W9-CJM-709

ENVIRONMENTAL CHEMODYNAMICS

ENVIRONMENTAL SCIENCE AND TECHNOLOGY

A Wiley-Interscience Series of Texts and Monographs

Edited by JERALD L. SCHNOOR, *University of Iowa*
ALEXANDER ZEHNDER, *Swiss Federal Institute for Water Resources and Water Pollution Control*

A complete list of the titles in this series appears at the end of this volume

ENVIRONMENTAL CHEMODYNAMICS

Movement of Chemicals in Air, Water, and Soil

Second Edition

LOUIS J. THIBODEAUX

A WILEY-INTERSCIENCE PUBLICATION
JOHN WILEY & SONS, INC.
New York • Chichester • Brisbane • Toronto • Singapore

This text is printed on acid-free paper.

Library of Congress Cataloging in Publication Data:

Thibodeaux, Louis J.

Environmental chemodynamics: movement of chemicals in air, water, and soil / Louis J. Thibodeaux.—2nd ed.

p. cm.—(Environmental science and technology)

Rev. ed. of: Chemodynamics. c1979.

"A Wiley-Interscience publication."

Includes index.

ISBN 0-471-61295-2 (cloth: alk. paper)

1. Environmental chemistry. I. Thibodeaux, Louis J. Chemodynamics. II. Title.

QD31.2.T47 1995

628.5'01'54—dc20 95-35279

Printed in the United States of America

10 9 8 7 6 5 4 3 2 1

To
Phoebe, Natalie, Whitney,
Bonnie, and Camille

SERIES PREFACE

Environmental Science and Technology

We are in the third decade of the Wiley Interscience Series of texts and monographs in Environmental Science and Technology. It has a distinguished record of publishing outstanding reference texts on topics in the environmental sciences and engineering technology. Classic books have been published here, graduate students have benefited from the textbooks in this series, and the series has also provided for monographs on new developments in various environmental areas.

As new editors of this Series, we wish to continue the tradition of excellence and to emphasize the interdisciplinary nature of the field of environmental science. We publish texts and monographs in environmental science and technology as it is broadly defined from basic science (biology, chemistry, physics, toxicology) of the environment (air, water, soil) to engineering technology (water and wastewater treatment, air pollution control, solid, soil, and hazardous wastes). The series is dedicated to a scientific description of environmental processes, the prevention of environmental problems, and to preservation and remediation technology.

There is a new clarion for the environment. No longer are our pollution problems only local. Rather, the scale has grown to the global level. There is no such place as "upwind" any longer; we are all "downwind" from somebody else in the global environment. We must take care to preserve our resources as never before and to learn how to internalize the cost to prevent environmental degradation into the product that we make. A new "industrial ecology" is emerging that will lessen the impact our way of life has on our surroundings.

In the next 50 years, our population will come close to doubling, and if the developing countries are to improve their standard of living as is needed, we will require a gross world product several times what we currently have. This will create new pressures on the environment, both locally and globally. But there are new opportunities also. The world's people are recognizing the need for sustainable development and leaving a legacy of resources for future generations at least equal to what we had. The goal of this series is to help understand the environment, its functioning, and how problems can be overcome; the series will also provide new insights and new sustainable technologies that will allow us to preserve and hand down an intact environment to future generations.

Jerald L. Schnoor
Alexander J. B. Zehnder

PREFACE

Writing in *Residue Reviews* (1974), R. Hague and V. H. Freed termed environmental chemodynamics an "emerging field": studies that demonstrate the relation of physicochemical principles in the behavior of chemicals in the environment. In other words, the studies attempt to answer the question: What happens to a chemical when it is introduced into the environment, and what are the factors that determine and influence its distribution? Within the last twenty years environmental chemodynamics has become a recognized subfield of environmental chemistry and environmental engineering and is also known as chemical fate and transport in the environment. Much new information has appeared on the subject since the appearance of the original edition in 1979. This new edition captures the essence of that material in textbook form.

Developed for senior and first-year graduate students in engineering and chemistry, the new material has been incorporated into the basic structure of the earlier book. Its organization and approach have been retained. The literature citations have more than doubled; other highlights of the second edition include:

- All chapters have been updated significantly.
- Chapter 5, on chemical exchange between the water column and bed sediment, and Chapter 7, on intraphase chemical movement, have been revised and expanded extensively.
- New sections on equilibrium models for environmental compartments, dry deposition of particles and chemical vapors onto water and soil surfaces, suspended solid behavior in flowing streams, chemical behavior in estuaries, behavior of particles in porous media, physical structure of the atmospheric boundary layer, and aspects of contaminant behavior in the subterranean (i.e., groundwater) have been added. The inclusion of these sections allowed the elimination of Chapter 8.
- Exercise and example problems have been increased by 50%.
- A solutions manual is available.
- The appendixes have been reformatted with more useful data for solution of the exercise problems and for practitioners.

Leveled at chemical engineering departments, environmental chemodynamics has been equally well accepted in university departments of civil and environmental engineering and environmental chemistry. It has served as the primary material or as the majority of the subject matter in numerous courses at the senior and first-year graduate levels. It has also served as one of many source materials in other environmental science–related courses. Only a working knowledge of algebra is needed for the interpretation and use of virtually all the mathematical models presented, whereas expertise in calculus and differential equations is necessary for the development of alternative models.

Practitioners similarly use the subject matter in environmental chemodynamics. Beginning in 1981, the American Institute of Chemical Engineers (AIChE) Continuing Education Department has offered the subject as a short course in conjunction with its national and annual meetings, this textbook being the primary source material. The course was expanded from two days to three days course in 1990; to date a total of 55 short courses have been presented, 38 of these under AIChE sponsorship. This record is indicative of its acceptance as a useful tool by practitioners, these being primarily engineers, chemists, and other scientists employed by consulting firms, governmental agencies, and manufacturing industries. Because of the potential environmental ills that result from the large-scale production and use of chemicals, a recent National Research Council report* contains the phrase statement: "Chemical engineers must take up the role of cradle-to-grave guardians for chemicals, ensuring their safe and environmentally sound manufacture, use and disposal..." Proper disposal requires a clear understanding and quantitative description of chemical processes in nature. As conceived and presented, this book is an extension of the practice of chemical processing in manufacturing. However, environmental chemodynamics presents a greater challenge to researchers and practitioners because these chemical processes occur within open "reactors" fashioned by the natural environment and controlled in large part by the capricious forces of nature. Despite these additional complexities, traditional engineering design predictions involving chemical behavior will need to be made. As a university-level course, the description of environmental chemodynamics involves interphase equilibrium, reactions, transport processes, mathematical models near natural interfaces (i.e., air–water–sediment-soil), and the associated boundary layer regions concerned with the fate of anthropogenic substances. Such a study quantifies the phase distribution, reaction rates, fluxes, half-lives, and concentration levels. These predicted quantities are necessary inputs into toxicological algorithms, procedures, and other formalisms that are used to assess the chemical hazards posed to humans and other life forms. The text material makes it clear that the level of environmental chemodynamic knowledge—and thus the reliability of predictions—is often inadequate. Much research remains to be done on the subject of chemical processes in nature.

LOUIS J. THIBODEAUX

Baton Rouge, Louisiana

* *Frontiers in Chemical Engineering* (National Academy Press, Washington, D.C.), 1987.

ACKNOWLEDGMENTS

My colleagues Danny D. Reible and Kalliat T. Valsaraj contributed in several ways to this book. Besides their many technical contributions, criticisms of and input on the topics to include was very valuable. They also contributed by covering my absence at numerous academic meetings while I tended to EC matters. Technical and topical contributions were also made by W. David Constant, James H. Clarke, Thomas R. Marrero, and Herbert E. Allen. Both Jim and Tom reviewed chapters of the manuscript and made suggestions.

Louisiana State University is acknowledged through the encouragement and support of the faculty students and staff. This includes the Dean of the College of Engineering, Edward McLaughlin, and the Chemical Engineering Department Chairmen, John R. Collier, Ralph W. Pike, Jr., and Arthur M. Sterling. Student feedback from using EC in numerous offerings of the technical elective course has been an enormously valuable contribution. Gregory Thoma, Jeffery Smith, and Brian Swift symbolize this group. Maureen Mitchem and Rajendran Subramanian did the word processing.

My family supported me in this effort in numerous ways. Joyce did the most, as usual. She processed words, supervised the production of the manuscript, and verified its accuracy.

Thanks to all of you.

L. J. T.

CONTENTS

LIST OF SYMBOLS

Symbols that appear infrequently or in one section only are not listed. Dimensions are given in terms of ampere (A), mass (M), amount of substance (mol), length (L), time (t), and temperature (T). Boldface symbols are vectors or tensors.

LATIN AND SCRIPT LETTERS

A	area, L^2
a	absorptivity, dimensionless; acceleration, L/t^2; or dispersivity, L
a_m	surface area per unit mass, L^2/M
a_v	interfacial area per unit volume, L^{-1}
B_1	Brunauer, Emmett, and Teller isotherm constant, dimensionless
C	an arbitrary constant, variable dimensions
C_d	drag coefficient, dimensionless
$\hat{C}_p$ and $\hat{C}_v$	heat capacity at constant pressure and volume respectively, per unit mass, L^2/t^2T
c	total molar concentration, mol/L^3
c_{ij}	molar concentration of species i in phase j, mol/L^3
c'_{ij}	fluctuating molar concentration, mol/L^3
D	characteristic length, L
D_{A31}	diffusion coefficient of chemical A in soil phase pore spaces filled with air, L^2/t
D_{2x}	longitudinal dispersion coefficient, L^2/t
D_{ij}	diffusion coefficient of species i in phase j, L^2/t
D_p	particle diameter, L, d_p also used
$\mathscr{D}_{ij}$	molecular diffusivity of species i in phase j, L^2/t
$\mathscr{D}^k_{ij}$	Knudsen diffusivity, L^2/t
$\mathscr{D}^T_{ij}$	Soret effect thermal diffusivity, L^2/tT
$\mathscr{D}^{(t)}_{ij}$	turbulent or eddy diffusivity of species i in phase j, L^2/t
d	diameter or zero-plane displacement, L
E	energy of interaction, ML^2/t^2 mol
e	2.71828..., or emissivity, dimensionless
F	Faraday constant, tA/mol
F_A	a fraction denoting efficiency, removal efficiency, and so on, dimensionless

f	fraction, or frequency and probability of occurence, dimensionless
f_{ij}	fugacity of species i in phase j, M/Lt^2
f^0_{ij}	pure component, M/Lt^2
$\mathbf{g}$	gravity vector, L/t^2
g	gravitational acceleration, L/t^2
H	diffusion hindrance factor, dimensionless
H_A	Henry's law constant, M/Lt^2
H_R	relative humidity, dimensionless
H_0	enthalpy, ML^2/t^2 mol
h_j	heat transfer coefficient for phase j, M/t^{3}T
h	Planck's constant, ML^2/t
h	elevation or depth, L
i	slope of water table, L/L
i_i	mass input loading rate per area, M/L^{2}t
$\mathbf{J}_i$	molar flux of species i relative to mass average velocity, mol/tL^2
$\mathbf{j}$	mass flux of species i relative to mass average velocity, mol/tL^2
j_D and j_H	Chilton–Colburn j-factor for mass and heat transfer respectively, dimensionless
K_3	hydraulic conductivity, L/t
${}^1K_{i2}$	overall liquid-phase mass transfer coefficient for species i across a gas–liquid interface, mole fraction driving force, mol/tL^2
${}^1K'_{i2}$	overall liquid-phase mass-transfer coefficient for species A across a gas–liquid interface, concentration driving force, L/t
$\mathscr{K}^*_{ijk}$	partition or distribution coefficient for species i between phase j and phase k, volume ratio, L^3/L^3
k	thermal conductivity, ML/t^{3}T
k_B	Boltzmann constant, ML^2/t^2 mol
k_3	intrinsic porous media permeability, L^2
k''_i	heterogeneous chemical reaction rate constant; n is order of reaction, mol^{1-n}/L^{2-3n}t
k'''_i	homogeneous chemical reaction rate constant; n is order of reaction, mol^{1-n}/L^{3-3n}t
${}^3k_{i1}$	individual gas-phase mass transfer coefficient for species i across the air–soil interface, mole fraction driving force, mol/tL^2
${}^3k'_{i1}$	individual gas-phase mass transfer coefficient for species i across the air–soil interface, concentration driving force, L/t
L	characteristic length, L
l	length or fetch, L
$\mathscr{M}$	moles of material, mol
M_i	molecular weight of species i, M/mol
$\mathscr{M}_i$	moles of component i, mol
m	mass of material, M
m_i	mass of component i, M
N	Avogadro's number, mol^{-1}

$\mathbf{N}_i$	molar flux of species i with respect to stationary coordinates, mol/tL^2
N_i	number of biological species of type i, dimensionless
$\mathbf{n}$	unit vector normal to surface, dimensionless
$\mathbf{n}_i$	mass flux of species i with respect to stationary coordinates, M/tL^2
n	real number, dimensionless
P	probability, dimensionless
p	fluid pressure, M/Lt^2
p_i	partial pressure of species i, M/Lt^2
p_i^0	pure component vapor pressure of species i, M/Lt^2
Q	volumetric flow rate, L^3/t
Q_{12}	radiant energy flow from phase 1 to phase 2, ML^2/t^3
q	energy flux rates relative to mass average velocity, M/t^3
R	gas constant, ML^2/t^{2}T mol or radius of sphere or cylinder, L
R_{i3}	retardation factor, dimensionless
R_i	molar rate of production of species i, mol/tL^3, or washout ratio, L^3/L^3
r	radial coordinate, L
r_h	hydraulic radius, L
r_i	mass rate of production of species i, M/tL^3
S	residual saturation, L^3/L^3
S_R	heat of reaction, ML^2/t^2 mol
s	fractional surface renewal rate, t^{-1}
s_{ij}	Soret coefficient, T^{-1}
T	absolute temperature, T
T'	fluctuating temperature, T
t	time, t
V	volume, L^3
$\mathbf{v}$	velocity, vector, L/t
v	velocity with subscript denoting direction, L/t
v_d	deposition velocity, L/t
v_i	velocity of species i, L/t
v_j'	fluctuating velocity of phase j, L/t
$\bar{v}_j$	mass average velocity of phase j, L/t
v_*	$\sqrt{\tau_0/\rho_j}$ = reference or friction velocity, L/t
W	total molar flow rate, mol/t
W_i	molar flow rate of species i, mol/t
w	width, L
w_i	mass flow rate of species i, M/t
w_{3i}'	particle mass reworking rate of biological species i, M/t · individual
x	rectangular coordinate, longitudinal direction, L
x_i	mole fraction of species i in water, dimensionless
y	rectangular coordinate, vertical direction, L
y_i	mole fraction of species i in air, dimensionless
z	rectangular coordinate, lateral direction, L
z_e	electrochemical valence, dimensionless

GREEK LETTERS

α_j	thermal diffusivity for phase j, L^2/t
$\alpha_j^{(t)}$	turbulent or eddy thermal diffusivity for phase j, L^2/t
α_i^*	relative volatility (gas–liquid) of species i, dimensionless
β	cloud cover factor, dimensionless
β_j	coefficient thermal expansion for phase j, T^{-1}
Γ	mass flow rate of liquid film per unit width of wetted surface, M/Lt
Γ_a	dry adiabatic lapse rate, T/L
$\Gamma(x)$	gamma function of x
γ	existing, in general diabatic, lapse rate in the surrounding air, T/L
γ_{ij}	chemical activity coefficient of species i in phase j, dimensionless
Δ	denotes difference when precedes symbol, dimensionless
δ	boundary layer thickness, L
ε_j	fraction void space occupied by phase j in a solid, dimensionless
ζ	dimensionless elevation
ζ_{ij}	coefficient of mass expansion for species i in phase j, dimensionless
θ	angle in cylindrical or spherical coordinates, radians, or wave period, t
κ_1	von Kármán's constant, dimensionless
λ_i	latent heat of vaporization of species i, L^2/t^2
μ_{ij}	chemical potential of species i in phase j, dimensionless
μ_j	viscosity of phase j, M/Lt
ν	frequency, Brunt–Vaisala, t^{-1}
ν_j	μ_j/ρ_j = kinematic viscosity of phase j, L^2/t
π	3.14159...
ρ	reflectivity, dimensionless
ρ_{ij}	mass concentration of species i in phase j, M/L^3
ρ'_{ij}	fluctuating mass concentration, M/L^3
σ	mean displacement or dispersion coefficient, L
τ	residence time, half-life, time period, etc., t
τ_h	tortuosity factor, dimensionless
τ_o	shear stress at fluid–solid interface, M/t^2L
υ	volume fraction, L^3/L^3
Φ_i	fraction species i on particles in air, dimensionless
ϕ	similarity function, dimensionless
ϕ_e	electrical potential, ML^2/t^3A
ϕ_i	mass fraction of species i in water, dimensionless
ψ_i	mass fraction of species i in air, dimensionless
Ω	density normalized gradient, L^{-1}
ω	frequency of oscillation, t^{-1}
ω_i	mass fraction of species i in solid, dimensionless

OVERLINES

$\cdot$	local value
$-$	time smoothed

BRACKETS

$\langle a \rangle$	average value of a over a flow cross section

SUPERSCRIPTS

o	initial value
*	equilibrium condition, value or solubility
$'$	deviation from time-smoothed value
$'$	denotes mass transfer coefficient has dimensions of L/t
(t)	turbulent
(l)	laminar
T	thermal diffusivity
1	gas interface
2	liquid interface
3	solid interface
4	other interface (e.g., oil)

SUBSCRIPTS

A, B, C, etc.	species in multicomponent system
b	bulk or "cup-mixing" value
h	heat transfer
i	interface
i	arbitrary chemical species (A for chemical A, B for chemical B, etc.)
j	arbitrary phase (1 for air, 2 for water, 3 for soil, 4 for second liquid, etc.)
m	completely mixed system, momentum with similarity function
o	at origin of space dimension
o	quantity evaluated at a surface or interface
p	plug flow system
t	total mass
w	water vapor
x	variable has movement in x direction
y	variable has movement in y direction
z	variable has movement in z direction
1, 2, 3, 4	see subscript j above

COMMONLY USED DIMENSIONLESS GROUPS

Bi_i	Biot number
Da_i	Damköhler number
Fr	Froude number
Gr	Grashoff number for heat transfer
Gr_{ij}	Grashoff number for mass transfer
Nu	Nusselt number for heat transfer
Nu_{ij}	Nusselt number for mass transfer
Pe	Peclet number
Po	Power number
Pr	Prandtl number
Ra	Raleigh number
Re	Reynolds number
Ri	Richardson number
Sc_{ij}	Schmidt number
Sh_{ij}	Sherwood number for mass transfer
St	Stokes number

MATHEMATICAL OPERATIONS

$$\mathrm{erf}(x) = \frac{2}{\sqrt{\pi}} \int_0^x e^{-t^2}\,dt = \text{error function of } x$$

$\exp(x) = e^x =$ the exponential function of x
$I_0(x)$ and $K_0(x)$ = modified Bessel functions, of x, of the first and second kinds
$\ln(x)$ = the logarithm x to the base e
$\log(x)$ = the logarithm x to the base 10

$$\Gamma(x, u) = \int_0^u t^{x-1} e^{-1}\,dt = \text{the incomplete gamma function}$$

$$\Gamma(x) = \int_0^\infty t^{x-1} e^{-1}\,dt = \text{the complete gamma function}$$

∇ = the "del" or gradient vector operator
∇^2 = the Laplacian operator

ENVIRONMENTAL CHEMODYNAMICS

1

INTRODUCTION

1.1. INTRODUCTION TO ENVIRONMENTAL CHEMISTRY AND ENGINEERING

Chemodynamics

With an unparalleled surge in creativeness, the human race has produced hundreds of thousands of "unnatural" chemicals. Some are regarded as potential threats to humanity and its living environment. Many of these xenobiotics or anthropogenic substances have found their way into the biosphere and have been classified as toxic or potentially harmful chemicals. Figure 1.1-1 illustrates the pathways by which pesticides are transported between environmental compartments. Since we are going to continue using chemicals, it is important to be able to trace their transport in the natural environment. *Environmental chemodynamics* is the name given to a subject that deals with the transport of chemicals (intra- and interphase) in the environment, the relationship of their physical–chemical properties to transport, their persistence in the biosphere, their partitioning in the biota, and toxicological and epidemiological forecasting based on physical–chemical properties. A comprehensive and systematic study of chemical movements in the environment is an interdisciplinary undertaking and must utilize the principles of such disciplines as chemistry, physics, systems analysis, mathematical modeling, engineering, and earth, medical, and biological sciences.

This book is concerned with several topics of environmental chemodynamics. Specifically, the subject is the interphase transport of chemicals and energy among the air, soil, and water phases of the environment. We focus on the mechanisms and rates of movement of chemicals across the three interfaces: air–soil, soil–water, and water–air. Our focus is on the region near the interfaces and the natural forces that affect and control transport in that region.

Once an anthropogenic substance enters the natural environment, human-initiated forces aimed at controlling, manipulating, modifying, and attenuating are usually secondary to the existing natural forces. These natural forces derive their energy from the sun and are manifest mainly in the form of fluid movements (air and water) and solar radiation. Observables that quantify aspects of these natural forces include temperature, incident radiation, flow velocity, pressure, relative humidity, and concentration.

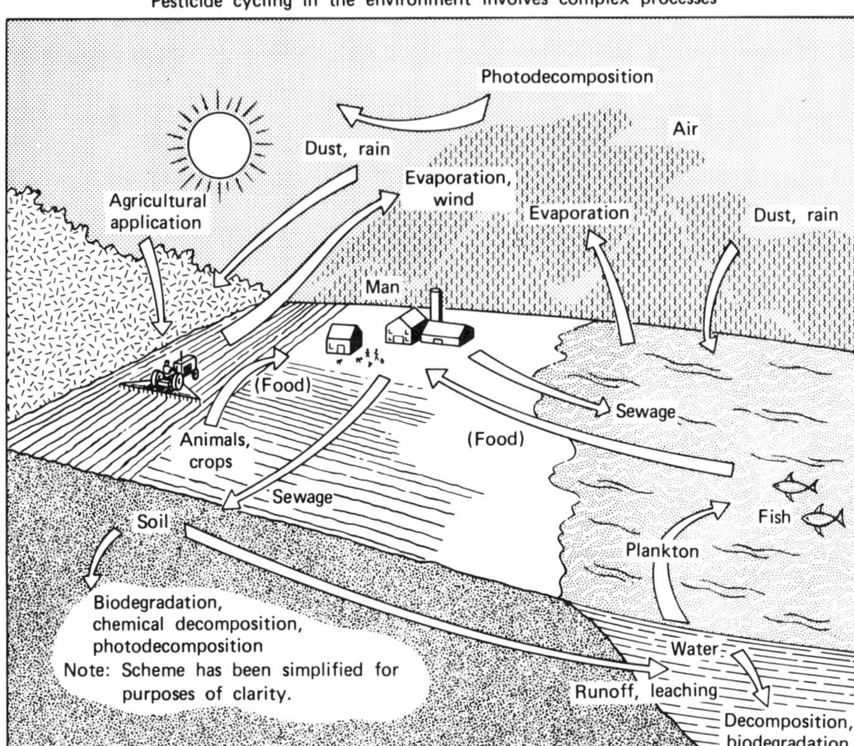

Figure 1.1-1. Pesticide pathways between environmental compartments. [Reprinted with permission of the copyright owner, American Chemical Society, from the April 22, 1974 issue of *Chemical and Engineering News* (Ref. 1).]

Chemical is used here in the broad sense and includes water, oxygen, carbon dioxide, sulfur dioxide, and DDT. Movements among the four ecosystems—atmosphere, hydrosphere, lithosphere, and biosphere—that constitute the environment are referred to as *interphase transfers*. The interphase transfer of water and oxygen is desirable, whereas that of sulfur dioxide and DDT is undesirable for the most part. The movement of chemicals within the environment has a profound effect upon environmental livability. The rates of transfer are important and the magnitude can also affect livability. The natural processes that promote these exchanges are ever present and are responsible for the magnitude and direction of the exchanges, both desirable or undesirable.

Once a chemical enters one of the mobile phases (i.e., air or water), it becomes dispersed rapidly because of fluid movements. Movement within a phase is termed *intraphase mass transfer*, *diffusion*, or *dispersion*. Interphase mass transfer is important to the movement of synthetic chemicals between the various phases of the ecosystem. People and the other organisms that constitute the biosphere reside, to varying degrees, within the other three spheres.

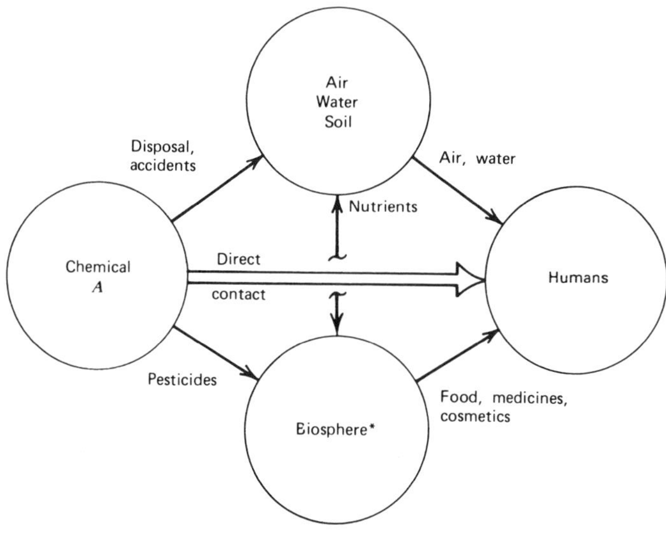

Figure 1.1-2. Routes of synthetic chemicals to humans.

Figure 1.1-2 traces the direct and indirect routes of synthetic chemicals through the environment and eventually to the human organism.

People encounter potentially harmful substances by direct contact with chemicals contained in food, food additives, medicines, cosmetics, the workplace, home, and so on. There are, however, several indirect contact modes of a more subtle nature. The continual intake of air and water is an indirect source of many chemical substances because of residuals in those phases of the environment. The pathways for entry of chemicals into these necessary elements of the ecosystem are shown in Fig. 1.1-3. The biosphere, in the form of natural foodstuffs, is another indirect source made even more critical by

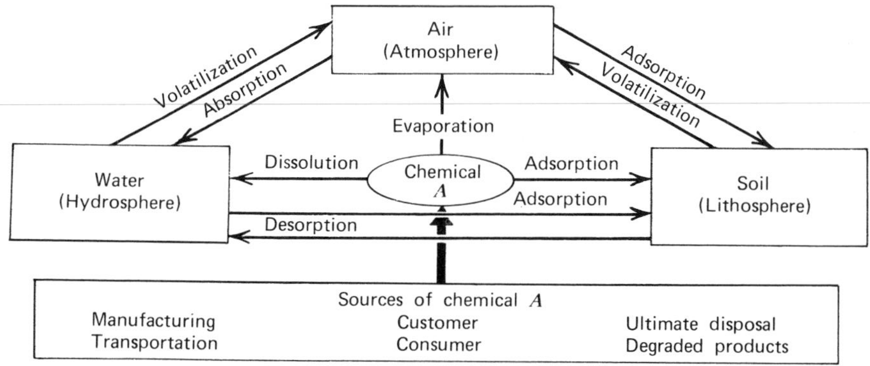

Figure 1.1-3. Movement of chemicals in the environment.

bioamplification processes. A large portion of our food consists of organisms at or near the ends of food chains all of which reside in some portion of the atmosphere, hydrosphere, or lithosphere. Medicines, cosmetics, and similar personal goods made of living matter constitute another indirect contact source.

This book is concerned with natural interphase transfer processes in general and specifically, with the movement of chemicals among the three portions of the ecosystems in which people find themselves. In general, chemicals are not always unnatural entities in the environment except when they are present in concentrated form. In isolated instances the movement of naturally occurring chemicals is addressed here (i.e., absorption of oxygen in water), but in the main we focus on the movement of synthetic chemicals.

Almost all activities of a progressive nature result in an upset in the natural status quo. The disturbance may be planned or unplanned. Chemicals and energy are inevitably placed into the environment in a nonnatural manner. The eventual assimilation of this disturbance is largely dependent on the interphase transport processes present in the natural setting. The discussion in this book provides (1) a qualitative understanding of these natural processes, and (2) quantitative tools by which to assess the response and/or recovery of the physical environment to chemical and energy stresses.

Chemical and energy stresses are eventually relieved but the rate is controlled by the natural exchange process. Interphase transport is usually the critical step. The phases involved are the atmosphere, hydrosphere, and lithosphere. We emphasize here the role of natural environmental transfer rates in assimilating stress. The question in many instances is the rate of assimilation, not whether the environment can assimilate the disturbance.

Intensity and lifetime are two important pieces of information with regard to environmental insults; intensity is measured as a concentration or temperature and lifetime is measured in real time. Both of these variables are intimately related to the interphase transport phenomena and their rates.

Transport science is quite advanced. The emphasis in this book is on application of chemical engineering transport concepts (i.e., heat, momentum, and mass transfer) to situations involving the natural environment. Chemical engineers readily deal with interphase transport; however, the application is typically inside process equipment of their design and under their control (i.e., reactors, mass transfer columns, heat exchangers, etc.). Except for system design and control, much of the body of chemical engineering knowledge can be applied to situations in a natural setting (i.e., out-of-doors) Here we redirect that knowledge onto environmental problems related to the production of chemicals, thereby completing the job of engineering chemicals for the benefit of humankind.

International System of Units

Public laws declare that the policy of the United States shall be to encourage educational agencies and institutions to prepare students to use the metric

Table 1.1-1. SI Base and Supplementary Units

Quantity	Name	Symbol
SI base units		
Length	meter	m
Mass	kilogram	kg
Time	second	s
Electric current	ampere	A
Thermodynamic temperature	kelvin	K
Amount of substance	mole	mol
Luminous intensity	candela	cd
SI supplementary units		
Plane angle	radian	rad
Solid angle	steradian	sr

system of measurement as part of the regular education program. Appendix A contains an interpretation and modification of the International System of Units (hereafter SI) for the United States.

The SI is based on seven fundamental or base units, as they are called, and the entire system is built from them. In rapid review the SI base unit names and symbols are given in Table 1.1-1. These units are combined by dividing, multiplying, or raising to negative or positive powers to produce the needed derived units. There are 19 of them, which have been given special names. Any one of the base or derived units can be made any size, very large or very small, by the addition of a prefix. The student is urged to consult Appendix A for a list of the derived units, prefixes, and additional information regarding SI.

Useful Interpretation of the Mole and Molecular Weight. Most of the SI units should be familiar to the student; however, some units important to the subject of chemistry need additional interpretation. The unit for the amount of substance, the mole, is defined as the amount of substance in a system that has the same number of entities, that is, molecules, atoms, ions, electrons, and particles, as there are atoms in 0.012 kg of carbon-12. There are 6.022E23 (Avogadro's number) atoms in 12 g of carbon-12. Since chemicals move about, react, exchange phase, and diffuse as well-defined entities, it is more useful and often easier to quantify the number of these entities rather than the mass. The most useful entity for the purposes of chemodynamics is the molecule.

Atomic weight is the relative mass of an atom based on a scale in which a specific carbon atom (carbon-12) is assigned a mass value of 12. A table of relative atomic weights for the chemical elements is given in Appendix C. Molecular weight is the sum of the atomic weights of all the atoms in a molecule. Atomic weight and molecular weight can be interpreted to have dimensions of g/mol, kg/kmol, mg/mmol, and so on. The atomic weight of carbon is 12.011 g/mol. The molecular weight of carbon dioxide is 43.999 kg/kmol. The molecular weight (also atomic and ionic) weight of species i is denoted by M_i. Review the example problems on the utility of the mole in environmental chemistry.

Scientific Notation and Machine Computation. With the advent of the computer and the electronic calculator, a shortcut form of scientific notation has been adopted. The essence of the change is to represent the power-of-10 multiplier with the letter E. For example, the standard scientific notation for Avogadro's number is 6.023×10^{23}; in electronic machine notation it is 6.023E23. Scientific notation for the numerical value of the natural log of $\frac{1}{2}$ is -6.93×10^{-1} and in machine notation it is -6.93E-01. The advantages of using machine notation, besides compatibility with electronic computing devices, are the absence of powers and a shortening of the written statement. The electronic computation machine form of scientific notation is used throughout this book.

Multicomponent and Multiphase Notation

It is necessary to introduce an intricate and descriptive system of notation in dealing with the multicomponent and multiphase nature of environmental chemistry. Chemical transport within the environment necessitates movement of the particular chemical species with respect to the other species. If the movement takes place within a single phase it is termed *intraphase transport*, and if it takes place between phases it is termed *interphase transport*. Next we discuss briefly the notation employed throughout the book.

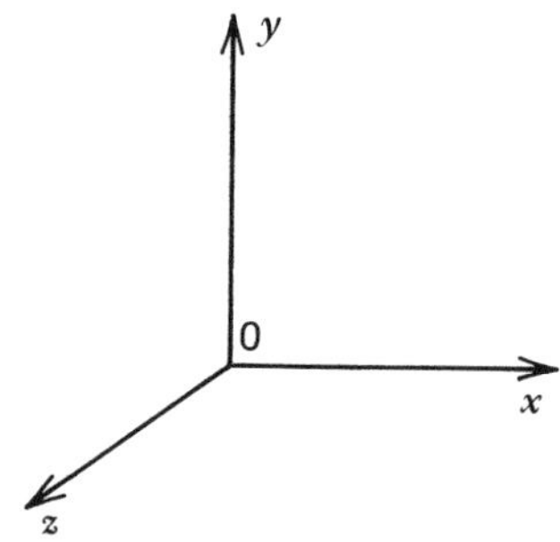

Space Variables and Directions. Chemical movement implies a species change of position with respect to a fixed point in space. Since most movements of concern in this book are at or near the surface of the Earth, the space variables and positive directions are defined with respect to that frame of reference. The dimension of the space variables is length (L).

$$0 \equiv (0, 0, 0)$$

The origin is a fixed, defined point in three-dimensional space. Normally throughout the book the x direction is horizontal and positive to the right, the y direction is vertical and positive upward, and the z direction is horizontal and positive toward the reader.

The surface of the land, sea, seabed, and so on, normally coincides with the x–z plane, so that the positive y direction is pointing away from the center of the Earth. Normally, fluid flow (i.e., water or air) is from left to right in the x direction. The "natural" space orientation is used whenever possible, but special orientations may be defined during the analysis of some problems.

Concentration and Phase Density. The quantitative intensity of the existence of a chemical species within any phase of the environment is an important piece of information. It is defined as concentration, expressed as an intensive measure, and has many forms of notational representation. It was recognized earlier when the mole was introduced that chemical species and such substances exist as well-defined entities. The most important entity with respect to environmental chemistry is the molecule; however, atomic and ionic entities are important also. The concentrations of chemicals are "naturally" expressed in molar units. Mole fraction of chemical species A represents the moles of A with respect to the total molar quantity of the phase. The mole fractions should represent the particular phase; specific definitions and notation are

$$x_i \equiv \text{mole fraction of chemical species } i \text{ in water (mol/mol)}$$

$$y_i \equiv \text{mole fraction of chemical species } i \text{ in air (mol/mol)}$$

The mol in the denominator is of the mixture. Other mole fractions, such as species i in oil, are defined when needed. Note that mole fractions must always be subscripted and the space variables not.

The simplest multicomponent system is a binary system consisting of species A and B. For this system the *law of the whole* (i.e., sum of the parts) requires that

$$x_A + x_B = 1 \qquad y_A + y_B = 1$$

In general, for an N-component system consisting of $A, B, \ldots, N$,

$$\sum_{i=A}^{N} x_i = 1 \qquad \sum_{i=A}^{N} y_i = 1 \qquad \text{and so on}$$

Molar concentration is also a convenient quantitative measure of chemical species intensity. The symbol c is used for concentration:

$$c_{i1} \equiv \text{molar concentration of species } i \text{ in air (mol/L}^3)$$

$$c_{i2} \equiv \text{molar concentration of species } i \text{ in water (mol/L}^3)$$

The volume (L^3) in the denominator is that of the mixture. The second subscript, when it is an integer, denotes phase j with 1 for air, 2 for water, and 3 for soil. Species concentrations in other phases are defined by employing additional integers in a similar fashion when they are needed. The sum of the molar concentrations in a phase is the molar phase density. For the case of air,

$$c_1 = c_{A1} + c_{B1} + c_{C1} + \cdots + c_{Nj} \tag{1.1-1}$$

Similar molar densities exist for water.

Mass units are also useful and convenient for expressing concentration of environmentally important chemical species. The mass fraction of species A represents the mass of A with respect to the total mass quantity of the phase. The mass fraction should also represent a particular phase. Specific definitions and notations are

$$\phi_i \equiv \text{mass fraction of species } i \text{ in water (M/M)}$$

$$\psi_i \equiv \text{mass fraction of species } i \text{ in air (M/M)}$$

$$\omega_i \equiv \text{mass fraction of species } i \text{ in soil (M/M)}$$

The mass (M) in the denominator is that of the mixture. Other mass fractions can be defined if needed.

Mass concentration is denoted by ρ; specifically,

$$\rho_{i1} \equiv \text{mass concentration of species } i \text{ in air (M/L}^3\text{)}$$

$$\rho_{i2} \equiv \text{mass concentration of species } i \text{ in water (M/L}^3\text{)}$$

$$\rho_{i3} \equiv \text{mass concentration of species } i \text{ in soil (M/L}^3\text{)}$$

The volume (L^3) in the denominator is that of the mixture.

Mass and molar concentrations are related by $c_i = \rho_i/M_i$. Just as for mole fraction and molar concentrations the law of the whole demands that

$$\sum_{i=A}^{N} \phi_i = \sum_{i=A}^{N} \psi_i = \sum_{i=A}^{N} \omega_i = 1$$

The sum of the mass concentration in a phase is the mass-phase density. In the case of air

$$\rho_1 = \rho_{A1} + \rho_{B1} + \rho_{C1} + \cdots + \rho_{N1} \tag{1.1-2}$$

A similar mass density exists for water. Table 1.1-2 contains a summary of the preceding concentrations and their interrelation.

Table 1.1-2. Notation for Concentration in Multicomponent Systems

Basic definitions of molar

$$y_A = \frac{c_{A1}}{c_1}, \quad x_A = \frac{c_{A2}}{c_2}, \quad y_B = \frac{c_{B1}}{c_1}, \quad x_B = \frac{c_{B2}}{c_2}, \quad \text{etc.} \tag{A}$$

$$c_{A1} = \frac{\rho_{A1}}{M_A}, \quad c_{A2} = \frac{\rho_{A2}}{M_A}, \quad c_{B1} = \frac{\rho_{B1}}{M_B}, \quad c_{B2} = \frac{\rho_{B2}}{M_B}, \quad \text{etc.} \tag{B}$$

$$y_A + y_B + y_C + \cdots + y_N = 1, \quad x_A + x_B + x_C + \cdots + x_N = 1, \quad \text{etc.} \tag{C}$$

$$c_1 = c_{A1} + c_{B1} + \cdots + c_{N1}, \quad c_2 = c_{A2} + c_{B2} + \cdots + c_{N2}, \quad \text{etc.} \tag{D}$$

Basic definitions of mass

$$\psi_A = \frac{\rho_{A1}}{\rho_1}, \quad \phi_A = \frac{\rho_{A2}}{\rho_2}, \quad \psi_B = \frac{\rho_{B1}}{\rho_1}, \quad \phi_B = \frac{\rho_{B2}}{\rho_2}, \quad \text{etc.} \tag{E}$$

$$\rho_{A1} = c_{A1} M_A, \quad \rho_{A2} = c_{A2} M_A, \quad \rho_{B1} = c_{B1} M_B, \quad \rho_{B2} = c_{B2} M_B, \quad \text{etc.} \tag{F}$$

$$\psi_A + \psi_B + \psi_C + \cdots + \psi_N = 1, \quad \phi_A + \phi_B + \phi_C + \cdots + \phi_N = 1, \quad \text{etc.} \tag{G}$$

$$\rho_1 = \rho_{A1} + \rho_{B1} + \rho_{C1} + \cdots + \rho_{N1}, \quad \rho_2 = \rho_{A2} + \rho_{B2} + \rho_{C2} + \cdots + \rho_{N2}, \quad \text{etc.} \tag{H}$$

Additional relations

$$x_A M_A + x_B M_B = M_2, \quad M_1 = \frac{\rho_1}{c_1}, \quad \text{etc.} \tag{I}$$

$$\frac{\psi_A}{M_A} + \frac{\psi_B}{M_B} + \cdots + \frac{\psi_N}{M_N} = \frac{1}{M_1}, \quad \text{etc.} \tag{J}$$

$$x_A = \frac{\phi_A/M_A}{\phi_A/M_A + \phi_B/M_B + \cdots + \phi_N/M_N}, \quad \text{etc.} \tag{K}$$

$$\psi_A = \frac{y_A M_A}{y_A M_A + y_B M_B + \cdots + y_N M_N}, \quad \text{etc.} \tag{L}$$

Trace Quantity Engineering and Very Dilute Solutions. We may define trace quantity engineering as the segment of engineering that covers the identification, control, and handling of trace components.[2] The limitation on what constitutes a trace quantity is somewhat arbitrary at this point. For trace quantities, parts per million (ppm) and parts per billion (ppb) have been conventional terms for expressing concentrations. Although continued use of these terms is anticipated, their inexactness should be recognized, especially when the trace quantity is not completely soluble in the diluent material and when the definition of the denominator is not clear.

A basis for defining a dilute solution was proposed by Thibodeaux.[9] Table 1.1-3 contains a convenient summary of the phase densities and definitions of molar concentration ranges for dilute chemical solutions and mixtures in the

Table 1.1-3. Phase Density and Dilute Solutions

	Pure Phase Density		Definition of dilute Solution[a] as a Maximum Concentration of Chemical A	
Phase	Molar (mol/L)	Mass (g/L)	Molar (mol/L)	Mass (g/L)
Air[b]	$c_1 = 0.0446$	$\rho_1 = 1.293$	$c_{A1} = 0.00223$	$\rho_{A1} = 0.06465$
Water				
Fresh[c]	$c_2 = 55.56$	$\rho_2 = 1000$	$c_{A2} = 2.778$	$\rho_{A2} = 50.00$
Sea[d]	$c_2 = 55.25$	$\rho_2 = 1019$	$c_{A2} = 2.763$	$\rho_{A2} = 50.95$

[a]Definitions are based on $0.05 \geqslant x_A, y_A > 0$.
[b]At 0°C, 760 mmHg.
[c]At liquid at 4°C.
[d]Of 3.5% salinity as NaCl.

air and water phases of the environment. Solid mixtures of interest in the area of chemodynamics differ drastically from fluid mixtures in that they are not homogeneous and are not usually true solutions. An example is a trace chemical A absorbed onto the surface of lumps of soil. In the case of chemicals on soils, dilute "loadings" are those of 50 g/kg or less.

The convenience of the preceding definition has other advantages for the quantitative treatment of dilute solutions. One advantage is that the relation between mole fraction and molar concentration is simplified. These relations are

$$y_A = \frac{c_{A1}}{c_1} \qquad x_A = \frac{c_{A2}}{c_2} \tag{1.1-3a, b}$$

where c_1 and c_2 are, respectively, the molar densities of air and water. The mole (mol) in the denominators is of air and water, respectively. The molar density of air is a strong function of temperature and pressure. From the ideal gas law, $c_1 = p/RT$. The molar density of water is a weak function of temperature (see Appendix D). The density of most earthen soils is constant over a relatively large range of temperature and pressure.

In a similar fashion the relation between mass fraction and mass concentration is simplified:

$$\psi_A = \frac{\rho_{A1}}{\rho_1} \qquad \phi_A = \frac{\rho_{A2}}{\rho_2} \tag{1.1-4a, b}$$

where ρ_1 and ρ_2 are the corresponding mass densities. The mass (M) in the denominators is of air and water, respectively. All the equations in Table 1.1-2 remain valid for dilute solutions, but several important ones are simplified drastically. The relations between mole fraction and mass fraction [i.e., lines (K) and (L) in Table 1.1-2] become

$$x_A = \frac{\phi_A M_B}{M_A} \qquad y_A = \frac{\psi_A M_B}{M_A} \tag{1.1-5a, b}$$

so the conversion involves only a ratio of molecular weights.

Parts per Million, Parts per Billion, etc. Parts per million, parts per billion, and so on, in weight, volume, and mole ratios, are used for chemical concentrations. Sometimes the ratios are unclear. The following are common usages. For gases, ppm is used on a volume basis (i.e., vol/vol). For liquids, both weight (i.e., wt/wt) and volume are used. For solids, a weight basis is normally used. Clearly, confusion can set in if these ratios are used other than for a general qualitative indication of concentration.

It is best if the vague ppm, ppb, and so on, usages are abandoned and chemical concentrations in fluids are placed on a mass (or mole) per unit volume basis. For gases (including air) temperature and pressure must be specified precisely so that the volume represents a known quantity of substance. The familiar standard conditions of chemistry and physics [i.e., standard temperature and pressure (STP)] are 0°C and 760 mmHg and the molar volume of any gas is 22.4 L/mol. All gas concentrations given either as c_{A1} or ρ_{A1} should be referenced to STP or some other standard. Although most liquid volumes, including water, change only slightly with temperature and pressure, a precise definition is needed. Water at 1.0 atm pressure and 4°C has a mass volume of 1.0 L/kg, and this is an excellent reference.

The use of ppm, ppb, and so on, for specifying trace quantity chemicals associated with soil or other solids is ideal, provided that weight ratios are used (i.e., g A/E6 g soil). This measure is imprecise unless the soil is on a water-free (unbound) basis. Reporting and measuring concentrations with the soil dried in air for a specified time at 1 atm and 100°C will establish a fairly precise standard for this case. Simple conversions exist between ppm, ppb, and ppt and concentration measured in SI units, provided that ppm, ppb, and ppt are defined precisely for each phase. See Example 1.1-3 for useful conversion factors.

More on Notation

Multicomponent and multiphase notation was introduced in defining species concentration within various phases. A mnemonic system is used throughout the book. This technique of notation makes reading and interpreting equations easier and is a direct aid in computer programming; however, writing them is somewhat more tedious. As an example c_{A2} can be read as the "molar concentration of species A in water."

By the use of overbars, superscripts, and subscripts along with the Latin and Greek alphabets, a fairly self-consistent system of notation can be developed. It works well most of the time but is not perfect. A complete list of notation appears in the front matter of the book. Examples of notation interpretation are as follows:

c^*_{A2i} = equilibrium molar concentration of species A in water at the interface.

y_A = mole fraction of species A in air

$\mathscr{D}^{(t)}_{A2}$ = turbulent diffusivity of A in a liquid phase

z = a distance from the origin in the z direction

n_{Az} = mass flux rate of species A in the z direction

${}^1k_{A2}$ = gas interface mass transfer coefficient for species A in water

The normal procedure is to read the superscripts and overbars first, then the basic symbol, followed by the reading of the subscripts. No more than three subscripts are used. The first subscript position denotes the species and is an uppercase Latin letter. The second subscript position denotes the phase ($1 \equiv$ gas, $2 \equiv$ liquid, and $3 \equiv$ soil) and is an integer. The third subscript usually specifies position within the phase (i.e., $i \equiv$ interface). When mole fraction and mass fraction symbols are used, a maximum of two subscript positions are employed. The phase subscript notation is superfluous since x implies water, y gas, and z solid. If the notation system seems confusing at this point, its usefulness and clarity should become evident as the text unfolds.

Example 1.1-1. Equal Molar Trace Quantity Concentrations of Chemicals in Water and Air. We desire to prepare standard solutions of low-molecular weight chlorinated hydrocarbons for laboratory use.

(a) Calculate the mass (kg) of chloroform ($CHCl_3$) and vinyl chloride (CH_2CHCl) in a 1 liter (L) aqueous solution in which the concentration of each is 1.0 mol/m^3 .
(b) Repeat the calculation for a 1-L air mixture at STP.

SOLUTION (a) Since the molar concentration of each chemical is equal, each liter of water contains

$$6.022\text{E}20 = \left(6.022\text{E}23\ \frac{\text{molecules}}{\text{mol}}\right)\left(\frac{1\ \text{mol}}{\text{m}^3}\right)\left(\frac{1\text{E}-3\ \text{m}^3}{\text{L}}\right)(1.0\,\text{L})$$

molecules of chloroform and 6.022E20 molecules of vinyl chloride. The mass

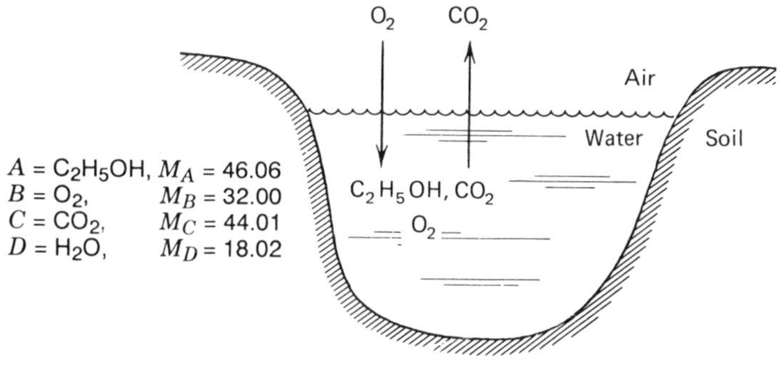

Figure E1.1-2

of chloroform ($A \equiv CHCl_3$), molecular weight 119.39, is

$$m_A = \left(\frac{1.0 \text{ mol}}{\text{m}^3}\right)\left(\frac{119.39 \text{ g}}{\text{mol}}\right)(1.0 \text{ L})\left(\frac{1\text{E-3 m}^3}{\text{L}}\right)\left(\frac{1\text{E-3 kg}}{\text{g}}\right)$$
$$= 119.39\text{E-6 kg}$$

The mass of vinyl chloride ($B \equiv CH_2CHCl$), molecular weight 62.5, is

$$m_B = (1.0)(62.5)(1.0)(1\text{E-3})(1\text{E-3}) = 62.5\text{E-6 kg}$$

(b) Since the molar concentration of each chemical is equal, each liter of air contains 6.022E20 molecules of each chemical. The mass of chloroform and vinyl chloride in air is 119.39E-6 and 62.5E-6 kg, respectively.

Example 1.1-2. Chemical Reaction and Movement in Molar Units. At a certain point in a river, ethanol in the water is being oxidized by a microbial enzyme reaction. The source of oxygen is the air above the river surface. Assume that the process is occurring at steady state. Atmospheric oxygen moves across the air–water interface at a constant rate. Once the oxygen is in the water the oxidation of ethanol occurs at a constant rate. The stoichiometry of the reaction is

$$C_2H_5OH + 3O_2 = 2CO_2 + 3H_2O$$

The evolution of carbon dioxide from the water also occurs at a constant rate. Although ethanol is volatile, assume that it remains in the water and its rate of disappearance ($-r_A$) is 5E-4 mg/L·s. Additional information is given in Fig. E1.1-2.

(a) Calculate the reaction rate of oxygen consumption (kmol/s) within a cubic meter of water located at the surface.

(b) Calculate the molar flux rate of carbon dioxide (kmol/s·m^2) as this trace chemical desorbs through a square meter of the surface.
(c) Convert the numerical results of parts (a) and (b) to mass (kg) units.

SOLUTION (a) *Reaction rate of oxygen:* First convert the ethanol reaction rate to molar units.

$$-R_A = 5\text{E-4}\ \frac{\text{mg}}{\text{L}\cdot\text{s}}\left|\frac{\text{kg}}{\text{E6 mg}}\right|\frac{1000\ \text{L}}{\text{m}^3}\left|\frac{\text{kmol}}{46.06\ \text{kg}}\right.$$
$$= 1.09\text{E-8 kmol/m}^3\cdot\text{s}$$

For each mole of ethanol 3 mol of oxygen is required; the rate of oxygen disappearance is

$$-R_B = 3(-R_A) = 3.27\text{E-8 kmol/m}^3\cdot\text{s}$$

The oxygen consumption rate is 1 m^3 of water is 3.27E-8 kmol/s.

(b) *Molar flux rate of carbon dioxide:* The stoichiometry demands that 2 mol of CO_2 be formed for each mole of C_2H_5OH rected; therefore,

$$R_c = 2(-R_A) = 2.18\text{E-8 kmol/m}^3\cdot\text{s}$$

A cubic meter of water will have 1 m^2 of exposed interface in contact with air, to yield

$$N_{Cz} = 2.18\text{E-8 kmol/m}^2\cdot\text{s}$$

(c) *Converting from molar to mass units:* Molecular weight is used in the conversion.

$$-r_B = 3.27\text{E-8}\ \frac{\text{kmol}}{\text{m}^3\cdot\text{s}}\left|\frac{32.00\ \text{kg}}{\text{kmol}}\right| = 1.05\text{E-6 kg/m}^3\cdot\text{s}$$

$$n_{Cz} = 2.18\text{E-8}\ \frac{\text{kmol}}{\text{m}^2\cdot\text{s}}\left|\frac{44.01\ \text{kg}}{\text{kmol}}\right| = 9.59\text{E-7 kg/m}^2\cdot\text{s}$$

Example 1.1-3. PPM, PPB, and PPT Converted to Concentrations Expressed in SI Units. Derive useful conversion factors for converting the following conventional concentrations to SI units:

(a) Parts per million (volume ratio) of chemical A in air.
(b) Parts per million (mass ratio) of chemical A in water.
(c) Parts per million (mass ratio) of chemical A within a soil.

Give the ppb and ppt result for each case also. The molecular weight of species A is M_A (g/mol).

SOLUTION (a) *1 ppm A in air:* Assume all components behave as ideal gases and that STP is used as the reference state.

$$1\ \text{ppm} = \frac{1\ \text{m}^3\ A}{\text{E6 m}^3\ \text{air}}\left|\frac{1000\ \text{L}}{\text{m}^3}\right|\frac{\text{mol}}{22.4\ \text{L}}\left|\frac{1000\ \text{mmol}}{\text{mol}}\right. = \frac{1}{22.4}\ \text{mmol}\ A/\text{m}^3$$

$$= \frac{1}{22.4}\ \frac{\text{mmol}\ A}{\text{m}^3}\left|\frac{M_A\ \text{mg}}{\text{mmol}}\right. = 1\times\frac{M_A}{22.4}\ \text{mg}\ A/\text{m}^3 \qquad \text{(E1.1-3A)}$$

In similar fashion,

$$1\ \text{ppb} = 1\times\frac{M_A}{22.4}\ \mu\text{g}\ A/\text{m}^3 \qquad \text{(E1.1-3B)}$$

$$1\ \text{ppt} = 1\times\frac{M_A}{22.4}\ \text{ng}\ A/\text{m}^3 \qquad \text{all at STP conditions} \qquad \text{(E1.1-3C)}$$

(b) *1 ppm A in water:* Use liquid water at 4°C, 1 atm as the reference state of volume.

$$1\ \text{ppm} = \frac{1\ \text{g A}}{\text{E6 g H}_2\text{O}}\left|\frac{1\ \text{g}}{\text{cm}^3}\right|\left(\frac{100\ \text{cm}}{\text{m}}\right)^3 = 1\ \text{g}\ A/\text{m}^3 \quad (=1\ \text{mg/L*}) \qquad \text{(E1.1-3D)}$$

$$1\ \text{ppb} = 1\ \text{mg/m}^3 \qquad \text{(E1.1-3E)}$$

$$1\ \text{ppt} = 1\ \mu\text{g/m}^3 \qquad \text{(E1.1-3F)}$$

L* ≡ liter is not a valid SI unit but is in common use.

(c) *1 ppm A within soil:* Since soil is not normally measured on a volume basis, mass is used.

$$1\ \text{ppm} = \frac{1\ \text{g}\ A}{\text{E6 g soil}}\left|\frac{1000\ \text{mg}}{\text{g}}\right|\frac{1000\ \text{g}}{\text{kg}}\right| = 1\ \text{mg}\ A/\text{kg soil} \qquad \text{(E1.1-3G)}$$

$$1\ \text{ppb} = 1\ \mu\text{g/kg soil} \qquad \text{(E1.1-3H)}$$

$$1\ \text{ppt} = 1\ \text{ng}\ A/\text{kg soil} \qquad \text{(E1.1-3I)}$$

Material Balance

This section may be skipped now. It will take on more concrete meaning when specific applications of chemical movement are addressed. A general form of the component material balance is presented for future use. The equation is simply a statement of the law of conservation of mass. It was proved

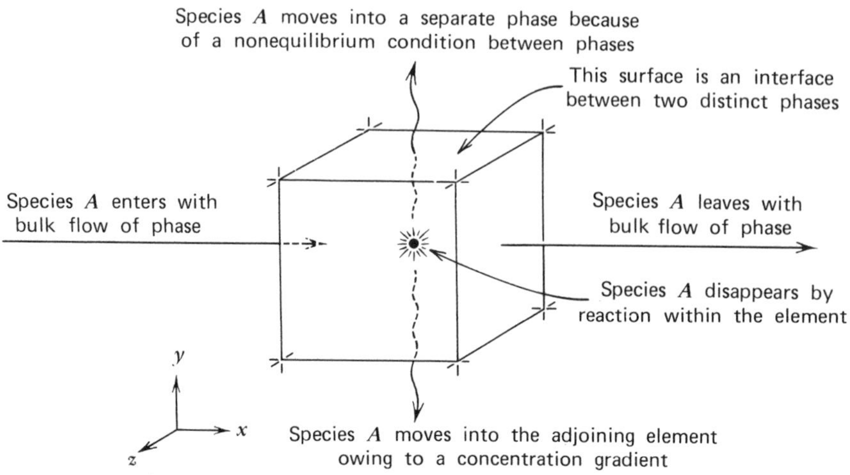

Species A moves into a separate phase because of a nonequilibrium condition between phases

This surface is an interface between two distinct phases

Species A enters with bulk flow of phase

Species A leaves with bulk flow of phase

Species A disappears by reaction within the element

y

x

z

Species A moves into the adjoining element owing to a concentration gradient

Figure 1.1-4. Material balance for a volume element fixed in space through which a fluid is flowing.

experimentally in 1777 when Antoine Lavoisier showed that there was no gain or loss of matter during the course of chemical and phase transformations.

Consider the volume element of arbitrary shape illustrated in Fig. 1.1-4. Where the composition within the system under study is uniform throughout (i.e., independent of position), the element may be defined as the entire system V and the accounting made for chemical A. Where the composition within the system is not uniform, a differential element of volume, dV, must be defined and the accounting of species A made for it. Thus, as illustrated in Fig. 1.1-4, we have

$$\begin{pmatrix} \text{rate of} \\ \text{mass } A \\ \text{flow into} \\ \text{volume} \\ \text{element} \end{pmatrix} = \begin{pmatrix} \text{rate of} \\ \text{mass } A \\ \text{flow out} \\ \text{of volume} \\ \text{element} \end{pmatrix} + \begin{pmatrix} \text{rate of mass } A \\ \text{loss due to} \\ \text{mass transfer*} \\ \text{from the} \\ \text{volume element} \end{pmatrix}$$

$$+ \begin{pmatrix} \text{rate of mass } A \\ \text{loss due to} \\ \text{chemical reaction} \\ \text{within the} \\ \text{volume element} \end{pmatrix} + \begin{pmatrix} \text{rate of} \\ \text{accumulation} \\ \text{of mass } A \\ \text{in volume} \\ \text{element} \end{pmatrix} \tag{1.1-6}$$

*Mass transfer includes interphase and intraphase movements.

A special form of Eq. 1.1-6 results if the volume element is assumed to move as a slug with the mean velocity of the fluid. In this case the flow into and flow out of terms are omitted. Additional assumptions can and will be placed on this general material balance as the needs of the problem dictate.

PROBLEMS

1.1A. Converting to the International System of Units from Other Systems

1. Using the conversion tables for SI in Appendix A, convert the following numerical entities to the equivalent SI units:

(a) Density of water, $\rho_2 = 62.3\,\text{lb}_\text{m}/\text{ft}^3$.

(b) Molecular diffusivity of oxygen in air, $\mathscr{D}_{O_2 1} = 0.206\,\text{cm}^2/\text{s}$.

(c) Temperature of a soil interface, $T_{3i} = 75°\text{F}$.

(d) Air pressure, $p_1 = 1\,\text{atm}\ (14.696\,\text{lb}_\text{f}/\text{in}^2)$.

(e) Latent heat of vaporization of water ($H_2O \equiv A$), $\lambda_A = 1051.5\,\text{Btu/lb}_\text{m}$.

(f) Mass flux rate of chemical A, $n_A = 4.1\text{E-8}\,\text{lb}_\text{m}/\text{s}\cdot\text{ft}^2$.

(g) Stefan–Boltzmann constant, $\sigma = 0.1712\text{E-8}\,\text{Btu/hr}\cdot\text{ft}^2\cdot\text{R}^4$.

(h) Water viscosity, $\mu_2 = 1\,\text{cP}$.

(i) Water heat capacity of $\hat{c}_{p2} = 1\,\text{cal/g}\cdot°\text{C}$.

(j) The gas constant, $R = 1.987\,\text{cal/mol}\cdot\text{K}$.

2. Using only base units, verify the dimensions of the following variables the notation for which appears in the list of symbols.

(a) $\hat{c}_\rho$ = heat capacity (L^2/t^2T).

(b) Q = rate of energy flow across a surface (ML^2/t^3).

(c) p = fluid pressure (M/Lt^2).

(d) F = force of a fluid on an adjacent solid (ML/t^2).

(e) W = rate of doing work on surroundings (ML^2/t^3).

1.2. ILLUSTRATION OF OBJECTIVES AND CONTENT: REAERATION OF NATURAL STREAMS

The objectives of chemodynamics were stated in Section 1.1. The scope and concept of this book were also described. In the absence of technical detail this discussion was somewhat abstract. In this section the objectives and content are redescribed in terms of a single, specific example—the movement of oxygen from air into the water of a natural stream. This is usually the first chemodynamic problem encountered by students beginning studies in environmental pollution.

Introduction

The discharge of organic impurities, such as municipal sewage and industrial waste, into natural streams presents a problem of primary importance in the field of environmental engineering. The decomposition of this organic matter by waterborne microorganism for metabolic processes results in the utilization of the dissolved oxygen in the stream.

$$C_xH_yO_z + nO_2 \xrightarrow[\text{enzymes}]{\text{microbial}} \text{products of } CO_2,\ H_2O, \text{ etc.}$$

The organic impurities denoted by the general formula $C_xH_yO_z$ are usually measured as an *oxygen demand*. This simplifies an otherwise complex stoichiometry. The biochemical conversion of 1 g of oxygen-demanding organic matter requires 1 g of molecular oxygen. The replacement of this molecular oxygen (i.e., O_2) by reaeration occurs through the water surfaces exposed to the atmosphere. The concentration of organic matter can be so great that there results a condition in which the receiving stream is completely devoid of dissolved oxygen. Because every stream has a limited capacity to assimilate organic wastes, evaluation of the natural purification capacity of a stream is of fundamental and practical value. It is a very old but still a very important problem, attracting the interest of environmental scientists. The mechanisms of stream reaeration is still undergoing investigation.[12]

Oxygen Equilibrium Between Air and Water

Molecular oxygen exists in the Earth's atmosphere and comprises 20.95% (volume). Oxygen is soluble in water to a small degree, reaching a maximum of 14.16 mg/L at 0.0°C and 760 mmHg. The solubility of oxygen in water decreases with increasing temperature, increasing chloride concentration (or concentration of ionic impurities in general), and decreasing partial pressure of oxygen in the atmosphere. High humidity and concentration of pollutants will also tend to slightly reduce O_2 solubility below the literature values. Relationships for dissolved oxygen solubility in equilibrium with dry air appear in Appendix C. These equations are represented by the general function f:

$$\rho_{A2}^* = f(T, \rho_{B2}, p_A^0)$$

where ρ_{A2}^* is the mass concentration of oxygen ($A \equiv O_2$) in equilibrium with air, T the water temperature, ρ_{B2} the salinity ($B =$ salts), and p_A^0 the partial pressure of oxygen in air.

Deoxygenation and Reoxygenation

The classical work of Streeter and Phelps in 1925[3] presented a mathematical analysis of the organic waste and oxygen content in water known as the *dissolved oxygen sag*. Consider the idealized stream shown in Fig. 1.2-1. A volume element of water V, moving at the mean velocity v_x, contains oxygen of concentration ρ_{A2} and organic material ($\equiv B$) of concentration ρ_{B2}. This

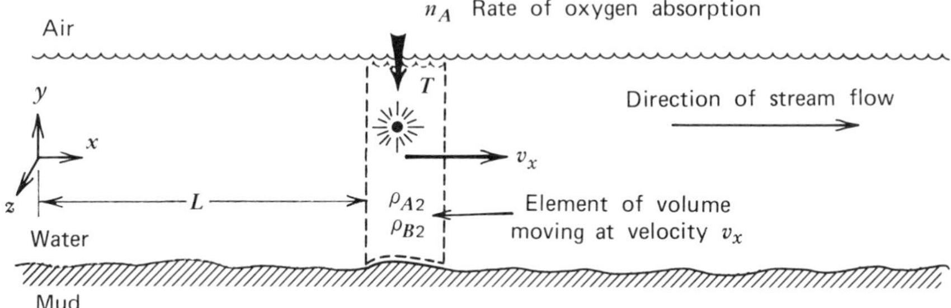

Figure 1.2-1. Natural stream reaeration.

volume element, located downstream from the organic waste point of entry at distance L, is assumed not to mix with elements upstream or downstream. Assume that only the two mechanisms biochemical oxidation and interphase mass transfer are occurring within the element. A material balance, according to Eq. 1.1-6, for component B in the volume yields the simple equation

$$0 = 0 + 0 + V(-r_B) + \frac{d}{dt}(V\rho_{B2}) \tag{1.2-1}$$

The rate of disappearance of B is usually assumed to be a first-order rate equation,

$$-r_B = k_B'''\rho_{B2} \tag{1.2-2}$$

where k_B''' is the rate constant. The negative sign appearing in front of the rate symbol denotes the disappearance of species B. The rate is normally defined positive for appearance of some reaction product.

Oxygen, component A, enters the volume element through the air–water interface. As shown in Chapter 4, the major controlling resistance to the absorption of oxygen resides in the water phase. The flux through the interface is represented by

$$n_A = {}^1k_{A2}'(\rho_{A2}^* - \rho_{A2}) \tag{1.2-3}$$

where ${}^1k_{A2}'$ is the liquid-phase mass transfer coefficient of component A associated with the air interface and $(\rho_{A2}^* - \rho_{A2})$ is the concentration departure from equilibrium for species A in water. It is common practice to measure organic concentrations in water as oxygen demands. The gross pollutant measures, such as *chemical oxygen demand* (COD) and *biochemical oxygen demand* (BOD), are measured in mg O_2/L. This makes the stoichiometric ratio, in an otherwise complex biochemical reaction, unity. A material balance on component A according to Eq. 1.1-6 results in

$$0 = 0 - n_A A_{xz} + \left(\frac{1}{1}\right) V(-r_B) + \frac{d}{dt}(V\rho_{A2}) \tag{1.2-4}$$

A_{xz} is the interfacial area in the x–z plane of air–water contact for the volume V. Using Eqs. 1.2-2 and 1.2-3 and constant V gives

$$^1k'_{A2} \frac{A_{xz}}{V} (\rho^*_{A2} - \rho_{A2}) - k'''_B \rho_{B2} = \frac{d\rho_{A2}}{dt} \tag{1.2-5}$$

In the interest of simplifying Eq. 1.2-5, the oxygen deficit is defined:

$$\Delta_A \equiv \rho^*_{A2} - \rho_{A2} \tag{1.2-6}$$

With this definition and the fact that ρ^*_{A2} is constant if stream temperature is constant, Eq. 1.2-5 becomes

$$\frac{d\Delta_A}{dt} = k'''_B \rho_{B2} - {}^1k'_{A2} \frac{A_{xz}}{V} \Delta_A \tag{1.2-7}$$

In the terminology used in stream pollution k'''_B is the *deoxygenation coefficient* with dimensions of (t^{-1}), and the group $(^1k'_{A2})(A_{xz}/V)$ is the *reaeration coefficient* with the same dimensions. Combining Eqs. 1.2-1 and 1.2-2 we get

$$\frac{d\rho_{B2}}{dt} = -k'''_B \rho_{B2} \tag{1.2-8}$$

These final two equations represent the quantitative result of the stream reaeration analysis. A simple integration of the two equations, with $\Delta_A = \Delta^0_A$ and $\rho_{B2} = \rho^0_{B2}$ at $t = 0$ as the initial condition representing in-stream oxygen deficit and organic concentration after mixing at the waste input point, results in

$$\Delta_A = \frac{k'''_B \rho^0_{B2}}{(^1k'_{A2}/h) - k'''_B} \left[\exp(-k'''_B t) - \exp\left(\frac{-{}^1k'_{A2} t}{h}\right)\right] + \Delta^0_A \exp\left(\frac{-{}^1k'_{A2} t}{h}\right) \tag{1.2-9}$$

where V/A_{xz} has been replaced by h, the average depth of a constant-cross-section stream. Figure 1.2-2 shows the shape of a typical oxygen sag curve. Time of flow can be related to the distance from the waste input point L and

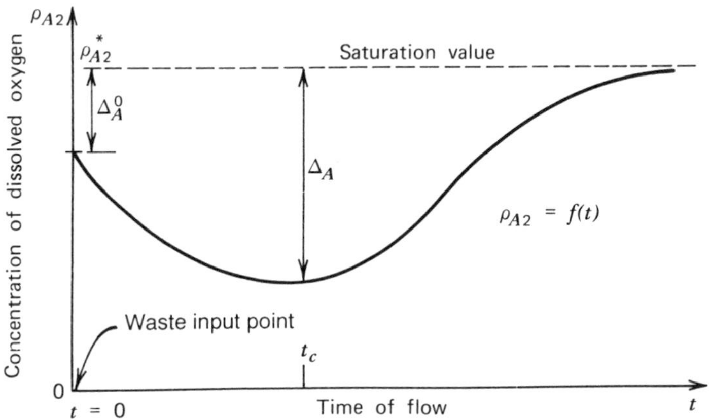

Figure 1.2-2. Dissolved oxygen sag curve.

the mean stream velocity v_x by

$$t = \frac{L}{v_x} \tag{1.2-10}$$

The minimum point in the sag curve is usually of special interest, being the location where the concentration of dissolved oxygen is a minimum. Typically, this minimum should not be less than the oxygen concentration desirable to sustain some higher life-forms within the stream. The minimum can be obtained by setting the first derivative of the deficit in Eq. 1.2-9 to zero and solving for t_c:

$$t_c = \frac{1}{({}^1k'_{A2}/h) - k'''_B} \ln \left\{ \frac{{}^1k'_{A2}}{hk'''_B} \left[1 - \Delta^0_A \frac{({}^1k'_{A2}/h) - k'''_B}{k'''_B \rho^0_{B2}} \right] \right\} \tag{1.2-11}$$

The downstream location of the minimum is $L_c = t_c v_x$.

Within its range of application this model of oxygen dynamics describes the major features of deoxygenation and reaeration. Equations 1.2-9 to 1.2-11 can be used for making calculations of the stream assimilation capacity for a particular waste input under ideal condition (see Problem 1.2B). This simple stream model has several important limitations. Oxygen concentration must be greater than zero. Most streams have other oxygen sources and sinks, including consumption at the mud–water interface and oxygen associated with algal respiration processes. Organic material may be released from deposits in bottom muds, and hydraulic backmixing usually occurs to some degree in all streams. Many streams change significantly in form within relatively short stretches (i.e., pool and ripple streams) so that the average depth h, velocity v_x, and other quantities affected by these variables also change significantly and the model is therefore inadequate.

Mechanism of Reaeration in Natural Streams

If there is no utilization of the dissolved oxygen within the body of water, the rate of reaeration may be obtained from Eq. 1.2-5.

$$\frac{d\rho_{A2}}{dt} = \frac{{}^1k'_{A2}}{h}(\rho^*_{A2} - \rho_{A2}) \tag{1.2-12}$$

which integrates to

$$\rho_{A2} = \rho^*_{A2} - (\rho^*_{A2} - \rho^0_{A2}) \exp\left(-\frac{{}^1k'_{A2}t}{h}\right) \tag{1.2-13}$$

These two equations show that the reaeration coefficient ${}^1k'_{A2}/h$ is critically important in predicting the oxygen uptake. This coefficient or ${}^1k'_{A2}$, the liquid-phase mass transfer coefficient, and the relationship to observable stream characteristics have had the attention of many investigators. O'Connor and Dobbins[4] have applied basic concepts of turbulence successfully to explain and quantify the reaeration coefficient. The following discussion is adopted from their work.

In any stream the ability to absorb oxygen from the atmosphere, the *reaeration capacity*, is a direct function of the degree of turbulent mixing. Atmospheric oxygen can be obtained only at the water surface, and the rate at which reaeration can take place is therefore directly limited by the rate of surface-water replacement in a flowing stream. Thus in a relatively still pool reaeration is a very slow process, whereas the reaeration capacity of a rapids section is very great. In turn, in any stream the rate of water surface replacement at the air–water interface is controlled by the stream's physical characteristics and is related to the associated water-flow properties.

O'Connor and Dobbins considered that the controlling factor in oxygen absorption was the resistance of a liquid film at the surface, through which oxygen must be absorbed by molecular diffusion. The film was assumed to be constantly renewed by unsaturated elements from the body of the stream through the mechanism of turbulence, as proposed originally by Higbie.[5] The rate of transfer through an element of the surface depends on the length of time it has been exposed to the atmosphere. The function describing the age distribution of surface elements was taken to be that of Dankwerts,[6] which is

$$\phi(t) = s\exp(-st) \tag{1.2-14}$$

in which $\phi(t)$ is the fractional part of the exposed surface area having ages between t and $t + dt$, and s is the rate of surface renewal, dimensions of t^{-1}. These ideas, along with Fick's law applied to molecular diffusion and the definition of a liquid-phase mass transfer coefficient, result in an equation for that coefficient:

$$^{1}k'_{A2} = (\mathscr{D}_{A2}s)^{1/2} \tag{1.2-15}$$

in terms of the molecular diffusivity of oxygen in water $\mathscr{D}_{A2}$ and the surface-renewal rate. Some basic theory of fluid turbulence is needed to understand and interpret s in Eq. 1.2-15. An in-depth presentation of turbulent flow is presented in a later chapter.

In turbulent flow a complex secondary motion is superimposed on the primary motion of translation. Turbulence is characterized by eddies that transport parcels of fluid from one layer to another with varying velocities. The eddy motion, which is erratic and seemingly unpredictable, can only be defined in terms of probability. Thus the principles of statistics are employed to define quantitatively parameters of turbulence, such as size of eddies and the velocity fluctuations.

The instantaneous velocity at any point in turbulent flow varies in magnitude and direction. Although there is no net flow in the vertical or y-direction, because of the eddies that exist there can exist a rapidly fluctuating (i.e., up and down) velocity. The intensity of this vertical velocity fluctuation can be calculated from direct measurements and is $\sqrt{\overline{(v'_y)^2}}$, where v'_y represents the instantaneous fluctuating velocity in the y-direction.

Although the velocity fluctuations define the intensity of turbulence, some linear measure is required to define the scale of turbulence. In this regard the mixing-length theory developed by Prandtl in 1925 is appropriate. By assuming that eddies move around in a fluid very much as molecules move about in a gas (actually, a very poor analogy), Prandtl developed an expression for the fluctuating velocity in a fluid in which the mixing length plays a role roughly analogous to that of the mean free path in gas kinetic theory. This way of thinking led Prandtl to the relation

$$\sqrt{\overline{(v'_y)^2}} = l\,\frac{dv_x}{dy} \tag{1.2-16}$$

where l is the mixing length and signifies a distance that a parcel of fluid moves from its point of departure from the mean motion until it mixes again with the main body of the fluid. It is doubtful whether it is possible to assign a more definite physical meaning to the mixing length, but it can be said that this length is a measure of the average size of the eddies responsible for the fluid mixing. The other term in Eq. 1.2-16 is the x-component velocity gradient in the vertical direction. Equation 1.2-16 applies to a turbulent condition in which there exists a velocity gradient.

In the cases of comparatively deep channels it is possible that the turbulence may approach an isotropic condition near the air–water interface. An isotropic turbulent condition is one in which the intensity of the velocity fluctuations in all three directions is very nearly the same. This type of turbulence, in which there is neither a shearing stress nor a velocity gradient, is approached in the flow downstream from screens, in a hydraulic jump, and in the center of a deep, wide-open channel. By contrast, nonisotropic turbulence is characterized by a significant correlation between the velocity fluctuations and by a velocity gradient and shearing stress. This type of turbulence is evidenced in flow in pipes and in comparatively shallow open channels.

In turbulent flow, momentum, mass, heat, or any inherent characteristic of the fluid can be transferred from one layer of fluid to another. The basic concepts of turbulence may therefore be used to determine the rate at which parcels of fluid at the surface layer can be replaced by parcels arising from the turbulent motion of the body of the fluid. The intensity of turbulence may be defined by some mean measure of the velocity fluctuations, such as $\sqrt{\overline{(v_y')^2}}$ and the scale of turbulence by the mixing length l. The mixing length signifies a distance a parcel moves from its point of departure from the mean forward motion until it mixes again with the main body of the fluid. Therefore, only parcels within a zone defined by a mixing length from the surface will affect the renewal of this surface. Furthermore, any parcel located at a distance greater than this length from the surface will be deflected from its vertical path before reaching the surface. It may be reasoned that vertical flow exhibiting small length and high-velocity characteristics will cause a greater rate of surface renewal than will flow of great length and low velocity. Therefore, parcels at the surface are replaced at a rate directly proportional to the intensity of turbulence and inversely proportional to the scale of turbulence. Surface renewal may be considered to take place in a period of time defined by

$$t = \frac{l}{\sqrt{\overline{(v_y')^2}}} = \frac{1}{s} \tag{1.2-17}$$

where the values of l and $\sqrt{\overline{(v_y')^2}}$ are those that prevail in the vicinity of the surface. At any point the period defined by Eq. 1.2-17 varies in time owing to the random nature of the variables involved. However, since the length and velocity are average values, t is then the average time during which such surface renewal takes place.

Although the velocity gradient may be zero at the stream surface in the case of isotropic turbulence, heat and mass can be transferred across this plane. Similarly, the mixing length and the vertical velocity fluctuations have finite values. The only way to obtain values of these parameters is by direct measurement. From published values of field observations of the mixing length and the vertical velocity fluctuation on the Mississippi River and its estuaries it was concluded that on the average the mixing length and the vertical velocity fluctuation were approximately 10% of the average depth and the mean flow velocity, respectively. These values may be used as an approximation to determine the rate of surface renewal:

$$s = \frac{\sqrt{\overline{(v_y')^2}}}{l} = \frac{0.1v_x}{0.1h} = \frac{v_x}{h} \tag{1.2-18}$$

Substitution of this value for s in Eq. 1.2-15 gives

$${}^1k'_{A2} = \left(\frac{\mathscr{D}_{A2}v_x}{h}\right)^{1/2} \tag{1.2-19}$$

This equation approximates the gas-interface liquid-phase mass transfer coefficient for the case of isotropic turbulence in natural streams. Equation 1.2-19 was verified by comparing values of the reaeration coefficient reported previously by others (see Problem 1.2C) on five natural streams and one bay.

A Summary of Reaeration Coefficients

Most equations developed for the prediction of this mass transfer coefficient are of the form

$${}^1k'_{A2} = \frac{\alpha v_x^{\beta}}{h^{\delta}} \tag{1.2-20}$$

Table 1.2-1 summarizes the results of several investigators. The O'Connor–Dobbins equation is applicable to rivers and nonstratified estuaries with depth range 1 to 30 ft and velocity range 0.5 to 1.6 ft/s. The Churchill et al. equation is for large, swift rivers with depth 2 to 11 ft and velocity 1.5 to 5.0 ft/s, while that of Owens et al. is for shallow, swift streams with depth 0.4 to 11 ft and velocity 0.1 to 5.0 ft/s.[11] Notable alternative theories of stream aeration have been developed by Thackston and Krenkel[7] and Tsivoglou and Wallace.[8] The former is based on the average eddy viscosity in the stream, and the latter is

Table 1.2-1. Summary of Liquid-Phase Mass Transfer Coefficients at Air–Water Interface of Natural Streams

Investigators	Limitations, Origin	α^a	β	δ
O'Connor and Dobbins[4]	Theoretical and experimental	$\mathscr{D}_{A2}^{1/2}$	0.5	0.5
Churchill et al.[13]	Regression analysis of field data	5.026^b	0.969	0.673
Owens et al.[14]	Field data	9.41^b	0.67	0.85

$^a\alpha$ at 20°C; ${}^1k'_{A2}(T) = {}^1k'_{A2}(20)\cdot 1.016^{T-20}$ (T in °C).
$^b v_x$ in ft/s, h in ft gives ${}^1k'_{A2}$ in ft/day.

Table 1.2-2. Reaeration Coefficient Values

Water Body and Example	Average Discharge (ft^3/s)	Average Depth (ft)	Average Velocity (ft/s)	Typical Reaeration Coefficient (day^{-1})
Pond or marsh				
(no example)	—	—	—	0.2
Lake				
(Lake Tahoe)	—	—	—	0.3
Rapids and waterfalls	10–100	0.5–1.0	1.0–2.0	10–20
(no example)				
Streams and rivers				
Pecos River	300	2	2.0	6.0
Raritan River	1500	3	2.0	4.0
Rio Grande River	7500	5	2.5	2.0
Wabash River	50,000	12	3.0	0.8
Columbia River	150,000	25	4.0	0.4
Mississippi River	500,000	50	5.0	0.2–0.3

Source: Ref. 10.

based on the energy expended when flowing water undergoes a change in elevation. Example 1.2-1 demonstrates the applicability of the material in this chapter.

Table 1.2-2 contains typical values of reaeration coefficients for various water bodies. Another summary is contained in Exercise Problem 1.2C.

Closure

This chapter has been concerned primarily with the absorption of oxygen into natural streams. The author has departed from the conventional approach and has emphasized the liquid-phase mass transfer coefficient at the air–water interface, ${}^1k_{A2}$, rather than the *reaeration coefficient* (normally noted as k_{21} in most reaeration literature). This approach was taken with the broader goals of chemodynamics in mind. There is another liquid-phase mass transfer coefficient in natural streams associated with the interphase movement of chemicals at the mud–water interfaces. This coefficient is noted by ${}^3k_{A2}$ for oxygen and details are presented later in the book. Although the emphasis in this chapter has been on the absorption of a single chemical species into water, the resulting coefficients presented in Table 1.2-1 have a much wider range of utility for quantifying the exchange and movement of chemicals through the air–water interface of natural streams.

The use of a mass transfer coefficient defined by Eq. 1.2-3 using a mass concentration difference seems more appropriate. This definition maintains a parallelism with Fick's first law of diffusion and is the most unrestrictive and generally used coefficient in the field of interphase mass transfer. Mass transfer coefficients are scalars and are for a particular chemical species (hence the subscript A for O_2). Under the proper transformation, the mass transfer coefficients represented in Table 1.2-1 can be used for estimating the desorption of chemical species from the stream into the overlaying air mass (see Eq. 3.1-25). Tsivoglou used this concept when he employed a krypton-85 tracer in natural stream desorption studies to obtain reaeration coefficients.

Example 1.2-1. Recovery Time of Stream Void of Oxygen. The time required for a stream to recover from a hypothetical state of no dissolved oxygen to some final degree of saturation gives an indication of speed of reaeration.

(a) Calculate the time in seconds required for a stream void of oxygen to reach a state of 50% saturation. Note that this is the oxygen absorption half-life for this particular stream. Compute the distance downstream (km) and the oxygen concentration (mg/L).

(b) Calculate the time (s) required to reach 95% saturation. This time (i.e. $t_{95\%}$) is a realistic measure of the recovery time. Compute the distance and concentration. Stream data: Average velocity 0.21 m/s, average depth 1.8 m, and $Đ_{A2} = 2.42\text{E-}5\ \text{cm}^2/\text{s}$ at 30°C, the water temperature.

SOLUTION The transient oxygen concentration in a stream undergoing recovery is represented by Eq. 1.2-13. For the conditions of the problem $\rho^0_{A2} = 0$ in oxygen-deficit terminology, the equation becomes

$$\frac{\Delta_A}{\Delta^0_A} = \exp\left(-\frac{{}^1k'_{A2}t}{h}\right) \tag{E1.2-1a}$$

where $\Delta_A \equiv \rho^*_{A2} - \rho_{A2}$ and $\Delta^0_A \equiv \rho^*_{A2} - \rho^0_{A2}$.

(a) The half-life is $\Delta_A = \Delta^0_A/2$ and Eq. E1.2-1a becomes

$$t_{1/2} = \frac{0.693h}{{}^1k'_{A2}} \tag{E1.2-1b}$$

From Eq. 1.2-19 the reaeration coefficient becomes

$$\frac{{}^1k'_{A2}}{h} = \frac{(Đ_{A2}v_x)^{1/2}}{h^{3/2}}$$

$$= \left(2.42E\text{-}5\ \frac{\text{cm}^2}{\text{s}}\left|\frac{0.21\ \text{m}}{\text{s}}\right|\frac{100\ \text{cm}}{\text{m}}\right)^{1/2}\Bigg/\left(1.8\ \text{m}\left|\frac{100\ \text{cm}}{\text{m}}\right.\right)^{3/2} = 9.33\text{E-}6\ \text{s}^{-1}$$

$$= 0.806\ \text{day}^{-1}$$

$$t_{1/2} = \frac{0.693}{9.33\text{E-}6\ \text{s}^{-1}} = 7.42\text{E}4\ \text{s} \quad (20.6\ \text{h})$$

From Eq. 1.2-10, $L = tv_2 = (7.42\text{E}4\ \text{s})(0.21\text{E-}3\ \text{km/s}) = 15.6\ \text{km}$.

$$\rho^*_{A2} = 7.50\ \text{mg O}_2/\text{L} \qquad \text{from Appendix C}$$

$$\rho_{A2} = \rho^*_{A2}\left(1 - \frac{\Delta_A}{\Delta^0_A}\right) = 7.50\left(1 - \frac{1}{2}\right) = 3.75\ \text{mg O}_2/\text{L}$$

(b) Ninety-five percent recovery; $\Delta_A = 0.05\Delta^0_A$.

$$t_{95\%} = \frac{3h}{{}^1k'_{A2}} = 3.22\text{E}5\ \text{s} \quad (89.3\ \text{h})$$

$$L = (3.22\text{E}5\ \text{s})(0.21\text{E-}3) = 67.6\ \text{km}$$

$$\rho_{A2} = 7.50(1 - 0.05) = 7.13\ \text{mg O}_2/\text{L}$$

PROBLEMS

1.2A. In-Stream Oxygen Concentration Below Outfall

A uniform natural stream 0.76 m deep flows with an average velocity of 0.12 m/s. At the point where the waste outfall enters the stream the after-mixing concentration of oxygen-demanding organic material is 25 mg/L (as oxygen). These oxygen-demanding substances decay and utilize oxygen according to Eq. 1.2-8. The integrated form of this equation is

$$\rho_{B2} = \rho^0_{B2}\exp(-k'''_B t) \tag{1.2A}$$

The rate constant k'''_B, where $B \equiv$ organic material, is 0.3 day^{-1}. Temperature is 30°C. Do the following:

1. Calculate and graph the oxygen concentration downstream of the discharge point for each 2 h of flow time for a period of 1 day. Assume that $\rho^0_{A2} = 6.17\ \text{mg O}_2/\text{L}$.
2. Calculate and graph to concentration of organic waste in mg/L for the same time intervals.
3. Calculate the critical time (days) and distance downstream (km).

1.2B. Stream Assimilation Capacity for Waste Material

The ability of a natural stream to oxidize oxygen-demanding organic (or inorganic) waste material and still maintain a dissolved oxygen content above some arbitrary minimum is termed the *stream assimilation capacity*. Figure 1.2B illustrates the idea of an aqueous waste entering a stream. The notation Q_w, ρ_{A2w}, and ρ_{B2w} represents the volumetric flow rate, dissolved oxygen

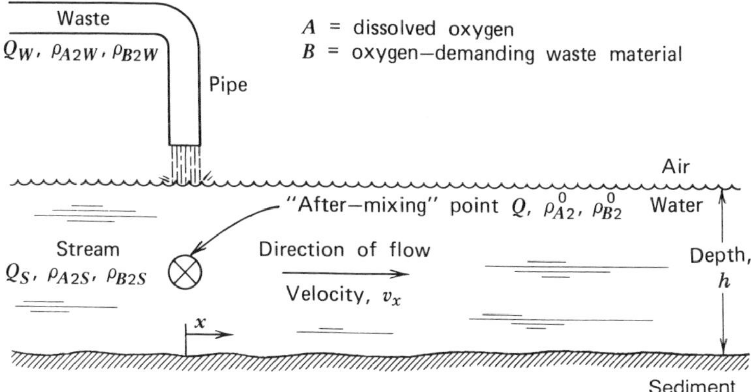

Figure 1.2B. Stream assimilation capacity.

concentration, and O_2-demanding matter concentration of the waste. The terms Q_s, ρ_{A2s}, and ρ_{B2s} represent the volumetric flow rate, dissolved oxygen concentration, and O_2-demanding matter concentration of the stream. The in-stream conditions after mixing is represented by similar notation: Q, ρ^0_{A2}, and ρ^0_{B2}. The assimilation capacity is the quantity represented by the product $Q_w \rho_{B2w}$.

Material balances for the mixing point result in three useful equations: For water,

$$Q = Q_s + Q_w \tag{1.2B-1}$$

For dissolved oxygen,

$$\rho^0_{A2} = \frac{Q_s \rho_{A2s} + Q_w \rho_{A2w}}{Q} \tag{1.2B-2}$$

For oxygen-demanding material,

$$\rho^0_{B2} = \frac{Q_s \rho_{B2s} + Q_w \rho_{B2w}}{Q} \tag{1.2B-3}$$

For many applications the waste flow rate is small compared to the stream flow rate, the dissolved oxygen concentration in the waste is zero, and the stream is practically void of oxygen-demanding matter prior to mixing with the waste. For this situation, Eqs. 1.2B-1 and 1.2B-2 become for dissolved oxygen,

$$\rho^0_{A2} \simeq \rho_{A2s} \tag{1.2B-4}$$

and for oxygen-demanding material,

$$\rho_{B2}^{0} = \frac{Q_w \rho_{B2w}}{Q_s} \tag{1.2B-5}$$

The assimilation capacity can now be represented by the product $Q_s \rho_{B2}^{0}$.

Do the following:

1. Verify Eqs. 1.2B-1 to 1.2B-5.
2. A chemical manufacturing company plans to build a plant on Clearwater River. Calculate the assimilation capacity of Clearwater River (kg oxygen-demanding material/day) if stream standards stipulate that the minimum dissolved oxygen in the river must not fall below 4 mg of dissolved oxygen/L under low-flow conditions of 1.8E5 m^3 day and a critical temperature of 30°C. It is assumed that the stream above the mixing point is fully saturated with oxygen and is free of oxygen-demanding organic matter. The following data apply: Average velocity 0.21 m/s, average depth 1.8 m, $k_B''' = (0.12)(1.047)^{T-20}$ day^{-1}, and T in °C.

1.2C. Comparing the O'Connor–Dobbins Reaeration Model with Stream Data

A portion of the field data employed by O'Connor and Dobbins in the verification of their proposed model appears in Table 1.2C.

1. Calculate a ${}^1k_{A2}$ ($A \equiv O_2$) value for each field data point in g/cm$^2\cdot$s. Note that ${}^1k_{A2} = {}^1k'_{A2}\rho_2$.
2. Prepare a parity plot for comparing the calculated coefficient with the coefficient observed. [*Hint:* Plot $(\log {}^1k_{A2})_{\text{obs}}$ versus $(\log {}^1k_{A2})_{\text{calc}}$.]
3. Discuss this graphical demonstration of verification.

Table 1.2C. Field Data and Observed Reaeration Coefficients

River	Depth (m)	Velocity (m/s)	Temperature (°C)	Renewal Time (s)	Observed ${}^1k'_{A2}/h$ (day^{-1})
Elk	0.27	0.30	12	1.1[a]	11.1
Clarion	0.58	0.17	13	1.7	6.00
Tennessee	1.2	0.22	23	5.3[a]	3.03
Illinois	2.8	0.42	27	6.7	0.62
San Diego Bay	3.7	.098	20	37.5	0.111
Ohio	2.1	0.18	24.5	13	0.44

[a]Time of renewal reduced by the percentage of rapids in the stretch.

1.2D. Location of Critical Oxygen Concentration

Demonstrate the details of obtaining Eq. 1.2-11 from Eq. 1.2-9.

1.2E. Comparing Reaeration Models with Stream Data

Using the appropriate equation from Table 1.2-1 calculate the reaeration coefficient (i.e., ${}^1k'_{A2}/h$ in day^{-1}) for the rivers and streams in Table 1.2-2. Assume that the water is at 25°C. Compare the calculated coefficients to the values in Table 1.2-2.

1.2F. Is the Reaeration Coefficient a Fundamental Stream Parameter?

The coefficient ${}^1k'_{A2}/h_2$ that results from modeling the oxygen processes in streams is a highly variable parameter whereas ${}^1k'_{A2}$ is not. Using the data in Tables 1.2-2 and 1.2C compute the statistics: mean, standard deviation and coefficient of variation of each parameter, as well as h_2 and v_x. Comment on the results.

1.2G. Putting SOD in the Stream Oxygen Model

Rederive the O_2-organic matter fate model accounting for bottom sediment oxygen demand (SOD) as a constant rate process; n_{A3} in g $O_2/m^2 \cdot s$.

REFERENCES

1. "Scientists Probe Pesticide Dynamics," *Chem. Eng. News*, Apr. 22, 1974, pp. 32–33.
2. J. R. Fair, B. B. Crocker, and H. R. Null, "Trace-Quantity Engineering", *Chem. Eng.*, Aug. 7, 1972, p. 60.
3. H. W. Streeter and E. B. Phelps, "A Study of the Pollution and Natural Purification of the Ohio River," *U.S. Public Health Service Bulletin 146,* 1925.
4. D. J. O'Connor and W. E. Dobbins, "The Mechanism of Reaeration in Natural Streams," *J. Sanit. Eng. Div., Proc. Am. Soc. Civ. Eng.*, **82**(SA6), 1115 (1956).
5. R. Higbie, "The Rate of Absorption of a Pure Gas into a Still Liquid During Short Periods of Exposure," *Trans. Am. Inst. Chem. Eng.*, **31**, 365 (1935).
6. P. V. Dankwerts, "Significance of Liquid-Film Coefficients in Gas Absorption," *Ind. Eng. Chem.*, **43**(6), 1469 (1951).
7. E. L. Thackston and P. A. Krenkel, "Reaeration Prediction in Natural Streams," *J. Sanit. Eng. Div., Am. Soc. Civ. Eng.*, **95**(SA1), 65–94 (1969).
8. E. C. Tsivoglou and J. R. Wallace, "Characterization of Stream Aeration Capacity," *EPA-R3-72-012*, U.S. EPA, Washington, D.C., 1972.
9. L. J. Thibodeaux, *Chemodynamics*, Wiley, New York, 1979, p. 11.

10. B. G. Liptak, Ed., *Environmental Engineers' Handbook*, Vol. 1, *Water Pollution*, Chilton, Radnor, Pa., 1974, p. 196.
11. Tetra Tech, *Rates, Constants and Kinetic Formulations in Surface Water Quality Modeling*, prepared for U.S. Environmental Protection Agency, 1978.
12. L. J. Thibodeaux, M. Poulin, and S. Even, "A Model for Enhanced Aeration of Streams by Motor Vessels with Application to the Seine," *J. Hazard. Mater.*, **37**, 459–473 (1994).
13. M. A. Churchill, H. L. Elmore, and R. A. Backingham, *J. Sanit. Eng. Div., Proc. Am. Soc. Civil Eng.* (SA4), 1 (July 1962).
14. M. Owens, R. W. Edwards, and J. W. Gibbs, *Int. J. Air Water Pollut.*, **8**, 469 (1964).

2

EQUILIBRIUM AT ENVIRONMENTAL INTERFACES

2.1. CHEMICAL EQUILIBRIUM AT ENVIRONMENTAL INTERFACES

Air, water, and earthen solids constitute the three major phases of the Earth's crust. The fourth phase, the biosphere, contains the living material and because of human and ecosystem health concerns it is the truly important one, however, on a mass basis the fourth phase is of minor importance; the bulk of the chemical residuals are usually found to be in the atmosphere or suspended in water or in surface soils, river silt, and similar earthen solids. The use of an equilibrium model is a way of connecting the biosphere to the other phases for estimating its concentration of hazardous substances.

As we begin to address the nature of chemical transfer between the three major phases it is necessary to consider some general physical and chemical properties of each phase. From a gross point of view, the atmosphere is a thin layer of a gaseous mixture surrounding the Earth's surface and remaining attached by the pull of gravity. The hydrosphere consists more than 99% of water, mainly in liquid form. The lithosphere is the earthen solid material underneath the air mass and the oceans. It is categorized as soil, rock, sand, mud, clay, and so on. A silt loam surface in good condition for plant growth is 20 to 30% (volume) air, 20 to 30% water, 45% minerals, and 5% organic matter. Tables 2.1-1 to 2.1-4 contain chemical data characterizing and typifying the three major phases: the sea level composition of the dry atmosphere, elements in solution in seawater, and those in rocks and soil in the lithosphere.

An *interface* is the place at which two different systems (or subsystems) meet and interact with each other; it is also the boundary between any two phases. Among the three phases (gas, liquid, and solid) there are five types of interface: gas–liquid, gas–solid, liquid–liquid, liquid–solid, and solid–solid. In considering chemical movements the last type is almost unimportant; however, the remaining four interfaces constitute important environmental planes through which a host of synthetic chemicals move, eventually becoming distributed throughout the ecosystem.

Table 2.1-1. Sea Level Atmospheric Composition for a Dry Atmosphere[a]

Constituent Gas	Molecular Fraction (%)	Molecular Weight (O = 16.000)
Nitrogen (N_2)	78.09	28.016
Oxygen (O_2)	20.95	32.0000
Argon (Ar)	0.93	39.944
Carbon dioxide (CO_2)	0.03	44.010
Neon (Ne)	1.8×10^{-3}	20.183
Helium (He)	5.24×10^{-4}	4.003
Krypton (Kr)	1.0×10^{-4}	83.7
Hydrogen (H_2)	5.0×10^{-5}	2.0160
Xenon (Xe)	8.0×10^{-6}	131.3
Ozone (O_3)	1.0×10^{-6}	48.0000
Radon (Rn)	6.0×10^{-18}	222.0

Source: Ref. 1. Reprinted with permission from *CRC Handbook of Chemistry and Physics*, 49th ed., Robert C. Weast, Ed., CRC Press, Cleveland, Ohio, 1968, p. F-147. © The Chemical Rubber Co., CRC Press, Inc.

[a]These values are taken as standard and do not necessarily indicate the exact condition of the atmosphere. Ozone and radon, particularly, are known to vary at sea level and above.

Table 2.1-2. Elements Present in Solution in Seawater (Excluding Dissolved Gases)

Element	Concentration (g/metric ton or ppm)
Cl	18,980
Na	10,561
Mg	1,272
S	884
Ca	400
K	380
Br	65
C (inorganic)	28
Sr	13
(SiO_2)	0.01–7.0
B	4.6
Si	0.02–4.0
C (organic)	1.2–3.0

Source: H. V. Sverdrup, M. W. Johnson, and R. H. Fleming, *The Oceans*, © 1942, renewed 1970, pp. 176–177. Printed by permission of Prentice Hall, Englewood Cliffs, N.J.

Table 2.1-3. Chemical Composition of Rocks (%)

Element	Average Igneous Rock	Average Shale	Average Sandstone	Average Limestone	Average Sediment
SiO_2	59.14	58.10	78.33	5.19	57.95
TiO_2	1.05	0.65	0.25	0.06	0.57
Al_2O_3	15.34	15.40	4.77	0.81	13.39
Fe_2O_3	3.08	4.02	1.07	0.54	3.47
FeO	3.80	2.45	0.30	—	2.08
MgO	3.49	2.44	1.16	7.89	2.65
CaO	5.08	3.11	5.50	42.57	5.89
Na_2O	3.84	1.30	0.45	0.05	1.13
K_2O	3.13	3.24	1.31	0.33	2.86

Source: "Chemical Composition of Average Rocks" (after Clarke), in F. J. Pettijohn, *Sedimentary Rocks, 2nd ed.* © 1949, 1957 by Harper & Row, Publishers, Inc. Used by permission of the publisher.

Table 2.1-4. Analysis of Representative U.S. Surface Soils (%)

Constituents	Norfolk Fine Sand, Florida	Sassafras Sandy Loam, Virginia	Ontario Loam, New York	Loam from Ely, Nevada	Hagerstown Silt Loam Tennessee	Cascade Silt Loam, Oregon	Marshall Silt Loam, Iowa	Summit Clay from Kansas
SiO_2	91.49	85.96	76.54	61.69	73.11	70.40	72.63	71.60
TiO_2	0.50	0.59	0.64	0.47	1.05	1.08	0.63	0.81
Fe_2O_3	1.75	1.74	3.43	3.87	6.12	3.90	3.14	3.56
Al_2O_3	4.51	6.26	9.38	13.77	8.30	13.14	12.03	11.45
MnO	0.007	0.04	0.08	0.12	0.44	0.07	0.10	0.06
CaO	0.01	0.40	0.80	5.48	0.37	1.78	0.79	0.97
MgO	0.02	0.36	0.75	2.60	0.45	0.97	0.82	0.86
K_2O	0.16	1.54	1.95	2.90	0.91	2.11	2.23	2.42
Na_2O	Trace	0.58	1.04	1.47	0.20	1.98	1.36	1.04

Source: Ref. 2. Copyright © 1974 by Macmillan Publishing Co., Inc.

A chemical introduced into the environment on one side of an interface will eventually become present, by a spontaneous process, in the other phases. As molecules of the chemical begin to accumulate on the other side, a certain level is reached after which gross accumulation ceases. At this time definite and possibly drastically different concentrations are in evidence on opposite sides of the interface. A system is in *equilibrium* when its state is such that it can undergo no spontaneous or unaided changes. When no further observable changes in concentration of the chemical occur on either side of the interface, equilibrium between phases has been attained. This description of phase equilibrium is strictly phenomenal.

A kinetic explanation of chemical equilibrium in an air–water system maintains that molecules of the chemical are crossing the interface continually but that the rate at which gas-phase molecules condense on the liquid surface is equal to the evaporation rate of the chemical from the liquid surface into the gas. Thermodynamic considerations require that the chemical potentials of the component in both phases of a multicomponent system are identical at equilibrium for constant pressure and temperature:

$$\mu_{A1} = \mu_{A2} \tag{2.1-1}$$

where $\mu \equiv$ chemical potential of constituent A. Since the fugacity of any component i is related directly to its chemical potential by the relation $(d\mu_A = RTd\ln f_A)_T$, fugacity can also be used as a criterion of equilibrium between phases. This is a more useful property than chemical potential for defining equilibrium, since fugacity can be expressed as an absolute value, whereas chemical potential can be expressed only relative to an arbitrary reference state. Conceptually, fugacity is closely related to pressure. For equilibrium of component A between two phases

$$f_{A1} = f_{A2} \tag{2.1-2}$$

where the subscripts refer to the phases: $1 \equiv$ air and $2 \equiv$ water. Since fugacity (f) can be expressed as a simple function of activity coefficient (γ), mole fractions (x_A and y_A), and a reference fugacity (f^0), a more useful form of the equation results:

$$(y_A\gamma_A f_A^0)_1 = (x_A\gamma_A f_A^0)_2 \tag{2.1-3}$$

This final equality is the starting point for calculations of equilibrium concentrations (i.e., x_A, y_A) of chemicals on both sides of an interface and can be generalized to include all four phase combinations.

One important and practical aspect needed to predict the movement of hazardous materials through the environment is phase equilibrium. This information can help answer such questions as: (1) If a liquid organic solution is in contact with the air, what is the maximum concentration of organics in the air?; or (2) If an oily organic phase is in contact with water, what is the maximum concentration of organics in water? In theoretical studies and practical calculations, a key parameter is the activity coefficient, γ_{ij}. This variable is a unifying concept that ties together questions of phase equilibria of all kinds: two-component, multicomponent, solid, liquid, and gaseous. It is the basis for calculation methods.

A thermodynamic text by Hougen et al.[3] contains details and extensions of the general criteria for equilibrium between phases. Extensions and applications of Eq. 2.1-3 for equilibria at environmental interfaces is treated in detail and quantitative results are made available in the following sections. Lyman et al.[4] treat the subject of equilibrium at natural environment interfaces in more detail. Their work is a reinterpretation and extension of the fundamentals of the subject as presented by Reid et al.[5]

Ideal Solutions

An *ideal solution* is defined as a solution in which the fugacity of each component i is equal to the product of its mole fraction and the fugacity of the pure component at the same temperature, pressure, and state of aggregation as those of the solution. An ideal liquid solution presupposes that when one component is mixed with another, mutual solubility results, that no chemical interaction with accompanying heat effects occur, that molecular diameters are the same, and that intermolecular forces of attraction and repulsion are the same between unlike as between like molecules. Most chemicals of concern in the environment form nonideal aqueous solutions as they exist in water. At low pressures gaseous mixtures exhibit nearly ideal behavior. For this reason chemicals in air as molecular dispersed species are assumed here to have ideal behavior.

Air–Water Equilibrium Occurrences

The interface between the atmosphere and hydrosphere is an important site for the transfer of chemicals because of the fluid nature of each phase. The material in this section is concerned with the equilibrium distribution of chemicals across gas–liquid interfaces. It includes air, pure gases, gas mixtures, water, nonaqueous liquids, and organic sludges where constituent partial pressure and solubility are the desired chemodynamic parameters.

Pure Gases in Contact with Water. When equilibrium is established between a pure gas and water at constant pressure and temperature, the solubility of the gas in water is manifest. We are concerned here only with atmospheric pressure data. For example, a vapor cloud of pure ethane gas from a ruptured cryogenic container on a windless day residing above a water body results in concentration levels approaching its solubility (see Example 2.1-1). Equation 2.1-3 becomes

$$1 = x_A \gamma_{A2} f^0_{A2} \tag{2.1-4}$$

where γ_{A2} is the activity coefficient, f^0_{A2} the pure component fugacity of A in the water, and x_A the mole fraction solubility of A in water. This equation is useful for calculating x_A for gases for which no solubility data are available, since γ_A and f^0_A can be obtained from further thermodynamic relations.

The system pressure can fall into one of three conditions. Some low-molecular-weight gases such as light hydrocarbons (butane and lighter) are unable to exert their full saturation vapor pressure at atmospheric temperature.

Solubility data are given at 1 atm where Eq. 2.1-4 applies, with f^0_{A2} being the saturation vapor pressure. If the vapor pressure is atmospheric pressure, $f^0_{A2} = 1$ in Eq. 2.1-4. If the system temperature exceeds the critical temperature of the gas, the saturation vapor pressure (i.e., f^0_{A2}) must be estimated by extrapolating to system temperature and the result used in Eq. 2.1-4. The activity coefficient, γ_{A2}, accounts for the nonideal behavior of A in liquid water. Activity coefficients for selected compounds are available in the literature; otherwise, they can be estimated by group contribution methods.[4,5]

Although some high-molecular-weight organics may have low pure-component vapor pressure and high boiling points relative to water, they nevertheless exhibit large relative volatilities, due to large activity coefficients in water. Activity coefficients of 10^3 to 10^7 for *n*-acids, *n*-primary alcohols, *sec*-alcohols, *n*-aldehydes, *n*-ketones, *n*-esters, *n*-ethers, *n*-chlorides, *n*-paraffins, and *n*-alkyl benzenes have been reported. Being equal the inverse solubility, see Eq. 2.1-38, explains the large values exhibited by sparingly soluble chemicals.

Solubility data are available for common gases. Some data are given in Appendix C. Gas solubility is highly dependent on pressure. The nature of the water also affects the magnitude of gas solubility, as indicated by the difference in oxygen solubility in fresh water and seawater (see Table C.2)

Example 2.1-1. Solubility of Ethane in Water. Estimate the solubility (g/m^3) of ethane ($C_2H_6 \equiv A$) in water at 25°C if it is present above water as a pure gas at 1 atm pressure. Here $\gamma_{A2} = 7.02\text{E}2$ at 25°C.

SOLUTION Since ethane in the gas is pure, $y_A = 1$; it forms an ideal gas mixture $\gamma_{A1} = 1$, and at 1 atm $f^0_{A1} = 1$, Eq. 2.1-3 applies. The critical pressure and temperature of ethane is 48.2 atm at 32.3°C. The specified temperature condition is below the critical temperature. The vapor pressure of ethane at 25°C is 39.4 atm $= f^0_{A2}$. This pressure is estimated from ethane vapor pressure data in a handbook:

$$x_A = \frac{1}{(7.02\text{E}2)(39.4)} = 3.62\text{E-}5$$

From lines (A) and (B)of Table 1.1-2,

$$\rho_{A2} = x_A M_A c_2 = 3.62\text{E-}5(30.07\,\text{g/mol})(55{,}400\,\text{mol}\,H_2O/\text{m}^3) = 60.3\,\text{g/m}^3$$

For the case of a gas mixture in contact with water, see Problem 2.1G.

Non-aqueous Liquids and Mixtures in Contact with Air. When equilibrium is achieved between a pure liquid chemical and the stagnant air space above at constant total pressure ($p_T = 1$ atm) and temperature, its pure component vapor pressure is manifest within the air space. An example of a pure liquid would be a layer of benzene, $p_A^0 = 0.125$ atm, floating on water. Equation 2.1-3 yields the air space mole fraction of chemical A:

$$y_A = \frac{p_A^0}{p_T} \tag{2.1-5}$$

where p_A^0 is the vapor pressure of liquid chemical A and p_T is the total pressure in the air space above the interface. Vapor pressure data for many pure liquids and solids at ambient temperatures can be found in *Perry's Chemical Engineers' Handbook*[63] and other chemical data handbooks. Some of the data for common liquids and solids are given in Appendix C. For some cases the chemical concentration or vapor density in air is more useful than mole fraction. These can be obtained through the ideal gas law to yield

$$\rho_{A1}^* = \frac{p_A^0 M_A}{RT} \tag{2.1-6}$$

where ρ_{A1}^* is the mass concentration of the volatile liquid in air and R is the universal gas constant, 82.1 atm $\cdot$ cm^3/ mol $\cdot$ K.

Effect of Temperature on Vapor Pressure. The forces causing the vaporization of a liquid are derived from the kinetic energy of translation of its molecules. An increase in this energy should increase the rate of vaporization and therefore the equilibrium vapor pressure, since it is directly proportional to the absolute temperature. This is found to be the case universally. It is the temperature of the liquid surface at the interface that is effective in determining the rate of vaporization and the vapor pressure.

If the temperature does not vary over wide limits and it is assumed that the latent heat of vaporization λ_A is constant, the volume of liquid is neglected and the ideal gas law applies; then the Clausius–Clapeyron equation is applicable:

$$\ln \frac{p_A^0}{p_{A0}^0} = \frac{\lambda_A}{R}\left(\frac{1}{T_0} - \frac{1}{T}\right) \tag{2.1-7}$$

where R is the universal gas constant and λ_A is the molal heat of vaporization. This equation permits calculation of the vapor pressure of a substance p_A^0 at a temperature T if the vapor pressure p_A^0 at another temperature T_0 is known. The latent heat of vaporization is the quantity of heat that must be added to transform a substance from the liquid to the vapor state at the same temperature. Values of the heats of vaporization at the normal boiling point of many compounds are listed in handbooks. Equation 2.1-7 gives rise to empirical formulas of the form $\ln p_A^0$ versus $1/T$ commonly used for correlating vapor pressure data (see Problem 2.1E).

When equilibrium is achieved between a mixture of nonaqueous chemicals and the vapor space above it at constant T and p the partial vapor pressure, p_A, of each chemical is manifest. Beginning with a gas phase ($1 \equiv$ air + vapors) and an organic phase ($4 \equiv$ organic) consisting of homologs, an equation analogous to Eq. 2.1-3 yields

$$p_A = x_{A4} p_A^0 \tag{2.1-8}$$

where x_{A4} is the mole fraction of constituents A in the nonaqueous liquid. This relation for obtaining the partial pressure, Raoult's law, is limited to mixtures that behave ideally. Examples of homologous mixtures that approach ideal behavior are the constituents in oil, gasoline, and oil fraction of sludges from oil–water separators (see Problem 2.1M). In the case of binary and multicomponent liquid mixtures that are nonideal, the equilibrium law at atmospheric pressure is

$$p_A = \gamma_{A4} x_{A4} p_A^0 \tag{2.1-9}$$

where γ_{A4} is the activity coefficient of constituent A in the organic phase. In general, activity coefficients are a function of composition; however, for sparingly soluble materials, $x_{A4} <$ E-3, it is almost independent of concentration. The limiting value as $x_{A4} \to 0$ is the activity coefficient at infinite dilution, γ_{A4}^{∞}, which is constant. An example of a nonideal binary mixture is 1,3,5-trinitronaphthalene, 0.05 g in 100 g of ethyl ether at 28°C. Munz and Roberts[16] report that cosolvents in water reduce the solute's γ_{A2} only at cosolvent mole fractions $x_{B2} >$ 5E-3 (A = solute, B = cosolvent).

Partition Coefficients for the Air–Water System. The need for equilibria information on the distribution of traces of chemicals between the air and water phases is of primary importance, since this is the more common case, rather than the preceding cases, where one or the other phase is a pure chemical. For example, the existence of traces of PCBs in air will involve all aquatic environments in the northern hemisphere, whereas a slick of benzene on a stream is a local matter.

Distribution Law (Definition). If a substance is added to a system of two mutually, totally, or nearly immiscible phases, in both of which it is soluble, it distributes itself between the two by dissolving in each in fixed proportions, independent of the quantity of the solute, the ratio of its concentrations in the two phases being constant at a constant temperature; this ratio (called the *distribution coefficient* or *partition coefficient*) is constant and equals the ratio of the values of the solubility of the substance in each of the two solvents. Other names are *partition law* and *Nernst theorem* or *law of distribution.* This definition appears in the 1964 edition of the *Concise Dictionary of Science* by Frank Gaynor. In principle it applies to the air–water, water–soil, and air–soil phases; however, for a given chemical it is constant only in a narrow range of concentrations (i.e., traces near zero) well away from the saturation concentration in the respective phases.

The distribution law definition for air–water is

$$\mathscr{K}^*_{A12} \equiv \frac{y_A}{x_A} \tag{2.1-10}$$

which can be reduced to fundamental thermodynamic parameters. Equation 2.1-3 can be transformed to yield

$$\mathscr{K}^*_{A12} = \frac{\gamma_{A2} p_A^0}{p_T} \tag{2.1-11}$$

which gives the partition coefficient of moles of A in air to moles of A in water, and is constant when $\gamma_{A2} = \gamma_{A2}^{\infty}$.

Henry's law applies for dilute solutions of chemicals, either a gas, liquid, or solid at ambient conditions, in water. Equilibrium is defined by giving the Henry's law constant and the temperature

$$p_A = H_A x_A \tag{2.1-12}$$

where H_A is the Henry's law constant for species A. Normally, p_A is the partial pressure of chemical A in the gas phase (air) measured in atmospheres and x_A is the mole fraction of chemical A in solution; therefore, H_A has dimensions of atmospheres. Now by using $p_A = y_A p_T$ along with Eq. 2.1-3, the Henry's law constant is also seen to be a function of the activity coefficient of A in water;

$$H_A = \frac{\gamma_{A2} p_A^0}{1} \tag{2.1-13}$$

Numerical values of Henry's law constant taken from handbooks and other sources should be used with care. The dimensions of the Henry's law constant depend on the particular definition and units used (see Problem 2.1K).

In summary, there are three ways of expressing air–water equilibrium for dilute solutions of chemical A: the partition coefficient, relative volatility, and Henry's law. As is evident from Eqs. 2.1-11 and 2.1-13, the partition coefficient and Henry's law reflect the same basic thermodynamic parameters: γ_{A2} and p_A^0. The third method is the relative volatility of chemical A to water. Correspondence of the numerical values of the relative volatility to unity quantify the preference of the chemical for water as opposed to air (see Thibodeaux[17] for definition and uses).

The choice of which one to use depends in part on the application and preferences of the user. Table 2.1-5 contains Henry's law constants and relative volatility values for trace quantities of common gases and liquids. Henry's law

Table 2.1-5. Vapor–Liquid Equilibriums of Selected Gases and Liquids in Water at 25°C

Component	Normal Boiling Point (°C)	Henry's Law Constant,[a] H_A	Relative Volatility,[b] α_A^*
Propylene	−48	5,690	182,100
Acetylene	−84	1,330	42,560
Bromine (Br_2)	−58.8	73.7	2,358
Acetylaldehyde	20.2	5.88	188
Acetone	56.5	1.99	63.7
Isopropanol	82.5	1.19	38.1
n-Propanol	97.8	0.471	15.1
Ethanol	78.4	0.363	11.6
Methanol	64.7	0.300	9.60
n-Butanol	117	0.182	5.82
Acetic acid	118.1	0.0627	2.01
Formic acid	100.8	0.0247	0.790
Propionic acid	141.1	0.0130	0.416
Phenol	181.4	0.0102	0.326

[a]See defining equation 2.1-12; H_A in atm/mol fraction.
[b]Defining equation $\alpha_A^* = (y_A x_B)/(y_B x_A)$, $B \equiv H_2O$.

constants of other common gases are given in Appendix C. Consult chemical handbooks as sources for additional data. Faced with no data, one must obtain activity coefficients and vapor pressures. Procedures are available for obtaining estimates of activity coefficients.[4,5] Problem 2.1F gives a simple method of estimating Henry's law constant based on vapor pressure and water solubility. This method has its limits, as is demonstrated in the problem solution. Wolfenden[18] observes that molecule self-association is likely to be encountered with polar compounds near their limits of solubility in water, so that such estimates should be viewed with skepticism.

All three air–water equilibrium parameters—$\mathscr{K}^*_{A12}$, H_A, and relative volatility—are truly constant only for dilute solutions when $\gamma_{A2} = \gamma_{A2}^\infty$. Activity coefficients are also a function of cosolvents in the aqueous phase. These other constituents in the aqueous phase can have a significant effect on the γ_{A2} value of the volatile chemical of interest.[16] See the discussion "Pure Liquids in Contact with Water" later in this section for further information on the cosolvent effect on activity coefficients.

Example 2.1-2. Vinyl Chloride in Air. The Henry's law constant for vinyl chloride ($A \equiv CH_2CHCl$) at 25°C has been reported to be 50, where H_ρ is defined as the concentration of A in air (mg A)/(L at 25°C, 760 mmHg) divided by concentration of A in water (mg A)/(L of water). Vinyl chloride appears to have a 50-fold greater distribution in air than in water under equilibrium conditions; however, the dimensions on H_ρ hamper this interpretation.

(a) Convert H_ρ to the partition coefficient given in Eq. 2.1-10 to effect a realistic partitioning of A between phases in terms of molecules and/or moles.

(b) Obtain the equilibrium concentration of A in air in mg A/L at 25°C, 760 mmHg above a wastewater treatment vessel in which the aqueous-phase concentration is 2.5 g/m^3.

SOLUTION (a) From Eq. 2.1-10,

$$\mathscr{K}^*_{A12}\left(\equiv\frac{y_A^*}{x_A}\right)=\frac{50\,\text{mg}\,A}{\text{L air}}\left|\frac{\text{L H}_2\text{O}}{\text{mg}\,A}\right|\frac{1000\,\text{g}}{\text{L H}_2\text{O}}\left|\frac{22.4\,\text{L air}}{\text{mol}}\right|\frac{\text{mol}}{18.01\,\text{g}}\left|\frac{298\,\text{K}}{273\,\text{K}}\right.$$

$$= 67{,}900$$

(b) From the definition of H_ρ

$$\rho^*_{A1} = H_\rho \rho_{A2} = 50(2.5\,\text{mg/L}) = 125\,\text{mg}\,A/\text{L in air}$$

Henry's constants are reasonably accurate in quantifying the chemical distribution between air and water for large flat interfaces such as aquatic bodies. Recent findings have identified unexpectedly high concentrations of certain pesticides and other hydrophobic organics in fog waters. This gives rise to significant deviations from the Henry's law predictions. Enrichment factors, defined as the ratio of the Henry's constant divided by the chemical concentration ratio observed air/water, several hundred to a few thousand have been compiled and reported by Valsaraj et al.,[57] who propose that the adsorption of hydrophobic compounds at the air–water interface of fog droplets of small diameters contributes significantly to the enhancement factors observed. Small droplets 2 to 8 μm in diameter have very large area/volume ratios.

Equilibria in the Natural Environment. Unfortunately, the behavior of chemicals in natural environments is rendered more complex by a number of factors. Most of the equilibrium data presented in handbooks and those reproduced in the appendixes were obtained under ideal laboratory conditions. These handbook data may not be the most desirable but must be used in the absence of data representing the natural environmental system and its accompanying synergistic effects.

Mackay and Shiu[6] review some of the factors that render more complex the equilibrium behavior of hydrocarbons in natural aquatic environments. The presence of electrolytes generally increases hydrocarbon activity coefficients and reduces solubility (i.e., the salting-out effect). Appendix C contains some data on the solubility of hydrocarbons in distilled water and seawater. Generally, the seawater solubility is about 70 to 80% of that in distilled water.

Groves[19] has developed a method of predicting cycloparaffin solubility in water of marine salinity levels from distilled water solubility data. It has been well established that hydrocarbons can exist in colloidal, micellar, or particulate form in appreciable quantities.[60,61] Filtration was found to reduce the apparent solubility. The presence of surface-active organic compounds (many of which occur naturally) increases the amount of colloidal hydrocarbons.[56] It appears that a typical oil, which may be truly soluble to the extent of 200 μg/L, may be solubilized to double this concentration by the presence of a few mg/L of dissolved organic surfactant and may be also present as particulate hydrocarbon to about 10 times the true solubility, or several mg/L.

Earthen Solid–Water Equilibrium Occurrences

The contact of earthen solids and water creates a complex interface with a large capacity for chemical uptake and eventual release. Describing the equilibrium state of chemicals between these two apparent phases in order to predict the environmental parameters that affect and regulate uptake and release is an ongoing challenge. A confounding factor in a proper thermodynamic description of the equilibrium process is the complex physicochemical nature of earthen solids. The earthen solids of primary concern from a chemical contamination point of view are surface soils, subsurface soils, and geologic formations containing groundwater plus suspended particles and bottom sediment in water bodies. For the purposes of chemical equilibrium, a more thorough description of earthen solids is needed than that presented in Tables 2.1-3 and 2.1-4.

Surface soil is probably the most complex of the earthen solids. It contains at least four separate phases. If the pore spaces are not completely saturated with water, they contain atmospheric gases; this constitutes two phases. Soil air usually contains considerably more carbon dioxide and slightly less oxygen than does atmospheric air. The soil water is possibly not unlike liquid water except for the few monomolecular layers, 0.1 to 0.5 μm in thickness, sorbed onto and near the solid surface. The solid phases are the natural organic matter (1 to 5%) and the mineral fractions (95 to 99%). The ubiquitous swelling phyllosilicate minerals known as smectite clays are components of many soils and sediments. Because of their small particle size ($<2\,\mu$m) and unusual intercalcation properties, they afford an appreciable surface area for the absorption of organic molecules.[20] Estuarine colloidal matter, a form of suspended earthen solids, has been characterized as a proteinaceous–carbohydrate polymer that is associated in varying amounts of poorly crystallized clay minerals and trace metals.[21] These two solid phases apparently interact in a synergistic manner and are a key to the absorptive character of soils and sediment for hydrophobic compounds. Additional information concerning soils and related earthen solids appear throughout the book and the index should be consulted.

In this section we present information on the various equilibrium scenarios necessary to describe most contact between earthen solids and water. These include pure solids and mixtures of solid chemicals in contact with water, nonaqueous liquids in contact with earthen solids, and partition coefficients for the distribution of trace contaminants between earthen solids and water.

Solid Chemicals in Contact with Water: Pure and Mixtures. The extent to which a solid substance mixes with a liquid to form a homogeneous system at a given temperature is defined as *solubility*. Following a maritime accident at the mouth of the Mississippi River, pentachlorophenol (PCP), a solid with density greater than water, was spread on the bottom sediment in a layer of the pure solid material. The water layers in intimate contact with the PCP constitute an equilibrium solution, 80 mg/L at 20°C, which is its solubility limit. For the case of pure solid chemicals (also gases and liquids) water solubility is an equilibrium state. Equation 2.1-3 for the case of a solid–liquid system becomes

$$f^0_{A3} = x_{A2}\gamma_{A2}f^0_{A2} \tag{2.1-14}$$

where 2 ≡ water phase and 3 ≡ solid phase. For chemicals that are solid at the temperature in question the situation is complex since the correct reference fugacity (f^0_{A2}) is the fugacity (or vapor pressure) of the pure solid in a hypothetical liquid state. This vapor pressure can be estimated by extrapolating the liquid vapor pressure curve below the triple point. When a solid chemical is in equilibrium with an aqueous solution of concentration x_A the fugacity f^0_{A3} equals the vapor pressure of the solid; thus $\gamma_{A2}f^0_{A2}$ can be calculated as f^0_{A3}/x_A. The activity coefficient can be estimated by techniques mentioned earlier. It can be obtained from experiment. For example, solid naphthalene has a vapor pressure of 1.14E-4 atm at 25°C. Extrapolation of the liquid vapor pressure curve gives a value of $f^0_{A2} = 3.07\text{E-}4$ atm. The solubility of solid naphthalene of 34.4 g/m^3 corresponds to a mole fraction x_A of 4.83E-6; thus, $\gamma_{A2}f^0_{A2}$ is 23.60 and γ_A is estimated to be 4.69E-4. Published aqueous solubility data for solid and liquid hydrocarbons under environmental conditions are available.[6] These data and those for other substances are given in Appendix C.

In the case of a mixture of solids a complication arises. The equilibrium concentration of any one chemical in the aqueous mixture may not be equal to its solubility in water. This is because in general the individual activity coefficients are a function of the other substances (i.e., cosolvent) present and the concentrations; however, Banerjee[22] found that for chlorobenzenes, a mixture of solids that do not interact, the components tend to behave independent of one another and their solubilities are approximately additive. (See the discussion "Multicomponent Liquids in Contact with Water" later in this section for further information on the cosolvent effect on activity coefficients).

Nonaqueous Liquids in Contact with Earthen Solids. It often occurs, particularly in the case of chemical spills on or in the ground, that the liquid phase is nonaqueous ($4 \equiv$ nonaqueous liquid phase). The introduction of this new phase into earthen solids involves yet another in an already complex system. An example of such an occurrence was the release of large quantities of benzene from damaged railroad tankcars in Dartmund, West Germany, as a result of air raids in World War II. Leaks of gasoline from storage tanks is a common occurrence. Nonaqueous liquids are, or contain, toxic substances that become sources for chemical transport to air (volatilization) or water (leaching). In this subsection techniques are developed for estimating chemical partitioning, under equilibrium conditions, between a nonaqueous liquid and water within soil pore spaces. The more general problem of chemical mass distribution between the various phases in soil, for determining concentration levels, is considered in Section 2.3.

Soils and sands have a capacity for retaining nonaqueous liquids. Freeze and Cherry[23] introduced the concept of residual saturation to quantify the volume of soil required to immobilize a volume of applied liquid. The definition of residual saturation, S_{43}, is

$$S_{43} \equiv \frac{V_4}{\varepsilon V_B} \tag{2.1-15}$$

where V_4 is the volume of nonaqueous liquid applied (cm^3), ε the porosity of dry soil (void fraction), and V_B the volume of bulk soil required to immobilize the liquid (cm^3). Measured values for eight liquids, including water, with a sand and a surface soil yielded S_{43} values of 0.26 to 0.75.[24] These numerical values indicate that a volume of nonaqueous liquid can contaminate up to four volumes of bulk soil.

Obviously, the residual saturation is not a thermodynamic quantity, but it does indicate that earthen solids have a large capacity for retaining these fluids. Doshi[25] observed little or no water loss upon the introduction and movement of immiscible liquids through sand containing 10% water; however, some water was lost if the liquids were miscible. It often occurs that a sufficient quantity of the nonaqueous liquid exists to form a separate phase. Equilibrium relationships are needed to relate chemical concentrations in the adjoining liquids within soil pore spaces.

The equilibrium partition coefficient can be defined in terms of mass concentration ratios,

$$\mathscr{K}^*_{A42} \equiv \frac{\rho_{A4}}{\rho_{A2}} \tag{2.1-16}$$

where ρ_{A4} and ρ_{A2} are the concentrations (gA/cm^3) in phases 4 and 2, respectively. For the case of equilibrium solutions in both phases a relationship such as Eq. 2.1-3 is used to obtain a thermodynamic equation for the partition coefficient:

$$\mathscr{K}^*_{A42} = \frac{\gamma_{A2} c_4}{\gamma_{A4} c_2} \tag{2.1-17}$$

where c_4 and c_2 are molar densities of the respective liquids (mol/cm^3). Generally, the activity coefficients are subject to cosolute effects (see the discussion "Multicomponent Liquids in Contact with Water" later in this section.) Example 2.1-3 is a simple but realistic application of the material discussed above.

Example 2.1-3. Landfill Leachate Concentration from Oily Sludge. Estimate the equilibrium concentration of benzene ($\equiv A$) in the water ($\equiv 2$)-filled pore space within a landfill cell containing an oily sludge ($\equiv 4$) with 1400 ppm (wt) benzene at 25°C. Assume that $\bar{M}_4 = 111$ g/mol, $\rho_4 = 0.9$ g/cm^3, and $\gamma_{A4} \simeq 1.0$, since benzene probably forms an ideal solution in the sludge.

SOLUTION The leachate concentration can be obtained from Eq. 2.1-16:

$$\rho_{A2} = \frac{\rho_{A4}}{\mathscr{K}^*_{A42}} \tag{E2.1-3A}$$

where $\rho_{A4} = 1440(0.9) = 1263$ mg A/L sludge. At its solubility, x^*_A, benzene in water forms a dilute solution and $\gamma_{A2} = 1/x^*_A$ (see Problem 2.1L), so that Eq. 2.1-17 simplifies to

$$\mathscr{K}^*_{A42} = \frac{c_4}{x^*_A c_2} \tag{E2.1-3B}$$

$c_4 = 0.9/111 = 0.00811$ mol/cm^3 and $c_2 = 0.0555$ mol/cm^3. Benzene solubility is 1780 mg/L, giving $x^*_A = 4.11\text{E-}4$. Substitution yields $\mathscr{K}^*_{A42} = 355$L H_2O/L sludge. Using Eq. E2.1-3A yields

$$\rho_{A2} = \frac{1263}{355} = 3.56 \text{ mg/L}$$

for the concentration of benzene in the leachate. This example illustrates that benzene enjoys a preferential partitioning in the oil phase, so that at the 1400 mg/kg level only 3.56 appears in the leachate.

Partition coefficients for Sediment–Water and Soil–Water System. In this section we are concerned with trace levels of chemicals in soil and sediment. A more precise definition of trace levels is needed than the value of 50,000 mg

A/kg soil or less introduced in Chapter 1. In this case a trace level of chemical contamination is one that allows the soil or sediment to retain its natural physicochemical characteristics, except possibly for aspects of the biology, and the contaminants exert none of their own properties nor change those of the soil. Applying a few parts per million of ethylene biomide to eradicate nematodes is an example of a trace level concentration of a chemical in surface soils.

In an analogous fashion to the air–water system, chemicals in low levels or residual amounts are partitioned between soil and water to effect an equilibrium state between the phases. Equation 2.1-3 points the way for obtaining the partition between the phases:

$$(\mathscr{K}^*_{A32} \equiv)\ \frac{x_{A3}}{x_{A2}} = \frac{\gamma_{A2} f^0_{A2}}{\gamma_{A3} f^0_{A3}} \tag{2.1-18}$$

where 2 ≡ water phase and 3 ≡ soil phase. The product $\gamma_{A3} f^0_{A3}$ is dependent on what material is assumed to comprise the soil phase or phases. As noted above, natural soils and sediment contain both organic and inorganic fractions. A two-phase model for simultaneous sorption onto the soil fractions has been proposed.[26] The overall partition coefficient can be expressed as

$$\mathscr{K}^*_{A32} = \mathscr{K}^*_{AC2}\omega_C + \mathscr{K}_{AI2}\omega_I \tag{2.1-19}$$

where ω_C and ω_I are the weight fractions of natural organic matter and inorganic matter in soil. Considering the complexity of the two phases, the partition coefficients for the organic matter, $\mathscr{K}^*_{AC2}$, and inorganic matter, $\mathscr{K}_{AI2}$, are normally redefined as concentration ratios rather than mole fraction ratios:

$$\mathscr{K}^*_{AC2} \equiv \frac{\omega_{AC}}{\rho_{A2}} \quad \text{and} \quad \mathscr{K}^*_{AI2} \equiv \frac{\omega_{AI}}{\rho_{A2}} \tag{2.1-20}$$

These coefficients have dimensions of volume per unit mass. Typical units are L/kg and cm^3/g. It is problematic whether these partition coefficients are true equilibrium constants in the classical sense. Nevertheless, they must serve this purpose. Prior to further theoretical considerations it is instructive to review some partition coefficient data.

Mud. By definition, mud is a slimy, sticky mixture of solid material with water. This mixture is present on the bottom of streams, ponds, lakes, estuaries, and so on, and a fair number of partition coefficient studies have been made for certain chemicals.

Sediment–Water Partition Coefficients. In many instances a specific undesirable substance such as kepone has been discharged into the water for extended periods of time, rendering the stream or river polluted. Once the

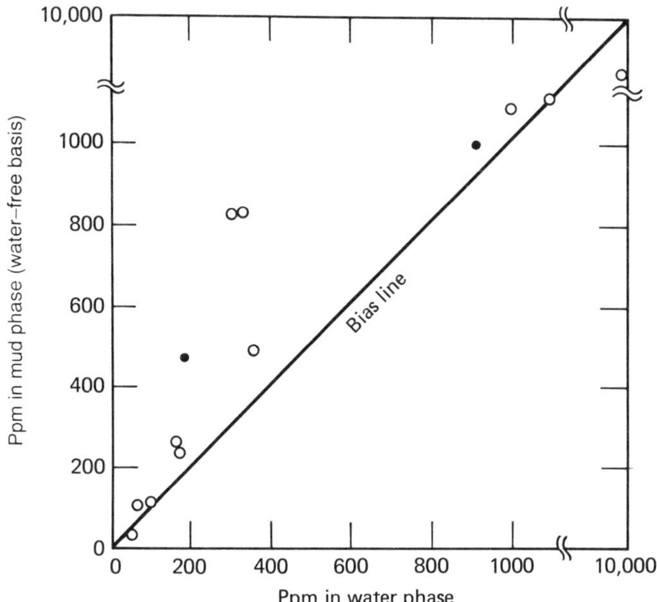

Figure 2.1-1. Equilibrium data for sodium sulfate between the soil and water at 77°F. (Reprinted by permission from Ref. 7.)

health hazard is discovered and the pollutant no longer discharged, the water is proclaimed clean and healthy. However, after the pollutant faucet is turned off, the health state of the stream bed is not advanced as quickly as that of the water itself. It is important to know the equilibria of various pollutants, such as kepone, phenol, sulfates, and detergents, between the water and soils comprising the muds, especially in the concentration range 5 to 10,000 ppm.

Greskovich[7] has published some water–mud equilibrium data for sodium sulfate and phenol. In acid-mine problem areas, waters are contaminated with sulfate ion forms such as sulfuric acid and ferric sulfate. Equilibrium studies were performed involving the sulfate system over the concentration range 40 to 10,000 ppm by weight. Various aqueous solutions were contacted with typical central Pennsylvania clayey-silt soil until equilibrium was established. The equilibrium data shown in Fig. 2.1-1 are reported as equilibrium concentration of sulfate in the water versus the concentration of sulfate in the soil (mud) on a water-free basis. It can be seen that the soil did exhibit a slight enhanced absorptive capacity for sulfate between 150 and 1000 ppm.

A second species study was performed with phenol (See Fig. 2.1-2). Little or no preferential absorption of phenol on soil was observed for the concentration range evaluated in these studies. It is often convenient to report values of the partition coefficient, if it is indeed constant, over a range of concentrations. Paris, Steen, and Baughman[8] obtained experimental values of the partition coefficients for two polychlorinated biphenyls (PCBs): Aroclors 1016 and 1242.

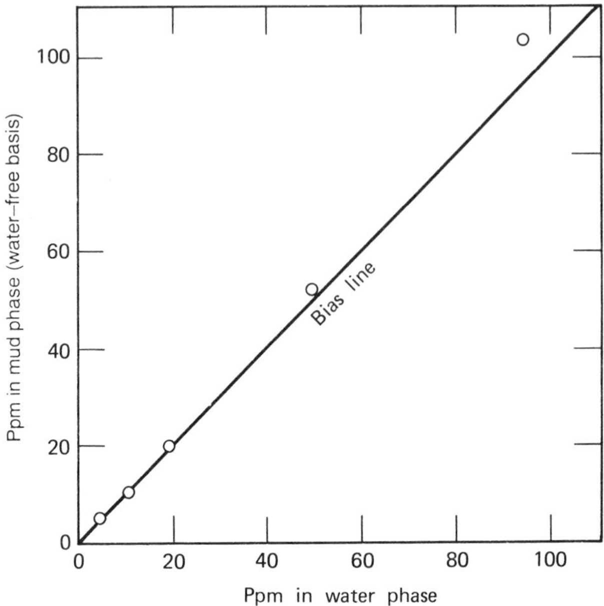

Figure 2.1-2. Equilibrium data for phenol between the soil and water at 77°F. (Reprinted by permission from Ref. 7.)

Table 2.1-6. Partition Coefficients ($\mathscr{K}^*_{A32}$)[a] of Aroclors 1016 and 1242[b]

Sediment Type	$\mathscr{K}^*_{A32}$ of A-1016	$\mathscr{K}^*_{A32}$ of A-1242
Doe Run Pond	(1.29 ± 0.8)E3	(1.09 ± 0.16)E3
Hickory Hills Pond	(1.30 ± 0.9)E3	(1.25 ± 0.11)E3
USDA pond	(1.37 ± 0.14)E3	(1.21 ± 0.12)E3

Source: Ref. 8.
[a] $\mathscr{K}^*_{A32} \equiv \omega_A/\phi_A$ in L/kg.
[b] Values are the mean of six determinations.

The experimental measured values are listed in Table 2.1-6. It appears that the values of the partition coefficients are independent of the Aroclor. The similarity in values for 1016 and 1242 is reasonable because the properties of these mixtures are so similar; however, the fact that all ponds display similar coefficients is surprising.

Sediment–water equilibrium was obtained for phosphorus release from moist Upper Klamath Lake, Oregon, mud.[9] For the release experiment the sediments were incubated at 10 and 23°C for a period of 45 days and total inorganic phosphorus in the external water was measured. The equilibrium data, shown in Table 2.1-7, indicate that phosphorus is strongly absorbed onto or within the soil solid matrix. Aerobic or anaerobic water conditions have a

Table 2.1-7. Sediment–Water Equilibrium for Upper Klamath Lake

Temperature (°C)	Inorganic Phosphorus Sediment (μg/g)	Water (μg/mL)
10	413	0.23
23	406	0.64
10	793	1.66
23	723	6.08
10	504	0.19
23	499	0.58

Source: Ref. 9.

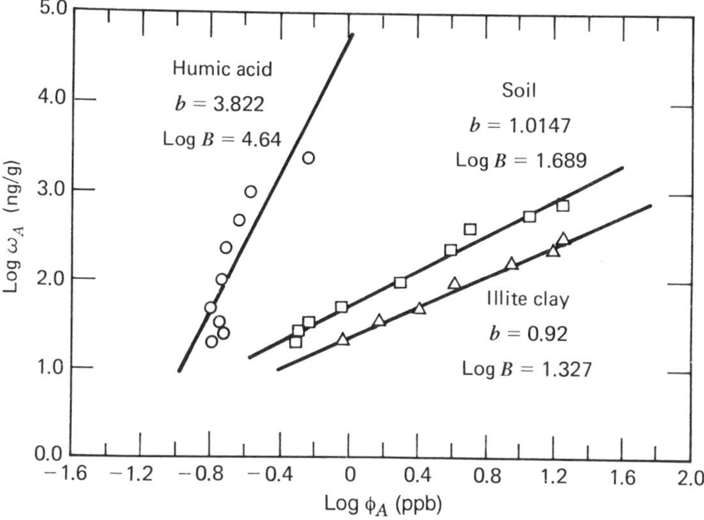

Figure 2.1-3. Freundlich plot for the adsorption of 2,4,2′,4′-tetradichlorobiphenyl on illite, humic acid, and Woodburn soil surface. (Reprinted by permission from Ref. 10.)

pronounced effect on the release and sorption of phosphate in soils and sediment (see Problem 2.1D).

Soil–Water Partition Coefficients. Surface soils are the recipient of chemicals both by intent and accident. In either case the partitioning between the soil–water and soil–solid phases is of interest since this environmental compartment is a primary point of contact between humans and their ecosystem. Hague[10] presents some results of the absorption of pesticides in a soil environment. The majority of data representing the absorption of pesticides on soil are represented by Freundlich-type isotherms (see Fig. 2.1-3). Langmuir and Brunauer–Emmett–Teller (BET) isotherms are little used for soil–pesti-

Table 2.1-8. Freundlich Isotherm Constant B and b for Adsorption of 2,4-D on Surfaces[a]

Surface	Temperature (°C)	b	log B
Illite	0	0.685	1.11
	25	0.719	1.02
Montmorillonite	0	0.925	−0.063
	25	1.004	−0.186
Sand	0	0.671	−0.984
	25	0.827	−1.454
Alumina	0	0.97	−0.06
	25	1.01	−0.08
Silica gel	0	0.90	0.58
	25	0.95	0.11
Humic acid	0	0.86	2.01
	25	0.931	1.9

Source: Reprinted by permission from Ref. 10.

[a]Constants to be used with ω_A in ng/g and ϕ_A in ppb.

cide systems. In practice, the logarithm of the chemical absorbed is plotted against the logarithm of chemical concentration in water at equilibrium. Such a plot usually results in a straight line and its slope and intercept give the value of the constants b and B. The constant B represents the extent of absorption, whereas b throws much light on the nature of the absorption as well as the role of solvent (water) in the absorption. Table 2.1-8 gives some absorption data typical of the soil system (see p. 63 for a review of theoretical adsorption isotherms).

Organic Chemicals. Organic chemicals include organic cations and neutral organic compounds and therefore cover essentially all organic pollutants for which sorption might be expected to contribute to environmental fate and impact in soil and sediment. In this section the equilibrium uptake and release of organic chemicals by natural sorbents are summarized. As noted above, natural sorbents include a phase synthesized by the association between clay minerals and natural organic matter. The word *sorption* is used with the intent of implying either chemical absorption onto or chemical solution into this phase since the exact process remains unresolved at this time.

Nonpolar Organic Chemicals. The nonpolar organic compounds, which include many pesticides, polychlorinated biphenyls (PCBs), and 2,3,7,8-tetrachlorodibenzodioxin (TCDD), are among the most important environmental pollutants. The hydrophobic (i.e., water avoiding) nature of this class of compounds highlights their preference for solid surfaces and the natural organic phase. The thermodynamic driving force for hydrophobic interactions is the increase in entropy resulting from the removal, or decrease, in the amount of hydration water surrounding an organic solute in water. Both adsorption equilibrium concepts and solution equilibrium concepts are presented to explain aspects of the partitioning mechanism.

EQUILIBRIUM SURFACE ADSORPTION. Dexter and Pavlou[27] present this as an approach to defining the accumulation of stable organic molecules on marine particulate interfaces within a coherent theoretical framework. The process can be conceptualized in terms of the following simplified physical model.

The system is comprised of porous solid particles (phase $\equiv$ 3) suspended in an aqueous solution of solute A. Molecules of A impinge on the surface of 3 at a rate proportional to the aqueous activity of A and the amount of surface area available for impingement. The molecules remain on the particles for a certain period of time, depending on the strength of their interaction with the surface. Eventually, any one molecule may obtain sufficient energy to overcome the binding and thus reenter the solution. At equilibrium, the number of molecules colliding with the surface per unit time is equal to the number of molecules leaving the surface. If it is assumed that the molecules of A are limited to relatively nonspecific bonding generated by London dispersive forces, the interaction energies of A with either the surface or other A molecules (multilayer formation) should not be markedly different. The strong bonding between water molecules and the lack of strong water–nonpolar solute intermolecular bonding gives rise to hydrophobic interactions. This condition can be characterized by an anomalously high molar free-energy change associated with the hydration–dehydration process of the solute, which is reflected in the low solubility of the compounds. The hydrophobic interaction provides considerable energy to the absorption process, which is independent of the nature of the adsorbent. From the foregoing considerations it appears that multilayer formation for nonpolar compounds absorbed from their aqueous solutions should be energetically feasible. In fact, such behavior has been observed for a number of systems using nonpolar and low-polarity organic molecules.

Since the BET multilayer adsorption theory contains the essential aspects of the physical model described above, it was used by Pavlou[28] to obtain a theoretical expression for the equilibrium partition coefficient, $\mathscr{K}^*_{A32}$, of stable organic chemicals in sediment. A modification accounts for variations in natural particle sizes and the content of organic matter, because there is much evidence that these variables contribute significantly.[26] Assuming low-ambient concentrations of A ($\leqslant 5\%$ of ρ^*_{A2}) yields a linear form for the BET isotherm manipulated to give

$$\mathscr{K}^*_{A32} = \frac{AM_A[1 - \exp(-A_C\omega_C/A)]\exp(\Delta E_{A3}/RT)}{A_A\rho^*_{A2}} \qquad (2.1\text{-}21)$$

where A is the specific surface area of the particulate matter (cm^2/g), A_C the surface area occupied by a unit mass of natural organic matter (cm^2/g), A_A the molar area of A (cm^2/mol), and ΔE_{A3} the difference in the energy of interaction of A as first layers and multilayers on the adsorptive surface (cal/mol). All the terms in Eq. 2.1-21 are specific for the compound and the natural environment under consideration.

Model parameters have been evaluated by Pavlou and Dexter. A representative value of 50 m^2/g was obtained for A based on N_2 adsorption on fine sediment samples collected in a deep basin of Hood Canal. For rough estimates they recommend $A = a \cdot 4\pi r_3^2$, with $a = 100$, as the empirical constant obtained by comparing actual particle surface area (N_2-BET adsorption) of sediment. The particle radius from sieve mesh size is r_3 (cm). ΔE_{A3} values for adsorbed chlorobiphenyls on surfaces of representative polarities were computed based on Fowke's equilibrium film and spreading pressure and surface tension data. The value for sorption on silica is approximately -20 kcal/mol, implying unfavorable adsorption due primarily to the strong water–silica interactions and necessitating very low water solubility (high activity coefficient) for significant adsorption. On the other hand, the exponential value in the BET equation is virtually constant over a wide range of organic surface polarities. The surface of the particulate matter for natural marine systems consists primarily of organic polyelectrolytes (aromatic and carbohydrate polymers) with significant numbers of moieties containing active oxygen (e.g., —OH, —COOH) with variable proteinaceous and lipoid residues. If propanol is chosen to represent a typical polar organic surface, the ΔE_{A3} term computed for chlorobiphenyl molecules is -2.8 kcal/mol. The following equation is recommended for computing A values:

$$A_A = \left[\frac{3M_A(\pi N)^{1/2}}{4\rho_A}\right]^{2/3} \tag{2.1-22}$$

where N is Avogadro's number (mol^{-1}), and ρ_A is the density of A (g/cm^3). This equation can also be used to estimate A_C values. Using several relatively simple alkanes, alcohols, and acids which can reasonably be expected to represent component moieties of natural polyelectrolytes yields 0.7E7 to 3.0E7 cm^2/g, and a rough average of 2E7 cm^2/g for A_C is recommended.

The Dexter–Pavlou development is a theoretical model that allows numerical estimates of the partition coefficient based on a few physicochemical parameters. Equation 2.2-21 is useful for estimating $\mathscr{K}^*_{A32}$; it contains the two dominant independent variables ω_C and ρ^*_{A2} in the correct functional form. Laboratory measurements have confirmed this functional dependence.

Example 2.1-4. Chlorinated Biphenyl Partition Coefficient. Estimate the soil–water partition coefficient for $C_{12}H_7Cl_3$, the dominant congener of Aroclor 1242, with molecular weight 258 g/mol, density 1.4 g/cm^3, and water solubility

0.4 mg/L at 25°C. The sediment has a particle surface area of 50 m^2/g and contains 5 wt% organic matter.

SOLUTION From Eq. 2.1-22 $A_A = 3.39\text{E}9\ \text{cm}^2/\text{mol}$; A_C and ΔE_{A3} are those suggested in the text. The values substituted in Eq. 2.1-21 are

$$\mathcal{K}^*_{A32} = 5\text{E}5\ \frac{\text{cm}^2}{\text{g}}\bigg|\ 258\ \frac{\text{g}}{\text{mol}}\left[1 - \exp\left(-2\text{E}7\ \frac{\text{cm}^2}{\text{g}}\bigg|\ 0.05\ \bigg|\ 5\text{E}5\ \frac{\text{g}}{\text{cm}^2}\right)\right]$$

$$\cdot\exp\left(-2.8\text{E}3\ \frac{\text{cal}}{\text{mol}}\bigg|\ \frac{\text{mol}\cdot\text{K}}{1.987\ \text{cal}}\bigg|\ 298\ \text{K}\right) 3.39\text{E}9\ \frac{\text{g}}{\text{cm}^2}\bigg|\cdot 4\text{E-}6\ \frac{\text{L}}{\text{g}}$$

$$= 727\ \text{L/kg}$$

This value is approximately half the measured values for Aroclor 1242 given in Table 2.1-6.

EQUILIBRIUM SOLUTION PARTITIONING. Chiou et al.[30] suggest that solubility in the organic matter is possible and is an appropriate mechanism to explain the soil-water distribution of nonionic organic compounds. The theoretical equilibrium partition coefficient expression developed by Dexter and Pavlou contains a term, ω_C, that quantifies the fraction natural organic matter content of the sediment (or soil). Expanding the $[1 - \exp(-A_C\omega_C/A)]$ group with Maclaurin's series and truncating after the second term yields

$$\mathcal{K}^*_{A32} \simeq \omega_C \mathcal{K}^*_{AC2} \tag{2.1-23}$$

where $\mathcal{K}^*_{AC2} \equiv A_C M_A \exp(\Delta E_{A3}/RT)/\rho^*_{A2}$. This result yields approximate numerical values for $\mathcal{K}^*_{A32}$ as Eq. 2.1-21 for $\omega_C \leqslant 0.05$ and clearly demonstrates the importance of this factor. Numerous investigators have observed that although $\mathcal{K}^*_{A32}$ tends to be variable over several orders of magnitude for a given solute with different soils, it generally correlates with ω_C of the sediment or soil (see Problem 2.1I). Typically, it is measured as the mass fraction of organic carbon. Equation 2.1-23 also contains the other two important variables for pollutant sorption: surface area and solubility in the aqueous phase.

Karickhoff[26] and Curtis et al.[31] present the details of the natural organic matter phase partitioning thermodynamics model as the theoretical basis for $\mathcal{K}^*_{AC2}$. Karickhoff demonstrates that $\mathcal{K}^*_{AC2}$ is a ratio of solute activity coefficients in the aqueous and organic matter phases and argues that one would expect highly nonideal behavior of the solute in the aqueous phase with large variations in γ_{A2} and solute interactions with the organic matrix to be similar in type and magnitude with smaller variations in γ_{AC}. This argument leads to water solubility being a primary dependent variable for expressing $\mathcal{K}^*_{AC2}$. Correlations of this form have been applied to the experimental data. For 47 compounds, including triazines, carbamates, organophosphates, and chlorinated hydrocarbons, the fitted equation with $r^2 = 0.93$ is

$$\log \mathscr{K}^*_{AC2} = -0.83 \log x^*_A - 0.01(T_{\rm mp} - 25) - 0.93 \qquad (2.1\text{-}24)$$

where $T_{\rm mp}$ is the melting point of a solid (°C) and x^*_A is the mole fraction solubility in water.

Extensive data exist for solute partitioning between octanol and water. These data have accumulated as a result of studies involving pharmaceuticals and the need for suitable analogs (i.e., octanol) for modeling drug distribution within the body. Applying Eq. 2.1-3 twice, once for the equilibrium distribution of A between natural organic matter (phase C) and water (phase 2), then again for A between octanol (phase 4) and water, with the definition of the respective partition coefficients, the following equation can be derived:

$$\mathscr{K}^*_{AC2} = \frac{(\gamma_{A2})_C}{(\gamma_{A2})_4} \left(\frac{\gamma_{A4}}{\gamma_{AC}}\right)_2 \frac{c_4}{c_C \rho_C} \mathscr{K}^*_{A42} \qquad (2.1\text{-}25)$$

where c_C and c_4 are the molar densities of organic matter and octanol (mol/cm^3) and ρ_C and is the mass density of organic matter (g/cm^3). The first term on the right-hand side is a ratio of solute activity coefficients in water saturated with dissolved organic matter and octanol, respectively. The next term is a ratio of solute activity coefficients in octanol and natural organic matter saturated with water. Apparently, water has a significant effect on the latter ratio, so that Curtis et al.[31] find it necessary to correlate the data of $\mathscr{K}^*_{AC2}$ versus $\mathscr{K}^*_{A42}$. The data were from several studies involving 40 observations that included halogenated alphatics and aromatics, alkyl-substituted aromatics, and polynuclear aromatic hydrocarbons. The correlating equation is

$$\log \mathscr{K}^*_{AC2} = 0.92 \log \mathscr{K}_{A42} - 0.23 \qquad (2.1\text{-}26)$$

with $r = 0.94$. This equation provides an alternative means of estimating $\mathscr{K}^*_{AC2}$. Consult Lyman et al.[4] for $\mathscr{K}^*_{A42}$ data and estimation techniques.

The paragraph above summarizes solution thermodynamics theory for obtaining partition coefficients. Three choices are available. One utilizes solute solubility in water data and another utilizes octanol–water partition coefficient data. Without these data on either x^*_A or $\mathscr{K}^*_{A42}$, these can be estimated from fragment constants and chemical structure factors. Finally, the fraction organic matter is needed to obtain the partition coefficient on the basis of soil or sediment mass.

Closure. Much of the information available on earthen solids–water partition coefficients is the result of adsorption experiments. Partition coefficients indexed to the fraction carbon content are relatively independent of other sediment–soil characteristics or geographic region. Pollutant solubility in water is also a prime correlating variable. Equations for $\mathscr{K}^*_{A32}$ based on either theoretical model, absorption or solution, contain these variables. The models also have a temperature-dependence term; however, in each case it is weak. Karickhoff[26] observes that the sorption coefficient is expected to decrease slightly with increased temperature and gives an equation for estimating the numerical magnitude. In addition, Dexter and Pavlou[27] observed using the adsorption model that increased salinity could be expected to increase $\mathscr{K}^*_{A32}$. A general model based on quantitative structure-activity relationships similar to Equation 2.1-24 containing corrections for soil pH and the dissociation constant, pK_a, of the chemical shows promise for estimating the sorption behavior of both ionized and nonionized chemicals.[64]

There are a number of other important aspects of adsorption partition coefficients in which coverage here is beyond the scope of this introductory treatment. Those wishing further details should consult Tinsley's chapter on chemodynamics.[29] The reviews of Karickhoff[26] and Curtis et al.[31] have been mentioned previously but contain much additional information. A most disconcerting aspect of these partition coefficients is the nonequilibrium behavior observed when desorption experiments are performed. Information on this aspect is contained in Chapter 7; see Fig. 7.3.4 and the subsection "desorption–dissolution of the solid-phase fraction" in Section 7.3.

Metals and Inorganic Chemicals. The equilibrium processes between dissolved metal species or other inorganic species and sediment or soil, under normal physicochemical conditions, generally leads to a large fraction of these substances being associated with the solid fraction. Some metals of concern are cadmium, copper, lead, zinc, cobalt, and mercury. Other types of inorganics include the ammonium and phosphate ions. There is ample evidence in the literature to indicate that these substances do associate with particulate matter; however, theoretical approaches have not been applied as widely as have empirical approaches to quantifying the process in natural systems. One result of this is that algorithms for estimating partition coefficients, such as the hydrophobic organics, are not available. In this section we track the reviews of Allen[58] and Delos et al.[32] in presenting some theoretical considerations that have led to simple parameterization of metal partitioning commonly used and factors that influence partitioning on aquatic solids.

In general the behavior of metals on earthen solids is akin to their association with aquatic solids, which can be characterized by the following five mechanisms: (1) absorptive bonding on fine-grained surfaces, (2) precipitation of discrete metal compounds, (3) coprecipitation of metals by hydrous Fe and Mn oxides and by metal carbonates, (4) association with organic molecules (i.e., chelates), and (5) incorporation into crystalline minerals. In most cases encountered in river systems it would seem that the absorption of metals onto inorganic surfaces is a dominant mechanism. In lakes where there is a large fraction of biological solids, sorption into biomass or binding by organic surface functional groups is important. For these reasons most of the models describing the interaction of adsorbate ions and surfaces are implicitly assumed to be adsorption as a two-dimensional surface phenomenon.

Theory-based adsorption models are generally a composite of surface complex formation and electrostatics ideas. The application of these models to well-characterized laboratory metal–ligand–surface systems results in excellent agreement with experimental observations. Despite the general lack of experimentation with natural aquatic sediment, which would allow the application of theory with confidence, there has arisen general agreement on many characteristic features of adsorption reactions.

Metal adsorption is considered to be analogous to the formation of soluble complexes, the principal difference being that the ligand in the reaction is attached to a surface site on the solid. Therefore, the same factors that affect soluble complex formation also affect the interactions at solid surfaces. This being the case, pH is one of the most influential parameters in governing metal adsorption. It affects both the type of surface sites and the speciation of the metal ions in solution through hydrolysis reactions. In general, as pH increases, metal ion adsorption increases; however, in situations where complexing ligands are present in the adsorbing system, the foregoing generalization is not always true. Competition for adsorption sites by Ca and Mg also occurs but is a smaller effect than pH and ligand content. Of course, metal adsorption is also highly dependent on the type, relative amount, size distribution, and concentration of aquatic sediment solids.

In a review of trace metal speciation for water, sediment, and soils, Allen[58] concludes that modeling frameworks are available to permit the evaluation of metal binding in sediments that are oxic. For the sorption of metals by iron oxide (FeO_x), manganese oxide (MnO_x), and organic matter (OM), which are generally considered to be the most important metal-binding materials, the quantity of metal (denoted by M) that will be bound by each is summed as the total sorbed metal. The equilibrium concentration of divalent metal ions present in porewater or solution $[M^{2+}] \equiv \rho_{M2}$ can be expressed as

$$\rho_{M2} = \frac{\omega_m}{\sum_s K_{s-m}[s]}$$

where $\omega_m \equiv [M]_{ads}$ is the quantity of sorbed metal and the denominator is the summation of the product of the conditional stability constant K times the quantity of the site, s, that is available for binding metal. This equation describes the binding of metal in oxic sediments, both suspended and surficial. It is descriptive of the situation that may be present in many fast-flowing rivers that are well oxygenated.

The denominator in the foregoing equation can be characterized as $\mathscr{K}^*_{A32}$. Figure 2.1-4 summarizes $\mathscr{K}^*_{A32}$ results for some metals data obtained for streams. The partition coefficients are termed *conditional* equilibrium constants

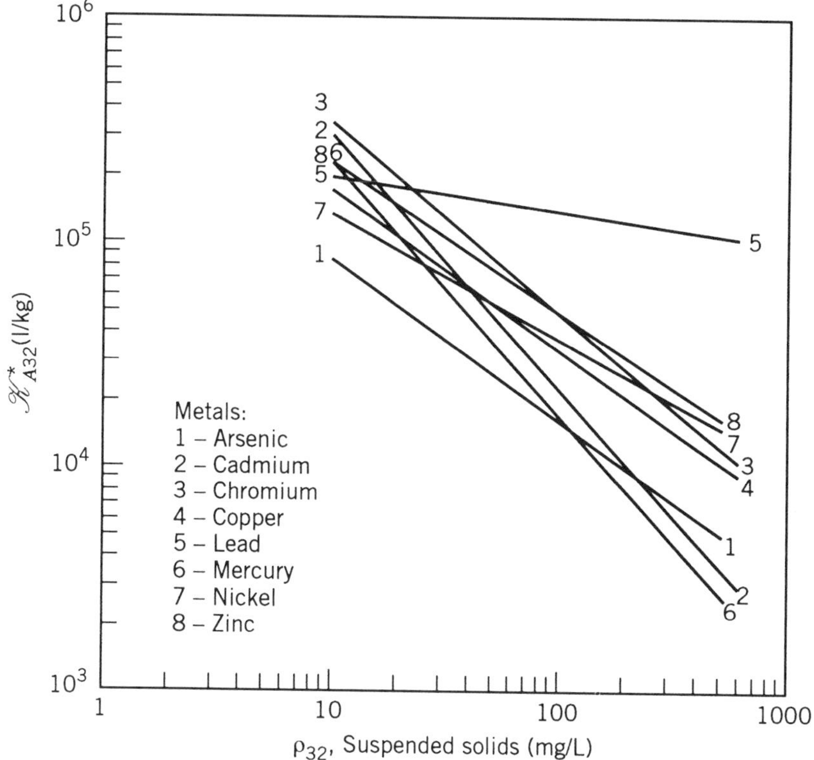

Figure 2.1-4. Metal partition coefficients as a function of suspended solids in streams. (From Ref. 32. Reprinted by permission, J. V. DePinto, 1995.)

because they are constant only for specific surface and bulk solution conditions. The lines in the figure do not reveal the wide variations in the data bases. As is also apparent in the figure, $\mathscr{K}^*_{A32}$ decreases with increased solids concentration, a commonly observed laboratory and field behavior for metals and organics. This behavior seems thermodynamically indefensible and is the subject of continuing theoretical study.

For hydrophobic organic compounds the normalizing factor is the organic carbon content of the sediment. For many metals the principal factor mitigating metal transport is sulfide in the sediment. As noted above, other phases that affect the partitioning include metal oxides and organic matter. Marine sediments often contain a large reservoir of sulfide, present largely as iron monosulfide. The sulfide present in sediments is quantified as acid volatile sulfide (AVS). The AVS is the gaseous sulfide evolved upon addition of cold acid to the sediment.[59] In the case of Cd and Ni toxicity, a consequence of mobility was not observed until the molar concentration of metal on the sediment exceeded the molar concentration of the sulfide. It is postulated that the iron sulfide could be reacting with the added cadmium to form highly insoluble cadmium sulfide.

It is important that the metals liberated during the evolution of gaseous sulfide of the AVS procedure also be determined. This molar metal content in the sediment is termed *simultaneous extracted metal* (SEM). Based on the measurements of AVS and SEM, a means exist to evaluate the potential mobility and toxicity of trace metals in sediments. If the SEM–AVS molar ratio is $\leqslant 1$, there will not be mobility, as the metals would be expected to be present as their sulfides. If SEM/AVS is >1, there may be mobility. The equation immediately above also describes porewater concentration of metal ion in sediments containing sulfide if the concentration of trace metal exceeds the concentration of AVS. This is the situation commonly encountered in contaminated sediments such as those present in harbors.

The disposal of hazardous waste, both chemical and radioactive, containing metals and elements usually involves placement into underground workings of soil at a few meters depth to 600 to 900 m in geologic media such as salt, basalt, granite, sandstone, or tuff. The possibility of migration of these elements in this portion of Earth's crust must include contact with groundwater. Subsoils and geologic media contain little or no natural organic matter as the term is used above. The distribution coefficient, $\mathscr{K}^*_{A32}$, is a very useful concept here also in that it represents sorption, a general term that covers all aspects of interaction between the mobile fluid phase and the immobile phase (solid plus trapped liquid): this may include ion exchange, ion adsorption, filtration, and precipitation. It cannot be considered an equilibrium, reversible phenomenon in the classical sense since the desorption isotherm often does not track the adsorption isotherm. (See p. 527 for details on this behavior.)

In the case of nuclear waste the three major elements of concern are iodine-129, neptunium-237, and plutonium-239. These elements are present in significant quantities in the waste and have a very wide range of physicochemical properties that are of importance in their interaction with rock formations. De Marsily et al.[33] give values of and discuss the distribution coefficients for each of these elements in Savannah River plant soil and typical desert soil as follows. For iodine, $\mathscr{K}^*_{A32} = 0$. It is well known and usually accepted that iodine is not (or very little) adsorbed. This property is valued in hydrology, where it is used as a "perfect" tracer. For neptunium $\mathscr{K}^*_{A32} = 15\,\text{mL/g}$. It is a relatively small distribution coefficient. For plutonium, two values are given: $\mathscr{K}^*_{A32} = 2000\,\text{mL/g}$ and $\mathscr{K}^*_{A32} = 0$. The first value is representative of ph $\simeq 7$ and valences IV or VI for plutonium (values for valence III could be higher). The chemistry of plutonium is rather complex, and in some circumstances it may react with other elements present in the water to generate complex molecules that are electrically neutral and that might therefore not be adsorbed.

The sorption of other radionuclides in the biosphere is also of interest. Table 2.1-9 contains the distribution coefficients of the three major radionuclides discussed above and several others measured between Hanford basalt and simulated groundwater. As with most metals, the sorption distribution coefficients exhibit a fairly wide range of values.

Table 2.1-9. Sorption Distribution Coefficients $\mathscr{K}^*_{A32}$ for Basalt

Element	Observed Range (mL/g)	Element	Observed Range (mL/g)
Cesium	70–2100	Neptunium	3–16[a] (150)
Strontium	58–240	Radium	48–150
Selenium	0–10[a] (17–80)	Uranium	0–70[a]
Iodine	0–3	Plutonium	10–10,000
Technetium	0–40[a] (4000)	Americium	10–100,000

Source: Ref. 34.

[a]Data obtained under oxidizing conditions. Values in parentheses are for anoxic conditions.

Earthen Solid–Air Equilibrium Occurrences

Many synthetic chemicals exist in the air as dispersed molecules and sorbed onto aerosols. As air masses containing these chemicals move over the land they become adsorbed onto earthen solids and vegetable surfaces. Chemicals placed directly onto the soil can and do vaporize into the overlying air mass. The air–earthen solid interface is large in areal extent. The size is slightly less in order of magnitude than the water–earthen solid and the air–water interfaces.

Pure Gases in Contact with Earthen Surfaces. Accidents involving vessels containing gaseous or liquefied substances may produce vapor clouds more dense than the surrounding air. This material, in gravity-driven flow or aided by light winds, subjects the earthen material underneath to its pure presence. Larry Moree suggests that the subsurface injection into the soil of methyl bromide, a pesticide used for nematode control, is a realistic example involving a pure gaseous chemical in direct contact with soil solids. While in contact some adsorption does occur. Situations involving such releases are highly transient, and it is unlikely that an equilibrium condition with pure gas can be realized since the gaseous material becomes quickly dissipated and diluted by air.

Pure Solid Chemicals in Contact with Air. In a fashion similar to pure liquids in contact with air, pure solids exert their vapor pressure in the space above. Equations 2.1-5 and 2.1-6 apply to pure solids in air. (See the section on nonaqueous liquids and mixtures in contact with air for further information.)

Chemical Adsorption onto Solids. Adsorption theory provides the basis for understanding the equilibrium association of substances in solution and the adjoining solid surfaces. Traditionally, the subject has been studied from the point of view of the vapor phase. It is introduced here in that context; however, there is no difference in principle between adsorption from liquid and vapor phases since, thermodynamically, the adsorbed phase concentration in equilibrium with a liquid must be precisely the same as that which is in equilibrium with the saturated vapor. For the purposes of this introduction the solid fractions of soil or sediment are considered to be the adsorbent. In previous sections we have described aspects of the solid nature of soils and sediment that provide the sites for chemical adsorption from either the vapor phase or aqueous solutions.

In what follows the classical approach to adsorption theory is presented. It is a surface phenomenon in that molecules become associated with sites on the solid surface in a reversible fashion that governs the equilibrium partitioning between phases. In considering the adsorption of a molecule from solution onto a solid, both physical and chemical characteristics of the adsorbent will influence the process. The actual surface area of the solid will have a profound effect, primarily through the availability of adsorption sites. The nature of the binding sites on the surface may be charged, having potential for hydrogen bonding or containing hydrophobic areas. This will influence adsorption characteristics. Tinsley[29] provides an overview of the types of binding sites available on soil and sediment surfaces.

Adsorption of Gases and Vapors onto Solids. Gaseous chemical molecules in air are adsorbed onto soil as soon as they come in contact with the surface. The adsorption depends on the energy of the surfaces representing various constituents of the soil. Factors affecting the adsorption of gases (and liquids) on soils include water content, water solubility, chemical structure, nature of the surface, organic matter, and so on. Although adsorption experiments are relatively simple to perform, the results are usually difficult to interpret. The adsorption behavior of a chemical on a soil surface can be visualized with the aid of a Leonard-Jones potential energy diagram (Fig. 2.1-5). As the distance of the adsorbate to the surface is decreased, there is attraction; with a further decrease the potential energy increases. Here ΔH_{I} is the heat of adsorption. Line I is for a physical or van der Waals type of adsorption, and line II is for chemisorption. The point where both the graphs intersect is the activation energy. Usually, the chemisorption process requires higher activation energy, and the heat of adsorption for chemisorption, ΔH_{II} is always greater than the heat for physisorption. For physisorption the activation energy as well as the heat of adsorption are quite low. The model of adsorption as a set of distinct localized sites is in general more appropriate to chemisorption than to physical adsorption since in many cases a physically adsorbed layer is highly mobile and resembles more closely a two-dimensional gas. It is likely that volatile organic chemicals, including the pesticide–soil system, generally follow a physical type of adsorption.

Adsorption data are generally represented by an equation known as an isotherm. An *isotherm* represents a relation between the amount of chemical on

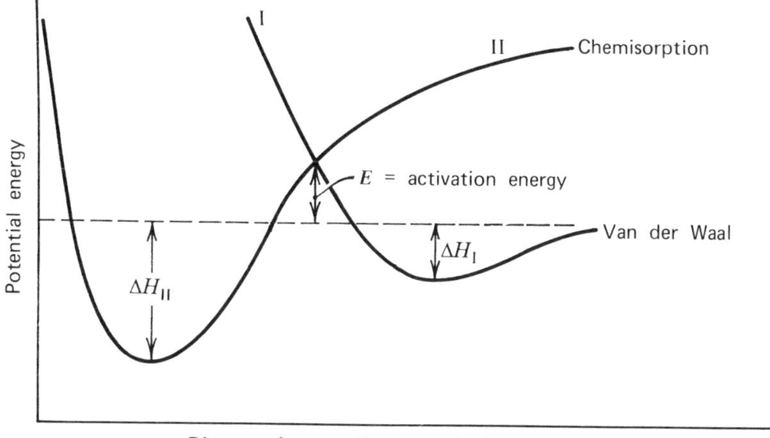

Figure 2.1-5. Adsorption process as represented by a potential energy diagram.

the soil and the chemical left in the solution (gas or liquid) after equilibrium at a constant temperature. The three well-known isotherms are Freundlich, Langmuir, and BET (Brunauer, Emmett, and Teller). The context and nomenclature used in the remainder of this section will be those of adsorption from the vapor phase.

In many cases the adsorption isotherm is satisfactorily represented by the empirical equation proposed by Freundlich:

$$\omega_A = B(p_A)^b \tag{2.1-27}$$

where ω_A is the mass of chemical A adsorbed on the material or soil, p_A the partial pressure in the gas phase, and B and b are empirical constants.

When $b = 1$ a linear relationship results and this is commonly referred to as *Henry's law*, by analogy with the limiting behavior of solutions of gases in liquids (see Eq. 2.1-12) and the constant of proportionality, which is simply the adsorption equilibrium constant B and is referred to as the *Henry constant*. For physical adsorbed species there is no change in molecular state and it follows that for adsorption on a uniform surface at sufficiently low concentrations such that all molecules are isolated from their nearest neighbor, the equilibrium relationship between fluid-phase and adsorbed-phase concentrations will be linear. Adsorption from aqueous solution represented by Eq. 2.1-19 is such an example.

The temperature dependence of the Henry constant obeys the *van't Hoff equation*,

$$B = B_0 \exp\left(-\frac{\Delta H_0}{RT}\right) \tag{2.1-28}$$

where ΔH_0 represents the difference in enthalpy between adsorbed and gaseous states. Differences in heat capacities between the phases was neglected in the development, and plots of $\ln B$ versus $1/T$ are often found to be essentially linear over a wide range of temperature.[35]

The simplest type of theoretical model occurs when adsorption is restricted to a single molecular layer on the solid. On this basis Langmuir developed an equation for this isotherm, assuming that at any pressure below saturation the amount of chemical A adsorbed is proportional to the partial pressure of the gas and the fraction of the surface left uncovered. He assumed further that the adsorbed gas molecules do not dissociate or interact on the surface. The result was

$$\omega_A = \omega_A^* \frac{b_1 p_A}{1 + b_1 p_A} \tag{2.1-29}$$

where ω_A^* is the amount of chemical A adsorbed per unit mass of adsorbent to cover the surface with a layer one molecule thick, and b_1 is an empirical constant. This expression shows the correct asymptotic behavior for monolayer adsorption since at saturation $p_A \to \infty$ and $\omega_A \to \omega_A^*$, while at low sorbate concentration Henry's law is approached. ω_A^* represents a fixed number of surface sites and should be temperature independent. The temperature dependence for the isotherm is through the equilibrium constant b_1 and should follow van't Hoff just as in Eq. 2.1-28.

Brunauer, Emmett, and Teller showed how to extend Langmuir's approach to multilayer adsorption, and their equation has become known as the *BET equation*. The basic assumption is that the Langmuir equation applies to each layer, with the added postulate that for the first layer the heat of adsorption, ΔH_1, may have a special value, whereas for all succeeding layers it is equal to λ_A, the heat of condensation of the liquid adsorbate. Evaporation and condensation can occur only from exposed surfaces to be covered by single, double, and multiple layers of molecules.[36] The BET equation is

$$\omega_A = \omega_A^* \frac{B_1(p_A/p_A^0)}{[1 - (p_A/p_A^0)][1 + (B_1 - 1)(p_A/p_A^0)]} \tag{2.1-30}$$

where p_A^0 is the pure component vapor pressure. The constant B_1 is an empirical constant but is related to the heat of adsorption of the first layer and the normal heat of condensation of the gas (or vapor) molecules. In equation form this is

$$B_1 \simeq \exp\left(\frac{\Delta H_1 - \lambda_A}{RT}\right) \tag{2.1-31}$$

It is approximate but, nevertheless, is used in interpreting the constant B_1. The BET equation can approximate three isotherm shapes: types I, II, and III. For large B_1 values (i.e., $\Delta H_1 \gg \lambda_A$) it reduces to the Langmuir Equation 2.1-29. This is a type I isotherm.

The theoretical models of Langmuir and BET are linearized for convenience in testing data (see problems 2.1D). In the case of Langmuir a plot of p_A/ω_A versus p_A should give a straight line, and the two constants ω_A^* and b_1 may be evaluated from the slope and intercept. Similarly, ω_A^* and B_1 can be obtained by plotting $(p_A/p_A^0)/\omega_A(1 - p_A/p_A^0)$ versus p_A/p_A^0 for the linearized BET. In either case it is theoretically possible to obtain the specific surface area of the solid, a_m:

$$\omega_A^* = \frac{a_m c_A^0}{N\sigma^0} \tag{2.1-32}$$

where c_A^0 is the molar density of the adsorbate on the surface and σ_0 is the area of a site.

Since its appearance in 1938, the BET equation has become the general method for obtaining surface areas from adsorption data. In the case of multilayer adsorption it seems reasonable to take σ^0 as an adsorbate (as opposed to an adsorbent site) area, based on either the solid or liquid density, depending on the temperature. The equation has become a standard one for practical surface area determinations, usually with nitrogen at 77 K as the adsorbate, but in general with any system giving sigmoidal-shaped isotherms. The effective molecular area for N_2 is usually 16.2 Å.[36] Equation 2.1-22 is such a derivative of Eq. 2.1-32.

The multicomponent extension of the BET equation, given by Hill,[37] appears to be in fair agreement with the data for oxygen and argon on chromia gel. For several competing adsorbates the Langmuir model can also be extended to binary or multicomponent systems. The resulting expression for the isotherm is[35]

$$\frac{\omega_A}{\omega_A^*} = \frac{b_1 p_A}{1 + b_1 p_A + b_2 p_B + \cdots} \tag{2.1-33a}$$

$$\frac{\omega_B}{\omega_B^*} = \frac{b_2 p_B}{1 + b_1 p_A + b_2 p_B + \cdots} \tag{2.1-33b}$$

For thermodynamic consistency $\omega_A^* = \omega_B^*$, but for physical adsorption of molecules of widely different size this assumption is unrealistic. If the extended Langmuir is regarded as an analytical description rather than a physical model, the use of different values of ω_i^* for each component becomes permissible, although extrapolation over large concentration range should be done with caution. A limited amount of binary equilibrium data supports the use of different ω_i^* values for two components.

It is pointed out again that the theoretical concepts also apply to the adsorption of chemicals from aqueous solutions (see the discussion "Sediment–water Partition Coefficients" earlier in this section).

Soil Particles and Moisture. A brief treatis on surface soils is presented in Section 6.3, pages 398–401. There, general information is provided on the nature of soil particles, size distributions, porosity, and water content. Additional information on the nature of the inorganic and organic constituents is also given on pages 415–420.

Surface soils are rarely absolutely dry, and water is derived primarily from rainfall. Dew and fog are secondary sources of soil moisture. Surface soils are saturated with water for short periods of time during a rain or when it is being irrigated. Water soon drains out of the larger macropores, to approximately 20 g H_2O per 100 g of dry soil for a silt loam and achieves a condition termed *field capacity*. Further drying occurs until the *wilting coefficient* is achieved at approximately 10 g per 100 g. The water remaining at this point is found in the smallest of the micropores and around individual soil particles. Upon further drying it will loose liquid water held in even the smallest of the micropores. The remaining water, <8 g per 100 g, will be associated with the surfaces of soil particles, particularly colloids as adsorbed moisture. It is held so tightly that much of it is considered nonliquid and can move only in the vapor phase. At this point the *hygroscopic coefficient* is reached. As might be expected, soils high in colloidal materials will hold more water under this condition than will sandy soils and those low in clay and humus content.

The BET adsorption model represented by Eqs. 2.1-30 to 2.1-32 provides the necessary theoretical framework to estimate the surface area of dry soils. A brief review of the standard methodology for determining surface areas of solids is given by Smith.[38] Although soil is a mixture of organic and inorganic fractions, the BET model assumes that all constituents are made up of similar 'solid' surfaces and the molecules are physically adsorbed and arranged in close two-dimensional packing without interaction. Orchiston[39] investigated the adsorption of water vapors on clays and found that the BET equation provided a good correlation over the usual range of application from 0.05 to 0.35 relative pressure. ω_i^* for H_2O was about 12 g per 100 g of dry montmorillonite, 2.4 for illite, and 0.14 for kaolinite. The corresponding values for B_1 were 21, 12, and 38, respectively. A value of $10.8A^2$ was assumed as the area available per water molecule for hexagonal close spacing on the surface of the adsorbent. This and other studies suggest that montmorillonite has an internal adsorbing surface of about 80%. Kaolinite has relatively small internal adsorption, and illite has some but much less than montmorillonite. The available data indicate that the B_1 average for water is 19 with $\sigma = 5.5$.

Bailey and White[40] review adsorption of organic pesticides by soil colloids and observe that the mineral fraction is composed of crystalline clay minerals and crystalline and amorphous oxides and hydroxides. They summarized soil constituent surface area determinations. Table 2.1-10 shows that the organic matter, vermiculite, and montmorillorite have the highest surface areas. Fuller[41] presented surface-area information on the silt, sand, and gravel fractions of soils and these data also appear in Table 2.1-10.

Table 2.1-10. Selected Surface Area of Soil Constituents

Soil Constituent	Surface Area, a_m (m^2/g)[a]	Surface Area, a_m (cm^2/g)[b]	Diameter[b] (mm)
Organic matter	500–800		
Vermiculite	600–800		
Montmorillonite	600–800		
Dioctahedral vermiculite	50–800		
Illite	65–100		
Chlorite	25– 40		
Kaolinite	7– 30		
Oxides and hydroxides	100–800		
Silt		454	0.05–0.002
Very fine sand		227	0.10–0.05
Fine sand		90.7	0.25–0.10
Medium sand		45.4	0.50–0.25
Coarse sand		22.7	1.0 –0.50
Fine gravel		11.3	2.0 –1.0

[a]From Bailey and White[40] (1964).
[b]From Fuller[41] (1980).

The inundation of surface and subsurface soils by large single spills or periodic small liquid or solid spills can result in a drastically altered system. The extent of alteration is such that the soil is changed from its natural state, and the adsorption models may not apply and alternatives need to be invoked. It is not uncommon to find that such inundation processes will form a fourth phase in addition to the air, soil, and water. Such is the case presented in the discussion "Nonaqueous Liquids in Contact with Earthen Solids" earlier in this section.

Equilibrium Partitioning of Organic Chemicals Between Air–Soil and Air–Particles. Potentially hazardous organic chemicals that exist in air and soil enjoy a wide range of pure-component vapor pressures. The availability of theoretically sound thermodynamic models is needed to quantify the relative distribution of these chemicals whether it be a toxic such as benzene ($p_A^0 = 95.2\,\text{mmHg}$ at 25°C) released to air from a manufacturing facility or the pesticide Dieldrin ($p_A^0 = 1.8$ E-7 mmHg at 25°C) applied onto an agricultural soil. In this section three models with algorithms are presented from which quantitative estimates can be made for the concentrations in air, soil, and aerosols under equilibrium conditions.

Because of their extremely high vapor pressure at ambient temperatures, the major constituents of air are not significantly adsorbed onto soil solids.

Therefore, when considering the equilibrium behavior of organic chemicals, oxygen and nitrogen can be treated as nonabsorbing species. Water vapor ($B \equiv H_2O$) cannot be so easily dismissed. The water content of soils can be large. Due to its pure component vapor pressure at ambient conditions, it can be present in air at concentrations up to approximately 7 vol% in vapor form. Since the molecules are known to compete with other adsorbates for sites on soil surfaces, it will be used as a criterion for model selection. The relative mass fraction for complete monolayer coverage, ω_B/ω_B^*, will be the quantitative criterion.

Dry Soils and Particles. If the water content of the earthen solid is less than 5% of that required for single-monolayer coverage (i.e., $\omega_B/\omega_B^* < 0.05$), the soil is considered to be dry. For volatile organic chemical adsorption modeling, this low level of site coverage by water molecules allows the use of the single-component BET isotherm (Eq. 2.1-30). A recent review of the available BET adsorption data for volatiles on soils shows that B_1 ranges from 2 to 80.[42] The adsorption is dominated by the mineral and organic matter content and is limited to the external surface. The areas presented in Table 2.1-10 are total areas. The external surface area comprises only about 10 to 20% of the total.

The total surface area of a dry soil, a_m, can be estimated based on the mass fractions of the constituents in an additive fashion:

$$a_m = \sum_{i=1}^{n} a_{mi}\omega_i \tag{2.1-34}$$

where n is the total number of constituents, a_{mi} is the total surface area of the constituents, and ω_i the mass fractions. In reviewing the data in Table 2.1-10, it becomes readily apparent that the montmorillonite fraction dominates. The organic matter contribution to the surface area can be significant if the soil clay minerals are dominated by the illite and kaolinite fractions.

The use of Eq. 2.1-30 for estimating the mass ratio, ω_A, of a volatile adsorbed onto dry soil or particles requires the amount of A needed to cover the adsorption sites with a layer one molecule thick, ω_A^*. The surface density combined with the soil surface area per unit mass yields

$$\omega_A^* = 0.917 F a_m \left(\frac{\rho_A^2 M_A}{N}\right)^{1/3} \tag{2.1-34a}$$

where ρ_A is the mass density of the adsorbate on the surface and F is the fraction of the total surface area that is external and available to accommodate sorbed molecules. The numerical constant in Eq. 2.1-34a has been subject to considerable investigation.[38] It is related to the projected area of a molecule on the surface when the arrangement is a close two-dimensional packing. This value is slightly smaller than that obtained by assuming that the adsorbed molecules are spherical and that their projected areas on the surface are circular. The mass density is normally taken as that of the condensed gas, pure liquid, or solid at the temperature of adsorption.

Junge[43] used the single-component BET model to obtain a theoretical expression for the fraction of a volatile component adsorbed to aerosol particles. Cupitt[44] applied the model, compared the results to the available data, and noted for atmospheric aerosols that gaseous toxic materials are probably adsorbed onto particles rather than in the vapor phase if the saturation vapor pressure is E-7 torr or less.

Damp Soils and Particles. The soil solids are considered damp if the water content is such that between 5 and 95% of the adsorption sites are covered with a monolayer of water. In this moisture range the water molecules compete with the volatile organic molecules for absorption sites. Valsaraj and Thibodeaux[42,45] used the multicomponent extension of the BET model to develop the theoretical effect for competing water molecules and demonstrated that soil moisture will increase the partial pressure of an organic chemical in the vapor space adjoining the solid particles. Using the competing adsorbate Langmuir model (Eq. 2.1-33), the pressure versus soil loading relationship is

$$p_A = \frac{p_A^*}{B_1} \frac{\omega_A/\omega_A^*}{1 - \omega_A/\omega_A^* - \omega_B/\omega_B^*} \tag{2.1-35}$$

where ω_B and ω_B^* are the soil moisture loading and moisture monolayer loading, respectively. As the ω_B/ω_B^* ratio approaches $1 - \omega_A/\omega_A^*$, the partial pressure, p_A, increases dramatically. This equation is obviously limited to ω_A and ω_B conditions such that $p_A \leqslant p_A^*$. There is a lack of data on adsorption in this soil moisture region; however, the data reported by Spencer et al.[48] support the general features of the predictions of the competitive adsorbate theory.

Wet Soil and Particles. If the moisture level is $\geqslant 95\%$ of ω_B^* the soil surface sites are nearly covered with water and the term *wet* can be used. In this case the volatile chemical can be viewed to exist at equilibrium in a three-phase system. Using air–water and water–solid as in the form of Eqs. 2.1-12 and 2.1-18 yields

$$p_A = \frac{\omega_A H_p}{\mathscr{K}_{A32}^*} \tag{2.1-36}$$

where H_p is Henry's constant in atm·L/mg A and $\mathscr{K}_{A32}^*$ is the particle–water partition coefficient in L/kg. Coupled with realistic expressions for H_p and $\mathscr{K}_{A32}^*$, Eq. 2.1-36 appears to be valid for many organics on soils.[45]

Example 2.1-5 p,p′-DDT Loading on Soil for Saturation Vapor Pressure. For a chemical in wet soil at the critical loading, ω_{AC} (mg/kg), or above, the air is saturated. From Eq. 2.1-36,

$$\omega_{AC} = \frac{p_A^0 \mathscr{K}^*_{A32}}{H_p} \tag{E2.1-5}$$

Based on the following data for *p,p'*-DDT at 25°C and a 0.6% organic matter soil, compute ω_{AC} in μg/g; $\mathscr{K}^*_{AC2} = 100{,}700$ L/kg, $\rho^*_{A2} = 0.015$ mg/L, and $p_A^0 = 1.7\text{E-}7$ mmHg.

SOLUTION For immiscible, sparingly soluble chemicals, Henry's constant is the ratio of vapor pressure to solubility:

$$H_p = \frac{1.7\text{E-}7}{760}\,\text{atm} \Big/ \frac{\text{L}}{0.015\,\text{mg}} = 1.49\text{E-}8\,\text{atm}\cdot\text{L/mg}$$

The soil-water partition coefficient is obtained from Eq. 2.1-23:

$$\mathscr{K}^*_{A32} = \omega_C \mathscr{K}^*_{AC2} = (0.006)100{,}700 = 600\,\text{L/kg}$$

The critical loading is

$$\omega_{AC} = 2.24\text{E-}10\,\text{atm} \left| \frac{600\,\text{L}}{\text{kg}} \right| \frac{\text{mg}}{1.49\text{E-}8\,\text{atm}\cdot\text{L}} = 9.0\,\mu\text{g/g}$$

It appears that a very low loading is required and the saturation vapor pressure is exerted above the soil. The reported experimental value at 30°C is 15 μg/g.

Closure. Experimental data show that the vapor pressures of pesticides are greatly influenced by their interaction with soils, due primarily to adsorption.[10] This class of compounds has been the subject of much study because of their application to soils for pest control purposes, and as a consequence, many data are available compared with other organic chemicals. How much adsorption reduces the vapor pressure of a pesticide in soil depends mainly on the nature of the pesticide, pesticide concentration in soil, soil water content, and soil properties such as organic matter and clay content.

Absorption reduces the chemical activity, or fugacity, below that of the pure compound. This is then reflected in changes in vapor pressure of the chemical. For weakly polar or nonionic pesticides on wet soil the amount of organic matter is the most important soil factor for increasing absorption (see Problem 2.1I) and, consequently, for decreasing vapor pressure or potential volatility affects. Most of the more volatile pesticides are only weakly polar or nonionic; thus their absorption by soils is closely related to organic matter content. For example, dieldrin vapor pressure in five soils varied inversely with soil organic matter content. The vapor pressure of weakly polar pesticides in soil increases greatly with increases in pesticide concentration and reaches saturation vapor densities equal to that of the pesticide without soil at relatively low soil pesticide concentration. In moist Gila silt loam, saturation vapor densities for dieldrin, *o,p′*-DDT, trifluralin, lindane, and *p,p′*-DDT were reached at soil concentrations of 25, 39, 73, 55, and 15 μg/g, respectively. Typical applications of pesticides are, for example, 1 kg/ha, which is equivalent to 10 $\mu g/cm^2$, or approximately 150 ppm in the top 0.5 mm of soil. It appears that in the case of wet soils immediately after application, some pesticides exert the vapor pressure of the pure chemical. Problem 2.1H illustrates the effect of moisture, organic content, and the soil mineral fraction on the vapor pressure of dieldrin.

Chemical Adsorption onto Altered Solid. As noted above, soils may be altered from their natural state by the massive application of chemical and wastes. The creation of an additional nonaqueous liquid phase necessitates an additional equilibrium relationship. In the section "Nonaqueous Liquids in Contact with Earthen Solids" we reviewed the available expressions for chemical equilibrium between the two liquids. In the case of unsaturated, altered soils, the air phase is present. The case involving three or more phases in soil is presented in Section 2.3.

Equilibrium Occurrences Between Water and Other Liquids

The contact of water with nonaqueous liquid chemicals, which involves two similar phases (i.e., liquid–liquid), occurs with some frequency. This environmental interface is present at the underside of an oil slick on the surface of water, where the hydrocarbon phase contacts the water phase. The inadvertent placement of high-density liquids in water bodies is another example of the creation of a liquid–liquid environmental interface.

The examples above are called *floaters* and *sinkers*, respectively; they involve a continuous layer of bulk liquid and are termed *macrophases*. This is in contrast to a microphase system, which involves the liquid in a dispersed state. An example of a microphase liquid–liquid system is that of oil globules trapped within the pores of soil or bottom sediment. Natural colloids and micells in the aqueous phase are also well-known examples but are not treated in this book. In the section "Nonaqueous Liquids in Contact with Earthen Solids" we considered the case of chemical equilibrium with oil as the microphase adjacent to porewater. The natural organic matter in soil and sediment may be considered a nonaqueous liquid microphase. Because of the ubiquitous nature of natural organic matter it is given special treatment (see the section "Partition Coefficients for the Sediment–Water and Soil–Water Systems" earlier in this section. Although the material noted above, concerns microphases, the results apply to the macrophase problem since the same thermodynamic principles are involved. The following subsections contain additional information on the chemical equilibrium between water and the other liquid (phase 4).

Pure Liquids in Contact with Water. It is obvious in the case of a pure liquid in contact with water that its solubility is the equilibrium concentration. A relationship such as Eq. 2.1-3 applies for two liquid phases,

$$x_{A2}\gamma_{A2}f_A^0 = x_{A4}\gamma_{A4}f_A^0 \tag{2.1-37}$$

where 4 ≡ nonaqueous liquid. For liquids that are sparingly soluble, the result is particularly simple

$$x_{A2} = \frac{1}{\gamma_{A2}} \tag{2.1-38}$$

The pure component fugacity, f_A^0, cancels since both are liquids of the same chemical. If water is also sparingly soluble in the organic phase, $x_{A4} = 1.0$. Also, for a practically pure organic phase, $\gamma_{A4} \simeq 1.0$ and γ_{A2}^{∞} is the activity coefficient at infinite dilution. The solubility, expressed as a mole fraction, is simply the reciprocal of the latter activity coefficient. Some data on pure liquid solubilities in water are provided in Appendix C and elsewhere in the book. Many data are available in chemical handbooks. In the absence of experimental data, activity coefficients can be estimated from the Margules and UNIFAC equations[22,46] for an approximate value of the solubility.

For systems that exhibit significant solubilities, the more complicated equilibrium law holds:

$$x_{A2}\gamma_{A2} = x_{A4}\gamma_{A4} \tag{2.1-39}$$

A prediction of solubility is still possible by means of this law if the dependence of the activity coefficients on composition are known. For example, water may be soluble in the organic phase and this factor must be accounted for in determining γ_{A4}. In addition, γ_{A2} may not be the infinite dilution value. We consider this very situation in the next section.

Multicomponent Liquids in Contact with Water. The simplest case of a multicomponent liquid is that of two miscible organics that are sparingly soluble in water. The benzene (A), toluene (B), and water (2) system is an example. What will be the benzene and toluene mole fractions in the aqueous phase at equilibrium? For A, Eq. 2.1-39 applies. For B, it becomes

$$x_{B2}\gamma_{B2} = x_{B4}\gamma_{B4} \tag{2.1-39a}$$

Certain facts about this system lead to simplifications. Benzene and toluene form an ideal solution; hence $\gamma_{A4} = \gamma_{B4} = 1.0$. The solubilities of each are very low and there will be no interaction while in aqueous solution, so that Eq. 2.1-38 can be used to approximate the respective activity coefficients. If $x_{A4} = 0.8$ and pure material mole fraction solubilities at 25°C are 4.00E-4 and

1.01E-4 for A and B respectively, computations with Eqs. 2.1-39 and 2.1-39a will yield $x_{A2} = 3.2\text{E-4}$ and $x_{B2} = 2.02\text{E-5}$. The aqueous phase is $1 - x_{A2} - x_{B2} = 0.9997$ mole fraction water, so the system is sparingly soluble.

In the general case when the organics are soluble in water and/or form nonideal solutions, an iterative scheme is necessary for solution of the equations since the activity coefficients are concentration and species dependent. Groves[46] has developed a simple approximate procedure that gives satisfactory accuracy for predicting hydrocarbon solubilities. Crude oil is usually a multicomponent liquid at ambient conditions. Leinonen et al.[11,12] have studied this system in some detail. For this equilibrium situation Eq. 2.1-39 applies for each constituent in the mixture. The equation also applies to nonaqueous liquid phases other than hydrocarbons.

The values of γ_{A4} in hydrocarbon mixtures are fairly close to unity; however, it is desirable to obtain its actual value particularly at a low concentration, where it tends to be highest. As noted earlier, the aqueous-phase activity coefficients γ_{A2} are very high (e.g., 10^3 to 10^7), and methods are now available to predict the influence one dissolved hydrocarbon has on another. A solubility enhancement of from 1 to 25% may occur, owing to the presence of other hydrocarbon components, which apparently reduces the hydrocarbon activity coefficient in the aqueous phase. Few data are available on solubility of hydrocarbon mixtures under environmental conditions.

Phase Equilibrium, Phase Nonequilibrium, and Interface Equilibrium

Phase equilibrium has been the topic thus far in this section. Equilibrium represents a static condition. The net movement of a chemical between phases (i.e., interphase transfer) ceases at equilibrium. Viewed from another perspective, it establishes the potential for a chemical to move between phases. Besides establishing the fact, the foregoing material allows one to calculate the magnitude of chemical concentrations when equilibrium has been achieved.

Nonequilibrium conditions must exist between the phases for the net movement of a chemical species to occur. If the concentrations of chemical A existing in adjoining phases (x_A, y_A) are different from the equilibrium concentrations (x_A, y_A^*), there will be interphase movement (see Problems 2.1A and 2.1B).

Even though the concentrations of the bulks of each phase may be in a nonequilibrium condition, equilibrium is assumed always to exist at the interface! The equilibrium interface (by definition) is assumed to be a hypothetical physical region two molecules thick, one monomolecular layer in each phase. This physical region is sufficiently small in scale so as to respond extremely quickly and remain at equilibrium at all times under conditions when the adjacent bulk phases are undergoing rapid fluctuations in concentration. The existence of equilibrium at the interface does not preclude interphase mass transfer when the bulk phases are not in equilibrium (see Fig. 2.2-1). Further developments of these concepts are presented in Section 3.1 under the topic "Binary Mass Transfer Coefficients in Two Phases and Two-Resistance Theory of Interphase Mass Transfer."

Note the following on equilibrium notation: An equilibrium condition expressed in either concentration or mole fraction units should be denoted by * as a superscript. For example, the set of concentrations (c^*_{A1}, c_{A2}) signifies gas-phase concentration in equilibrium with a liquid-phase concentration, whereas (y_A, x^*_A) signifies a liquid mole fraction in equilibrium with a gas mole fraction. These superscripts were omitted in the earlier section on phase equilibrium for reasons of simplicity and clarity; however, from this point forward the designation is used. The following examples demonstrate the use of the equilibrium notation in phase equilibrium calculation.

PROBLEMS

2.1A. Equilibrium, Nonequilibrium, and Direction of Chemical Movement

Concentrations of chemicals on either side of environmental interfaces are given for the following situations. Do these concentrations represent equilibrium conditions? Give the direction of chemical movement, if any.

1. A sample of seawater (18,980 ppm chloride concentration) at 18.5°C has a dissolved oxygen concentration of 8.0 mg/L. The other phase is air.
2. A sample of pure water in contact with pure *n*-butane gas at 1 atm pressure and 25°C has a concentration of 61.4 g *n*-butane/m^3.
3. The concentration of toluene in air above a pool of pure toluene liquid was 102 g/m^3 at 25°C and 1 atm pressure.
4. The concentration of NH_3 above a septic water body at pH = 7, T = 80°C, y_A = 2E-4, and x_A = 1.5E-9 (see Table 4.1-4).
5. A water sample contains 60 ppm phenol, and the adjoining mud contains 100 ppm (use Fig. 2.1-2).

2.1B. Graphical Representation of Nonequilibrium

The graphical representation in Fig. 2.1B shows the equilibrium curve and a nonequilibrium point (x_A, y_A) for chemical A. This point represents the concentrations across a certain environmental interface. In what direction is chemical A moving: G to L or L to G? Explain your answer.

2.1C. Applying Criteria for Equilibrium to Specific Cases

State and justify the transformation of Eq. 2.1-3 to: 1. Eq. 2.1-4; 2. Eq. 2.1-5; 3. Eq. 2.1-8; 4. Eq. 2.1-11; 5. Eq. 2.1-13.

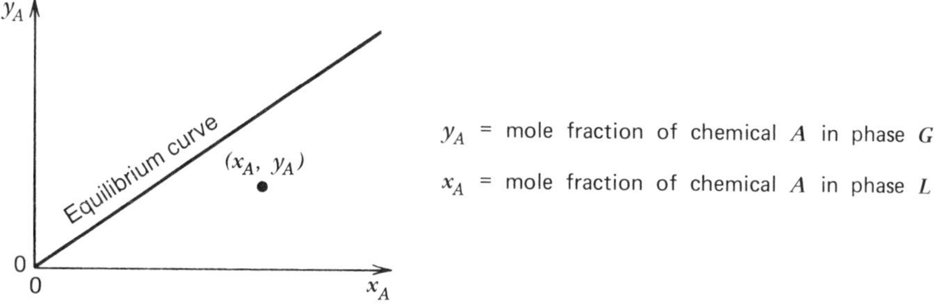

Figure 2.1B. Nonequilibrium condition.

2.1D. Distribution of Phosphorus Between Soil and Water

Samples of Crowley and Mhoon soils were incubated under aerobic and anaerobic conditions, and the distribution of added soluble phosphorus between the solution and solid phases was determined.[13] Each sample (300 g) was kept in suspension by use of a magnetic stirrer in 1500 mL of water in a sealed 2-L flask. Slow streams of air for the aerobic treatments and argon for the anaerobic treatments were continuously bubbled through the suspensions. The samples were incubated for 17 days at 30°C before the incremental additions of phosphorus as $Ca(H_2PO_4)_2$. Samples of the suspension were removed 24 hr after each addition of phosphorus and filtered prior to analyzing for phosphorus. The chemical analysis at equilibrium is shown in Table 2.1D.

1. Transform these data so that plots like Figs. 2.1-1 and 2.1-2 can be constructed. Make the vertical axis μg P/g soil and the horizontal axis μg P/mL.
2. Is phosphorus partitioned more strongly in the soil or in the water?
3. What effects do anaerobic conditions have on the release of phosphorus from the soil?
4. Is the shape of the experimental absorption isotherms in part 1 similar to that of the Freundlich (Eq. 2.1-27), Langmuir (Eq. 2.1-29), or BET (2.1-30)?

2.1E. Vapor Pressure of *p,p'*-DDT

The abbreviated name for 1,1,1-trichloro-2,2-bis(4-chlorophenyl)ethane is *p,p'*-DDT.

1. Zhdamirov et at.[47] give the following pure-component vapor pressures for *p,p'*-DDT: T(°C), p_A^0(mmHg) = (10, 2.61E-8), (20, 1.48E-7), (30, 7.18E-7), and (40, 3.23E-6). Using a relationship such as $\log p_A^0 = A + B/T$ between p_A^0 (mmHg) and T(K) find A and B values that best fit the data.
2. Using the ω_{AC} result for wet soil from Example 2.1-5, estimate the partial pressure at this loading level if the soil is dry. Assume that the soil has

Table 2.1D. Soil–Water Equilibrium for Phosphate in Soils

Anaerobic		Aerobic	
Added (μg P/g soil)	Equilibrium Solution (μg P/mL)	Added (μg P/g soil)	Equilibrium Solution (μg P/mL)
Crowley soil			
15	0.025	15	0.025
30	0.05	30	0.04
45	0.12	45	0.095
60	0.8	60	0.4
180	4.0	180	5.0
420	10.6	420	30
780	30	780	90
1080	40	1080	100.2
Mhoon soil			
10	1.5	10	0.07
20	2.0	20	0.15
40	2.5	40	0.35
80	3.0	80	4.0
160	8.0	160	15
360	30	360	40
660	70	660	85
1020	80	1020	120

Source: Ref. 13.

90 m^2/g total surface area but that only 15% is available as adsorption sites. Give the answer in mmHg.

2.1F. On Calculating Henry's Law Constants[14]

Henry's law constants can be calculated by

$$H_\rho \equiv \frac{\rho_{A1}^*}{\rho_{A2}^*} = 16.04 \frac{p_A^0 M_A}{T\rho_{A2}^*} \tag{2.1F}$$

where ρ_{A1}^* and ρ_{A2}^* are the equilibrium concentrations in the air and water phases, respectively, p_A^0 the vapor pressure of the pure solute (mmHg), M_A the molecular weight of the solute, T the temperature (K), and ρ_{A2}^* the solubility of the solute in water (mg A/L). The ρ_{A2}^* must be finite.

1. Derive Eq. 2.1F. (*Hint:* Use Eq. 2.1-3 twice. Use it once for pure A vapor above an aqueous solution of A to obtain an expression for γ_{A2}^∞. Equation 2.1-13 is the second use.)

Table 2.1F. Equilibrium Data for Chlorohydrocarbons in Water at 25°C, 1 atm

Compound	Solubility ρ_{A2}^* (mg/L)	Vapor Pressure, p_A^0 (mmHg)	Henry's Law Constant, H_ρ (found)
$CH_2Cl\ CH_2Cl$	8700	82	0.040
CH_3CCl_3	730	99	1.4
$CH_2{=}CHCl$	60	760	50
$CH_2{=}CCl_2$	400	497	6.3
$CCl_2{=}CCl_2$	140	18.6	0.50

Source: Ref. 14.

2. Calculate the Henry's law constant given by Eq. 2.1F for the chlorohydrocarbon data in water compiled by Dilling[14] in Table 2.1F. Compare calculated value to other values in the literature.

3. M. M. Claith (Ph.D. thesis, University of California–Riverside, June 1978) measured H_ρ for *S*-ethyl *N,N*-dipropylthiocarbamate as 7.29E-3 (cm^3 air/cm^3 water) at 30°C. If the solubility and vapor pressure reported are 320 mg/L and 2.97E-2 mmHg at 30°C, calculate H_ρ with Eq. 2.1F.

2.1G. Gas Mixtures Above Water

1. For a binary mixture of gases A and B, show the development of the following: $x_A = y_A/\gamma_{A2} f_{A2}^0$, $x_B = (1 - y_A)/\gamma_{B2} f_{B2}^0$, and $x_A + x_B + x_2 = 1$. Note that γ_{i2} values are functions of concentration: x_A and x_B.

2. Calculate the equilibrium concentrations in water (mg/L) for the gas mixture; 75 vol% vinyl chloride 25% ethane at 25°C. The solubility of vinyl chloride is 30.6 mg/L at 25°C. Is the water mixture a dilute solution?

3. The sea level composition of dry air is shown in Table 2.1-1. Compute the equilibrium concentration of nitrogen, oxygen, and carbon dioxide in water at 20°C. Give the numerical results in mole fraction (x_A) and concentration (ρ_{A2} in mg A/L).

2.1H. Dieldrin Vapor Pressure with Soil

The chemical formula for dieldrin is $C_{13}Cl_6O$, the molecular weight 381, density 1.75 g/cm^3, and melting point 175°C. Spencer et al.[48] give the vapor pressure above pure crystals as 2.6-E6, 1.05E-5, and 3.47E-5 mmHg at 20, 30, and 40°C, respectively. It was applied to Gila silt loam, a desert soil containing 18.4% clay (predominately montmorillonite) and 0.6% organic matter.

1. Estimate the total surface area from the soil fractions. The measured value is 90 m^2/g.

Table 2.1I. Sorption of 4-Amino-3,5,6-Trichloropicolinic Acid by Soils at pH 2 (wt %)

Soil	Organic Matter Content of Soil	Acid in Solution at Equilibrium	Acid sorbed on Soil at Equilibrium
D_1	1.0	51	49
N_1	2.7	23	77
B_2	4.1	11	89
B_1	10.7	5.8	94
Q_1	32.2	2	98

Source: Ref. 15.

2. Estimate the partial pressure (mmHg) above dry Gila silt loam for a loading of 100 μg/g at 40°C. Assume that 18.4% of the total surface is available for adsorption of dieldrin and $B_1 = 40$. The measured value was approximately 8.8E-8 mmHg.

3. Compute the single-monolayer coverage water content, ω_B^* (g H_2O/g soil). Assume 100% surface availability for water.

4. Estimate the partial pressure of dieldrin on soil at 40°C at a loading of 100 μg/g for $\omega_B/\omega_B^* = 0.05, 0.7, 0.95$, and 0.995. Answer in mmHg.

5. Calculate ω_{AC} in μg/g. The reported value at 20 and 30°C for wet soil is 25 μg/g.

2.1I. Soil–Water Partition Coefficient for an Organic Pesticide

The organic content of the soil strongly affects the partitioning of organic pesticides between the soil and water phases. Table 2.1I contains experimental results of an equilibrium study with soils of organic matter $\geqslant 1\%$. For each test 4 mL. of a 1.0 ppm solution was added to each 1.0 g of soil. The pH was adjusted to 2 with HNO_3 and the slurry incubated for 1 hr.

1. Determine the soil–water partition coefficient, $\mathscr{K}^*_{A32}$ (L/kg soil) for each.

2. Determine the organic matter–water partition coefficient, $\mathscr{K}^*_{AC2}$ (L/kg OM), for each.

3. Perform simple statistics (i.e. $\bar{x}$ and σ) on each coefficient.

Which coefficient is nearly constant? Does this suggest anything about the adsorption process?

2.1J. Runoff Contamination from Organic Pesticide

The managers of a large farming operation plan to use 4-amino-3,5,6-trichloro-picolinic (ATP) acid to control a certain pest. It is expected that the concentration of ATP at the soil surface will be 15 μg/g. The soil contains 5% organic matter. The state environmental protection department is concerned about the water runoff contamination. Calculate the maximum water concentration of ATP likely to occur in the runoff in mg ATP/L. Additional data: $\mathscr{K}^*_{A32} = 546$ L/kg.

2.1K. The Many Forms of Henry's Law

Henry's law for expressing the equilibrium distribution of a chemical between dilute gas and liquid phases enjoys several formulations. The mole fraction form is dimensionless and can be expressed as

$$y_A = H_{Ax} x_A \tag{2.1K-1}$$

A common form is given by Eq. 2.1-12:

$$p_A = H_A x_A$$

where H_A is in atm/mol fraction. Other forms are

$$p_A = H_p \rho_{A2} \tag{2.1K-2}$$

where H_p is in atm$\cdot$cm^3/g, and

$$\rho_{A1} = H_\rho \rho_{A2} \tag{2.1K-3}$$

where H_ρ is in cm^3 water/cm^3 air STP. In the preceding expressions y_A and x_A are in mole fractions, p_A in atm, ρ_{A1} and ρ_{A2} in g/cm^3 air STP and g/cm^3 water, respectively. Find the multipliers a, b, and c for the equality

$$H_{Ax} = aH_A = bH_p = cH_\rho \tag{2.1K-4}$$

2.1L. Cumene Partitioning Between Water and Oil Slick Phases

Beginning with Eq. 2.1-39, show that the concentration of cumene in water is given by

$$c_{A2} = c^*_{A2} x_{A4} \gamma_{A4} \tag{2.1L}$$

where c^*_{A2} is the solubility of cumene in water, x_{A4} the mole fraction of cumene within the oil slick, and γ_{A4} the activity coefficient of cumene within the oil slick. (*Hint:* In your development, apply Eq. 2.1-39 twice. Apply it to a two-phase system consisting of pure cumene and water. Apply it to a two-phase system consisting of the oil slick and water.)

Table 2.1M. Constituents in the Oil Fraction

Chemical	M_A (g/mol)	ω_A (μg/g)
Benzene	78	1,400
Toluene	92	5,600
Ethylbenzene	106	4,300
m,p-Xylene	106	15,000
o-Xylene	106	32,000
Sat. hyd.[a]	85	24,000
Indane	118	28,000
C_3-benzene	120	160,000
Dimethylpentene	98	1,900
Sat. hyd[a]	57	5,200

[a]Saturated hydrocarbon, exact species unknown.

2.1M. Oily Waste Vapor Pressures

A refinery waste contains 27.74% oil; the remainder is inorganic solids of a soil nature. Table 2.1M contains the organic chemical analysis of the oil fraction.

1. Compute the average molecular weight of the oil fraction.
2. Compute the partial pressure of each known species at 25°C. Answer in mmHg.
3. Compute the total pressure exerted by the mixture at 25°C.

2.2. THERMAL EQUILIBRIUM AT ENVIRONMENTAL INTERFACES

The movement of thermal energy is an important aspect of the subject of chemodynamics as an effect in itself or as it affects the movement of chemicals within environmental phases and at interfaces. Thermal energy is mechanical, potential, and kinetic energy of random motion on a molecular or microscopic scale. Heat is thermal energy in the process of being added to or removed from a given substance or moving from one portion of material substance to another by a temperature gradient. In describing thermal phenomena we find that we need an additional fundamental quantity besides mass, length, and time and that quantity is temperature.

The sensations of hotness and coldness of a given body are determined by what is called its *temperature.* Add heat to a body and its temperature ordinarily rises. Remove heat from a body (i.e., let it give thermal energy to some other body) and its temperature ordinarily goes down. Two bodies have the same temperature if when placed in contact, no heat flows from one to the other. Body A is at a higher temperature than body B if when they are placed in contact, heat flows from A to B. A group of objects, two phases of substance,

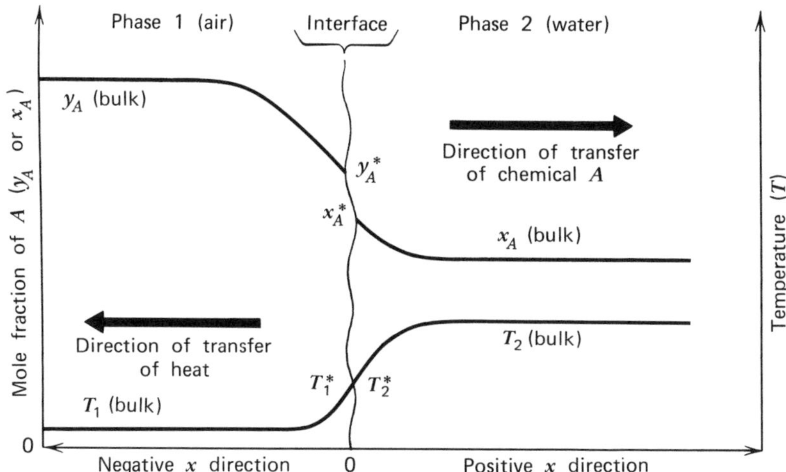

Figure 2.2-1. Heat and mass transfer through environmental interfaces.

and so on, placed within a well-insulated enclosure (i.e., net heat flow is zero) transfer heat in such a way that eventually they come to the same temperature. At this stage *thermal equilibrium* is said to be established, and no rearrangement of the objects, phases, and so on, can result in further transfer of heat.

The similarities and contrasts of chemical and thermal equilibrium and nonequilibrium concepts are noteworthy from the viewpoint of environmental interfaces. Figure 2.2-1 shows an environmental interface consisting of two fluids in which both heat and chemical transfer is occurring. Profiles of concentration and temperature in the region near the interface represent time-averaged values. Equilibrium, both chemical and thermal, exists at the interface. Whereas the chemical concentration profile is usually discontinuous at the interface, the thermal profile is continuous, since thermal equilibrium is satisfied when

$$T_{1i}^* = T_{2i}^* \tag{2.2-1}$$

Temperature of phase 1 equals temperature of phase 2 at the interface, where equilibrium is always assumed to exist. The bulk phases (i.e., that region far back from the interface) are not at a thermal or chemical equilibrium condition, so interphase movement of heat and mass can take place. The mass of material associated with the interface regions where temperature and concentration gradients exist is a vanishingly small fraction of the mass of the respective phases. Temperature and concentration measurements made by sampling and analyzing a particular phase yield the bulk phase temperature and concentration. Special techniques are necessary to detect the interface concentrations and temperatures and these are rarely known.

2.3. CHEMICAL EQUILIBRIUM MODELS FOR ENVIRONMENTAL COMPARTMENTS

Equilibrium models are used to address aspects of the potential for chemical redistribution in the natural environment. Concepts concerned with equilibrium between two phases in contact, as presented earlier in this chapter, are a key factor in these models. Other elements of the modeling include the definition of compartments and mass balances. Presented in this section are the fundamental concepts of chemical equilibrium models which include the assumptions, development techniques, and some simple applications.

In its simplest form a chemical equilibrium model is a mathematical representation of the distribution of a species between environmental compartments based on the thermodynamics of the system. It is convenient to view the natural environment as a multimedia system consisting of macro- and microcompartments. For example, one of the simplest multimedia systems is one consisting of the three macrocompartments: atmosphere, water, and soil. A slightly more complex system is the model ecosystem shown in Fig. 2.3-1, chosen to represent a slice of the environment. The contents within each compartment are homogeneous, and chemical equilibrium is assumed to be achieved between compartments. This means that there are no concentration gradients within a compartment and that the rates of chemical movement between compartments has ceased. Production, consumption and transformations are not considered.

Hamaker[49] first used such an approach in attempting to calculate the percentage of a chemical in the soil for the microcompartments consisting of air, water, and solids in a surface soil. Baughman and Lassiter[50] introduced the concept of the evaluative model, in which no attempt is made to simulate the real environment. The aim is, rather, to provide behavioral information characteristic of the substance; for example, the relative amounts of substance that will partition into each compartment and the relative concentrations.

Mackay[51,52] approached the development of equilibrium modeling using the concept of fugacity directly. When a system is at equilibrium, the fugacity in each compartment matches that in any other compartment. Equation 2.1-2 is an example of this for the air and water compartments. In Mackay's development a slice of the Earth is selected as a unit world or model ecosystem, (see Fig. 2.3-1). Fugacities are calculated for each compartment of the ecosystem, and the overall distribution patterns of a given chemical are predicted. Mackay observes that fugacity is merely a surrogate for concentration and that, in effect, the two are related by a constant, the fugacity capacity of the phase, for a chemical. The fugacity approach is proving to be a useful concept for developing and presenting multimedia environmental models; for full details, see the textbook by Mackay.[62] In an alternative approach, McCall et al.[53] define a model ecosystem that also represents a unit world; however, the development incorporates chemical equilibrium expressions as concentration ratios. Theoretically, both approaches are essentially the same and the numerical results are the same.

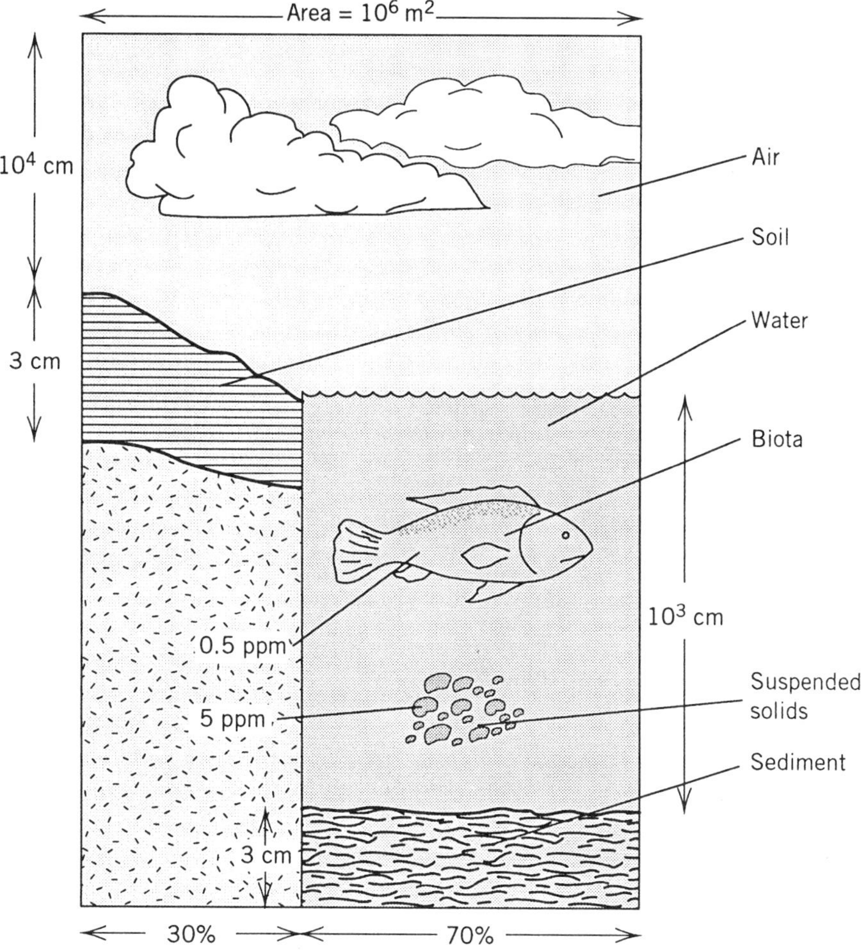

Figure 2.3-1. Idealized ecosystem for equilibrium modeling. (From Reference 51. Reprinted with permission, D. Mackay, 1995.)

Equilibrium models can thus be obtained which predict the distribution patterns of chemicals in a simulated environment representative of a segment of the world. The goal is not to predict actual expected concentrations, but to predict expected behavior. The models can address such general questions as to which phase a substance will tend to migrate; whether a pesticide applied to a soil will leach or be volatile; and whether a chemical will accumulate in the biotic compartment. In the following sections two equilibrium models are presented. One represents a macrosystem and the other a microsystem.

Two-Compartment Equilibrium Model

This section covers the assumptions and development of a two-compartment equilibrium model constructed for a chemical in the sediment and water compartments.[55] Although developed for this case, one should realize the generality of the final results and that by changing the phase notation it can be used for any two compartments.

Consider only the water (2) and sediment (3) compartments shown in Fig. 2.3-1. In addition to the mass of the chemical, m_A, the system contains the masses of water, m_2, and sediment, m_3. The distribution of A in any single compartment is assumed to be uniform. The concentration of A in water is designated ρ_{A2} and that in sediment is ω_A. The sediment compartment consists of two fractions, suspended sediment and bottom sediment, both with the same concentration of A. It is now assumed that A has been introduced into the system, and that net chemical transport between phases has occurred but has now ceased, so that the two compartments are at chemical equilibrium. The final equilibrium state is independent of how A is introduced. It may be introduced as a pure substance, such as in laboratory batch apparatus containing sediment and water, or it may enter with a phase such as a pesticide sorbed onto suspended sediment runoff into a pond. The chemical has distributed itself between the water and sediment phases such that:

$$A(2) \rightleftharpoons A(3) \tag{2.3-1}$$

where $A(2)$ denotes A dissolved in water and $A(3)$ denotes A on the sediment. The equilibrium concentration ratio is denoted by the partition coefficient (Eq. 2.1-18) in mass units,

$$\mathscr{K}^*_{A32} = \frac{\omega_A}{\rho_{A2}} \tag{2.3-2}$$

The total mass of A is

$$m_A = m_{A2} + m_{A3} \tag{2.3-3}$$

where m_{A2} and m_{A3} are the mass in water and sediment, respectively. Also note that the masses in Eq. 2.3-3 can be expressed in terms of the concentrations:

$$m_A = \frac{\rho_{A2} m_2}{\rho_2} + \omega_A m_3 \tag{2.3-4}$$

where ρ_2 is the density of water.

The development above represents the primary equations for the two-compartment equilibrium model. These equations can be manipulated to yield various relationships. One useful result is the mast fraction of A in each compartment. That in the water is m_{A2}/m_A, so that using the equations yields

$$\frac{m_{A2}}{m_A} = \frac{1}{1 + \mathscr{K}^*_{A32}\rho_{32}} \tag{2.3-5}$$

where $\rho_{32} \equiv m_3\rho_2/m_2$ is the apparent concentration of sediment in the water. This result shows the fraction to be independent of chemical concentration. The remaining fraction is on the sediment. As illustrated in Problem 2.3A, even though $\mathscr{K}^*_{A32}$ and ω_A are large for many chemicals, ρ_{32} is small, so that the model predicts that the water contains a significant fraction of A.

Other end uses can be made of the model equations. The equilibrium scenario can be a batch or a continuous-flow process. A batch process is implied in the development above, in that fixed quantities of A, water, and sediment were used. Problem 2.3B illustrates the use of the model in a continuous-flow process.

Three-Compartment and Higher Equilibrium Models

In this section the specific focus is on a three-compartment, microscale environmental model. Several things will be achieved. The section begins with an introduction to the possible chemical association with various phases in a soil system. Then an example problem is given to illustrate how equilibrium fundamentals can simplify the understanding of chemical associations between phases. Finally, the algorithm for the three-compartment equilibrium model is presented.

From the introduction presented in the discussion "Nonaqueous Liquids in Contact with Earthen Solids," it should be clear that five individual phases may be present in an apparent earthen solid phase to challenge the skills of a thermodynamicist in predicting concentrations. Fortunately, if consideration is given to how phase interaction is likely to occur in a few realistic contact scenarios, the equilibrium analysis can be simplified somewhat. In the case of an episodic spill involving a nonaqueous liquid, a likely layering of phases for a soil system is: mineral, natural organic matter, water, nonaqueous liquid, and soil gas. If the mineral and natural organic phases are combined, a four-compartment system results: gas (1), water (2), mineral–organic matter (3), and nonaqueous liquid (4), so that the equilibrium relationship for A may be viewed as

$$A(3) \rightleftharpoons A(2) \rightleftharpoons A(4) \rightleftharpoons A(1) \tag{2.3-6}$$

It is possible to define equilibrium distribution coefficients for each two-phase association analogous to Eqs. 2.1-10 and 2.3-2. These coupled with a mass balance yields the algorithm to compute the concentration of A in each phase. In another scenario of preceived phase interaction it may be assumed that the nonaqueous liquid may not completely cover the water. In this case the equilibrium relationship is

$$\begin{array}{c} A(3) \rightleftharpoons A(2) \rightleftharpoons A(4) \\ \diagdown\!\!\diagdown \quad \diagup\!\!\diagup \\ A(1) \end{array} \tag{2.3-6a}$$

since both liquids contact the soil gas. The need for an additional distribution coefficient is suggested in this relationship.

Soil that has been repeatedly contaminated by oil, chemicals, solvents, sludges, and so on, is a realistic phase association scenario. Such long-term contact, which prevails around fuel terminals, chemical plants, refineries, and waste disposal sites, produces a situation where the natural organic matter is overwhelmed, so that the soil solids may become "oil wet" rather than "water wet". The two possible equilibrium relationships are

$$A(4) \rightleftharpoons A(2) \rightleftharpoons A(1) \tag{2.3-7}$$

and

$$\begin{array}{ccc} A(4) & \rightleftharpoons & A(2) \\ & \nwarrow\!\!\searrow \quad \nearrow\!\!\swarrow & \\ & A(1) & \end{array} \tag{2.3-7a}$$

Here phase 4 is the combined nonaqueous liquid and original natural organic matter plus mineral. Equation 2.3-7 applies to the envisioned layered association, where oily solids are coated with water and a continuous layer of water and soil–air exist in the remaining spaces (Case A, Fig. E2.3-1). Equation 2.3-7*a* applies when it is assumed that water forms an incomplete coating over the oily solids, so that soil–air contacts both water and the nonaqueous liquid (Case B, Fig. E2.3-1). Identical concentrations will result in either case because the same number of phases are present regardless of position. The following example illustrates this principle.

***Example 2.3-1. Equilibrium Concentration in the Leachate and Vapor Space for an Oily Sludge*[54].** If the benzene distribution problem of Example 2.1-3 is extended to include a vapor space, three phases are present. Show that Eqs. 2.3-7 and 2.3-7*a* are equally valid representations and that an arbitrary ordering of phases is not necessary.

SOLUTION Consider the following phase layering cases for air, water, and sludge around an inert core (Fig. E2.3-1).

The binary-phase partition coefficients are $\rho_{A1} = \mathscr{K}_{A12}\rho_{A2}$ (air–water), $\rho_{A2} = \mathscr{K}_{A24}\omega_A\rho_4$ (water–sludge), and $\rho_{A1} = \mathscr{K}_{A14}\omega_A\rho_4$ (air–sludge). For case A, Eq. 2.3-7 suggests that two partition coefficients are needed. In matrix notation,

$$\begin{vmatrix} 1 & -\mathscr{K}_{A12} & 0 \\ 0 & 1 & -\mathscr{K}_{A24} \end{vmatrix} \cdot \begin{vmatrix} \rho_{A1} \\ \rho_{A2} \\ \omega_A\rho_4 \end{vmatrix} = \begin{vmatrix} 0 \\ 0 \\ 0 \end{vmatrix}$$

An additional relationship is needed for a unique solution to the unknown vector. For case B, Eq. 2.3-7*a* suggest that all three partition coefficients are

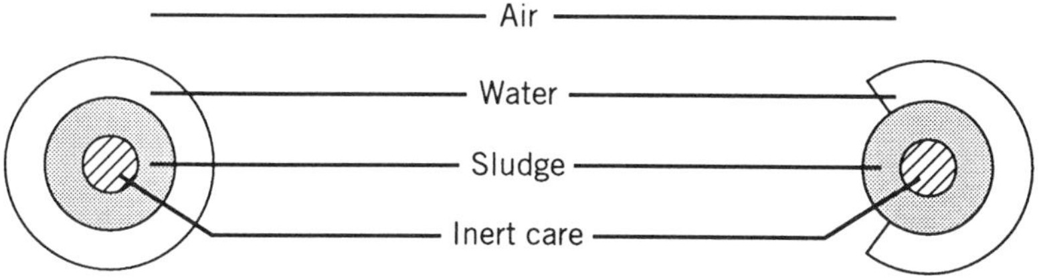

Figure E2.3-1

needed. Expressing the system in matrix format followed by triangulation yields

$$\begin{vmatrix} 1 & -\mathscr{K}_{A12} & 0 \\ 0 & 1 & -\mathscr{K}_{A24} \\ 0 & 0 & \mathscr{K}_{A14} - \mathscr{K}_{A12}\mathscr{K}_{A24} \end{vmatrix} \cdot \begin{vmatrix} \rho_{A1} \\ \rho_{A2} \\ \omega_A \rho_4 \end{vmatrix} = \begin{vmatrix} 0 \\ 0 \\ 0 \end{vmatrix}$$

Since $\omega_A \neq 0$, the entry in column 3 row 3 of the coefficient matrix must equal zero; this yields

$$\mathscr{K}_{A14} = \mathscr{K}_{A12}\mathscr{K}_{A24} \tag{E2.3-1}$$

The third partition coefficient is therefore a product of the other two. This being the case, the matrix representation for cases A and B are identical, so that Eqs. 2.3-7 and 2.3-7*a* must be identical representations of the same equilibrium chemical distribution. This implies that phase ordering or layering is *not* important to the final equilibrium state.

The principles involved in developing the equilibrium relationships are well established and straightforward. Difficulties arise in the application. It must be decided how many realistic phase associations apply in a given situation. Equations 2.3-6 and 2.3-7 offer two possibilities, but others can be formulated. A slightly less difficult problem is that of arriving at the various partition coefficients, since much guidance is provided in this chapter.

A component mass balance is needed to account for the quantity of chemical in each phase. The three phases in Eq. 2.3-7 used to represent the distribution of A in soil is the basis for the three-compartment model. The mass of A at equilibrium, m_A, in a unit volume of soil is distributed

$$m_A = V_1\rho_{A1} + V_2\rho_{A2} + V_4\rho_4\omega_A \tag{2.3-8}$$

The equilibrium partition coefficient relationships for A in phases 1 and 2 and in phases 2 and 4 provide the additional equations. Solution of this set of equations yields:

$$\omega_A = \frac{m_A \mathscr{K}_{A42}}{\rho_4(V_1 \mathscr{K}_{A12} + V_2 + V_4 \mathscr{K}_{A42})} \tag{2.3-9}$$

where V_1, V_2, and V_3 are the respective phase volumes in soil. This equation gives the concentration in soil; the other two concentrations are obtained from the respective partition coefficient relationships.

The three-compartment equilibrium model presented above was developed for a soil microsystem, but the resulting equations have general applicability for use in any system of three phases. For example, on the macroscale level it can be used for any three of the phases shown in Fig. 2.3-1. Applications involving more compartments, which include the "unit world," "terrestrial" and "pond" simulations, are available.[52,53] These and others exist in computer software forms.

PROBLEMS

2.3A. Equilibrium Fraction of Two Pesticides in Water or Aquatic Environments

Methoxychlor [2,2-bis(*P*-methoxyphenyl)-1,1,1-trichloroethane] is a popular substitute for DDT [2,2-bis(*P*-chlorophenyl)-1,1,1-trichloroethane] in the control of many insects. In aquatic systems these chemicals are distributed between the water and sediment phases at equilibrium according to Eq. 2.3-1.

1. Beginning with the definition of the mass fraction of A in the water show details of the derivation of Eq. 2.3-5.
2. The sediment–water partition coefficients for methoxychlor and DDT are 620 and 8600 L/kg, respectively.[55] Compute the expected distribution of each chemical, as mass fractions, in the water and sediment of Beaver Reservoir in northwestern Arkansas. The reservoir has a suspended sediment concentration $\rho_{32} = 6.5\,\text{mg/L}$, a volume of 1.65E6 acre·ft, and occupies 28,200 acres. Assume that the chemically active bottom sediment is 1 mm thick and has a bulk density of 1.5 g/cm^3.

2.3B. Equilibrium Model Calculation for Steady-Flow System

The Ignocare Chemical Corporation discharges 1000 kg of benzene into the Mississippi River daily. The river flow is 5E5 ft^3/sec and has 2400 mg/L suspended sediment. Neglecting benzene's sorption onto bottom sediment, biodegradation, volatilization, and so on, calculated the expected concentration in water at equilibrium.

2.3C. Microscale Equilibrium Model

This problem concerns the three compartments soil–sludge, soil–air, and soil–water. The sludge described in Example 2.1-3 has 1400 ppm (wt) benzene initially; some must move to the other phase to achieve equilibrium.

1. Present the three equations with unknown concentrations and show the steps in the derivation of Eq. 2.3-9.
2. A soil contains a sludge residual saturation of 0.75. The original soil had 0.5 porosity when dry but now contains 10 vol% water. The sludge density is 0.9 g/cm^3. Calculate the equilibrium concentration of benzene in the sludge (mg/kg), water (mg/L), and air (mg/L).
3. What is the pore–air benzene concentration if the soil contains no water?

REFERENCES

1. R. C. Weast, Ed., *CRC Handbook of Chemistry and Physics*, 49th ed., Chemical Rubber Co., Cleveland, Ohio, 1968.
2. N. C. Brady, *The Nature and Properties of Soils*, Macmillan, New York, 1974, p. 24.
3. O. A. Hougen, K. M. Watson, and R. A. Ragatz, *Chemical Process Principles, Vol. II, Thermodynamics*, Wiley, New York, 1959, pp. 849–856, 862–879.
4. W. J. Lyman, W. R. Reehl, and D. H. Rosenblatt, *Handbook of Chemical Property Estimation Methods: Environmental Behavior of Organic Compounds*, McGraw-Hill, New York, 1982.
5. R. C. Reid, H. M. Prausnitz, and T. K. Sherwood, *Properties of Gases and Liquids*, McGraw-Hill, New York, 1977.
6. D. Mackay and W. Y. Shiu, "The Aqueous Solubility and Air–Water Exchange Characteristics of Hydrocarbons under Environmental Conditions", in *Chemistry and Physics of Aqueous Gaseous Solutions*, ASTM, Philadelphia, 1974, pp. 93–110.
7. E. J. Greskovich, "Equilibrium Data for Various Compounds between Water and Mud," *Am. Inst. Chem. Eng. J.*, **20**(5), 1024 (1974).
8. D. F. Paris, W. C. Steen, and G. L. Baughman, "Role of Physico-chemical Properties of Aroclors 1016 and 1242 in Determining Their Fate and Transport in Aquatic Environments", *Paper PEST 110*, 172nd American Chemical Society National Meeting, San Francisco, 1976.
9. R. E. Wilding and R. L. Schmidt, "Phosphorus Release from Lake Sediments," *EPA-RE-73-024,* U.S. EPA, Washington, D.C., 1973.
10. R. Hague, "Role of Adsorption in Studying the Dynamics of Pesticides in a Soil Environment," in R. Hague and V. H. Freed, Eds., *Environmental Dynamics of Pesticides*, Plenum Press, New York, 1975, pp. 97–114.
11. P. J. Leinonen, D. Mackay, and C. R. Phillips, "A Correlation for the Solubility of Hydrocarbons in Water," *Can. J. Chem. Eng.*, **49**, 288–290 (Apr. 1971).

12. P. J. Leinonen and D. Mackay, "The Multicomponent Solubility of Hydrocarbons in Water," *Can. J. Chem. Eng.*, **51**, 230–233 (Apr. 1973).
13. W. H. Patrick, Jr., and R. A. Khalid, "Phosphate Release and Sorption by Soils and Sediments: Effect of Aerobic and Anaerobic Conditions," *Science*, **186**, 53–55 (Oct. 1974).
14. W. L. Dilling, "Interphase Transfer Processes. II. Evaporation Rates of Chloro Methanes, Ethanes, Ethylenes, Propanes, and Propylenes from Dilute Aqueous Solutions: Comparison with Theoretical Predictions," *Environ. Sci. Technol.*, **11**(4), 405–409 (1977).
15. J. W. Hamaker, C. A. I. Goring, and C. R. Youngson, "Sorption and Leaching of 4-Amino-3,5,6-trichloropicolinic Acid in Soils," in R. F. Gould, Ed., *Organic Pesticides in the Environment*, Advances in Chemistry Series 60, American Chemical Society, Washington, D.C., 1966, pp. 23–37.
16. C. Munz, and P. V. Roberts, "Effects of Solute Concentrations and Cosolvents on the Aqueous Activity Coefficient of Halogenated Hydrocarbons," *Environ. Sci. Technol.*, **20**, (8) (1986).
17. L. J. Thibodeaux, *Chemodynamics: Environmental Movement of Chemicals in Air, Water and Soil*, Wiley, New York, 1979, p. 45.
18. R. Wolfenden, "Waterlogged Molecules," *Science* **222**, Dec. 9, 1983, pp.1037–1094.
19. F. R. Groves, Jr., "Solubility of Cycloparaffins in Distilled Water and Salt Water," *J.. Chem. Eng. Data*, **33**, (2) (1988).
20. T. J. Pinnavaia, "Intercalated Clay Catalysts," *Science*, **220** (4595), Apr. 22, 1983, pp. 365–371.
21. J. C. Means and R. Wijayaratne, "Role of Natural Collides in the Transport of Hydrophobic Pollutants," *Science*, **215**, Feb. 19, 1982, pp. 968–970.
22. S. Banerjee, "Solubility of Organic Mixtures in Water," *Environ. Sci. Technol.*, **18**(8), 587–591 (1984).
23. R. A. Freeze and J. A. Cherry, *Groundwater*, Macmillan, New York, 1979, pp. 444–447.
24. T. Altenbaumer, R. Kew, B. Mathews and L. Wright, "A Study of Contaminated Soil Volumes Resulting from Organic Liquid Spills," Senior Project Final Report, Department of Chemical Engineering, University of Arkansas, Fayetteville, Ark., Oct. 1982.
25. D. V. Doshi, "'Modeling Vertical Migration of Non-aqueous Phase Liquid Waste in Unsaturated Soils," Thesis, Louisiana State University, Baton Rouge, La., May, 1987.
26. S. W. Karickhoff, "Organic Pollutant Sorption in Aquatic Systems," *Hydraul. Eng.*, **110**(6), 707–735 (June 1984).
27. R. N. Dexter and S. P. Pavlou, "Distribution of Stable Organic Molecules in the Marine Environment: Physical Chemical Aspects, Chlorinated Hydrocarbons," *Marine Chem.*, **7**, 67–84 (1978).
28. S. P. Pavlou, "Thermodynamic Aspects of Equilibrium Sorption of Presistent Organic Molecules at the Sediment Seawater Interface: A Framework for Predicting Distributions in the Aquatic Environment," Chapter 17 in R. A. Baker, Ed., *Contaminants and Sediments*, Ann Arbor Science, Ann Arbor, Mich. 1980.

29. I. J. Tinsley, *Chemical Concepts in Pollutant Behavior*, Wiley, New York, 1979, pp. 9–31.

30. C. T. Chiou, L. J. Peters, and V. H. Freed, *Science*, **206**, 831 (1979).

31. G. P. Curtis, M. Reinhard, and P. V. Roberts, Chapter 10 in J. A. Davis and K. F. Hayes, Eds., *Geochemical Processes at Mineral Surfaces*, American Chemical Society, Washington, D.C., 1986.

32. C. G. Delos, J. V. DePinto, W. L. Richardson, P. W. Rodgers, and K. Rygwelski, *Technical Guidance Manual for Performing Waste Load Allocations*, Appendix B, Chapter 3, Book II, U.S. EPA, Office of Water Research, Washington, D.C., Aug., 1984.

33. G. de Marsily, E. Ledoux, A. Barbeau, and J. Margat, "Nuclear Waste Disposal: Can the Geologist Guarantee Isolation?" *Science*, **197**, 519 (1977).

34. B. J. Wood, "Backfill Performance Requirements, Estimates from Transport Models," *RHO-BWI-SA-58*, USDOE Contract DE-AC06-77RL01030, Rockwell Hanford Operations, Richland, Wash., Nov. 1980.

35. D. M. Ruthven, *Principles of Absorption and Adsorption Processes*, Wiley, New York, 1984, pp. 43–61.

36. A. W. Adamson, *Physical Chemistry of Surfaces*, Wiley, New York, 1976, pp. 548–634.

37. T. L. Hill, *J. Chem. Phys.*, **14**, 46, 268 (1946).

38. J. M. Smith, *Chemical Engineering Kinetics*, 2nd ed., McGraw-Hill, New York, 1970, pp. 296–302.

39. H. D. Orchiston, *Soil Sci.*, **78** 463–479 (1954).

40. G. W. Bailey and J. L. White, *Agric. Food Chem.*, **12**(4), 324–332 (1964).

41. W. H. Fuller, "Soil Modification to Minimize Movement of Pollutants from Solid Waste Operations," in *CRC Critical Reviews in Environmental Control*, CRC Press, Boca Raton, Fla., Mar. 1980, p. 216.

42. K. T. Valsaraj and L. J. Thibodeaux, "Equilibrium Adsorption of Chemical Vapors on Surface Soils, Landfills and Landfarms: A Review," *J. Hazard. Mater.*, **19**, 79–99 (1988).

43. C. E. Junge, "Basic Consideration About Trace Constituents in the Atmosphere as Related to the Fate of Global Pollutants," in I. H. Suffett, Ed., *Fate of Pollutants in the Air and Water Environments*, Part I, Wiley, New York, 1977.

44. L. T. Cupitt, "Fate of Toxic and Hazardous Materials in the Air Environment," *EPA-600/3-80-08U*, U.S. EPA, Research Triangle Park, N.C., Aug. 1980.

45. K. T. Valsaraj and L. J. Thibodeaux, "Equilibrium Adsorption of Chemical Vapors onto Surface Soils: Model Predictions and Experimental Data," in J. Schnoor, Ed., *Fate and Transport of Pesticides and Chemicals in the Environment*, Wiley, New York, 1992, pp. 155–174.

46. F. R. Groves, Jr., "Effects of Cosolvents on the Solubility of Hydrocarbons in Water," *Environ. Sci. Technol.*, **22**(3), 282–286 (1988).

47. G. G. Zhdamirov, E. E. Popov, and N. F. Lapina, *Proceedings of the Symposium on Environmental Transport and Transformation of Pesticides*, U.S. EPA, Athens, Ga, Feb. 1978, p. 53.

48. W. F. Spencer, M. M. Claith, and W. J. Farmer, *Soil Sci. Soc. Am. Proc.*, **33**, 509 (1969).
49. J. W. Hamaker, "Diffusion and Volatilization," in C. A. I. Goring and J. W. Hamaker, Eds., *Organic Chemicals in the Soil Environment*, Marcel Dekker, New York, 1972.
50. G. L. Baughman and R. Lassiter, *ASTM STP 657*, ASTM Philadelphia, 1978, p. 35.
51. D. Mackay, *Environ. Sci. Technol.*, **13**, 1218–1223 (1979).
52. D. Mackay and S. Patterson, *Environ Sci. Technol.*, **15**, 1006–1013 (1981).
53. P. J. McCall, R. L. Swann and D. A. Laskowski, 'Partition Models for Equilibrium Distribution of Chemicals in Environmental Compartments," in R. L. Swann and A. Eschenroeder, Eds., *Fate of Chemicals in the Environment*, ACS Symposium Series 225, American Chemical Society, Washington, D.C., 1983.
54. L. J. Thibodeaux and S. M. Withers, "Hazardous Chemicals in the Natural Environment: Mobility of Constituents," in S. L. Sandler and B. A. Finlayson, Eds, *Chemical Engineering in a Changing Environment*, AIChE, New York, 1988.
55. N. L. Wolfe, R. G. Zepp, D. F. Paris, G. L. Baughman, and R. C. Hollis, *Environ. Sci. Technol.*, **11**, 1077-1081 (1977).
56. K. T. Valsaraj and L. J. Thibodeaux, *Water Res.*, **23**(2), 183–189 (1989).
57. K. T. Valsaraj, G. J. Thoma, D. D. Reible, and L. J. Thibodeaux, *Atmos. Environ.*, **27A**(2), 203–210, (1993).
58. H. E. Allen, "The Significance of Trace Metal Speciation for Water, Sediment and Soil Quality Standards," *Sci. Total Environ.*, Part 1, 23–45 (1993).
59. H. E. Allen, G. Fu, and B. Deng, "Determination of Acid-Volatile Sulfide (AVS) and Simultaneously Extracted Metals (SEM) for the Estimation of Potential Toxicity in Aquatic Sediments," *Environ. Toxicol. Chem.*, **12**, 1441–1453 (1993).
60. K. T. Valsaraj, A. Gupta, L. J. Thibodeaux, and D. P. Harrison, *Water Res.*, **22**(9), 1173–1183 (1988).
61. K. T. Valsaraj and L. J. Thibodeaux, *Water Res.*, **23**(2), 183–189 (1989).
62. D. Mackay, *Multimedia Environmental Models: The Fugacity Approach*, Lewis Publishers, Chelsea, Mich., 1991, p. 257.
63. R. H. Perry and D. Green, Eds., *Perry's Engineers' Handbook*, 6th ed., McGraw-Hill, New York, 1984.
64. S. Bintein and J. Devillers, "QSAR Model for Organic Chemical Sorption in Soil and Sediments," *Chemosphere*, **28**(6), 1171–1188 (1994).

3

TRANSPORT FUNDAMENTALS

On the topic of the fate and tracking of elusive pollutants in an aqueous medium, it has been flippantly written that the "physical mechanisms are all expressions of the pressure to escape from the aqueous to a gaseous or solid phase." Presumably, the author was referring to the expression

$$\text{rate of movement of chemical } A = K(\Delta p_A)$$

If one interprets *pressure* to mean the departure from equilibrium, some appropriate rate constant K is needed to complete the equality. This chapter is concerned with presenting fundamental concepts to provide a basis for the proper quantification of chemical transport rates in the region near and at environmental interfaces. The rate equations may be as homely as the above, must evolve from a sound grasp of processes or mechanisms, and be couched in an exact mathematical form for employment in a mass balance.

The phrase *fate of chemical A* is often used in conjunction with other phrases, such as *chemical mobility*, *partitioning*, *dissipation rate*, *concentration level*, *residence time*, and so on, to describe behavior. Quantifying the fate of chemical A is rooted in the mass balance. Antoine Lavoisier is primarily responsible for placing the chemical mass balance in general scientific use.[45] It was, however, an accepted theoretical concept for decades before he, together with his wife, Marie Paulze, developed it as an experimental tool for the laboratory using a "balance sheet" method to account for chemical quantities in various phases, reaction vessels, and so on, to explain losses or gains in chemical reactions. The law of conservation of matter is a necessary tool in environmental chemodynamics and is presented as the continuity equation for species A.

A place to begin is with the integral mass balance for species A in a single-phase system

$$\frac{d}{dt}\iiint_V \rho_A \, dV = \iint_A \rho_A \mathbf{V}\cdot(-\mathbf{n})\,dA + \iint_A \mathbf{j}_A\cdot(-\mathbf{n})\,dA + \iiint_V r_A \, dV \qquad (3.0\text{-}1)$$

In words, this equation tells us that the time rate of change of mass of species A in an open system, fixed with respect to Earth's surface, is equal to the net rate at which mass of species A is brought into the systems by convection, the net rate at which the species A "diffuses" into the system (relative to the mass average velocity), and the net rate (sources minus sinks) at which the mass of species A is produced by chemical reactions. The velocity of the fluid, $\mathbf{v}$, is relative to the fixed boundaries of the system. Here the system has a volume V and closed bounding surface A as in Fig. 1.1-5 and 1.2-1. The unit spatial vector $\mathbf{n}$ is outwardly directed to the surface so that only those components of $\mathbf{v}$ and $\mathbf{j}_A$ so oriented are accounted for. It is common to use such integral balances in situations where we wish to make a statement about a single-phase system as a whole without worrying about a detailed description of the motions of the fluids within the system.

With the aid of the generalized transport theorem[26] and Green's transformation, the integral equation for species A above results from integrating the differential equation of continuity for species A. This latter integral species mass balance is

$$\iiint_V \left(\frac{\partial \rho_A}{\partial t} + \nabla \cdot (\rho_A \mathbf{V}_A) - r_A \right) dV = 0 \qquad (3.0\text{-}2)$$

At every point in the system the integrand is zero. The quantity $\rho_A \mathbf{V}_A$ is called the mass flux of species A, $\mathbf{n}_A$, with respect to a fixed frame of reference (Earth's surface). For environmental chemodynamic applications the continuity equation for A may be rearranged to

$$\underset{\text{accumulation}}{\frac{\partial \rho_{Aj}}{\partial t}} + \underset{\substack{\text{convection} \\ \text{or advection}}}{\nabla \cdot (\rho_{Aj} \mathbf{V}_j)} + \underset{\substack{\text{constitutive} \\ \text{term}}}{\nabla \cdot \mathbf{j}_{Aj}} = \underset{\text{reaction}}{r_{Aj}} \qquad j = 1, 2, 3 \qquad (3.0\text{-}3)$$

where $\mathbf{j}_{Aj}$ is the mass flux of species A with respect to the mass average velocity, $\mathbf{V}_{Aj} - \mathbf{V}_j$, of the media. The equation has been modified to denote its application to the prime medias: air (1), water (2), and soil solids (3). Choosing the proper constitutive equations for the mass flux vector $\mathbf{j}_A$, which quantifies the chemodynamics processes of species A, is one of the primary goals of this chapter. The constitutive equations that represent the mass-flux vector for every species in a multi-component body are determined by the history of the motions of all species that the body has undergone as well as the thermal history; however, the temperature distribution and motions of individual species outside an arbitrarily small neighborhood of species A may be ignored. Also, the constitutive equations must be invariant under changes of the frame of (time and space) reference. The term accounts for numerous phenomelogical processes that affect the transport of A through space. The interface surface that separates the environmental media are important locations for the entrance or exit of chemical A. For this interface Eq. 3.0-3 reduces to $\nabla \cdot \mathbf{j}_{Aj}|_i = 0$ when advection is absent. Applying this result for the case of A in the vertical y-direction yields

$$j_{Aj}|_i = Z_{Ay}(t, x, o, z) \tag{3.0-4}$$

where i denotes the interface plane at $y = 0$. Here the zeta term, Z_{Ay}, denotes the flux rate expression; it is independent of y but can be a function of time, t, and location on the surface, x and z. Equation 3.0-4 is one coupling boundary condition at the interface for Eq. 3.0-3 and it also represents the inviolable condition of the equality of normal components of the fluxes in the two adjacent media. For the bounding surface of the media the zeta expression is a vector, $\mathbf{Z}_A \equiv (Z_{Ax}, Z_{Ay}, Z_{Az})$. As noted earlier, considerable attention in this book is devoted to attempts at obtaining realistic expressions for $\mathbf{j}_{Aj}$, the flux within a phase, and $\mathbf{Z}_A$, the flux at the interface that separates two phases. The zeta symbol will not be used beyond this point. The j symbol will be used for both fluxes since the two can be distinguished from the context of use. Often the flux symbol $\mathbf{n}_{Aj}$ will also be used; in this case, some implicit or explicit determination that advection is negligible is implied.

The expressions above reflect the entire fate of chemical A within any phase of the environment. They are written in vector notation to depict the three-dimensional nature of the movement terms. Rarely are they employed in the general form shown. When applied to solids, $\mathbf{v} = 0$, for example.

Convection (also called *advection*) is the term commonly employed to account for movement due to bulk flow of the phase. In the field of meteorology advection denotes the process of transport of an atmospheric property solely by the mass motion of the atmosphere. In the field of oceanography it denotes the process of transport of water, or any aqueous property, solely by the mass motion of the oceans, most typically via horizontal currents. The velocity concentration products (e.g., $\rho_{Aj}\mathbf{v}$) give rise to the preceding interpretations.

In the case of laminar fluid flow and solids the constitutive equation is Fick's first law, since this mass flux expression accounts for the movement of chemical A within a phase due to a concentration gradient. If turbulence is present, the constitutive equation should be one that quantifies the mass flux due to the randomlike exchanges of particles between regions of the phase. When applied to an interfacial region it accounts for the interface transfer of chemical A and enters Eq. 3.0-3 as a boundary condition. The reaction term accounts for the generation or depletion of A by a homogeneous chemical reaction within the phase. As the reaction term appears in Eq. 3.0-3, it is positive for the net production of A within the phase.

Depending on the magnitude and polarity (i.e., positive, zero, or negative), the total effect of advection, mass flux, and reaction of chemical A may result in it increasing, remaining constant, or decreasing within the phase. Therefore, the magnitude and polarity of the accumulation term reflects the quantitative fate of chemical A within the respective phase of the environment.

Equation 3.0-3 applies to a single media phase. If the chemical is present in two phases, it must be applied twice. For example, a study of the chemodynamics of polynuclear aromatics (PNAs) in the atmosphere requires Eq. 3.0-3 for the aerosol phase ($j = 3$) and again for the air phase ($j = 1$). The examination of contaminants within bed sediment requires three uses: porewater ($j = 2$), bed solids ($j = 3$) and suspended organic colloids ($j = 4$). Clever order-of-magnitude analysis results in paring many terms from the general equations to yield a manageable set while retaining realism for the intended model use.

Besides mass balances on the chemical constituents of interest, balances on momentum and energy complete the set of so-called equations of change. These equations provide the fundamentals for chemical transport and provide a basic structure for developing realistic mathematical models for quantifying aspects of behavior. The material in this chapter also addresses processes and mechanisms that lead to realistic constitutive equations for $\mathbf{j}_A$ in Eq. 3.0-3 and similar relationships for the other equations of change. The treatment of the subject of transport phenomena presented here is a review with the emphasis on application to environmental chemodynamics. Those wishing a more detailed and definitive approach should consult such traditional works as Refs 3 and 26.

The style and order of presenting the material in this chapter is one of bootstrap induction. Some aspects of molecular diffusion transport are commonly known. Being ever present and active in the environment, diffusion and mass transport are introduced first to induce the reader toward considering the next level of environmental chemodynamic realism. In a style that emphasizes real-world environmental applications rather than the traditional abstract deductive approach, the inductive approach arrives at the same place: a toolkit containing concepts, equations and a methodology for a qualitative understanding and quantitative description.

Figure 3.0-1 illustrates the three dominant environmental interfaces and the interface exchanges sites that are of concern. Beginning with basic ideas in molecular diffusion, topics of mass transfer and turbulence in the environment are developed here and related to the natural physical forces that are the prime movers behind the transport of chemicals across the interfaces of the geospheres. Each interface has at least one fluid phase associated with it. The specific type of movement of the fluid near the interface regulates to some degree the rate of interphase transfer through the particular boundary. For example, the fluid movements near the air–water interface of the oceans are only remotely related to the fluid movements near the same interface of a fast-moving landlocked stream. The physical flow characteristics on the bottom of an estuary are unlike those at the bottom of a stream, and airflow above the ocean does not compare directly to that above earthen solids and vegetation. As to the extent to which information is available on the particular

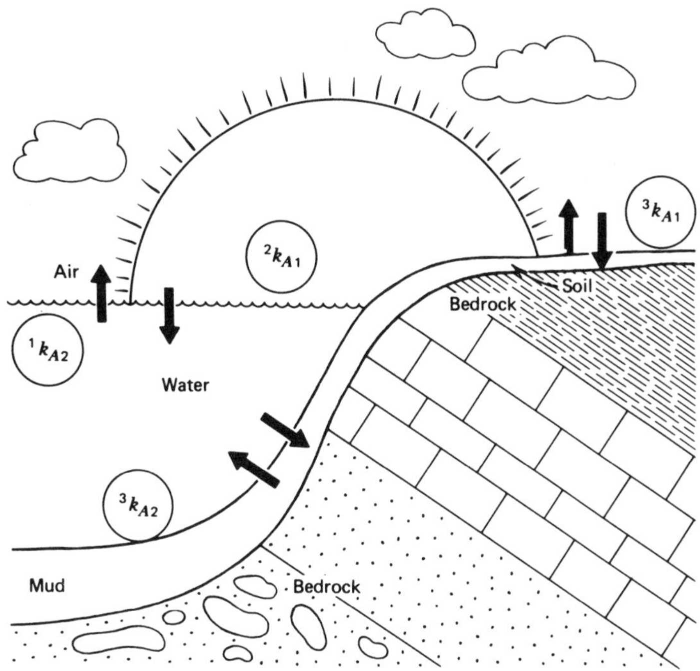

Figure 3.0-1. Interfaces in the geospheres.

aspects of the interphase transport and fluid movement, that body of knowledge, with its special adaptations, are presented for the most part in Chapter 4 and following chapters. In this chapter we attempt to unify the transport concepts into a somewhat general framework so that each example application presented will not be interpreted as an isolated case.

From the undergraduate point of view of practical application, this chapter is theoretical. Those students who wish to get on to the meat of the subject of chemodynamics may skip this chapter now and return to it at various times when insight into particular mechanisms of transport is needed.

3.1. DIFFUSION AND MASS TRANSFER

Advective movement of a chemical, either within the atmosphere or within the hydrosphere, is a process of transport affected solely by the mass motion of the atmosphere or the hydrosphere. Nonequilibrium conditions of a chemical species, between phases or within a single phase, must exist for diffusion movement of the chemical to occur. Nonequilibrium conditions between phases has been presented in Chapter 2. Within a single phase, nonequilibrium conditions are usually characterized by a concentration gradient (i.e., $\partial c_A/\partial_x$).

An all-encompassing theory capable of explaining and quantifying diffusion movement is incomplete. The incompleteness is due in large part to the complexity of the fluid turbulence associated with the mass motion of fluids. Turbulence is almost always present in both the atmosphere and hydrosphere. A diverse body of knowledge has been developed to explain and quantify diffusion chemical movements in both the human-influenced and the natural environment. The following is a review of fundamental diffusion processes with restrictions to dilute solutions and environmental temperatures and pressures.

General Description of Molecular Diffusion

At any temperature above absolute zero, the individual molecules of a gas move incessantly and at random, apparently independent of each other. Frequent collisions occur between particles, so a single particle follows a zigzag path. However, an aggregation of diffusing particles has an observable drift, from places of higher to places of lower concentration. For this reason diffusion is known as a transport phenomenon.

Molecular diffusion in liquids can be observed by placing a layer of iodine solution, avoiding any convection or stirring, underneath pure water. At first only the lower part, containing the iodine, is colored, but then the color is observed to spread slowly upward. Eventually, the upper part becomes colored, too, the intensity of the color decreasing from bottom to top. After a long time the entire mass is uniformly colored. Diffusion in gases can be observed similarly. In gases the process is much more rapid than in liquids. Diffusion in solids is generally the slowest type.

The average path traveled by a molecule in the interval between collisions is known as the *mean free path.* It decreases with increasing concentration. Another quantity characterizing the diffusing substance is the displacement. By displacement is meant the distance between the original position of a particle and its position after a certain period of time t. The mean displacement is zero, since in the absence of a difference in concentration, positive and negative displacement are equally probable. For this reason the mean-square displacement $\bar{x}^2$ is introduced.

Consider a hypothetical horizontal plane passing through the iodine–water system described earlier. Iodine molecules move in both directions through the plane. If one considers all the particles in a layer of a certain thickness on either side of a plane, it will be observed that there are more iodine molecules in the same thickness below than above the plane. Since the same percentage of molecules crosses the hypothetical boundary per second, this results in more molecules moving upward than downward, until there is no longer a concentration difference across the plane. There results an overall flow from positions of higher to positions of lower concentrations. In this way the concentration is equalized. This rush to equalization, which is observable macroscopically is called *diffusion.* It is obvious that the mean-square displacement and the diffusion time are a measure for the rate of diffusion. The ratio of the mean-square displacement and the diffusing time is the diffusion coefficient as follows:

$$\mathscr{D} = \frac{\bar{x}^2}{2t} \tag{3.1-1}$$

The mathematical description of the process of diffusion is accomplished with this quantity $\mathscr{D}$, as shown below. On a larger scale, the process of an oil slick spreading on water can be quantified by dispersion coefficients using Eq. 3.1-1 (see Problem 3.1F).

Fick's First Law. Qualitative observations of diffusion preceded quantitative descriptions. Robert Brown provided in 1827 the closest thing to direct, visual evidence for the motion of molecules. He observed that very minute particles suspended in a gas or liquid and viewed under a microscope were seen to be in a state of continual random motion. Random molecular motion is sufficient to bring about diffusion, as noted earlier. Thomas Graham (1805–1869) quantified that the relative rates at which gases diffuse is inversely proportional to the square roots of their respective densities or molecular weights.

In a paper in 1855, Fick finally put Graham's experiments on a qualitative basis. Fick's introduction of his basic idea is that "the diffusion of the dissolved material ... is left completely to the influence of the molecular forces basic to the same law ... for the spreading of warmth in a conductor and which already have already been applied with great success to the spreading of electricity." In other words, diffusion can be described on the same mathematical basis as Fourier's law of heat conductance or Ohm's law for electrical conduction. He defined a one-dimensional flux

$$J_{Az} = -c_2 \mathscr{D}_{A2} \frac{dx_A}{dz} \tag{3.1-2}$$

of component A through an area A across which diffusion is occurring, with J_{Az} the molar flux per area, c_2 the molar density of the fluid, x_A the mole fraction of A in the fluid, and z the distance. The quantity $\mathscr{D}_{A2}$, which Fick called the "constant depending on the nature of the substances," is the diffusion coefficient or binary diffusivity of A in 2. The units of the mass diffusivity $\mathscr{D}_{Aj}$, $j = 1$, 2, and 3 are cm^2/s.

Table 3.1-1 contains typical diffusivities. A more complete table of diffusivities appears in Appendix C. Relationships, based in part on theory and in part on experimental observations, are available for estimating diffusivities.[1] Gas diffusivities are dependent on the chemical species diffusing, temperature, and pressure. Liquid diffusivities are dependent on the species, temperature, and viscosity of the mixture. More on this later.

Table 3.1-1. Experimental Diffusivities

System	Temperature (K)	$\mathscr{D}_{ij}$ (cm^2/s)
Gas pair, CO_2–air	293	0.153
Liquid O_2–H_2O	293	1.8E-5
Solid state, H_2 in SiO_2	773	0.6–2.1E-8

Diffusion of Trace Chemicals in Stagnant Media. Molecular diffusion is a slow mechanism for the movement of chemicals through the atmosphere and the hydrosphere. Consider the simplistic problem of the diffusion of carbon dioxide into a stagnant layer of air lying above a water body containing a sufficient quantity of the dissolved gas so that the air interface is always saturated at 20°C.

This particular diffusion problem requires Fick's second law for adequate description:

$$\frac{\partial c_{A1}}{\partial t} = \mathscr{D}_{A1} \frac{\partial^2 c_{A1}}{\partial y^2} \tag{3.1-3}$$

and the well-known semi-infinite solid solution provides a realistic model of the process. The region of air enriched with CO_2 grows upward from the interface as the air mass remains in contact with the water body. The solution of Eq. 3.1-3 yields an equation for the CO_2 distribution in the stagnant air. The term $y_A = c_{A1}/c_1$ is the mole fraction concentration of CO_2 ($\equiv A$) for a given distance of penetration y and time t:

$$\frac{y_A - y_A^*}{y_A^0 - y_A^*} = \frac{2}{\sqrt{\pi}} \int_0^{y/\sqrt{4\mathscr{D}_{A1}t}} e^{-n^2}\, dn \tag{3.1-4}$$

and y_A^* and y_A^0 are the saturation mole fraction at the air–water interface and the initial mole fraction, respectively. $\mathscr{D}_{A1}$ is the diffusivity of CO_2 in air at 20°C, 1 atm. The right-hand side is commonly known as the Gauss error integral or probability function. Tabularized values of this function appear in Appendix B. Equation 3.1-4 is often written as

$$\frac{y_A - y_A^*}{y_A^0 - y_A^*} = \operatorname{erf}\left(\frac{y}{\sqrt{4\mathscr{D}_{A1}t}}\right) \tag{3.1-5}$$

Table 3.1-2 contains calculated CO_2 concentrations within the stagnant air mass. The table entries indicate the slowness of the molecular diffusion process in air.

Table 3.1-2. Mole Fraction CO_2 (y_A) in the Stagnant Air mass[a]

Time, t (s)	Penetration Distance, y (cm) 0.001	0.01	0.10	1.00	10.0
1	0.0657	0.0654	0.0606	0.0326	0.0300
60 (1 min)	0.0658	0.0657	0.0651	0.0592	0.0307
3600 (1 h)	0.0658	0.0658	0.0657	0.0649	0.0574

[a] $y_A^* = 0.0658$, $y_A^0 = 0.030$, $\mathscr{D}_{A1} = 0.153$ cm^2/s at 20°C, 1 atm.

Table 3.1-3. O_2 Concentration (mg/L) in the Stagnant Water[a]

Time, t (s)	Penetration Distance, y (cm) 0.001	0.01	0.1	1.0	10.0
300 (5 min)	0.069	0.70	6.1	9.17	9.17
36,000 (10 h)	<0.001	0.064	0.64	5.69	9.17
172,800 (2 days)	<0.001	0.028	0.29	2.87	9.17

[a] $\mathscr{D}_{O_2,H_2O} = 1.80E\text{-}5$ cm^2/s at 20°C.

Now consider a stagnant layer of oxygen-rich water, concentration 9.17 mg/L, suddenly placed above an anaerobic mud layer capable of consuming and depleting the water of its oxygen content. Due to molecular diffusion, the water layers adjacent to the mud become void of oxygen and this voided region gradually increases in thickness with time. Table 3.1-3 contains oxygen concentrations within the overlaying water. The entries indicate the extreme slowness of the molecular diffusion process in water.

The O_2 and CO_2 transfer processes in these two natural locales are in general much faster than is reflected in Tables 3.1-2 and 3.1-3. Although present, molecular diffusion is dominated by a more rapid transport mechanism. Fluid mixing is always occurring in nature to some degree. Mixing near environmental interfaces speeds up these interphase mass transfer processes. It is due in large part to this flow-induced mixing or turbulence that regions of the environment, such as rivers, can assimilate large quantities of some waste organic chemicals and retain some degree of diversity of biological life (see Section 1.2).

Eddy Diffusion

As a fluid moves parallel to a fixed surface, as, for example, air across the surface of the soil, it can be in laminar flow or turbulent flow. Only very slow-moving air is laminar, the connotation being that adjacent layers of fluid remain distinct and identifiable and do not intermix. As the wind speed increases, fluid layers become irregular and eddies develop. An eddy is thought of as an irregular but somehow identifiable material and wind structure, perhaps similar to a "puff of wind" or to a "cat's paw" over open water, having the ability to transfer air properties across the flow in a way that can conveniently be thought of as analogous to transfer by the air molecules on a much smaller scale.

Eddy Diffusion Coefficient. Mass transfer in a turbulent stream is essentially a mixing process, whereby mass is transported by the mixing and blending of the eddies. For brevity it is referred to as *eddy diffusion*, although its similarity to molecular diffusion lies only in the form of expressing the flux proportional to the concentration gradient. It is usually very rapid, although near a phase boundary where the eddy motion is damped it may be unimportant compared with the parallel process of diffusion by molecular motion.

The eddy diffusion coefficient $\mathscr{D}_{A2}^{(t)}$ is defined to follow the pattern of molecular diffusion as relating the molar flux of a species to the concentration gradient of the same species. Including the contributions of molecular diffusion, the total flux relationship is

$$J_{Ay} = -c_2(\mathscr{D}_{A2}^{(l)} + \mathscr{D}_{A2}^{(t)}) \frac{dx_A}{dy} \tag{3.1-6}$$

The difficulty in this approach to practical environmental problems lies not only in the complex dependence of $\mathscr{D}_{A2}^{(t)}$ on the properties of the turbulent flow field but also in the fact that the flux is not always proportional to the concentration gradient.

The magnitude of the eddy diffusion coefficient in the natural environment is usually many times larger than the molecular diffusivity. Table 3.1-4 contains typical ranges for eddy diffusivities observed in nature. *A* similarly defined eddy diffusion coefficient exists in the air (i.e., $\mathscr{D}_{A1}^{(t)}$).

Table 3.1-4. Diffusion Coefficients in Various Aqueous Environments

Environment Region	$\mathscr{D}_{A2}^{(l)}$ or $\mathscr{D}_{A2}^{(t)}$ (cm^2/s)
Dispersion: horizontal surface waters	100–1,000,000
Eddy diffusion: in pipes and flat ducts, normal to flow	0.1–100
Turbulent diffusion: vertical thermocline, and deeper regions in lakes and oceans	0.01–1.0
Molecular diffusion: salts and gases in water	10^{-4}–10^{-5}
Molecular diffusion: solutes in sediments and soils	10^{-6}–10^{-8}

Source: Ref. 2.

Eddy diffusion is initiated and propagated by fluid turbulence. It is possible to define an eddy viscosity in a fashion similar to Eq. 3.1-6. In the interest of understanding turbulent shear in flow near fixed boundaries, Prandtl developed eddy movements into a quantitative model by employing *mixing lengths.* A mixing length is roughly related to eddy diameters. One outcome of the Prandtl mixing-length model is that eddy viscosity increases linearly with distance from the fixed boundary. Details of the Prandtl mixing-length model are presented later in this chapter.

Eddy diffusion coefficients are impossible to use in many cases where diffusion chemical movements are of interest. This is particularly true near phase boundaries. Here $\mathscr{D}_{Aj}^{(t)}$, $j = 1, 2$, values are generally not estimable and concentration gradients are likewise difficult to obtain or estimate. This approach is abandoned for many complex situations near environmental interfaces. Faced with these difficulties a simpler flux equation has been created similar to Newton's law of heat conduction.

Mass Transfer Coefficient. Based on a previous argument concerning the difficulty of obtaining the relative contributions of $\mathscr{D}_{A2}^{(l)}$ and $\mathscr{D}_{A2}^{(t)}$ near the interface, it has proved to be convenient to replace the diffusion term with the product of a mass transfer coefficient k'_{A2} and some characteristic composition difference Δx_A between the fluid at the interface and the fluid far removed from the interface. For dilute aqueous solutions near environmental interfaces, the rate expression equivalent to Eq. 3.1-6 is

$$J_{Az} = c_2 k'_{A2} \Delta x_A \tag{3.1-7}$$

Here the flux rate, J_{Az}, is in mol $A/\mathrm{L}^2 \cdot \mathrm{t}$, the coefficient in L/t, and the concentration difference in mole fraction A in water. The mass flux rate expression employs the same mass transfer coefficient, except that the rate j_{Az} is in $\mathrm{M/L^2 \cdot t}$, the concentration difference in the mass fraction A in water, $\Delta\phi_A$. Another useful form of the mass flux rate is

$$j_{Az} = k'_{A2} \Delta\rho_{A2} \tag{3.1-8}$$

where the concentration difference (in $\mathrm{M/L^3}$) is a typical measurement used in a water chemistry analysis.

The coefficients defined above are useful for most environmental chemodynamic applications. These are typical for low mass transfer rates and dilute solutions. In Section 1.1 a dilute solution was defined as one containing 5% or less of the constituent of interest. For high concentrations and high mass transfer rates, Eqs. 3.1-7 and 3.1-8 are not applicable. Significant errors in the flux rate can result and Problem 3.1H illustrates this for the effect of the water vapor flux on chemical A vaporizing at the air–soil interface. In the remainder of this section, more general mass transfer coefficient rate expressions are developed for binary mixtures.

It is instructive and convenient to rewrite Eq. 3.0-3 as

$$\frac{\partial \rho_{Aj}}{\partial t} + \nabla \cdot \mathbf{n}_{Aj} = r_{Aj} \tag{3.1-9}$$

where $\mathbf{n}_{Aj}$ combines the advective and other flux contributions for species A. These other contributions include the diffusive flux under consideration here. Deleting the subscript that denotes the phase, n_{Ax} is the total flux of A in the x direction with respect to a vertical y–z plane fixed to the surface of the Earth. Similar components of the flux exist in the y and z directions, and this general definition has some inherent advantages in environmental chemodynamic applications involving the multicomponent fluids air and water.

Diffusion causes advection. The two processes always occur together: one cannot occur without the other. For example, the evaporation of water from the sea surface and subsequent steady-state transport through the adjacent boundary air sublayer creates an effective upward advective velocity, v_y. This upward velocity hinders the deposition of both gaseous- and particle-bound chemical species A to the sea surface. Equation 3.0-3 can be rewritten in molar concentration form, c_{Aj}. The material immediately following continues the discussion of the total flux expression in molar form, $\mathbf{N}_{Aj}$.

The historical form of Fick's law, Eq. 3.1-2, applies for the flux rate of A through a plane moving at v_z. The z component of the flux with respect to a coordinate system fixed to the surface of the earth is

$$N_{Az} = c_2 v_z x_A - c_2 \mathscr{D}_{A2} \frac{dx_A}{dz} \tag{3.1-10}$$

The total flux is the result of two quantities; the first term accounts for the bulk motion of the fluid and the second term results from diffusion superimposed on the bulk flow. As written, the equation is for molecular diffusion. For transport across an interface a mass transfer coefficient form of the equation is more convenient:

$$N_{Az} = c_2 v_z x_{Ai} + \dot{k}_{A2}(x_{A\infty} - x_{Ai}) \tag{3.1-11}$$

As written, the equation is for the flux from a concentration $x_{A\infty}$ to the interface with concentration x_{Ai}, the subscript i denoting the interface. The reader should realize that the existence of a concentration gradient induces a velocity in what otherwise is a motionless fluid and that v_z in Eqs. 3.1-10 and 3.1-11 accounts for this effect. In addition, the mass transfer coefficient itself is a function of the mass transfer rate and this is denoted by an overdot; see Ref. 3 for details. The coefficient has dimensions mol/L$^2\cdot$t.

The molar average velocity for a binary mixture of A and B moving in the z direction is defined

$$v_z \equiv \frac{N_{Az} + N_{Bz}}{c_2} \tag{3.1-12}$$

where $N_{Az} = c_2 v_{Az}$ and $N_{Bz} = c_2 v_{Bz}$, so that v_{Az} and v_{Bz} are the velocities of A and B in L/t. Equations 3.1-10 and 3.1-11 become

$$N_{Az} = (N_{Az} + N_{Bz})x_A - c_2 \mathscr{D}_{A2} \frac{dx_A}{dz} \tag{3.1-13}$$

and

$$N_{Az} = (N_{Az} + N_{Bz})x_{Ai} + \dot{k}_{A2}(x_{A\infty} - x_{Ai}) \tag{3.1-14}$$

This form of Fick's first law and the mass transfer coefficient equivalent are very general flux expressions of which Eqs. 3.1-2 and 3.1-8 are special cases. The latter forms are simpler and commonly used for most environmental chemodynamic applications. The corrections necessary for the coefficient at high mass transfer rates, $\dot{k}_{A2}$, are also negligible for most applications and the overdot is eliminated hereafter. Equivalent mass flux rate expressions and expressions for the gas phase or other fluid phases can be written by analogy, but this is left to the user. Note that the relationship between the primed coefficient in Eq. 3.1-8 and that without in Eq. 3.1-14 involves the phase density as a multiplier. For the molar flux the relationship is

$$k_{A2} = k'_{A2} c_2 \tag{3.1-15}$$

where c_2 is the molar density of the aqueous phase.

With respect to the movement of chemicals across the major interfaces of the geospheres, there are four individual mass transfer coefficients. These coefficients and their approximate interface locations are shown in Fig. 3.0-1. Note that coefficients are associated only with the fluid sides. There are two gas-phase and two liquid-phase coefficients, as noted by the subscripts 1 and 2. The superscript denotes the coexisting phase of the interface. Extensions to other fluids and even fluidized solid particles are also possible.

Equations such as 3.1-14 have proved to be extremely utilitarian for computing flux rates across phase boundaries. Conventional techniques of qualitative and quantitative chemistry along with phase equilibrium principles (Chapter 2) allow x_A's, y_A's, and hence Δx_A and Δy_A to be determined. Many experimental observations, in both the laboratory and the field, along with much theoretical work, have resulted in a host of semiempirical equations from which fairly accurate values of the individual mass transfer coefficients can be obtained. Specific correlations and applications are presented in later chapters.

No notational distinction is being made between point values of mass transfer coefficients and average values, but variations can occur from point to point on an interface. For example, the water-side coefficient for the wavy surface of a sandy streambed varies from crest to trough. Only average coefficients are usually available, and point values should be area averaged. In either case the common form for the chemical rate is

$$w_A = k'_{A2} A \, \Delta\rho_{A2} \tag{3.1-16}$$

where w_A is in m/t and A is the plane surface area of the interface in L^2. For those cases in which the area is difficult to define, such as that in a swarm of gas bubbles or other irregular geometric forms occupying a volume V, the modification of Eq. 3.1-16 is

$$w_A = k'_{A2} a_v V \, \Delta\rho_{A2} \tag{3.1-17}$$

Here the coefficient and the area per unit volume product, $k'_{A2} a_v$, is retained as a single term. The term a_v, the interfacial area per unit volume, has dimensions of (L^2/L^3) or L^{-1}.

Equations 3.1-11, 14, 16 and 3.1-17 do not contain a concentration gradient, so a mathematical analysis cannot yield spatial concentrations of chemicals in the various fluid phases. The mass transfer coefficients are usually associated with the region of the phase near the interface where concentration gradients are presumed to be greatest.

As noted earlier, the coefficients k'_{A1} and k'_{A2} have dimensions of velocity. This has led to these coefficients being called *piston velocities*. Apparently, the interpretation is that a piston of this velocity moves through the fluid and sweeps all the species A molecules ahead of it.

Mass Transfer Theories

The mass transfer coefficients introduced in Eqs. 3.1-11 to 3.1-17 are nothing more than a proportionality constant at this point. The coefficients are termed conductances and the reciprocals, $1/k'_{A2}$, are conventionally termed *resistances*. These coefficients are important quantities for calculating the rate of chemical transfers at environmental interfaces and require a degree of theoretical development for identifying measurable parameters that realistically reflect environmental processes.

It is important to understand what natural forces and occurrences within the environment influence the magnitude of these coefficients to the same degree that chemical potential and the displacement from equilibrium aided understanding of the driving force Δx_A.

Flow happenings near environmental interfaces are of paramount importance, but they are difficult to describe mathematically because of the complexities of turbulent flow; however, several conceptual models have been developed to aid our understanding. Although simple in concept, these models have been highly successful in helping us gain insight into natural mechanisms. They are all speculations and are continually being revised as more observational data become available. The most relevant facts pertaining to mass transfer into or away from a turbulent stream are that the resistance is confined to a region adjacent to the interface where the concentration gradients are steep.

Stagnant-Film Model. The film model is over 70 years old, having been proposed by Nernst[4] in 1904 and applied to gas absorption by Whitman[5] in 1923. When a fluid flow is turbulent past a solid surface, the velocity is nevertheless zero at the surface; there must be a viscous layer or film in the fluid very near the surface. Beyond this film fluid mixing is very rapid so that the rate of mass transfer is limited by the rate of molecular diffusion through the viscous layer.

The film idea can also be applied to both fluid surfaces near a gas–liquid or liquid–liquid interface. There is assumed to be a stagnant film in each phase. The rate equation for the film model at low mass transfer rates across a liquid film is

$$J_{Az} = \frac{\mathscr{D}^{(l)}_{A2} c_2}{\delta_{A2}} (x_{A\infty} - x_{Ai}) \tag{3.1-18}$$

where c_2 is the average molar density of the liquid mixture (mol/cm^3) and δ_{A2} is the effective film thickness (cm). Therefore in theory the mass transfer coefficient k_{A2} is equal to the group $\mathscr{D}^{(l)}_{A2} c_2/\delta_{A2}$ (mol/s·cm^2). In some special cases where δ_{A2} is known, this group can be used to estimate k_{A2}, but this is usually an exception. Whereas the stagnant-film model yields an excellent conceptual picture for mass transfer near interfaces, it has little practical value for obtaining numerical results.

A prediction of this theory is that mass transfer coefficients for different solutes being transferred under the same fluid flow conditions are directly proportional to the molecular diffusivities $\mathscr{D}^{(1)}_{A2}$ of the solutes. Various experimental observations have shown that the mass transfer rate is proportional to $\mathscr{D}^{n}_{A2}$, $0.1 \lesssim n \lesssim 0.9$. It was recognized very early that the film concept was a gross oversimplification of the actual conditions near a phase boundary. Although still useful in teaching elementary mass transfer concepts, the film theory has been largely discredited, except as a limiting case.

One useful result of the film model has been the development of the Sherwood number, Sh. A dimensionless ratio can be created by forming a quotient of the Eqs. 3.1-7 and 3.1-18 to yield the defined Sherwood number,

$$\mathrm{Sh}_{A2} \equiv \frac{k'_{A2} L}{\mathscr{D}^{(l)}_{A2}} \tag{3.1-19}$$

where L, which replaces the fictitious film thickness δ_{A2}, is an appropriate length that reflects the geometry of the system. The Sherwood number has been found to be a convenient dimensionless group useful in correlating experimental data. Based on an analysis of the dimensions of the variables that influence mass transfer, the Sherwood number may be expected to depend on the Reynolds number $\mathrm{Re} \equiv Lv/\nu_2$ and the Schmidt number $\mathrm{Sc}_{A2} \equiv \dfrac{\nu_2}{\mathscr{D}_{A2}^{(l)}}$ as

$$\mathrm{Sh} = f(\mathrm{Re}, \mathrm{Sc}, \text{and geometry}) \tag{3.1-20}$$

v is average fluid velocity (cm/s) and ν_2 is the kinematic viscosity of the fluid (cm^2/s). Numerous correlations for different geometric shapes and flow conditions take this form. The Sherwood number, Sh_{A2}, and the Nusselt number for mass transfer, Nu_{A2}, are different names for the same dimensionless group (see Table 3.3-2).

Penetration Theory. In 1935, Higbie[6] developed a basic idea about the mechanism of mass transfer near fluid–fluid interfaces that established what is known as *penetration theory*. Turbulent flow produces eddies that may be visualized as parcels of fluid consisting of an enormous number of molecules that exist together for a while as a well-defined entity and move about continuously, then eventually "dissolve" into the surrounding fluid. The parcels are continuously coming into being and dissolving throughout the entire fluid body. Only molecular movement occurs within the parcels. As a parcel of uniform concentration $c_{A\infty}$ contacts an interface, one side is in direct contact with the other phase for a short period of time (i.e., the average exposure time). The equilibrium concentration at the interface is c_{Ai}. During this exposure time molecular diffusion results in the penetration of chemical A into (or out of) the parcel. The depth of penetration is small compared to the dimensions of the parcel.

Fick's second law (i.e., Eq. 3.1-3) describes the mathematics of the process. With the semi-infinite solid boundary conditions, Eq. 3.1-4 is the solution that gives the concentration of A as a function of penetration depth y and exposure time $\bar{t}$. The average flux rate over the exposure time is

$$J_{Ay} = 2\sqrt{\frac{\mathscr{D}_{A2}^{(1)}}{\pi\bar{t}}}\,(c_{A2\infty} - c_{A2i}) \tag{3.1-21}$$

This theory predicts that the mass transfer coefficient k'_{A2} is equal to $2\sqrt{\mathscr{D}_{A2}^{(l)}/\pi\bar{t}}$. The k'_{A2} should vary as the square root of the molecular diffusivity, whereas the film model indicates the first power. The square root is nearer the truth in many instances. In practice the average exposure time $\bar{t}$ is seldom known, so that the model cannot be used to predict mass transfer rates except in special cases; however, its qualitative aspects explain how mass transfer can occur to and from an interface without the need of a stagnant film.

Surface-Renewal Theory. The assumption that all parcels had the same exposure time was revised by Dankwerts[7] in 1951. He assumed that parcels can remain in contact with the surface for variable times that may be any value from zero to infinity. He created the fractional renewal rate, s (s^{-1}), and used it in the surface-age distribution function $\phi = s\exp(-st)$. Here ϕ represents the probability that a parcel will be exposed for time t before being replaced by fresh mixed fluid from the bulk, a useful concept in stream oxygenation (see Eq. 1.2-15). The mean steady-state flux normal to the phase boundary was shown to be

$$J_{Az} = \sqrt{\mathscr{D}^{(l)}_{A2}\, s}(c_{A\infty} - c_{Ai}) \tag{3.1-22}$$

Since values of s are not generally available, its appearance in the model presents the same problems as do δ_{A2} and $\bar{t}$ of the film and penetration models, respectively.

Fluid turbulence enters each model in an ad hoc fashion. It is assumed to exist in the bulk fluid region beyond the film and exerts negligible resistance to mass transfer. Turbulent eddies from the bulk fluid region place fresh parcels of fluid adjacent to the interface for finite, uniformly short exposure time periods in the penetration theory. In surface-renewal theory, the same eddies create an interface plane with a known surface-age distribution function and surface renewal rate constant. So from a process perspective, environmental factors that enhance fluid turbulence decrease δ_{Az}, $\bar{t}$ and increase s. According to Equations 3.1-18, 3.1-21, and 3.1-22 this will increase k'_{Az} and hence the rates of mass transfer. A less contrived structure of turbulence is presented in Section 3.2.

Boundary Layer Theory. In fluid flow parallel to a smooth, flat solid object, boundary layers are formed adjacent to the surface. For an isothermal fluid with mass transfer to or from the wall, momentum and concentration gradients form in thin layers next to the solid (i.e., boundary layers) and provide the site of resistances to chemical transport. Theoretical and experimental studies with this simple but realistic model system can be utilized effectively as a first approximation at most solid–fluid interfaces. This flow and geometric configuration, shown in Fig. 3.1-1, is an elementary version used in the science of boundary layer theory.[27]

Useful results include expressions for the transport coefficients between the bulk flow region and the solid wall. For laminar flow within the boundary layer, $\mathrm{Re}_L \equiv v_\infty L/\nu_2 \leqslant 1\mathrm{E}5$ and

$$\mathrm{Sh}_{A2} = 0.664\mathrm{Re}_L^{1/2} \cdot \mathrm{Sc}_{A2}^{1/3} \tag{3.1-23}$$

where Re_L is the Reynold's number, v_∞ the approach velocity, and L the surface length (see Fig. 3.1-1). For turbulent flow within the boundary layer,

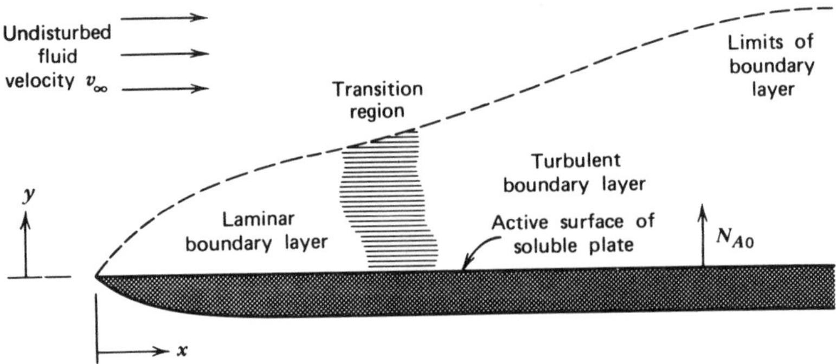

Figure 3.1-1. Boundary layer on a flat plate.

$Re_L > 1E5$ and

$$Sh_{A2} = 0.036 Re_L^{0.8} \cdot Sc_{A2}^{1/3} \tag{3.1-24}$$

The Sherwood number defined in Eq. 3.1-19 is used to express the mass transfer coefficient in dimensionless form. Both of the relations above result from analytical solutions of the equations of change. The laminar equation is from an exact solution of the momentum and mass equations while approximate velocity and concentration profiles are used for the turbulent flow case.[3] Both contain the Schmidt number modification which in effect adjusts for the relative thickness of the momentum and mass transfer boundary layers.[28]

Extensions and unification of the preceding four models have resulted in qualitative understanding and quantitative expressions capable of explaining many experimental observations. Reviews of these further developments can be found in textbooks by Treybal[8] and Sherwood et al.[9]

The preceding models help explain some of the occurrences at natural fluid–fluid and fluid–solid interfaces that affect the magnitude of the individual phase coefficients. Although the expressions are given for the liquid side of the interface, similar expressions apply for cases in which the fluid is a gas.

Generality of Mass Transfer Coefficients

Once an experimental value of a mass transfer coefficient has been obtained for a certain chemical in a particular phase and location, it may be used for other chemical species in the same phase and location by employing the proper transformation techniques based on the best available mass transfer theories. Individual phase mass transfer coefficients are scalars and should be therefore independent of the direction of the concentration driving force. This means that a coefficient measured under absorption conditions can be transformed and used in a desorption application (see, e.g., Problem 4.1C). The equation most commonly used to transform a coefficient for species A to species B is

$$k_{B2} = k_{A2}\left(\frac{\mathscr{D}_{B2}^{(l)}}{\mathscr{D}_{A2}^{(l)}}\right)^n \tag{3.1-25}$$

All other factors included in k_{A2} for A, such as fluid type, temperature, flow rate, location, depth, and so on, must remain constant for B. Here $n = 1$ for film theory, $n = \frac{1}{2}$ for the penetration and surface-renewal theories, and boundary layer theory suggests that $n = \frac{2}{3}$. Personally, the penetration and surface-renewal theories seem more appropriate for fluid–fluid interfaces, and the boundary layer theory is more appropriate for solid–fluid interfaces; however, the variations in concentration differences and gradients of the chemical between nearby locations along environmental interfaces likely imparts much greater uncertainties in the flux rate estimate than does the value of n chosen to adjust the mass transfer coefficient.

When using published coefficients and correlations, care must be taken that the geometry and boundary conditions are identical. In some cases the mass transfer coefficient is reported in the form of dimensionless number correlations of the form of Eqs. 3.1-23 and 3.1-24. In this case the diffusivity transformation is inherent in the correlating equation and Eq. 3.1-25 is not needed. These type correlations should also be phase independent. For example, if experimental measurements are made on a liquid-phase coefficient above a flat plate and correlated correctly with appropriate dimensionless groups, it is logical to assume that one can obtain reasonable estimates of gas-phase coefficients from this same correlation. Using the appropriate velocity, molar density, and other fluid properties, as specified for the phase, we obtain the appropriate individual phase coefficient.

Estimating Molecular Diffusivities. Molecular diffusivity emerges as an important variable for use in the constitutive equations. This is evident from the preceding section. Good experimental data and estimation techniques are therefore needed. This section provides a means to both. Consult Appendix C for data on molecular diffusivities in air and water.

Marrero and Mason[29] show that the empirical correlation suggested by Fuller, Schettler, and Giddings is reliable for estimating molecular diffusion coefficients for binary gas systems from moderate to high temperatures at low pressure. The correlation is based on special diffusion volumes and is based on an extensive amount of data.

$$\mathscr{D}_{A1} = \frac{0.001\,T_1^{7/4}(1/M_A + 1/M_1)^{1/2}}{P[(\Sigma_A V_i)^{1/3} + 2.72]^2} \tag{3.1-26}$$

where T_1 is in K, P in atm, and $\mathscr{D}_{A1}$ in cm^2/s. The volumes are listed in Table 3.1-5. The upper part of the table contains increments to be summed when

Table 3.1-5. Special Atomic Diffusion Volumes

Atomic and Structural Diffusion Volume Increments				
C = 16.5	O = 5.48	Cl = 19.5	Aromatic or heterocyclic rings = −20.2	
H = 1.98	N = 5.69	S = 17.0		
Diffusion Volumes of Simple Molecules				
H_2 = 7.07	O_2 = 16.6	Kr = 22.8	N_2O = 35.9	Cl_2 = 37.7
D_2 = 6.70	(Air = 20.1)	Xe = 37.9	NH_3 = 14.9	Br_2 = 67.2
He = 2.88	Ne = 5.59	CO = 18.9	H_2O = 12.7	SO_2 = 41.1
N_2 = 17.9	Ar = 16.1	CO_2 = 26.9	CCl_2F_2 = 114.8	SF_6 = 69.7

molecules are other than simple gases. The values for the simple gases are listed in the lower part. This correlation cannot distinguish between isomers. Thus far, special volumes have not been determined for F, Br, and I. The aromatic ring volume is subtracted to account for bound shortening observed for this type of molecular structure. The reliability of Eq. 3.1-26 for compounds of complex structures such as heptachlor, chlordane, and dieldrin has not been determined.

In 1827, Robert Brown observed that very minute particles suspended in a gas or liquid viewed under a microscope were seen to be in a state of continual random motion. The particles did not show any indication of settling to the bottom, nor did they show signs of slowing down their motion. The smaller the particle size and the higher the temperature of the medium, the more vigorous was the movement of the particles. This phenomenon is known as *Brownian motion* and the movement of the particles is termed *Brownian diffusion.*

The Stokes–Einstein equation placed the observed Brownian diffusion on a quantitative basis. Here a hydrodynamic theory was used to relate the particles or solute molecules mobility under the influence of a chemical potential gradient. In terms of a concentration gradient the flux is

$$\mathbf{J}_A = -\frac{k_B T}{6\pi\mu_j r}\nabla c_{Aj} \tag{3.1-27}$$

where $\mathscr{D}_{Aj} \equiv k_B T/6\pi\mu_j r$ is the Stokes–Einstein equation. In this equation k_B is Boltzmann's constant 1.3805E-16 $\text{g}\cdot\text{cm}^2/\text{s}^2\cdot\text{K}$, T is in K, μ_j is the solution viscosity in g/cm·s, and r is the particle radius (cm). With these units the diffusion coefficient is in cm^2/s. This equation has been shown to be fairly good for describing the diffusion of large spherical particles, colloidal particles or large spherical molecules, under which conditions the solvent appears to the diffusing species as a continuum. Cussler[31] notes that when the particle size is less than five times that of the solvent, the Stokes–Einstein equation breaks down.

For liquids the Stokes–Einstein equation has also been shown to be fairly good for describing the diffusion. The simple hydrodynamic approach gives expressions for the diffusion coefficient for spherical molecules in dilute solutions and also for the coefficient of self-diffusion. Wilke and Chang[12] have developed an approximate analytical relation based on the Stokes–Einstein equation that gives the diffusion coefficient in cm^2/s for small concentrations of A in water (H_2O liquid denoted by subscript 2):

$$\mathscr{D}_{A2} = 7.4\text{E-}8 \frac{(\psi M_2)^{1/2} T_2}{\mu_2 \tilde{V}_A^{0.6}}, \tag{3.1-28}$$

Here $\tilde{V}_A$ is the molar volume of the solute A in cm^3/mol as liquid at its normal boiling point, μ_2 the viscosity of the solution in centipoises, ψ the *association parameter* for the solvent, and T_2 the absolute temperature in K. Recommended values of ψ are 2.26 for water and 1.0 for heptane and hexane. This equation is for dilute solutions of nondissociating solutes and is usually good within $\pm 10\%$.

Temperature corrections of diffusivities for gases and liquids can be inferred from Eqs. 3.1-26 and 3.1-28, respectively. The $\frac{7}{4}$ power of the absolute temperature is used for gases. For liquids, since viscosity is strongly temperature dependent, the following proportionality should be used:

$$\mathscr{D}_{A2} \text{ at } T_2 = \mathscr{D}_{A2} \text{ at } T_1 \frac{T_2 \mu_1}{T_1 \mu_2} \tag{3.1-29}$$

where the subscripts 1 and 2 on the viscosity refer to temperatures T_1 and T_2.

According to Thomas Graham, at constant temperature the relative rates at which gases diffuse is inversely proportional to the square roots of their respective densities or molecular weights. Stated mathematically, Graham's law in terms of diffusivities is

$$\frac{\mathscr{D}_{A1}}{\mathscr{D}_{B1}} = \sqrt{\frac{M_B}{M_A}} \tag{3.1-30}$$

where M_A and M_B are the molecular weights of species A and B, respectively. The molecular weight correction for liquid diffusivities can be approximated by

$$\frac{\mathscr{D}_{A2}}{\mathscr{D}_{B2}} = \left(\frac{M_B}{M_A}\right)^{6/10} \tag{3.1-31}$$

according to Eq. 3.1-28. All the preceding relations apply to molecular diffusivities only and not turbulent diffusion coefficients or individual mass transfer coefficients. In light of Eqs. 3.1-26 and 3.1-28 the two forms of Graham's law, Eqs. 3.1-30 and 3.1-31, will provide rough estimates. However, fair estimates can result if elemental makeup and molecular structure are similar.

Example 3.1-1 Gas-Phase Transport Coefficients for Naphthalene

(a) Beginning with the diffusivity of anthracene at 0°C in Appendix C, use Graham's law to estimate it for naphthalene ($\equiv A$).

(b) If the mass-transfer coefficient for carbon dioxide at 25°C is 3000 cm/h, estimate it for naphthalene using Eq. 3.1-25 with $n = 1, \frac{2}{3}, \frac{1}{2}$ in turn.

SOLUTION (a) Anthracene ($\equiv B$) $\mathscr{D}_{B1} = 0.0421$ cm^2/s and $M_B = 178.22$. $M_A = 128.16$ and $\mathscr{D}_{A1} = 0.0421(178.22/128.16)^{1/2} = 0.0496$ cm^2/s. This is 3.3% less than the tabulated value, 0.0513.

(b) From Eq. 3.1-26, $\mathscr{D}_{A1} = 0.0513(298/273)^{7/4} = 0.0598$ cm^2/s, Eq. 3.1-25 yields: ${}^2k'_{A1} = 3000(0.0598/0.164)^n = 1090$ cm/h for $n = 1$, 1530 for $n = 2/3$, and 1810 for $n = 1/2$.

Binary Mass Transfer Coefficients in Two Phases and Two-Resistance Theory of Interphase Mass Transfer

In many environmentally important mass transfer systems there is an interface with concentration gradients on both sides. Consider, as an example, the contacting of liquid water, which contains species A, with air into which A is desorbing at steady state and constant temperature. The general shapes of the concentration gradients in the two phases are sketched in Fig. 3.1-2. Concentrations appear as mole fractions and, as is typical of most environmental mass transfer applications, dilute solutions of A exist in both phases. In this development the major resistances to mass transfer reside in the gas phase and in the liquid phase. If the interface resistance is negligible and if equilibrium is assumed here, $y^*_{Ai} = f(x_{Ai})$, where the functional form of the relationship is derived solely on the thermodynamics of gas–liquid equilibriums as presented in Chapter 2.

The molar flux of species A across each phase can be expressed in terms of the individual mass transfer coefficients and the concentration difference in each phase. For slow mass transfer rates in a dilute solution, Eq. 3.1-14 applies for the liquid phase:

$$N_{Az} = {}^1k_{A2}(x_A - x_{Ai}) \tag{3.1-32}$$

and similarly for the gas phase:

$$N_{Az} = {}^2k_{Ai}(y_{Ai} - y_A) \tag{3.1-33}$$

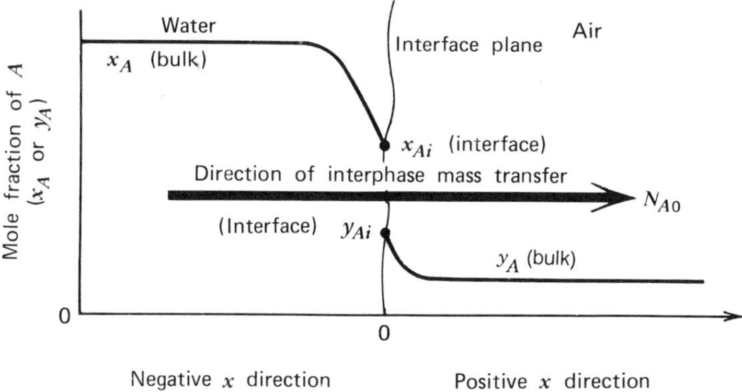

Figure 3.1-2. Interphase mass transfer.

Local average flux rates are used so that average individual coefficients may be used.

For convenience one may define overall mass transfer coefficients based on an overall concentration difference:

$$N_{Az} \equiv {}^2K_{A1}(y_A^* - y_A) \tag{3.1-34}$$

and

$$N_{Az} \equiv {}^1K_{A2}(x_A - x_A^*) \tag{3.1-35}$$

where ${}^2K_{A1}$ is the overall gas-phase mass transfer coefficient, ${}^1K_{A2}$ the overall liquid-phase mass transfer coefficient, y_A^* a nonexistent gas-phase mole fraction concentration in equilibrium with x_A, and x_A^* a nonexistent liquid-phase mole fraction concentration in equilibrium with y_A. In application either Eq. 3.1-34 or 3.1-35 may be used; the choice is left to the student.

The relationship between the overall coefficients and the individual phase coefficients are easily derived. This derivation is left as an exercise for the student (see Problem 3.1B). The relationships are

$$\frac{1}{{}^2K_{A1}} = \frac{H_{xA}}{{}^1k_{A2}} + \frac{1}{{}^2k_{A1}} \tag{3.1-36}$$

and

$$\frac{1}{{}^1K_{A2}} = \frac{1}{{}^1k_{A2}} + \frac{1}{{}^2k_{A1}H_{xA}} \tag{3.1-37}$$

where H_{xA} is the slope of the equilibrium line in the neighborhood of (x_{Ai}, y_{Ai}). For dilute solution H_{xA} is the Henry's law constant in mole fraction form (i.e.,

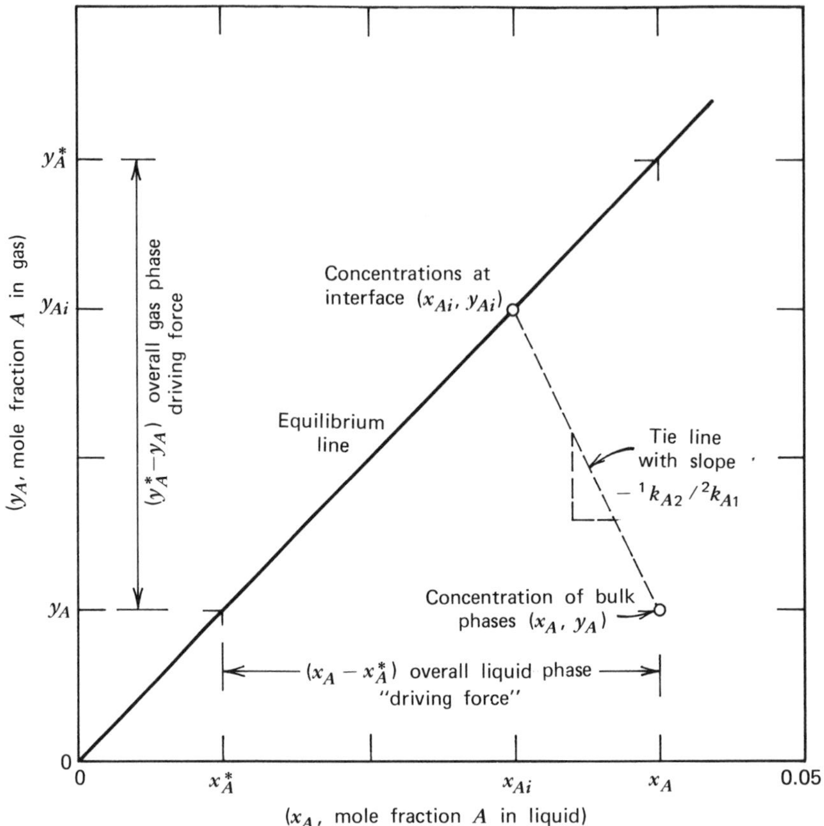

Figure 3.1-3. Graphical interpretation of the two-resistance theory.

$y_A^* = H_{xA}x_A$). If concentration driving forces such as that in Eq. 3.1-8 are used instead of mole fraction driving forces, Eqs. 3.1-36 and 3.1-37 should not be used. (See Problem 3.1I for the correct relationship in this case.)

The preceding discussion is the essence of the two-resistance theory for interphase mass transfer. The overall resistance for mass transfer across a fluid-fluid interface is the sum of the resistance in each phase as given by Eq. 3.1-36 or 3.1-37. Figure 3.1-3 contains a graphical interpretation of the theory. The equilibrium line is shown and the point set representing the bulk phase concentrations (x_A, y_A) indicates that a potential for mass transfer exists. The interface concentration point set (x_{Ai}, y_{Ai}) is obtained from a tie-line, the equation for which can be obtained from the ratio of Eqs. 3.1-32 and 3.1-33.

An analysis similar to that given above holds for other fluid–fluid interfaces through which a third constituent is moving. Examples include an organic phase in contact with air or water. If the organic phase is positioned on the air–water interface, the interface resistance created may be nonnegligible and an overall resistance containing three terms is needed. An example of this is an oil slick on the surface of water in waste impoundment or the surface of the ocean.

Controlling Resistance for Interphase Mass Transfer. The reciprocal of the mass transfer coefficient is conventionally termed a *resistance*. Equation 3.1-36 can be interpreted as the overall resistance ($1/{}^2K_{A1}$), and it is equal to the sum of the resistance in the liquid phase ($H_{xA}/{}^1k_{A2}$)and the gas phase ($1/{}^2k_{A1}$). It frequently occurs that one phase or the other dominates the overall resistance and therefore controls the process. A variation in individual coefficients ${}^2k_{Ai}$ and ${}^1k_{A2}$ can cause this; however, the numerical values of H_{xA} the Henry's law constant, cover a wide range, and it is this term that usually determines which phase resistance controls the rate of mass transfer. For example, if H_{xA} is large (i.e., $H_{xA} > 1$), the liquid-phase resistance dominates. If H_{xA} is small ($H_{xA} < 1$), the gas- phase resistance dominates. In the air–water system nearly insoluble substances such as the PCBs and oxygen are liquid-phase controlling, whereas soluble chemicals, such as phenol and propionic acid are gas-phase controlling. Some chemicals, such as methanol, encounter significant resistances in each phase. (See Tables 2.1-5 and 4.2-2 for the range in H_{xA} values.)

PROBLEMS

3.1A. Derivation of Film Theory Result

Show that Eq. 3.1-13 may be transformed to Eq. 3.1-18 for the region near an interface between phases where a stagnant film of thickness δ_{A2} is assumed to exist. (*Hint:* Assume that N_{Az} in Eq. 3.1-13 is constant through the film and see Fig. 3.1A. Note that there is a similar development for the gas film.)

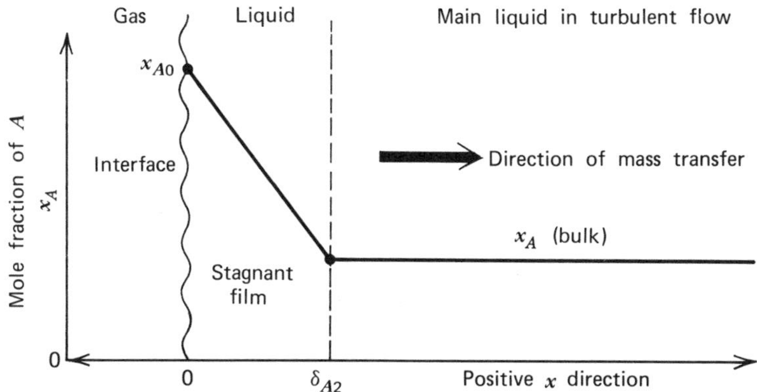

Figure 3.1A. Diffusion through a stagnant film.

3.1B. Completion of Two-Resistance Theory Derivation

Show that Eqs. 3.1-32, 3.1-33, and 3.1-35 can be combined to yield Eq. 3.1-37. [*Hint:* Start with the tautology

$$x_A - x_A^* = (x_A - x_{Ai}) + (x_{Ai} - x_A^*) \qquad (3.1B)$$

and refer to Fig. 3.1-3 as needed.]

3.1C. Effective Film Thickness, Penetration Time, and Surface-Renewal Rate

The mass transfer coefficient for the dissolution of furfural into water was found to be 0.44 lb mol/hr-ft^2. Using 1.31E-5 cm^2/s for the diffusivity of furfural in water, calculate the following:

1. The effective thickness of the stagnant film (cm).
2. The average penetration time (s).
3. The surface-renewal rate (s^{-1}).

3.1D. Estimating Liquid-Phase Diffusivity

Calculate the molecular diffusivity of the following, all at 25°C and 1 atm for dilute solutions in cm^2/s:

1. Furfural in water.
2. A 1-μm-diameter colloidal particle in water.
3. Ozone in air.

3.1E. Estimating Individual Gas-Phase Mass Transfer Coefficients

Use the boundary layer theory correlations for the flat-plate geometry to obtain the air-side coefficient within an experimental chamber that simulates chemical evaporation from soils. Use $\mathscr{D}_{Ai} = 0.035$ cm^2/s, $v_\infty = 0.22$ cm/s, $L = 30$ cm, and $T = 25°C$.

3.1F. Dispersion Coefficient for Heavy Fuel Oil on Sea Surface

Failure of a storage tank at a harborside refinery resulted in the release of 270,000 bbl of heavy fuel oil into Mizushima harbor and a substantial part of Japan's Seto Inland Sea. Figure 3.1F shows the progressive spread of the oil slick.

1. Compute the dispersion coefficient for the oil slick on horizontal surface waters in ft^2/s. Note that Eq. 3.1-1 contains $\bar{x}^2$, the mean-squared displacement; however, the data in the figure are plume size (see Eq. 7.1-67).

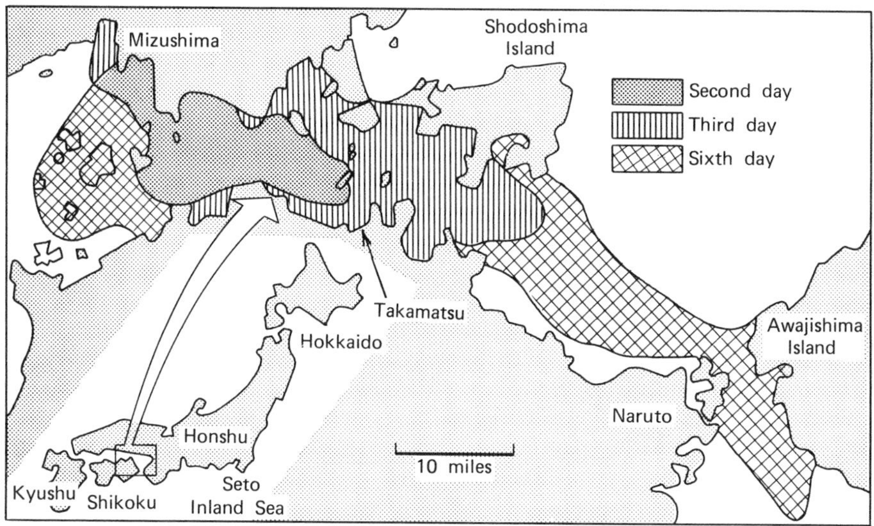

Figure 3.1F. Spread of oil slick. (Reprinted with permission of copyright owner, The American Chemical Society, from the June 2, 1975 issue of *Chemical and Engineering News*.)

2. Graph the Seto Sea slick data onto Fig. 7.1-23.
3. Determine A_L in Eq. 7.1-67 using the Seto data.
4. Is the behavior of this slick like other horizontal dispersion phenomena on waters? Why?
5. Using the $\frac{4}{3}$ power law, estimate the mean size of the slick, L (in miles), on the tenth day.

3.1G. Transient Molecular Diffusion of Benzene in Water from Sea-Surface Slick

1. Beginning with Eq. 3.0-3, show the development of Eq. 3.1-3, Fick's second law.
2. Using Eq. 3.1-5, obtain Eq. 3.1-21, the penetration theory flux expression.
3. Benzene was spilled on water to form a sea-surface slick. Assume that both the slick and the water underneath are perfectly still and that molecular diffusion is the only operative mass transfer mechanism. Calculate the benzene concentration in the water at each centimeter down to 10 cm underneath the slick after a 24-hr diffusion time. Assume that the water contained a background concentration of 10 g/m^3 benzene prior to the spill and is at 25°C.

4. How realistic is a 24-hr diffusion time for the water layers under a sea-surface slick? Discuss this in reference to sea-surface turbulence, benzene evaporation, and so on.

3.1H. Water Evaporation Effects on Volatile Chemical Transport

Water is a very volatile liquid and is usually present in air in concentrations which exceed that of many volatile contaminants. The flux of water vapor at environmental interfaces can enhance or suppress the flux of other volatile species in air.

1. Show that the simultaneous evaporation of water vapor and chemical A from a moist soil surface in contact with dry air containing no A results in the flux expression

$$N_{Az} = \left\{\frac{y_{Bi}[(M_A/M_B)^{1/2} - 1] + 1}{1 - y_{Bi}}\right\} {}^3k_{A1} y_{Ai} \tag{3.1H-1}$$

   where $A \equiv$ the volatile chemical and $B \equiv H_2O$.

2. Calculate the enhancement factor, the term in brackets, for a saturated soil at 40°C. Species A has a molecular weight of 300 g/mol.

3. Show that for condensation of water on the soil surface (i.e., dew formation) the flux expression is

$$N_{Az} = \left[1 - y_{B\infty}\left(\frac{M_A}{M_B}\right)^{1/2}\right] {}^3k_{A1} y_{Ai} \tag{3.1H-2}$$

4. Calculate the suppression factor. Assume that the air is saturated with moisture at 40°C and the soil surface is near 1°C. Use data from part 2 as needed.

3.1I. Two-Resistance Law with Concentration Driving Forces

If the phase concentrations are in mole fraction form, Eqs. 3.1-36 and 3.1-37 are valid for computing the overall mass transfer coefficient. If phase concentrations are in units of moles per volume (i.e., c_{A1} and c_{A2}), show that the correct forms of the two-resistance law are

$$\frac{1}{{}^2K'_{A1}} = \frac{1}{{}^2k'_{A1}} + \frac{H_{Ax}c_1}{c_2\,{}^1k'_{A2}} \tag{3.1I-1}$$

and

$$\frac{1}{{}^1K'_{A2}} = \frac{1}{{}^1k'_{A2}} + \frac{c_2}{H_{Ax}c_1\,{}^2k'_{A1}} \tag{3.1I-2}$$

Note that if mass concentrations are used, ρ_1 and ρ_2 replace c_1 and c_2 in the equations above.

3.1J. Continuity and Flux Expressions for Species A

1. Transform Eq. 3.0-3 to the form

$$\frac{\partial \rho_{A1}}{\partial t} + \nabla \cdot \mathbf{n}_A = r_A \tag{3.1J}$$

2. Using the definition of $\mathbf{n}_A$ from Eq. 3.1J, obtain the mass flux equivalent to Eq. 3.1-13.

3. Considering Eqs. 3.1-13 and 3.1-14, show under what conditions it can be assumed that $n_{Aj} \simeq j_{Aj}$.

3.2. TURBULENCE IN THE ENVIRONMENT

In the preceding section the transport processes of molecular diffusion were introduced. The slowness of this process for chemical movement was demonstrated by examples of the movement of CO_2 into stagnant air above the air–water interface and the depletion of oxygen from a stagnant water layer above the mud–water interface. Interphase chemodynamics are much more rapid than the molecular diffusion process. The turbulent (or eddy) diffusion process was mentioned briefly in Section 3.1. The idea that mass is transferred in turbulent flow by a mixing or blending of eddies was presented. Knowledge on turbulence is increasing and can be used to obtain eddy diffusion coefficients for several flow situations. The eddy diffusion approach was abandoned, and mass transfer coefficients were created. By this creation the unknown eddy diffusivity and unknown concentration gradient were replaced by a single constant and known concentration difference. Although highly utilitarian, the mass transfer coefficient lumps together and hides much of our ignorance of the turbulent transport processes. Section 3.1 contains a presentation of three ad hoc turbulent models that relate the mass transfer coefficient to molecular diffusivity and fluid dynamic parameters.

In Section 3.2 we readdress turbulence and turbulent processes in the environment. Fundamental turbulent flow concepts are presented and simple models are developed. The models are then extended to the derivation of useful analytical results. These results include reasonably accurate mathematical relationships for describing flow and turbulent transport of chemicals in the regions of environmental interfaces. Aspects of both mechanical- and fluid density–induced turbulence will be presented.

The topic of turbulence and turbulent flow in the environment cannot begin without introducing the idea of laminar flow, if for no other reason than to

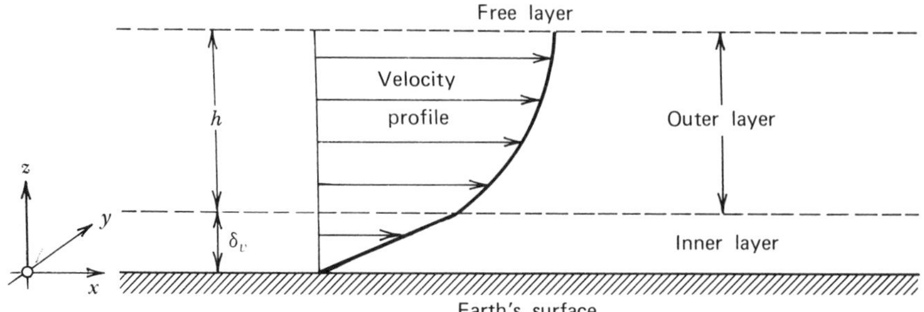

Figure 3.2-1. Atmospheric boundary layer.

create a contrast. In this section we are concerned with the flow of fluids in general and specifically, with the flow of air and water in the regions near solid surfaces. General equations embodying the laws of conservation of mass (the continuity equations) and Newton's second law of motion (the Navier–Stokes equation) are presented to provide a unified introduction to laminar flow and later turbulent flow. The equation of energy and the multicomponent continuity equation for dilute solutions are introduced also, to complete the presentation of the equations of change. Turbulent flow ideas and simple models are then extended to the development of useful results. The results include reasonably accurate mathematical relationships plus a better qualitative understanding of some fundamentals of turbulence.

In general, flow of air or water in the region near the earth's surface can be crudely divided into two layers, often referred to as *boundary layers*. Looking at Fig. 3.2-1, the atmospheric boundary layer is generally divided into two distinct layers: an inner layer of height δ_v on the order of a few centimeters in which the velocities relative to the earth's surface are close to zero and the viscous stresses dominate. The flow in this layer is laminar. In the outer layer of height h, on the order of a kilometer, Reynolds stresses (a turbulent flow stress to be defined later) dominate. Because of the relative importance of the various forces, the behavior of these two layers is different.

The movement of water near a solid surface results in a profile similar to that shown in Fig. 3.2-1. In the case of the benthic boundary layer, the three bottommost hydraulic layers are termed from bottom upward: *viscous layer*, *logarithmic layer*, and *outer layer*.

Equations of Change

In this section the *equations of change* are presented and not derived. Procedures of derivation from first principles may be found in any one of several textbooks on transport phenomena.[3,26] These equations provide a starting point for both qualitative discussions and quantitative formulations concerning interface regions.

Equation of Continuity. The equation of continuity is developed by writing a total mass (i.e., all chemical species) balance over a stationary volume element $\Delta x\,\Delta y\,\Delta z$ through which a fluid is flowing (see Fig. 1.1-4) and is

$$\frac{\partial \rho}{\partial t} = -\left[\frac{\partial}{\partial x}(\rho v_x) + \frac{\partial}{\partial y}(\rho v_y) + \frac{\partial}{\partial z}(\rho v_z)\right] \tag{3.2-1}$$

This is the *equation of continuity*, which describes the rate of change of density at a fixed point resulting from the changes in the mass fluxes at the point. It is frequently desirable to modify Eq. 3.2-1 by performing the operations indicated and collecting all derivatives of ρ on the left side:

$$\frac{\partial \rho}{\partial t} + v_x\frac{\partial \rho}{\partial x} + v_y\frac{\partial \rho}{\partial y} + v_z\frac{\partial \rho}{\partial z} = -\rho\left(\frac{\partial v_x}{\partial x} + \frac{\partial v_y}{\partial y} + \frac{\partial v_z}{\partial z}\right) \tag{3.2-2}$$

Remember that this equation is simply a statement of the conservation of total mass.

Equation of Motion. The *equation of motion* is the result of a momentum balance for the volume element $\Delta x\,\Delta y\,\Delta z$, such as that used in the preceding section. Momentum and forces acting on the volume element are summed for each space dimension, yielding three equations. The * component of the equation of motion is

$$\begin{aligned}\frac{\partial}{\partial t}(\rho v_*) = &-\left[\frac{\partial}{\partial x}(\rho v_x v_*) + \frac{\partial}{\partial y}(\rho v_y v_*) + \frac{\partial}{\partial z}(\rho v_z v_*)\right]\\ &-\left(\frac{\partial}{\partial x}\tau_{x*} + \frac{\partial}{\partial y}\tau_{y*} + \frac{\partial}{\partial z}\tau_{z*}\right) - \frac{\partial p}{\partial *} + \rho g_*\end{aligned} \tag{3.2-3}$$

Letting * be x, y, and z in turn generates all three. Collectively, the equations state that

$$\begin{pmatrix}\text{rate of}\\ \text{increase}\\ \text{of momentum}\\ \text{per unit volume}\end{pmatrix} = -\begin{pmatrix}\text{rate of}\\ \text{momentum gain}\\ \text{by convection}\\ \text{per unit volume}\end{pmatrix} - \begin{pmatrix}\text{rate of}\\ \text{momentum gain}\\ \text{by viscous forces}\\ \text{per volume}\end{pmatrix}$$

$$-\begin{pmatrix}\text{pressure forces}\\ \text{on element}\\ \text{per unit volume}\end{pmatrix} + \begin{pmatrix}\text{gravitational force}\\ \text{on element}\\ \text{per unit volume}\end{pmatrix}$$

The terms τ_{xx}, τ_{xy}, and so on, are the nine momentum fluxes known as *stresses*. Thus τ_{xx} is the normal stress on the x face (see Fig. 1.1-4), and τ_{yx} the x-directed tangential (or shear) stress on the y face resulting from viscous forces. These equations of motion state that a small volume of element moving with the fluid is accelerated because of forces acting on it. In other words, this is a statement of Newton's second law in the form: mass × acceleration = sum of forces.

Viscosity. For Newtonian fluids the stresses may be expressed in terms of Newton's law of viscosity; for example

$$\tau_{yx} = -\mu \frac{dv_x}{dy} \tag{3.2-4}$$

Atmospheric air and liquid water both behave as Newtonian fluids. Equation 3.2-4 is also a definition of viscosity. Tables D.1 and D.2 contain viscosity and kinematic viscosity, $\nu \equiv \mu/\rho$, data for water and air. The SI unit for viscosity is the newton-second per square meter.

The preceding equations (i.e., Eq. 3.2-3) in their complete form are seldom used to set up flow problems. Geometric considerations and simplifying assumptions yield a more manageable set of equations. If Newton's law is used for closure with constant density and viscosity, the three equations of motion become

$$\rho\left(\frac{\partial v_*}{\partial t} + v_x \frac{\partial v_*}{\partial x} + v_y \frac{\partial v_*}{\partial y} + v_z \frac{\partial v_*}{\partial z}\right) = -\frac{\partial p}{\partial *} + \mu\left(\frac{\partial^2 v_*}{\partial x^2} + \frac{\partial^2 v_*}{\partial y^2} + \frac{\partial^2 v_*}{\partial z^2}\right) + \rho g_* \tag{3.2-5}$$

Replacing * with x, y, and z generates those three equations. Here p is pressure, and g_* is the * component of gravitational acceleration. Equation 3.2-5 is the celebrated Navier–Stokes equation, and it has been widely used for describing flow systems in which viscous effects are dominant.

Equation of Energy. The *equation of energy* is the result of the conservation of energy or the first law of thermodynamics. The equation is obtained by writing the law of conservation of energy for a pure fluid contained within the stationary volume element $\Delta x\, \Delta y\, \Delta z$. The equation of energy in terms of the fluid temperature T is

$$\rho \hat{C}_v \left(\frac{\partial T}{\partial t} + v_x \frac{\partial T}{\partial x} + v_y \frac{\partial T}{\partial y} + v_z \frac{\partial T}{\partial z}\right) = -\left(\frac{\partial q}{\partial x} + \frac{\partial q}{\partial y} + \frac{\partial q}{\partial z}\right) - T\left(\frac{\partial p}{\partial T}\right)_{\hat{V}} \left(\frac{\partial v_x}{\partial x} + \frac{\partial v_y}{\partial y} + \frac{\partial v_z}{\partial z}\right) \tag{3.2-6}$$

This equation states that the temperature of a fixed fluid element changes with time due to advection of thermal energy, transport of thermal energy by conduction, and expansion effects.

A term accounting for the irreversible rate of internal energy increase by viscous dissipation has been omitted from the right-hand side of Eq. 3.2-6. This term is needed only in special situations such as extremely high shear rates. The pure fluid form of the equation of energy is usually as applicable to environmental situations as is the multicomponent form. The latter includes a heat of reaction term, but since most environmental chemical solutions are dilute, this factor does not usually influence fluid temperature significantly.

Thermal Conductivity. The local heat flow per unit area (heat flux) in the positive y direction is designated q_y. Fourier's law of heat conduction states that the heat flux by conduction is proportional to the temperature gradient, or mathematically,

$$q_y = -k \frac{dT}{dy} \tag{3.2-7}$$

This is also a definition of thermal conductivity k. In addition to the thermal conductivity, a quantity known as the thermal diffusivity α is convenient to use; it is defined as

$$\alpha \equiv \frac{k}{\rho \hat{C}_p} \tag{3.2-8}$$

Thermal conductivity data for air, water, and some solid materials are available in Appendix D. Density and heat capacity data are also available there. The SI units of thermal conductivity are the joule per meter-second-kelvin.

For most uses in chemodynamics, the fluids may be considered incompressible (ρ = constant), $\hat{C}_p = \hat{C}_v$, and $(\nabla \cdot \mathbf{v})$ is zero. Using Fourier's law for closure on the heat conduction terms, Eq. 3.2-6 becomes

$$\rho \hat{C}_p \left(\frac{\partial T}{\partial t} + v_x \frac{\partial T}{\partial x} + v_y \frac{\partial T}{\partial y} + v_z \frac{\partial T}{\partial z} \right) = k \left(\frac{\partial^2 T}{\partial x^2} + \frac{\partial^2 T}{\partial y^2} + \frac{\partial^2 T}{\partial z^2} \right) \tag{3.2-9}$$

For solids, velocities are zero, and Eq. 3.2-9 becomes

$$\rho \hat{C}_p \frac{\partial T}{\partial t} = k \left(\frac{\partial^2 T}{\partial x^2} + \frac{\partial^2 T}{\partial y^2} + \frac{\partial^2 T}{\partial z^2} \right) \tag{3.2-10}$$

Equations 3.2-9 and 3.2-10 are the starting points for most of our subsequent uses of heat transfer near environmental interfaces.

Multicomponent Continuity Equation for Dilute Solutions. The multicomponent continuity equation for dilute solutions was presented early in Chapter 3 and appears in vector form (see Eq. 3.0-3). This early presentation was used to define the concept of the fate of chemical A in the environment and to develop heuristic ideas concerning chemical movements by diffusion and mass transfer. If Fick's first law and constant diffusivity are used for closure, Eq. 3.0-3 in molar concentration expanded form for the case of chemical A in air becomes

$$\frac{\partial c_{A1}}{\partial t} + v_x \frac{\partial c_{A1}}{\partial x} + v_y \frac{\partial c_{A1}}{\partial y} + v_z \frac{\partial c_{A1}}{\partial z} = \mathscr{D}_{A1} \left(\frac{\partial^2 c_{A1}}{\partial x^2} + \frac{\partial^2 c_{A1}}{\partial y^2} + \frac{\partial^2 c_{A1}}{\partial z^2} \right) + R_A \tag{3.2-11}$$

Similar equations for chemicals in water and in a solid phase can be written. For solids, velocities are zero, and Eq. 3.2-11 becomes

$$\frac{\partial \omega_A}{\partial t} = \mathscr{D}_{A3} \left(\frac{\partial^2 \omega_A}{\partial x^2} + \frac{\partial^2 \omega_A}{\partial y^2} + \frac{\partial^2 \omega_A}{\partial z^2} \right) \tag{3.2-12}$$

where ω_A is the mass fraction of A on the solid.

Closure. This completes the presentation of the equations of change. As was pointed out earlier, these equations are usually simplified further. The process will be demonstrated throughout the book as they are transformed and applied to specific chemodynamic problems (see Problem 3.2A for an application.) The preceding set of general equations therefore serves as a unified starting point for the many specific chemodynamic applications to follow. This procedure often avoids the necessity of resorting to first principles and rederivation to effect a new problem analysis. Besides specific applications, the equations of change taken together as a group convey a certain amount of information.

An inspection and comparison of the equations of motion, energy, and multicomponent continuity reveal many similarities. This aspect is one of the important lessons conveyed by the transport phenomena concepts. The concern is with the transfer of momentum, energy and mass from one point in the universe to another. Specifically, it should be noted that the units of molecular diffusivity $\mathscr{D}_{Aj}$ are cm^2/s. Kinematic viscosity ν and the thermal diffusivity α also have the same units. The ways in which these three quantities are analogous can be seen from the following equations for the fluxes of mass, momentum, and energy in one-dimensional systems:

$$j_{Ay} = -\mathscr{D}_{A1} \frac{d}{dy} (\rho_{A1}) \qquad \text{(Fick's law)} \tag{3.2-13}$$

$$\tau_{yx} = -\nu_1 \frac{d}{dy} (\rho_1 v_x) \qquad \text{(Newton's law)} \tag{3.2-14}$$

$$q_y = -\alpha_1 \frac{d}{dy} (\rho_1 \hat{C}_{p1} T) \qquad \text{(Fourier's law)} \tag{3.2-15}$$

This set is incomplete without Ohm's law; the flux of charged species down a potential gradient gives rise to the phenomenon of electrical conductivity.[44] Although electron charge balances are used in environmental chemodynamic applications in a stoichiometric or equilibrium context such as oxidation-reduction reactions, chemical species flux limitations due to an Ohm's law resistance have not yet been identified to my knowledge.

It can now be appreciated that the similarity of the equations of change is due in large part to the mathematics of expressing the basic phenomenological processes. These equations state, respectively, that mass, momentum, and energy transports occur because of gradients in mass concentration, momentum concentration, and energy concentration. The three variables denoted by $\mathscr{D}_{A1}$, ν_1, and α_1 are fundamental transport properties dependent only on molecular processes (see Hirschfelder et al.[14]). The use of these for closure places the major limitation on the equations of change in that they do not apply if turbulence is present. They apply only to solids and fluids in laminar (i.e., viscous) flow.

Viscous Sublayer (or Inner Layer)

Except for a thin layer of air close to surfaces, the atmosphere is essentially always turbulent. There is also a viscous sublayer on the seafloor that is distinguished from the zone above in that turbulent transfer of momentum is less important than the molecular transfer of momentum. Figure 3.2-2 illustrates the flow characteristics occurring in the viscous sublayer next to a solid when the mass of fluid is moving in the positive x direction.

Simplifying the x component of the Navier–Stokes equation results in

$$\mu_1 \frac{d^2 v_x}{dy^2} = 0 \tag{3.2-16}$$

Since a real fluid is compelled by molecular attraction to adhere to a solid boundary, $v_x = 0$ at $y = 0$. The viscous sublayer is assumed to have a film

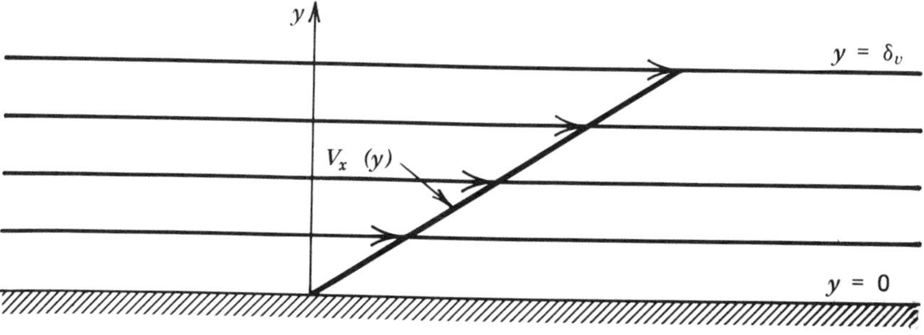

Figure 3.2-2. Flow in the viscous sublayer.

thickness of δ_v, at which $v_x = v_x(\delta_v)$. These two boundary conditions, combined with Eq. 3.2-16, yield the fact that the velocity profile is linear, as depicted in Fig. 3.2-2:

$$v_x(y) = \frac{v_x(\delta_v)y}{\delta_v} \tag{3.2-17}$$

The momentum transfer rate to the solid wall (τ_0) is obtained by use of Eq. 3.2-14 and is

$$\tau_0 = \frac{-\mu_1 v_x(\delta_v)}{\delta_v} \tag{3.2-18}$$

This τ_0 is also the shear stress exerted by the fluid onto the surface and has SI units of newton per square meter.

Because of increasing agitation of the molecules, there is a continuous transfer of momentum from regions of high bulk velocity to regions of low bulk velocity, and the rate of transfer of momentum across the unit area of a plane surface in the fluid is expressed by the product of the viscosity and the gradient (Eq. 3.2-18). From a molecular point of view, it should be observed that although there is no net transfer of molecules across a plane parallel to the direction of flow, the existence of a gradient of bulk velocity ensures that the random motion brings about a continuous cross-stream transfer of momentum.

Some Characteristics of Turbulent Flow

Turbulence is the property, easy to recognize but difficult to define, of irregular, chaotic motion possessed by almost all natural fluid flows. Nearly all natural motion, whether of air or water, is turbulent. Turbulent flow has been defined: motion of fluids in which local velocities and pressures fluctuate irregularly, in a random manner. Figure 3.2-3 shows a portion of an anemometer record of the natural wind velocity at an elevation of 35 m. In contrast, the laminar flow pictured in Fig. 3.2-2 is highly ordered. For chemodynamic practical purposes we can best define a turbulent fluid flow as one that has the ability to disperse particles and/or molecules embedded within itself quite rapidly, at a rate orders of magnitude greater than can be accounted for by molecular diffusion.

The air velocity increases with distance from the air–soil interface; the laminar flow pattern, with its steady advance in separate layers, is not maintained: the flow becomes unsteady, with chaotic movements of parts of the air in different directions superimposed on the main flow of the air, as in Fig. 3.2-3. Movement of any particular element of fluid is now very complicated, and it can be described in terms of averages.

In turbulent flow, transfers of momentum between neighboring pulses of the fluid are of primary importance, as is described in detail later. These inertial

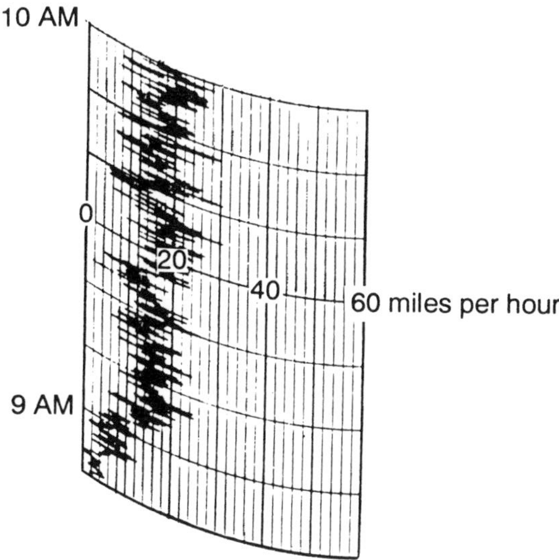

Figure 3.2-3. Wind velocity and velocity fluctuations.

(momentum) effects in turbulent flow (as contrasted with purely viscous effects in laminar flow) cause the velocity and density of the flowing fluid to assume great importance. Turbulent flow replaces laminar flow when these inertial effects are great compared with viscous effects. For the flow of fluid in a pipe of diameter d, Reynolds (1883) used the ratio of inertial force to viscous force to characterize the change of flow from laminar to turbulent and thus created the dimensionless ratio known as the *Reynolds number:*

$$\mathrm{Re} \equiv \frac{v_x d}{\nu} \tag{3.2-19}$$

Here ν denotes the ratio μ/ρ.

The effect of random disturbances (and there are many in the natural environment) on a fluid in laminar flow is shown in Fig. 3.2-4. In part (*a*) the flow is pure laminar with a streamline flow pattern in the viscous sublayer and well into regions beyond. In part (*b*) a random disturbance is shown, the density and velocity gradient effects (momentum effect) being shown by the arrow. In part (*c*) more momentum is being transferred into the disturbance than is being damped out, and an eddy is formed. Laminar motion is stable as long as there is no net transfer of energy from the primary flow into any superimposed random disturbances. All disturbances are damped by the viscosity μ, and the closeness of the solid wall. Figure 3.2-5 shows the turbulent eddies above the air–soil interface as viewed by an observer moving from left to right with the mean velocity. This is the case for fully developed turbulent flow. Near the wall, the strong velocity gradients within the fluid tear the fluid

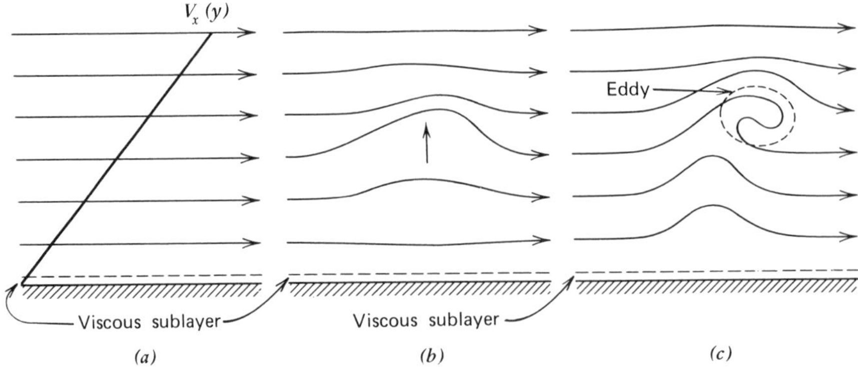

Figure 3.2-4. Origin of an eddy. (From Ref. 15.)

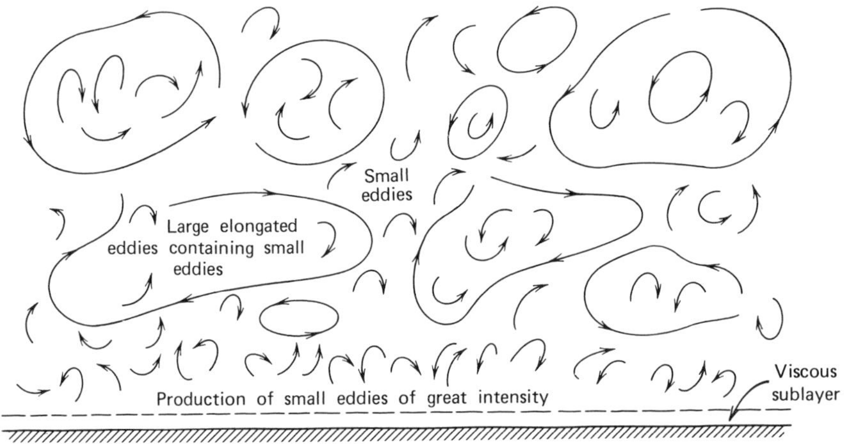

Figure 3.2-5. Eddy production patterns of turbulent flow above the air–soil interface. (From Ref. 15.)

into small eddies. Some of these migrate upward, where larger eddies are also to be found.

Experiments with many fluids in smooth, circular pipes of different diameters have confirmed that Re does indeed characterize the velocity of flow at which laminar breaks down to turbulent flow. At Reynolds numbers up to about 2000, the flow in a smooth pipe is always laminar. Between 2000 and 4000 (called the *transition region*), there is usually a gradual change to turbulent flow. Generally, turbulence is fully established when Re > 4000. For geometries other than that of a uniform circular pipe, the characteristic length d in Eq. 3.2-19 can be suitably assigned.

For the case of fully developed turbulent airflow above the air–soil interface, Slade[16] employs a semiquantitative development based on the five dimensional quantities (i.e., dv_x/dy, ν, y, ρ, and τ_0) and the well-known pi theorem of dimensional analysis to obtain the Reynolds number

$$\mathrm{Re} \equiv \frac{v_* \delta_v}{\nu} \tag{3.2-20}$$

where δ_v is some fixed elevation above which turbulence is present and v_* is the friction velocity, equal to $\sqrt{\tau_0/\rho}$. It has been found experimentally in numerous specific cases, pipe flow being one, that for the natural reference length used in Reynolds numbers, values above about 1E2 or 1E3 are always turbulent.

What length scale applies in the atmosphere? Because flows for which Re ≫ 1E2 are ordinarily turbulent, it follows that an upper limit to δ_v, the depth of the laminar sublayer for the atmosphere, will be on the order of a millimeter since v_* is known from observations to be on the order of 100 cm/s and v for air equals about 1E-1 cm^2/s. This means that blades of grass, grains of dirt, sticks, twigs, people, and so on, all protrude through the laminar sublayer. Anyone who has lost a hat in a sudden gust of wind has inadvertently experienced that δ_v is small and that one does indeed protrude through the viscous sublayer. People live at the air–soil interface and therefore have experienced many aspects of the fluid flow phenomena, including laminar motion, eddies, and convection. This experience, together with the qualitative treatment presented here, should provide a degree of understanding of the turbulent processes at work at all interfaces over which a fluid is moving, including ocean, lake, and river bottoms. These turbulent processes are important to the subject of chemodynamics.

Turbulent Equations of Change

Statistical Nature of Turbulent Flow. We must now focus our attention on the fluid behavior at one point above the air–soil interface where turbulent flow exists. Consider the anemometer record shown in Fig. 3.2-3, the average velocity in the x direction $\bar{v}_x$, and the instantaneous velocity v_x. We can imagine that while we are watching this one spot above the surface, the average wind velocity increases slowly from a value of 3 mi/h at 8:50 A.M. to a value of 20 mi/h at 10:00 A.M. The average velocity fluctuates on a time period much greater than that of the instantaneous velocity v_x. We define the time-smoothed (i.e., average) velocity $\bar{v}_x$ by taking a time average of v_x over a time interval t_0 large with respect to the time of rapid turbulent oscillations but small with respect to the larger period oscillations:

$$\bar{v}_x = \frac{1}{t_0} \int_t^{t+t_0} v_x \, dt \tag{3.2-21}$$

For the wind data in Fig. 3.2-3, 5 min seems reasonable for t_0. The instantaneous velocity may be written as the sum of the time-smoothed velocity $\bar{v}_x$

and a velocity fluctuation v_x':

$$v_x = \bar{v}_x + v_x' \qquad (3.2\text{-}22)$$

A similar expression can be written for the pressure, which is also fluctuating; clearly, $\bar{v}_x' = 0$ by the foregoing definitions. But $v_z'^2$ will not be zero and, in fact, $\sqrt{\overline{v_x'^2}}/\langle \bar{v}_x \rangle$ is a measure of the magnitude of the turbulent disturbance and is known as the *intensity of turbulence*. Notice that the fluctuating, or turbulent, component of the wind in Fig. 3.2-3 is of the same order of magnitude as the mean (i.e., intensity of about unity). This is a characteristic of atmospheric turbulence and distinguishes it sharply from wind tunnel turbulence, where the intensity is likely to be 0.01 to 0.001.

A convenient concept of the turbulent wind velocity at a fixed point is a velocity v_x in the x direction, representing the mean wind and three mutually perpendicular vectors in the x, y and z directions with time-varying components v_x', v_y', and v_z', to give

$$v_y = \bar{v}_y + v_y' \qquad (3.2\text{-}23)$$

$$v_z = \bar{v}_z + v_z' \qquad (3.2\text{-}24)$$

along with Eq. 3.2-22. The v_x' vector represents the fluctuation or eddy velocity in the direction of the mean wind, v_y' the vertical fluctuation, and v_z' the fluctuation across the mean wind. Using these definitions, measurements of turbulence in the atmosphere reveal that (1) the intensity of turbulence is essentially independent of wind speed but does depend on atmospheric stability; (2) at low levels turbulence is nonisotropic (i.e., $\sqrt{\overline{v_x'^2}}/\bar{v}_x \neq \sqrt{\overline{v_y'^2}}/\bar{v}_x \neq \sqrt{\overline{v_z'^2}}/\bar{v}_x$); and (3) some authorities have concluded that turbulence is isotropic at heights in excess of about 25 m.

Time Smoothing of the Equations of Change for Incompressible Fluid. In this section we present the equations that describe the time-smoothed velocity and pressure for an incompressible fluid. *A* brief outline of the definition is presented, and details are left as an exercise for the student (see problem 3.2C). Starting with the x component of the equation of motion with Newton's law for closure on the shear stresses yields

$$\frac{\partial}{\partial t}(\rho v_x) + \frac{\partial}{\partial x}(\rho v_x v_x) + \frac{\partial}{\partial y}(\rho v_y v_x) + \frac{\partial}{\partial z}(\rho v_z v_x) = \frac{-\partial p}{\partial x} + \mu\left(\frac{\partial^2 v_x}{\partial x^2} + \frac{\partial^2 v_x}{\partial y^2} + \frac{\partial^2 v_x}{\partial z^2}\right) + \rho g_x \qquad (3.2\text{-}25)$$

Replace v_x, v_y, and v_z by Eqs. 3.2-22, 3.2-23, and 3.2-34, respectively, and p by $\bar{p} + p'$ everywhere they appear. Now take the time average of the result according to Eq. 3.2-21 for v_x and a similar integral average for p, v_y, v_z and any products of these. The following equation results:

$$\frac{\partial}{\partial t}(\rho\bar{v}_x) + \frac{\partial}{\partial x}(\rho\bar{v}_x\bar{v}_x) + \frac{\partial}{\partial y}(\rho\bar{v}_y\bar{v}_x) + \frac{\partial}{\partial z}(\rho\bar{v}_z\bar{v}_x) + \frac{\partial}{\partial x}(\rho\overline{v'_x v'_x})$$

$$+ \frac{\partial}{\partial y}(\rho\overline{v'_y v'_x}) + \frac{\partial}{\partial z}(\rho\overline{v'_z v'_x}) = \frac{-\partial\bar{p}}{\partial x} + \mu\left(\frac{\partial^2\bar{v}_x}{\partial x^2} + \frac{\partial^2\bar{v}_x}{\partial y^2} + \frac{\partial^2\bar{v}_x}{\partial z^2}\right) + \rho g_x \qquad (3.2\text{-}26)$$

This is the time-smoothed x component of the equation of motion. All terms appearing in Eq. 3.2-25 appear in this time-smoothed equation replaced with average values, but in addition, new terms arise that are associated with the turbulent velocity fluctuations. These new terms are the components of the turbulent momentum flux and are usually referred to as the *Reynolds stresses:*

$$\tau_{xx}^{(t)} = \rho\overline{v'_x v'_x}; \qquad \tau_{xy}^{(t)} = \rho\overline{v'_y v'_x}; \qquad \tau_{xz}^{(t)} = \rho\overline{v'_z v'_x} \qquad (3.2\text{-}27\text{a,b,c})$$

With the preceding definitions and $\boldsymbol{\tau} = \bar{\boldsymbol{\tau}}^{(l)} + \bar{\boldsymbol{\tau}}^{(t)}$, Eq. 3.2-26 can be transformed to appear like Eq. 3.2-3. The equation of motion can therefore be used for turbulent flow problems provided that one changes all v_i to $\bar{v}_i$, ρ to $\bar{\rho}$, and τ_{ij} to $\tau_{ij}^{(l)} + \tau_{ij}^{(t)}$. Similar time-smoothing results can be obtained for the y and z-components of the equation of motion.

When considering the energy balance it is necessary to use a point temperature made up of an average temperature and a fluctuating component: $T = \bar{T} + T'$. Time smoothing the equation of energy results in additional fluxes also. The development is restricted to a fluid of constant ρ, $\hat{C}_p$, μ, and k. In vector form the timed-smoothed equation is

$$\rho\hat{C}_p\left(\frac{\partial\bar{T}}{\partial t} + \mathbf{v}\cdot\nabla\bar{T}\right) = k\nabla^2\bar{T} - \nabla\cdot\mathbf{q}^{(t)} \qquad (3.2\text{-}28\text{a})$$

(See Problem 3.2C, part 3 for directions of derivations.) Additional terms created have led to the definition of the turbulent energy fluxes with components:

$$q_x^{(t)} = \hat{C}_p\overline{v'_x T'} \qquad q_y^{(t)} = \rho\hat{C}_p\overline{v'_y T'} \qquad q_z^{(t)} = \rho\hat{C}_p\overline{v'_z T'} \qquad (3.2\text{-}28b, c, d)$$

The similarity between the components of $\bar{q}^{(t)}$ and those of $\bar{\tau}^{(t)}$ should be noted.

Time smoothing the multicomponent continuity equation for dilute solutions also results in new terms. (see Problem 3.2C, part 4). In vector form the equation is

$$\frac{\partial\bar{c}_{A1}}{\partial t} + \mathbf{v}\cdot\nabla\bar{c}_{A1} = \mathscr{D}_{A1}^{(l)}\nabla^2\bar{c}_{A1} - \nabla\cdot\mathbf{J}_{A1}^{(t)} - k_A'''\bar{c}_{A1} \qquad (3.2\text{-}29a)$$

and the new terms are

$$J_{Ax}^{(t)} = \overline{v'_x c'_{A1}} \qquad J_{Ay}^{(t)} = \overline{v'_y c'_{A1}} \qquad J_{Az}^{(t)} = \overline{v'_z c'_{A1}} \qquad (3.2\text{-}29b, c, d)$$

where c'_{A1} is the fluctuating component of concentration around the mean $\bar{c}_{A1}$. The new terms describe the turbulent mass transport. All fluxes are defined with respect to the mass average velocity.

Interpretation of the Reynolds Stresses

The meaning of the Reynolds stresses can be made clear by considering the case of airflow in the x direction near a soil surface. The mean velocity is distributed as shown in Fig. 3.2-6, with $\bar{v}_y$ and $\bar{v}_z$ zero. The quantity $\tau_{yx}^{(t)} = \rho\overline{v'_x v'_y}$ represents the Reynolds stress in the x direction acting on a plane perpendicular to the y direction. We would like to show that this term is different from zero and negative.

Suppose that a small pocket of fluid having a mean velocity $\bar{v}_{xA}$ at position A in Fig. 3.2-6 is transported, because of a negative v'_y, to a region B, where $\bar{v}_{xB}$ is smaller. Since the pocket will retain approximately its original velocity v_{xA} at position B, there is created a positive v'_x, and $v'_x v'_y$ is negative. On the other hand, if v'_y happens to be positive, the fluid is transported to position C, where the mean velocity is greater than v_{xA}; a negative v'_x is created, and $v'_x v'_y$ is again negative. It is thus easy to see that $\overline{v'_x v'_y}$ will be a negative quantity so that the mean turbulent stress $\bar{\tau}_{yx}^{(t)}$ shown in Eq. 3.2-26b is negative in value; this means momentum is being transferred to the ground.

Prandtl Mixing-Length Theory. By analogy with the kinetic theory of gases, Prandtl (1926) postulated that as the masses of fluid migrated laterally, they carried with them the mean velocity (and hence the momentum concen-

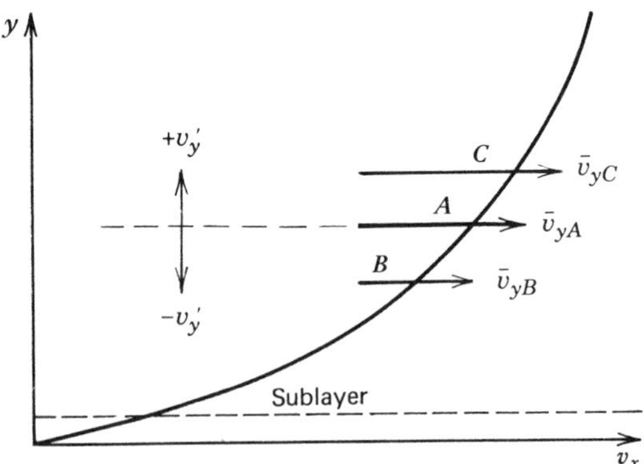

Figure 3.2-6. Time-smoothed velocity distribution above the air–soil suface.

tration) of their point of origin. Still considering a flow above a soil surface and referring to Fig. 3.2-6, we define the mixing length in the following way. Consider again a small pocket of fluid that is displaced from A to B in the y direction with a velocity v'_y. In reality, the lump of fluid will gradually lose its identity, but in the definition of the mixing length it is assumed to retain its identity until it has traveled a distance l defined as the *Prandtl mixing length.* For the small distance involved, we can write

$$\frac{d\bar{v}_x}{dy} = \frac{\bar{v}_{xB} - \bar{v}_{xA}}{l} \tag{3.2-30}$$

As mentioned earlier, the pocket of fluid can be assumed to retain its original velocity, so that $\bar{v}_{xB} - \bar{v}_{xA}$ is approximately $-v'_x$. Therefore, we have

$$v'_x = -l\,\frac{d\bar{v}_x}{dy}$$

and, in general,

$$\overline{|v'_x|} = l\left|\frac{d\bar{v}_x}{dy}\right| \tag{3.2-31}$$

Prandtl also assumed that $\overline{|v'_y|}$ was about the same absolute magnitude as $\overline{|v'_x|}$, so that

$$\overline{v'_x v'_y} = l^2\left|\frac{d\bar{v}_x}{dy}\right|^2$$

Since the sign of $\overline{v'_x v'_y}$ is negative and the opposite of $d\bar{v}_x/dy$ (i.e., the slope of $\bar{v}_x$ versus y in Fig. 3.2-6), we write the preceding equation for the turbulent shear stress as

$$\tau_{yx}^{(t)} = \rho\overline{v'_x v'_y} = -\rho l^2\left|\frac{d\bar{v}_x}{dy}\right|\frac{d\bar{v}_x}{dy} \tag{3.2-32}$$

At this point the overbar denoting average velocity will be omitted. This is a main result of the Prandtl mixing-length theory. Here l is a quantity having the dimension of length. According to Prandtl, l is the average distance traveled by fluid lumps involved in the turbulent mixing process. According to more recent views, however, l is to be interpreted as a quantity proportional to the average size of macroturbulent eddies and/or to the average length of their displacement during turbulent mixing. The average size of macroturbulent eddies and the average length of their displacements vary along the thickness of the flow; that is, l is a function of the distance from the wall. At the present state of knowledge, the exact form of the variation of l with y is not known.

By analogy with the molecular viscosity, an *eddy kinematic viscosity* or *turbulent coefficient of viscosity* can be defined for parallel flow as

$$\bar{\tau}_{xy}^{(t)} = -\rho \nu^{(t)} \frac{d\bar{v}_x}{dy} \tag{3.2-33}$$

Unlike molecular viscosity, the quantity $\nu^{(t)}$, introduced by Boussinesq (1877), is not a function of state but depends strongly on position. We have no way of calculating $\nu^{(t)}$ a priori, although it can be determined experimentally from a given distribution of $\bar{v}_x$ versus y.

Comparing Eqs. 3.2-32 and 3.2-33, it may appear that the only result of the Prandtl mixing length has been to replace one empirical, nondeterminable quantity with another, but the mixing length is easier that $\nu^{(t)}$ to visualize and estimate. In the next section it is shown that some valuable results for the velocity distribution for turbulent flow near environmental interfaces can be obtained with the simple relation

$$l = \kappa_1 y \tag{3.2-34}$$

where y is the distance from the surface and κ_1 is a universal constant, 0.4.

Velocity Distribution in Two-Dimensional Turbulent Flow with Free Surface. In this section the result of the Prandtl mixing-length theory is employed to obtain the velocity profile for water flowing in a stream. Figure 3.2-7 shows a section of a natural stream with a momentum transport-free surface at the air–water interface and a bottom made of a granular medium.

The left side of Fig. 3.2-7 shows schematically the distribution of the shear stress $\tau^{(t)}$ and $\tau^{(l)} = \tau - \tau^{(t)}$ along the depth of the flow. Here δ_v is the thickness of the laminar sublayer. It can be shown that above the laminar sublayer the

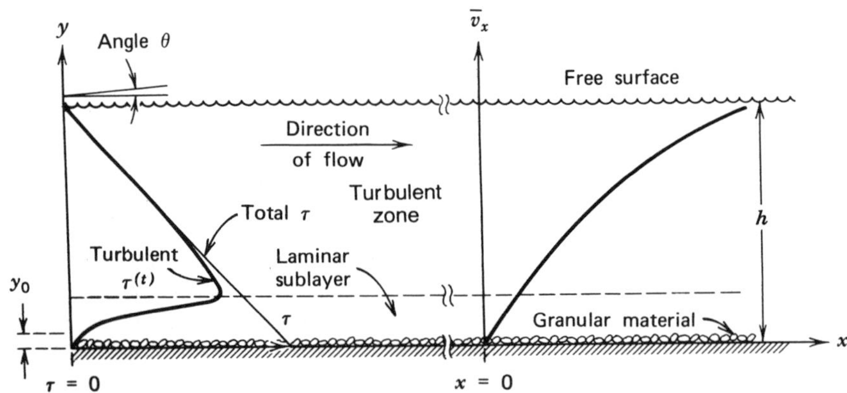

Figure 3.2-7. Flow of a natural stream.

shear stress is almost equal to the turbulent shear stress $\tau^{(t)}$, the component $\tau^{(l)}$ being negligible. Within the sublayer the opposite is true. The total shear stress within the fluid may be obtained from an x-momentum balance, and the result is

$$\tau_{yx} = \rho_2 h g_y \tan\theta \left(1 - \frac{y}{h}\right) \tag{3.2-35}$$

where $\tan\theta$ is the slope of the free surface (see Problem 3.2D). Note that the shear stress at the bottom, τ_0, is $\rho_2 h g_y \tan\theta$.

Now in the turbulent zone, that section of the stream from the laminar sublayer to the free surface, the total stress is equal (almost) to the turbulent stress. Using Eq. 3.2-34 for the mixing length and equating Eqs. 3.2-32 and 3.2-35, we obtain

$$\rho_2 \kappa_1^2 y^2 \left(\frac{dv_x}{dy}\right)^2 = \tau_0 \left(1 - \frac{y}{h}\right) \tag{3.2-36}$$

Prandtl made a mathematical simplification (physically indefensible) at this point by setting the right side equal to τ_0. This simplifies the mathematics somewhat, and it has been shown that the results differ very little from that obtained by integrating Eq. 3.2-36. After taking the square root, we get

$$\frac{dv_x}{dy} = \pm \frac{v_*}{\kappa_1} \frac{1}{y} \tag{3.2-37}$$

in which the plus sign will have to be used. Here v_* is the definition of $\sqrt{\tau_0/\rho_2}$, which has the dimensions of velocity and is commonly called the *shear* or *friction velocity*. If we integrate from the lower edge of the turbulent layer where $y = \delta_v$ to a distance y above the bottom, we get

$$v_x = v_x|_{\delta_v} + \frac{v_*}{\kappa_1} \ln \frac{y}{\delta_v} \qquad h \geqslant y \geqslant \delta_v \tag{3.2-38}$$

This is the relation between velocity and distance from the bottom. At the surface $y = h$ and v_x is the maximum, $v_x|_h$. For this case,

$$v_x|_h = v_x|_{\delta_v} + \frac{v_*}{\kappa_1} \ln \frac{h}{\delta_v} \tag{3.2-39}$$

and subtracting from Eq. 3.2-38, we obtain

$$\frac{v_x|_h - v_x}{v_*} = -\frac{1}{\kappa_1} \ln \frac{y}{h} \tag{3.2-40a}$$

which is a form of the *logarithmic distribution law*.

Logarithmic Velocity Distribution. Equation 3.2-40a and the more general form

$$\frac{v_x}{v_*} = \frac{1}{\kappa_1} \ln y + \text{constant} \tag{3.2-40b}$$

have been found to be a very good approximation for quantifying the velocity profiles in the turbulent regions above almost all environmental interfaces. This relation mimics the profiles in streams, in the air layers above both the soil surface and ocean surface, and in the benthic boundary layer (see Problems 3.2F and 3.2G).

The integration constant in Eq. 3.2-40b is usually defined so as to introduce the effect of surface roughness by requiring that $v_x = 0$ when $y = y_0$; y_0 is called the *roughness length* because it expresses the effect of varying interface roughness on the velocity profile:

$$v_x = \frac{v_*}{\kappa_1} \ln \frac{y}{y_0} \tag{3.2-41}$$

This equation is valid for $y \geqslant y_0$ and applies only above the laminar sublayer.

Sometimes y_0 is chosen so that $v_x = 0$ when $y = 0$. If this is done, the velocity profile equation takes the form

$$v_x = \frac{v_*}{\kappa_1} \ln \frac{y + y_0}{y_0} \tag{3.2-42}$$

Since, as a practical matter, interest is ordinarily centered on velocity at heights where $y \gg y_0$ the two forms are substantially equivalent. Values of y_0 and v_* found from field experiments appear in Table 3.2-1.

For the sake of simplicity, a two-layer fluid model consisting of a laminar layer adjacent to the solid surface with a turbulent zone beyond was presented above. It has become customary to think of three arbitrary zones: the laminar sublayer, in which Newton's law of viscosity is used to describe the flow; the buffer zone, in which the laminar and turbulent effects are both important; and the region of fully developed turbulence, in which purely laminar effects are of negligible importance. Although these regions should not be interpreted too literally, the thickness of the laminar sublayer for smooth surfaces is

$$\delta_v \simeq \frac{5\nu}{v_*} \tag{3.2-43}$$

and the distance from the wall to the outer fringes of the buffer zone is approximately $5\delta_v$.

Table 3.2-1. Universal Velocity Profile Parameters

Type of Surface	y_0 (cm)	v_* (m/s)
For wind near Earth's surface[a]		
Smooth mud, snow, sea surface, desert	0.001–0.01	0.16–027
Lawn grass to 5 cm	1–2	0.43
Lawn grass to 60 cm	4–9	0.60
Fully grown root crops	14	1.75
For the benthic boundary layer[b]		
Deep sea	2.0	0.001
Shelf	—	0.01
Natural streams		
Riverbeds	3.0–90	—

[a]From Ref. 16.
[b]From Ref. 17.

Figure 3.2-8. Three-dimensional sand ripples (Euphrates River). (From Ref. 18.)

Roughness of Environmental Interfaces. Since we live at the air–soil interface and frequently venture onto the air–water interface, we are familiar with the makeup and variations of the surface structures and therefore need no elaboration here. The other environmental interface is not so familiar. Figures 3.2-8 and 3.2-9 show photographs of a streambed and seabeds.

(*a*)

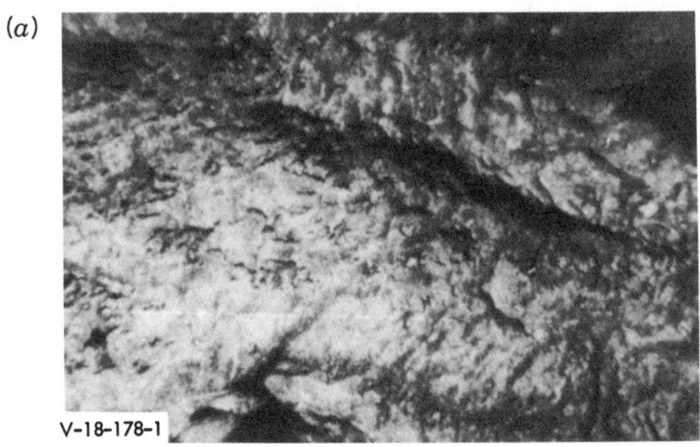

(*b*)

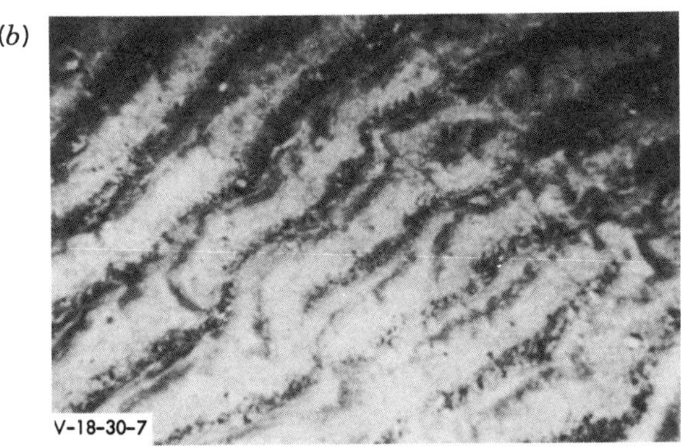

Figure 3.2-9. Four photographs taken on the eighteenth voyage of Columbia University's research vessel *VEMA*. (Photographs courtesy of Maurice Ewing, Lamont Geological Observatory.) (From Ref. 19.)

Transport by Free Convection

In forced-convection situations Eq. 3.2-3 applies; but in the case of free convection it is necessary to modify the equations of motion to account automatically for the buoyant effects in the fluid. Free thermal convection is defined as flow imparted to a fluid by temperature differences within. It may be considered to be a form of turbulence.

The transport of volatile pesticides from the surface of the soil, onto which they are applied, into the lower atmospheric boundary layer is enhanced significantly on sunny days. The soil absorbs solar radiation and converts it to thermal energy in the adjacent air layer. This layer is much warmer than those above and therefore less dense, resulting in the fluid instability that drives the free-convection process. Similar fluid instabilities occur in water due to thermal energy transport. Density instabilities also occur due to chemical concentration

(c)

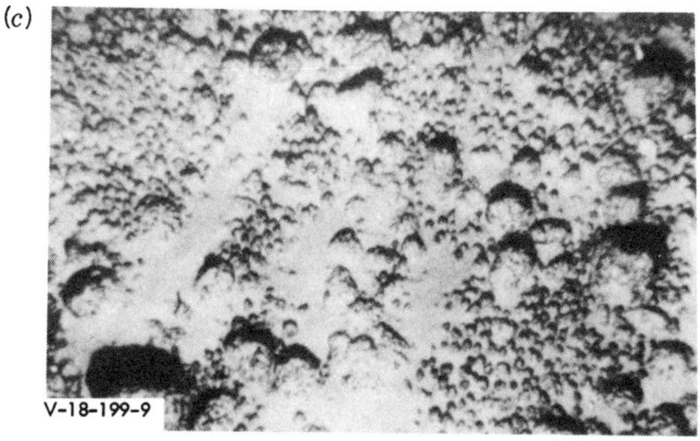

(d)

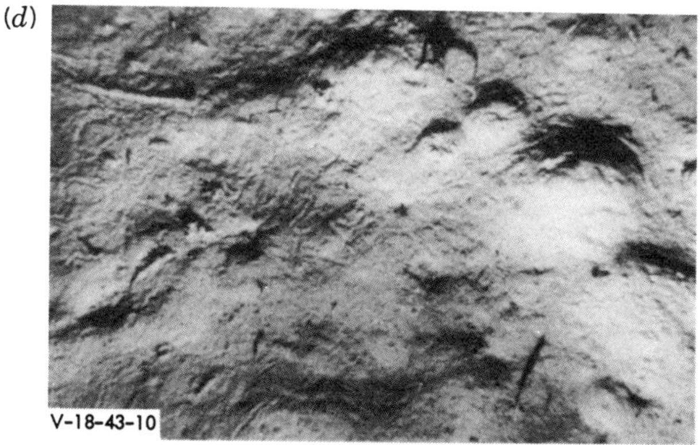

Figure 3.2-9. (*Continued*).

differences. For example, some hydrocarbon vapor fractions around leaks in underground storage tanks can move rapidly through the soil, due to their density being greater than the gas in the pore spaces.

Since most fluids usually exhibit a change in density when heated, it is convenient to define a thermal coefficient of volume expansion:

$$\beta_j \equiv \frac{-1}{V_j}\left(\frac{\partial V_j}{\partial T}\right)_p = \frac{-1}{\rho_j}\left(\frac{\partial \rho_j}{\partial T}\right)_p \tag{3.2-44}$$

Expanding the density, ρ_j, with a Taylor series in temperature, T, about some reference temperature $\bar{T}$ (as yet unspecified) and using Eq. 3.2-44 yields

$$\rho_j = \bar{\rho}_j - \bar{\rho}_j \beta_j (T - \bar{T}) + \cdots \tag{3.2-45}$$

a simple but useful equation of state.

In a flow system experiencing free convection the fluid temperature varies about some mean value $\bar{T}$. If the velocity gradients result entirely from temperature in equalities, the flow is usually quite slow. If the y-component of Eq. 3.2-3 is applied to a fluid for $v_y = 0$ then $dp/dy = \bar{\rho} g_y$; the pressure gradient is due entirely to the weight of the fluid. Of course, the velocity is not zero, but, the approximation that the pressure gradient is still $\bar{\rho} g_y$ applies. Using this and Eq. 3.2-45 in the y-component of Eq. 3.2-5 yields, in vector notation,

$$\bar{\rho}_j \frac{\partial v_y}{\partial t} + \mathbf{v} \cdot \nabla v_y = \mu_j \nabla^2 v_y - \bar{\rho}_j \beta_j g_y (T - \bar{T}) \qquad (3.2\text{-}46)$$

This is the y component of the Navier–Stokes equation and is used for problems of free convection in fluids when a mean temperature can be defined. It is limited to low velocities and small temperature variations.

Chemical concentration gradients also exist in fluids, and these can affect the flow field. If the simple equation of state for fluid density is now assumed to be a function of T and x_A, a double Taylor series expansion yields

$$\rho_j = \bar{\rho}_j - \bar{\rho}_j \beta_j (T - \bar{T}) - \bar{\rho}_j \xi_{Aj} (x_A - \bar{x}_A) + \cdots \qquad (3.2\text{-}47)$$

where $\bar{x}_A$ is a reference chemical mole fraction concentration and the mass coefficient of volume expansion is defined as

$$\xi_{Aj} \equiv -\frac{1}{\rho_j} \left(\frac{\partial \rho_j}{\partial x_A} \right)_{T,P} \qquad (3.2\text{-}48)$$

The once modified Navier–Stokes equation is now

$$\bar{\rho}_j \frac{\partial v_y}{\partial t} + \mathbf{v} \cdot \nabla v_y = \mu_j \nabla^2 v_y - \bar{\rho}_j \beta_j g_y (T - \bar{T}) - \bar{\rho}_j \xi_{Aj} g_y (x_A - \bar{x}_A) \qquad (3.2\text{-}49)$$

This equation states that the creation of temperature or concentration differences within a stagnant fluid will induce flow. These differences occur regularly near environmental interfaces.

Nondimensionalizing the respective equations yields the Grashoft numbers for heat and mass transfer. These nondimensional groups appear in Table 3.3-2 and play the same role in free convection that the Reynolds number does in forced convection. In this regard Eqs. such as 3.1-23 and 3.1-24 result for mass transfer correlations with forced convection, and one would then expect equations of the form

$$\mathrm{Sh}_{Aj} = f(\mathrm{Gr}_{Aj}, \mathrm{Sc}_{Aj}) \qquad (3.2\text{-}50)$$

for mass transfer coefficient correlations in the presence of free convection.

Such correlations are available, and several with specific application to environmental chemodynamics are presented in the following chapters.

PROBLEMS

3.2A Transformation and/or Simplification of the Equations of Change

Certain simplifications, assumptions, and restrictions are necessarily performed on the equations of change when they are applied to specific problems. List these when doing the following transformations.

1. Starting with Eq. 3.0-3, obtain Eq. 3.2-11.
2. Starting with Eq. 3.2-11, obtain Eq. 3.1-3.
3. Starting with Eq. 3.2-5, obtain Eq. 3.2-16.
4. Starting with Eq. 3.2-16, obtain Eqs. 3.2-17 and 3.2-18.

3.2B. Approximate Height of Viscous Sublayer

1. Verify that the height of the sublayer at the air–soil interface is of the order of a millimeter.
2. Compute the height of the sublayer to be found at the water–sediment interface.

3.2C. Time Smoothing the Equations of Change

1. Following the procedure outlined in the text, show details of the transformation from Eq. 3.2-25 to Eq. 3.2-26 that were omitted.
2. Using a procedure similar to that in part 1, time-smooth the continuity equation, Eq. 3.2-1.
3. Starting with the following form of the equation of energy:

$$\rho \hat{C}_p \frac{\partial T}{\partial t} = \nabla \cdot (\rho \hat{C}_p T \mathbf{v}) + k \nabla^2 T \tag{3.2C-1}$$

   and $T = \bar{T} + T'$, time-smooth and show the origin of Eqs. 3.2-28*a* to 3.2-28*d*.

4. Starting with the following form of the multicomponent equation of continuity

$$\frac{\partial c_{A1}}{\partial t} = -\nabla \cdot \mathbf{v} c_{A1} + \mathscr{D}_{A1} \nabla^2 c_{A1} - k_A''' c_{A1} \tag{3.2C-2}$$

and $c_{A1} = \bar{c}_{A1} + c'_{A1}$, time-smooth and show the origin of Eqs. 3.2-29*a* to 3.2-29*d*.

3.2D. Shear Stress Relation in Natural Flowing Stream

Show that Eq. 3.2-35 is valid for natural stream flow. Start by performing an x-component momentum (or force) balance on an element of fluid of thickness Δx, bounded by the free surface and bottom and extending a distance w in the z direction.

3.2E. Logarithmic Velocity Distribution for Turbulent Flow near Environmental Interfaces

Refer to text material starting with Eq. 3.2-32 and show all steps in the development of the logarithmic velocity distribution (Eq.3.2-40). List the assumptions needed.

3.2F. Velocity Profile near Natural Interfaces

The wind velocities listed in Table 3.2F were observed above a crop canopy near Davis, California, in July 1962.

1. Compute v_* (cm/s) and y_0 (cm) for each time.
2. Based on the values of v_* and y_0, what type of crop might be present?

3.2G. Viscous Sublayer at Seafloor

Caldwell and Chris[46] made measurements of currents near the seafloor at a site on the Oregon continental shelf in 199 m of water. The surface sediments are silty sand with a mean grain diameter of 0.042 mm. No bedforms were present to influence the flow significantly. Flow measurements with a heated thermistor, taken above the sediment, appear in Table 3.2G.

Table 3.2F

y (cm)	$\bar{v}_x$ (cm/s) at 1400 h	$\bar{v}_x$ (cm/s) at 1430 h
25	176	208
50	238	279
100	291	343
200	335	397

Table 3.2G

y (cm)	$\bar{v}_x$ (cm/s)	y (cm)	$\bar{v}_x$ (cm/s)
0.15	1.3	3.0	8.5
0.21	2.3	5.0	8.6
0.32	3.7	6.0	9.4
0.53	5.5	10.0	9.8
1.0	7.5	20.0	10.3
1.5	7.7	30.0	10.6
2.0	8.2		

1. Determine the thickness of the laminar sublayer, δ_v in mm.
2. Compute the shear stress exerted on the seafloor, τ_0 in g/cm·s^2, based on the laminar sublayer data.
3. Recompute τ_0 based on the turbulent profile data.
4. Compute y_0 in mm based on the turbulent profile data.

3.2H. Mixing-Length-Theory Expressions for Heat and Mass Transfer

Develop the mixing length theory expressions for heat and mass transfer by the following procedure.

1. *Heat transfer*. Beginning with Eq. 3.2-28*c*, parallel the Prandtl mixing-length theory for momentum transfer and show that the turbulent energy flux expression is

$$\bar{q}_y^{(t)} = \rho \hat{C}_p l^2 \left| \frac{d\bar{v}_x}{dy} \right| \frac{d\bar{T}}{dy} \tag{3.2H-1}$$

in which l is the Prandtl mixing length. (*Hint:* Draw a time-smoothed temperature profile to accompany the time-smoothed velocity shown in Fig. 3.2-6.) Implicit in this development is $v^{(t)}/\alpha^{(t)} = 1$. Prandtl recommends setting $l = \kappa_1 y$, in which y is the distance from the surface. Using $l = \kappa_1(y + y_0)$, where y_0 is a small roughness length, and Eq. 3.2-42, show that the turbulent thermal diffusivity is

$$\alpha^{(t)} = \kappa_1(y + y_0)v_* \tag{3.2H-2}$$

where $\alpha^{(t)}$, the turbulent thermal diffusivity, is defined by

$$\bar{q}_y^{(t)} \equiv -\alpha^{(t)} \frac{d}{dy}(\rho \hat{C}_p T) \tag{3.2H-3}$$

2. *Mass transfer*. Begining with Eq. 3.2-29c, parallel the Prandtl mixing length theory for momentum transfer, and show that the turbulent mass flux expression is

$$J_{Ay}^{(t)} = l^2 \left| \frac{d\bar{v}_x}{dy} \right| \frac{d\bar{c}_{A1}}{dy} \tag{3.2H-4}$$

in which l is the Prandtl mixing length. (*Hint:* Draw a time-smoothed concentration profile to accompany the time-smoothed velocity shown in Fig. 3.2-6.) Prandtl recommends setting $l = \kappa_1 y$, in which y is the distance from the surface. Using $l = \kappa_1(y + y_0)$, where y_0 is a small roughness length and Eq. 3.2-42, show that the turbulent diffusivity is

$$\mathscr{D}_{A1}^{(t)} = \kappa_1(y + y_0)v_* \tag{3.2H-5}$$

where $\mathscr{D}_{A1}^{(t)}$ is defined similar to Eq. 3.1-6. The turbulent diffusivities given by Eqs. 3.2H-2 and 3.2H-5, unlike the molecular counterparts α and $\mathscr{D}_{A1}^{(l)}$, which are properties of the fluids, are functions of external quantities y and v_*. Bird et al.[3] note that because of the analogy between heat and mass transfer, it is often assumed that $\mathscr{D}_{Aj}^{(t)}/\alpha_j^{(t)} = 1$.

3.3. OTHER TRANSPORT TOPICS

In this section the topic of heat transfer is extended and the analogy theories are presented.

Some Fundamentals of Heat Transfer

Heat transfer and temperature play a major role in the movement of chemicals near environmental interfaces, and the basic topics presented in Section 3.2, including the equation of energy and Fourier's law, need to be extended. A quantitative description of heat transfer processes in the environment requires several types of rate equations and several corresponding transfer coefficients.

Fourier's Law and Thermal Conductivity. Fourier observed that the heat flux by conduction is proportional to the temperature gradient, or, to put it somewhat pictorially, "heat slides downhill on the temperature versus distance graph." Fourier's law is given in Eq. 3.2-7. The thermal conductivity k, defined by this equation, applies to fluids at rest and to solids or fluids in laminar flow where the heat flux is by molecular movements. Thermal conductivity is a fluid property reflecting only molecular aspects, and for this reason Eq. 3.2-7 is of limited utility in fluids undergoing turbulent flow. However, heat transfer in soil is one application where the equation can be used.

The thermal diffusivity α, defined by Eq. 3.2-8, is a more useful quantity than the thermal conductivity for some purposes. Since density ρ and heat capacity $\hat{C}_p$ of the material are properties, thermal diffusivity is a property of the medium through which heat is moving. The thermal diffusivity of water at 20°C is 1.42E-3 cm^2/s.

In the fluid regions near environmental interfaces molecular movements play a minor role. The dominant mechanism by which heat is transferred in a flowing system (i.e., stream, lake, air, etc.) is by random bulk movements, known as *turbulent energy flux*. By analogy with Fourier's law of heat conduction, one may write

$$q_y = -(k^{(l)} + k^{(t)}) \frac{dT}{dy} \tag{3.3-1}$$

the quantity, $k^{(t)}$ being called the *turbulent coefficient of thermal conductivity* or *eddy conductivity*. It is not a physical property of the fluid like $k^{(l)}$, but depends on position, direction, and the nature of the turbulent flow. An *eddy thermal diffusivity* can be defined as

$$\alpha^{(t)} \equiv \frac{k^{(t)}}{\rho \hat{C}_p} \tag{3.3-2}$$

The eddy thermal diffusivity has the same dimensions as the eddy kinematic viscosity $\nu^{(t)}$ (Eq. 3.2-33) and eddy diffusion coefficient $\mathscr{D}_A^{(t)}$ (Eq. 3.1-6). These quantities may be compared for turbulent momentum and energy transport. The ratio $\nu^{(t)}/\alpha^{(t)}$ is on the order of unity. Values in the turbulence literature vary from 0.5 to 1.0.

Observations of heat conduction in water have yielded much information on the magnitude and character of this eddy thermal diffusivity. There is a wide spread in measured values, from as low as 4E-2 to as high as 200 cm^2/s. Table 3.3-1 summarizes some turbulent diffusivity observations from inland water bodies and the seas. Thermal gradients are very pronounced in the vertical direction in water bodies. Eddy thermal diffusivities reported for the horizontal direction are typically an order of magnitude greater than those in the vertical direction. Koh and Fan[20] have performed a literature review and present a summary of $\alpha^{(t)}$ values in equation and graphical form. The review closes with the following remarks (paraphrased): In general $\alpha_y^{(t)}$ has its maximum value in the surface layer. In the ocean $\alpha_y^{(t)}$ at the surface varies between 10 and 200 cm^2/s; in coastal areas between 10 and 50 cm^2/s; and in lakes about 10 cm^2/s. Below the surface mixed layer (or epilimnion) $\alpha_y^{(t)}$ drops to its minimum in the thermocline (on the order of 1 cm^2/s in open ocean); in lakes $\alpha_y^{(t)}$ may drop as low as 0.05 cm^2/s. Below the thermocline $\alpha_y^{(t)}$ may increase again.

Equation 3.3-1 is difficult to use. In general, the reasons for the difficulty are identical to those given for turbulent mass diffusivities. The thermal diffusivity

Table 3.3-1. Effective Vertical Thermal Diffusion Coefficients $\alpha_2^{(t)}$ in Water Bodies

	$\alpha_2^{(t)}$ (cm^2/s)		
Water Bodies	Surface	Thermocline	Depth
Lakes and reservoirs[a]			
Fontana Reservoir	0.81	0.14	1.15 at 29 m
Hungry Horse Reservoir	3.7	0.12	1.7 at 54 m
Lake Tahoe	9.2	0.1	1.2 at 90 m
Castle Lake	2.6	0.02	1.5 at 22 m
	$\alpha_2^{(t)}$ (cm^2/s)	Depth of Layer (m)	
Oceans and seas[b]			
Philippine Trench	2.0–3.2	5000–9788	
Mediterranean	42	0–28	
California Current	30–40	0–200	
Caribbean Sea	2.8	500–700	
Water in laminar flow	$\alpha_2^{(l)} = 0.00143$ cm^2/s		

[a]From Ref. 21.
[b]From Ref. 22.

is highly variable near environmental interfaces, and the thermal gradient is almost never known. For these reasons and others, Eq. 3.3-1 is not generally used for quantifying heat flux rates near or at environmental interfaces.

Sensible, Latent, and Radiant Energy Transfer at Interfaces. The three fundamental mechanisms of energy transfer at interfaces and their rate expressions are presented here. Topics concerned with specific interfaces are presented in those chapters.

Sensible Heat Transfer. Sensible heat is thermal energy given up or absorbed by a body on being cooled or heated as the result of the body's ability to hold thermal energy; this excludes latent heats of fusion or vaporization. An alternative form of Eq. 3.2-7 for expressing the sensible heat flux across an interface is by use of Newton's law of cooling:

$$q_{s0} = \dot{h}(T_0 - T_b) \tag{3.3-3}$$

Here q_{s0} is the sensible heat transfer rate per unit area of interface, $\dot{h}$ the local value of the heat transfer coefficient, T_0 the surface or interface temperature, and T_b the bulk or "cup mixing" fluid temperature. Newton's law of cooling is not really a law but, rather, a defining equation for h, which is called the *heat transfer coefficient*. Just as with mass transfer at phase boundaries (Eq. 3.1-8), the equivalent heat transfer relation finds major utility at these boundaries where the turbulent thermal conductivity and the thermal gradient are rarely known with certainty.

For surfaces of finite area A, one can parallel Eq. 3.3-3 and define an average heat transfer coefficient

$$Q_{s0} = hA(T_0 - T_b) \tag{3.3-4}$$

Here Q_{s0} is the sensible heat transfer rate and h is the average coefficient. Typical SI units for Q are joule per second (J/s) and for q_{s0} are joule per second square meter ($J/s \cdot m^2$). Just as for mass transfer coefficients, $\acute{h}$ and h are scalars. Q and q are vectors (i.e., they have magnitude and direction) and take the direction from the sign of $T_0 - T_b$. Heat transfer into a body is a positive number whereas heat transfer from a body to its surroundings is a negative number.

See Sections 4.3, 5.4 and 6.4 for specific heat transfer applications at the air–water, sediment–water and air–soil interfaces, respectively.

Latent Heat Transfer. *Latent heat* is defined as the amount of heat absorbed or evolved by 1 mol, or a unit mass, of a substance during a change of state (such as fusion, sublimation, or vaporization) at constant temperature and pressure. The latent heat transfer rate is important at air–water and air–soil interfaces. At these planes water is present and changes phase, either from a liquid to a gas (vaporization) or from a gas to a liquid (condensation). As one may expect, this relationship of energy movement rate is related to the water movement rate:

$$q_{10} = n_{A0}\lambda_A \tag{3.3-5}$$

Here q_{10} is the rate of latent heat transfer at the interface in units of $J/m^2 \cdot s$, n_{A0} is the mass flux rate of water ($A \equiv H_2O$) moving across the interface in units of $g/m^2 \cdot s$, and λ_A is the latent heat of vaporization of water in units of J/g. The latent heat of vaporization of water at 24°C is 2.44E + 6 J/kg. A relationship for computing the molar flux rate has been presented in Eq. 3.1-14; multiplying it by molecular weight A yields the mass flux rate n_{A0}.

Even though other chemicals may be moving across the interface with accompanying heat of vaporization or condensation effects, their contribution to the latent heat transfer is negligible because of their dilute nature. It is therefore necessary to account for the water only when quantifying latent heat effects across the air–water or the air–soil interface.

Radiant Energy Transfer. *Radiant energy* is defined as energy transmitted by electromagnetic waves through space or some medium. This energy, or electromagnetic radiation, becomes heat only after interaction with matter upon which it is absorbed. As a solid body exists at a temperature above absolute zero, some constituent molecules and atoms at the surface are raised to *excited states*. There is a tendency for the atoms or molecules to return spontaneously to lower energy states. When this occurs, energy is emitted in the form of electromagnetic radiation. Because the radiation emitted may result from changes in the electronic, vibrational, and rotational states of the atoms and molecules, the radiation is distributed over a range of wavelengths. Thermal radiation involves wavelengths primarily in the range 0.1E-6 to 10E-6 m. The preceding is a brief description of the emission of radiant energy. The reverse process, which is known as *absorption*, occurs when the addition of radiant energy to a molecule or atomic system causes the system to go from a low to a high energy state.

Radiation impinging on the surface of an opaque solid is either absorbed or reflected. The fraction of the incident radiation that is absorbed is called the *absorptivity* and is given the symbol a. *Kirchhoff's law* states that at a given temperature, the emissivity e and absorptivity a of any solid surface are the same when the radiation is in equilibrium with the solid surface. This allows us to conclude that

$$e = a \tag{3.3-6}$$

Emissivities are usually measured and reported. Typical values are: rough, unglazed silica brick, 0.80; candle soot, 0.952; planed oak, 0.895; and water, 0.95 to 0.963.

It has been shown experimentally that the total energy emitted from a real surface is

$$q_{e0} = e\sigma T^4 \tag{3.3-7}$$

in which T is the absolute temperature. The Stefan–Boltzmann constant σ has been found to have a value 5.67E-12 $\mathrm{J/s \cdot cm^2 \cdot K^4}$. The emissivity e must be evaluated at temperature T (absolute).

A reasonably accurate quantitative treatment of radiation between nonblack surfaces can be treated as follows. Consider the air–water interface of area A, in which the air is surface 1 and the water is surface 2. Since over a limited region A is flat, it intercepts none of its own rays. The rate of energy emission from surface 1 to surface 2 is given by

$$Q_{\vec{12}} = e_1 A\sigma T_1^4 \tag{3.3-8}$$

and the rate of energy absorption from surface 2 by surface 1 is

$$Q_{\vec{21}} = a_1 A\sigma T_2^4 \tag{3.3-9}$$

The net radiation rate from 1 to 2 is therefore

$$Q_{12} = \sigma A(e_1 T_1^4 - a_1 T_2^4) \tag{3.3-10}$$

where e_1 is the value of emissivity of surface 1 at T_1. The absorptivity a_1 is usually estimated as the value of e of surface 1 at T_2.

Thermodiffusional Transport. There is a tendency for molecules to diffuse under the influence of a temperature gradient. The phenomenon is known as the *Soret effect*. The flux contribution for an aqueous solution is

$$j_A = -\mathscr{D}_{A2}^{T} \rho_{A2} \frac{dT}{dz} \tag{3.3-11}$$

where $\mathscr{D}_{A2}^{T}$ is the thermal diffusion coefficient in dimensions of L^2/tT. The ratio of this coefficient to the molecular diffusivity is $s_{A2} \equiv \mathscr{D}_{A2}^{T}/\mathscr{D}_{A2}$, the Soret coefficient, with dimensions T^{-1}. This effect is an additional mechanism for mass transport and is handled quantitatively by adding it algebraically with Fick's law so that the total flux is the sum of the two.

Lerman[32] reports that the value of the Soret coefficient for common electrolytes and constituents in natural waters is between 0.001 to 0.02 K^{-1} and gives experimental data. He notes that the contribution is small in most cases where molecular diffusion fluxes are pronounced. However, in the vicinity of a heat source in geologic media, this mechanism can result in large-scale and relatively rapid fluxes of aqueous components from the hot zone toward the cool zone.[33]

Analogy Theories of Momentum, Heat, and Mass Transfer

The analogy theories extend the experimental data collected on one type of environmental transport parameter for use in others. For example, data on heat transfer coefficients between air and a soil surface can be extended by the use of the analogy theories to calculate mass transfer coefficients of chemical A between air and the soil surface. The first hint of analogy between the three transport phenomena was pointed out in Section 3.2 following the presentation of the equations of change (see specifically Eqs. 3.2-13, 3.2-14, and 3.2-15). In this section the analogy theories are presented and their utility in interphase chemodynamic problems is illustrated.

First, the Prandtl mixing-length theory is extended to heat and mass transfer to develop turbulent temperature and concentration profiles in that region above the laminar sublayer. Next, the Reynolds analogy is developed based on a proportionality. The von Kármán analogy theory, which considers not only the turbulent zone but the buffer layer and the laminar sublayer, is presented and discussed. Finally, utility and limitations of the analogy theories are considered.

Similar Turbulent Profiles. According to Prandtl's theory, momentum and energy are transferred in turbulent flow by the same mechanism. By employing an approach similar to that developed for momentum transfer, the turbulent energy flux expression (Eq. 3.2-28*c*) can be related to the time-smoothed velocity and temperature gradients in the fluid above a solid interface (see Eq. 3.2H-1 of Problem 3.2H, which is analogous to Eq. 3.2-32 at this point).

In the surface boundary layer, at steady state, it is assumed that the energy flux is independent of height, which then gives

$$\bar{q}_{\bar{y}}^{(t)} = q_0 \tag{3.3-12}$$

Recall that a similar simplification (i.e., $\tau_{yx}^{(t)} = \tau_0$) was made in the development of velocity profiles. If now Eqs. 3.2H-1 and 3.2-32 are divided one by the other, we get

$$\frac{q_0}{\tau_0} = \frac{\hat{C}_p \, d\bar{T}}{d\bar{v}_x} \tag{3.3-13}$$

which may be integrated from a point $y = \delta$ beyond the edge of the sublayers where $\bar{T} = \bar{T}|_\delta$ and $\bar{v}_x = \bar{v}_x|_\delta$ to yield

$$\frac{\hat{C}_p(\bar{T} - \bar{T}|_\delta)}{q_0} = \frac{\bar{v}_x - \bar{v}_x|_\delta}{\tau_0} \tag{3.3-14}$$

This result simply states that within the turbulent zone above the laminar sublayers the velocity and temperature profiles are similar; this agrees roughly with field observations.

Now Eq. 3.2-38 can be arranged in the form

$$\frac{\bar{v}_x - \bar{v}_x|_\delta}{\tau_0} = \frac{1}{\kappa_1 \rho v_*} \ln \frac{y}{\delta} \tag{3.3-15}$$

with the aid of the friction velocity. Clearly, from Eq. 3.3-14,

$$\bar{T} - \bar{T}|_\delta = \frac{-q_0}{\kappa_1 \rho \hat{C}_p v_*} \ln \frac{y}{\delta} \tag{3.3-16}$$

Hence one deduces that the temperature profile in the turbulent portion of the surface boundary layer will also be a logarithmic function.

By following an analogous development with time-smoothed velocity and concentration profiles (see part 2 of Problem 3.2H), we obtain the turbulent concentration profile

$$\bar{c}_A - \bar{c}_A|_\delta = \frac{-N_{A0}}{\kappa_1 v_*} \ln \frac{y}{\delta} \tag{3.3-17}$$

The height δ appearing in Eqs. 3.3-15 to 3.3-17 is well within the turbulent portion of the boundary layer. It is convenient to extrapolate the quantities to the surface; in doing so δ for the velocity profile becomes the roughness height y_0, where the velocity is zero. In the case of the temperature profile δ becomes y_T and $\bar{T}(y_T)$ is the extrapolated surface temperature. Similarly, for the concentration profile, δ becomes y_c and $\bar{c}_A(y_c)$ is the extrapolated surface concentration. The fluxes in Eqs. 3.3-16 and 3.3-17 are positive for transport into the boundary layer. In effect, y_0, y_T and y_c are chosen so that the profiles fit the data and any theoretical interpretation should be done with caution. In addition they should not be confused with the diffusive sublayers (see Problem 3.3C).

The three equations immediately above are valid for mechanical turbulence only. Recall that the Prandtl mixing-length theory used $v_x' = -l\, d\bar{v}_x/dy$ and similarly $\rho_{A1}' = l\, d\bar{\rho}_{A1}/dy$. The latter is not exactly correct because it can be argued that mixing lengths for mass and momentum are not the same numerical values. A correction factor must be applied. This and other important correction factors are included in Problem 4.1N and Eq. 6.2-16. If thermal turbulence is present, other modifications are necessary. See the appropriate sections in Chapter 6 on transport in the air above soil surfaces.

Reynolds Analogy. The following development of the Reynolds analogy is adopted from Bennett and Myers.[23] The proportionality of heat and momentum transfer can be stated in terms of four quantities, which we define with reference to a fluid at a bulk temperature T_b flowing past and losing heat to an interface, which is at temperature T_0. The four quantities are:

(a) The heat flux from the fluid to the interface, $h(T_b - T_0)$ (J/m$^2\cdot$s).

(b) The momentum flux at the wall, τ_0, (kg$\cdot$m/m$^2\cdot$s^2).

(c) The rate at which energy available for transfer as heat is transported parallel to the interface, $n\hat{C}_p(T_b - T_0)$ (J/m$^2\cdot$s), where n is the mass flow rate in kg/s$\cdot$m^2.

(d) The rate at which momentum is transported parallel to the interface nv_b, (kg$\cdot$m/m$^2\cdot$s^2).

It is then postulated that these four quantities can be described by the following proportionality: a/c = b/d. Substituting the respective quantities yields

$$\frac{h}{\hat{C}_p} = \frac{\tau_0}{v_b} \tag{3.3-18}$$

If the variables D, a characteristic length, k, thermal conductivity, and μ, fluid viscosity, are introduced to preserve the equality and create dimensionless groups, Eq. 3.3-18 becomes

$$\frac{\text{Nu}}{\text{Pr}} = \frac{\text{Re}\,\tau_0}{\rho v_b^2} \tag{3.3-19}$$

where Nu is the Nusselt number for heat transfer and Pr is the Prandtl number. The definitions of both of these dimensionless groups appear in Table 3.3-2. Equations 3.3-18 and 3.3-19 are a statement of the Reynolds analogy between heat and momentum transfer. A heat transfer coefficient adjacent to a surface, h, can be obtained directly from the fluid dynamic shear stress at the surface, τ_0. It gives best results when applied to gases, for which Pr is approximately 1. As an application, consider the problem of determining the heat flux from warm air to a cool soil surface. If temperature T_b and v_b are measured at a fixed height and τ_0 is estimated from velocity profile data (Eq. 3.2-42), the Reynolds analogy yields a value of the heat transfer coefficient. The heat flux can be estimated by use of Eq. 3.3-4 if the soil surface temperature T_0 is known.

Although Reynolds was concerned only with the analogy between heat and momentum transfer, the equations referred to as the Reynolds analogy can readily be extended to cover mass transfer. A similar development of mass and momentum transfer at an air–soil environmental interface results in

$$^3k_{A1} M_1 = \frac{\tau_0}{v_b} \tag{3.3-20}$$

where M_1 is the molecular weight of the air–chemical mixture. Introducing c_1, fluid molar density, $\mathscr{D}_{A1}$, molecular diffusivity, D, a characteristic length, μ_1, viscosity, and creating dimensionless groups yields

$$\frac{\text{Nu}_{A1}}{\text{Sc}_{A1}} = \frac{\text{Re}\,\tau_0}{\rho v_b^2} \tag{3.3-21}$$

where Nu_{A1} is the Nusselt number for mass transfer and Sc_{A1} is the Schmidt number. As with the heat transfer analogy result, Eqs. 3.3-20 and 3.3-21 are limited to the pure turbulent regions of the surface boundary layer except for cases where the Schmidt number is unity (Sc = 1.0). For fluid with Pr = Sc = 1, the Reynolds analogy applies not only to the turbulent region but also to the laminar sublayer.

Other Analogies. The restriction Pr = Sc = 1 limits the general utility of the Reynolds analogy. This restriction is critical in the case of environmental chemodynamic applications since the Prandtl number for water is 7.7 at 60°F and the Schmidt number for most chemicals in water ranges from 300 to 2700!

It is obvious that considering only the transport processes in the turbulent region and neglecting the laminar sublayer has its faults. Major improvements on the Reynolds analogy have been made, resulting in the Prandtl–Taylor equation, which was extended by von Kármán. Von Kármán considered the resistance of heat transfer to be composed of three parts: the laminar sublayer, the buffer zone, and the turbulent region. The von Kármán equation for mass and momentum transfer is

$$\frac{\mathrm{Nu}_{A1}}{\mathrm{Sc}_{A1}} = \frac{\tau_0}{\rho v_b^2} \frac{\mathrm{Re}}{1 + 5\sqrt{\tau_0/\rho v_b^2}\{\mathrm{Sc}_{A1} - 1 + \ln[(1 + 5\mathrm{Sc}_{A1})/6]\}} \tag{3.3-22}$$

The denominator on the right-hand side is seen to be a correction to Eq. 3.3-21, the Reynolds analogy. The von Kármán equation can also be expressed in terms of heat and momentum transfer. Both forms of the von Kármán equation reduce to the Reynolds analogy when Pr = Sc = 1. Equation 3.3-22 should be used to correct the deficiencies of the Reynolds analogy.

One final analogy worth discussing is the empirical *j*-factor relation proposed by Chilton and Colburn.[24] They found that a certain group of terms, called the *j*-factor, made possible excellent correlation of the data for any two transport processes near flat surfaces in the absence of form friction. The mass transfer *j*-factor is defined

$$j_D \equiv \frac{\mathrm{Nu}_{A1}}{\mathrm{Re} \cdot \mathrm{Sc}_{A1}^{1/3}} \tag{3.3-23}$$

so that the Chilton–Colburn *j*-factor analogy equation for mass transfer is

$$j_D = \frac{\tau_0}{\rho v_b^2} \tag{3.3-24}$$

The heat-transfer *j*-factor definition is

$$j_H \equiv \frac{\mathrm{Nu}}{\mathrm{Re} \cdot \mathrm{Pr}^{1/3}} \tag{3.3-25}$$

Put in the algebraic form of the two previous transport analogy theory results, Eq. 3.3-24 becomes

$$\frac{\mathrm{Nu}_{A1}}{\mathrm{Sc}_{A1}} = \frac{\tau_0}{\rho v_b^2} \frac{\mathrm{Re}}{\mathrm{Sc}_{A1}^{2/3}} \tag{3.3-26}$$

This final equation must be compared to Eqs. 3.3-21 and 3.3-22 for viewing all the analogous theory results on a common basis. They differ in the functional form of the denomination on the right-hand side of the equality sign. All forms reduce to the Reynolds analogy for Sc = 1.

The j-factor relation has been shown experimentally to have considerable merit in correlating heat and mass transfer data so that the cross analogy (i.e., $j_D = j_H$) is a particularly good one. It agrees closely with the predictions of boundary layer theory for the flat plate when Pr and Sc exceed 0.5 and appears to be fairly good for turbulent flow. Both j_H and j_D are frequently correlated as a function of Re, geometry, and boundary conditions. Due to its strong data base and simplicity of form, the Chilton–Colburn analogy is recommended.

Limitations of Using Analogies. The importance of analogies is that they permit one to obtain mass transfer correlations from heat transfer correlations for equivalent boundary conditions merely by substituting Nu_{Aj} for Nu and Sc_{Aj} for Pr. The same can be done for any flow geometry and for laminar or turbulent flow. Note that to obtain a valid analogy, one has to assume (1) constant physical properties, (2) a small rate of mass transfer, (3) no chemical reactions in the fluid, (4) no viscous dissipation, (5) no emission or absorption of radiant energy, and (6) no pressure diffusion, thermal diffusion, or forced diffusion. All of these assumptions apply for most chemodynamic applications. A summary of analogous quantities for heat and mass transfer is given in Table 3.3-2.

Example 3.3-1. Converting Mass Transfer Coefficient to Heat Transfer Coefficient. The liquid-phase mass transfer coefficient for phosphorus ($A \equiv \mathrm{P}$) at the bottom of Lake Washington has been estimated to be ${}^3k_{A2}/c_2 = 36$ m/yr. Estimate the heat transfer coefficient at the lake bottom in $\mathrm{J/m^2 \cdot s \cdot K}$. Assume that the diffusivity for phosphorus is 1.0E-5 cm²/s and the water is at 20°C.

SOLUTION In obtaining a heat transfer coefficient from a mass transfer coefficient, it is important to realize that a major message of the analogy theory is

$$j_H = j_D \tag{E3.3-1}$$

Substitution for the definitions of the j-factors:

$$\frac{\mathrm{Nu}}{\mathrm{Re} \cdot \mathrm{Pr}^{1/3}} = \frac{\mathrm{Nu}_{A2}}{\mathrm{Re} \cdot \mathrm{Sc}_{A2}^{1/3}}$$

At these conditions $\mathrm{Pr} = 7.0$ and $\mathrm{Sc}_{A2} = 890$. Substituting the definitions of the respective Nusselt numbers, we obtain

$$\frac{hD}{k_2(7.0)^{1/3}} = \frac{{}^3k_{A2}D}{c_2\mathscr{D}_{A2}(890)^{1/3}}$$

Table 3.3-2. Heat and Mass Transfer Analogies at Low Mass Transfer Rates

Group or Parameter	Heat Transfer Quantities	Mass Transfer Quantities
Profiles	T	x_A
Diffusivity	$\alpha \equiv \dfrac{k}{\rho \hat{C}_p}$	$\mathscr{D}_{Aj}$
Effect of profiles on density	$\beta \equiv -\dfrac{1}{\rho}\left(\dfrac{\partial \rho}{\partial T}\right)_{p,xA}$	$\zeta_A = \dfrac{-1}{\rho}\left(\dfrac{\partial \rho}{\partial x_A}\right)_{p,T}$
Transfer coefficient	$h = \dfrac{q}{A\,\Delta T}$	${}^{j}k_{A1} = \dfrac{w_A}{A\,\Delta x_A}$
Groups that are the same	$\mathrm{Re} \equiv \dfrac{Dv\rho}{\mu}$	$\mathrm{Re} \equiv \dfrac{Dv\rho}{\mu}$
	$\dfrac{L}{D}$	$\dfrac{L}{D}$
Groups that are different	$\mathrm{Nu} \equiv \dfrac{hD}{k}$	$\mathrm{Sh}_{Aj} \equiv \dfrac{k_{Aj}D}{c_j \mathscr{D}_{Aj}}$
	$\mathrm{Pr} \equiv \dfrac{\hat{C}_\rho \mu}{k} = \dfrac{\nu}{\alpha}$	$\mathrm{Sc}_{Aj} \equiv \dfrac{\mu}{\rho \mathscr{D}_{Aj}} = \dfrac{\nu}{\mathscr{D}_{Aj}}$
	$\mathrm{Gr} \equiv \dfrac{D^3 \rho^2 g \beta\,\Delta T}{\mu^2}$	$\mathrm{Gr}_{Aj} \equiv \dfrac{D^3 \rho^2 g \zeta_A\,\Delta x_A}{\mu^2}$

Source: Ref. 10, p. 646.

Solving for h using $k_2 = 0.59\ \mathrm{J/m \cdot s \cdot K \cdot}$ yields

$$h = \left(\frac{7.0}{890}\right)^{1/3}\left(\frac{36\,\mathrm{m}}{\mathrm{yr}}\right)\left(\frac{0.59\,\mathrm{J}}{\mathrm{m \cdot s \cdot K}}\right)\left(\frac{\mathrm{s}}{1\mathrm{E\text{-}5\,cm^2}}\right)\left(\frac{1\mathrm{E4\,cm^2}}{\mathrm{m^2}}\right)\left(\frac{\mathrm{yr}}{3.51\mathrm{E\text{-}7\,s}}\right)$$
$$= 120\ \mathrm{J/m^2 \cdot s \cdot K}$$

Particles and Porous Media

The transport principles in previous sections have been concerned primarily with chemical species or thermal energy within a single phase. The emphasis was on the fluid phases. Here the attention shifts to the solid phase. *Particles,* as the term is used here, includes suspensions of solid matter in water, fine liquid and solid matter in air (i.e., aerosols), hydrometers and dust. Porous media include saturated and unsaturated (with respect to liquid water) surface soils, subsoils, deep geologic formations, and aquatic bed-sediment.

Interest in processes involving particles and porous media assumes that this solid phase is important in chemical transport to some degree, along with one or more contiguous fluid phases. In Chapter 2 material was presented on the role that solid surfaces play in the equilibrium partitioning of chemicals; in addition, these surfaces provide sites for molecules to react or undergo transformations and for ionic species to be exchanged. Here the concern is with some general processes involving the behavior of particles and porous media as they exist in and with the natural fluids. Aspects of chemical transport are not addressed. The importance of this subject deserves a chapter of its own, but space limitations require that specific details concerning chemicals be presented in the appropriate chapters concerned with interphase or intraphase transport.

Particles can exist within the pore spaces of porous media; organic colloidal particles in bed–sediment porewaters is an example. Also, the internal structure of most particles is porous. For these reasons transport related aspects involving porous media are presented first.

Porous Media. The following examples of chemical movement within soil and sediment highlight the need for information on these natural porous media. Soil layers provide the primary barrier to chemical vapors moving from buried waste to the atmosphere. Natural clay liners may be the primary barriers to the transport of chemicals in leachate generated as water enters and perculates through the buried waste. Further movement of contaminants is aided by groundwater flow in both the saturated and unsaturated zones. Chemical desorption and movement from bed sediment maintains an ongoing contaminant source to the overlying water body in addition to the groundwater. Porous media of chemodynamic interest must contain voids that are at least partially interconnected, so that the material is permeable to air or water. Typical pore sizes may vary in diameter from 1 μm. to a few millimeters or more.

The continuity equation of A can be used for the porous media, but the definition of some terms must be modified. The concentration, ρ_{Aj}, in Eq. 3.0-3 represents an average value over a region within the homogeneous medium involving numerous individual pores. This region is large compared to individual pore diameters but small compared to the size of the contaminant plume. For the fluid phase $\bar{v}_j$ is the average pore or interstitial velocity (defined as $Q/A \cdot \varepsilon$) and $\bar{j}_{Aj}$ is expressed as $-\mathscr{D}_{A3}^{(t)} \cdot \nabla\rho_{Aj}$; where $\mathscr{D}_{A3}^{(t)}$ is the coefficient of (hydrodynamic) dispersion, a second-order tensor that combines the mixing effects due to molecular diffusion and advection effects. The ad hoc generalization of the coefficient of dispersion allows for its known nonisotropic behavior in porous media. Dullien[34] and Bear[35] should be consulted for further information.

One idealized model for porous media is a bundle of capillary tubes with laminar flow within the tubes. This simple system provides the basis for an understanding of the dispersion process as a result of fluid flow in porous media. The Taylor–Aris theory for the effective axial coefficient of hydrodynamic dispersion in capillary tubes is

$$\mathscr{D}_{A3x}^{(t)} \equiv \mathscr{D}_{Aj}^{(l)} + \frac{r^2 \bar{v}_x^2}{48 \mathscr{D}_{Aj}^{(l)}} \tag{3.3-27}$$

where r is the tube radius and $\bar{v}_x$ is the mean velocity in the tube. This theoretical result is valid for large times and effectively combines the mixing effects due to molecular diffusion and advection. Expressed as a dispersion coefficient, axial mixing increases rapidly with tube diameter and velocity, at least when the second term dominates.

Dullien[34] notes that an additional important difference between dispersion in a uniform capillary tube and a porous medium is introduced by the nonuniformity in the capillaries. This gives rise to dispersion by another mechanism, different velocities in pores in parallel connection. Due to these and other geometric factors in addition to possible dynamic factors, generalized correlations of $\mathscr{D}_{A3x}^{(t)}$ with average pore or interstitial velocity are used with field data. This is for the sake of simplicity, since rigorous justification is lacking. The general form is

$$\mathscr{D}_{A3x}^{(t)} = a\bar{v}_x^n \tag{3.3-28}$$

where a, the dispersivity, is one value for the longitudinal direction and another for the transverse; n is also obtained by curve-fitting field or laboratory data.

Knudsen Diffusion. If the pores are very small, the molecules within will collide most often with the walls and not with other molecules, because the distance between molecular collisions is greater than the pore diameter. For liquids the mean free path is a few angstroms (1 Å = 1E-10 m) and this is the case. For gases or vapors within pores, this may not be the case. Based on the simple kinetic theory of gases, Smith[36] gives the following equation for evaluating the Knudsen diffusivity:

$$\mathscr{D}_{A1}^{k} = 9700 d \left(\frac{T}{M_A} \right)^{1/2} \tag{3.3-29}$$

where $\mathscr{D}_{A1}^{k}$ is in cm^2/s, d is diameter in cm, and T is in K. If the value of $\mathscr{D}_{A1}^{k}$ is less than $\mathscr{D}_{A1}$ computed from Eq. 3.1-26; it should be used for the molecular diffusivity within the pores. Unlike $\mathscr{D}_{A1}$, the Knudsen diffusivity is independent of pressure.

Particles. The mass of chemical species sorbed onto solid particles suspended in water constitutes a significant fraction of these contaminants in some streams and lakes. Dust from a contaminated surface soil suspended into air by the winds may be a major off-site transport mode. Similarly, fine particles deposited onto water surfaces may carry nonnegligible quantities of chemical into aquatic systems. Colloidal particles exist within the water filled pores in soils and sediment and may contribute significantly to chemical transport. The airborne and waterborne particles of concern are typically very small. In the case of aerosols a typical range is 0.03 to 10 μm radius. Aquatic colloids are 0.4 μm in diameter or smaller. Larger aquatic suspended matter ranges up to 1 mm in diameter.

Due to their large sizes in comparison to the molecules of air or water, particles present some unique challenges in formulating flux rate expressions. Of course, the continuity equation, Eq. 3.0-3, still applies to particles; however, the forms of $\mathbf{j}_{Aj}$ and the z-expression, denoted in Eq. 3.0-4, are complicated beyond those for molecular species. Very small particles may behave as large molecules, but the larger ones enjoy other mechanisms of transport at natural interfaces. In addition, the nature of the interface exaggerates the complexity, and therefore the following material is presented by interface type.

Air–water. The primary mechanism by which aerosols are produced in the marine environment is through bursting of bubbles at the sea surface. Water, salts, organic matter, and a net electric charge are transferred to the air. The production of sea-salt aerosols is due to the agitation of the sea surface by the wind force, and in this regard its formation is similar to that of dust aerosols. Bubbles are most numerous in the white caps associated with breaking waves, where bulk air is entrained into the surface waters. As air bubbles reach the interface they burst. The bubble-surface collapse process ejects a jet of water that produces 1 to 10 drops above the interface. Additional drops are formed from the bursting water film that covers the bubble. The distinction between jet and film drops is of some importance, because surface-active materials are associated with the latter.

By the process noted above, sea-salt particles enter the atmosphere. Here they can absorb or otherwise accommodate chemical constituents from the air. The original particles, coagulated particles, transformed particles, those that have become condensation nuclei or products of atmospheric gas-phase reactions, and land-originating particles can reenter the sea. (The information above was extracted primarily from Warneck[37] and Pruppacher and Klett.[38]; the following, which is concerned with the reentry of aerosols into the sea, is abstracted from Slinn.[39])

The governing transport processes of atmospheric trace constituents through the air–water interface are generally referred to as precipitation scavenging, dry deposition, and resuspension. The latter was described above and the other two are referred to as *wet* and *dry removal processes*, respectively. The same terms are used by meteorologist when referring to the transport of gases and vapors to the air–soil interface.

As usual, the continuity equation and coupling boundary conditions describe the entire problem of interest, but this formalization exceeds present analysis capabilities. Consequently, the objective here is to develop approximations, or parameterizations, for terms in these equations and for the boundary conditions in forms suitable for applications at space and time scales of interest. The emphasis is on particles and those transport aspects relevant to describing air-pollution fluxes to water bodies through the air interface and on parameterizing the microscale aspects. This scale, in which the terms (i.e., concentration, velocity, etc.) in the continuity equation are averaged, includes not only the molecular-induced motions of the trace constituents but also the motions induced by high-frequency ($\geqslant 1$ cycle/min) turbulence.

The term *precipitation scavenging* is used here to describe trace constituent removal from the atmosphere by various types of precipitation. It refers to hydrometers (raindrops, snowflakes, ice pellets, etc.) whose gravitational terminal velocity exceeds local updrafts by 10 cm/s. Incorporation of air pollutants into cloud droplets is not precipitation scavenging, since it does not involve removal from the atmosphere. For aerosols containing the constituents A, the wet flux to the air–water (or air–soil) interface can be estimated by

$$n_A = R_A v_2 \bar{\rho}_{A1} \qquad (3.3\text{-}30)$$

where v_2 is the precipitation rate (L^3/L^3t), $\bar{\rho}_{A1}$ the concentration of A in bulk air (air + particles) measured at a convenient reference height (e.g., 1 or 10 m) and R_A the washout ratio for A (L^3 air/L^3 water). For raindrops, if chemical equilibrium is achieved, $R_A \simeq 1/H_p$ where H_p is Henry's constant.

The *washout ratio* is defined as the trace constituent concentration in surface-level precipitation to its concentration in surface-level air. It is therefore a convenient term for comparison with experimental field data. Also, theoretically, it is related to the removal rate, which can be obtained from an analysis of the interaction of particles and hydrometers as the latter moves through clouds. It includes the precipitation rate, mean drop radius, and hydrometer-particle collection efficiencies. The latter is also a function of particle radius and involves Brownian diffusion, interception, and impaction processes. Similar processes and relationships exist for snowflakes. Some measured washout ratios were found to be generally in the range E5 to E6 (L air)/(L water).[39]

In the case of the dry deposition of aerosols and the trace constituents they contain, the flux can be parameterized as

$$n_A = v_{dA} \bar{\rho}_{A1} \qquad (3.3\text{-}31)$$

where v_{dA} is the deposition velocity (L/t). Within the region above the surface where $\bar{\rho}_{A1}$ is typically measured, two atmospheric layers are present: the constant flux layer and the deposition layer. Particle transport in the former is dominated by turbulence and in the latter by Brownian diffusion. Particle deposition velocities can be roughly estimated using data from wind–water tank measurements. It appears that turbulence enhanced gravitational settling and Brownian diffusion are the two important processes.

Brownian for the small particles and gravity settling for the large. Gravity settling under Stokes' law conditions is the origin of the deposition velocity term, and in the limit for large particles the settling velocity approximates the deposition velocity. For a particle diameter range of 0.03 to 1.0 μm depositing on a water surface v_d varies from 0.005 to 0.04 cm/s.[39]

Air–soil. On the continents, coarse particles in the air are due to dust from the wind-driven erosion of soils and released plant material. The origin of coarse particles is similar to that of sea salt over the ocean in that it is surface derived. The fine particles arise mainly from condensation processes, but a mineral component is still present in the submicrometer size range. These small solid and liquid particles arise from the condensation of vapors when the vapor pressure exceeds the saturation point. For example, smoke from the open burning and often incomplete combustion of wood or agricultural refuse arises at least in part from the condensation of organic vapors. Urban and industrial sources also include particles and vapors primarily from combustion processes.

Continental and marine aerosols are blended to some extent in coastal regions to the effect that, depending on wind direction, a continental dust plume may travel far over the ocean, or sea salt may be carried inland for hundreds of kilometers. Miscellaneous sources of continental aerosols include biogenic particles released from plants in the form of seeds, pollen, spores, leaf waxes, resins, and so on. Volcanism constitutes another source of particulate matter in the atmosphere. Like volcanos, forest fires tend to emit particles into a buoyant plume reaching far into the troposphere.

Whereas most continental sources of aerosols involve direct injection (i.e., smokestacks, exhaust pipes, fires, volcanos, etc.) into the atmospheric boundary layer, the z_{Ay} function in Eq. 3.0-4 is handled as a constant or a time- and position-varying numeric relationship. Dust emissions from soils caused by wind erosion has been investigated extensively by D. A. Gellette and collaborators. Soil-derived particle fluxes are dependent on local conditions and the prevailing wind force. It is the only vertical mass flux of soil-derived aerosols amenable to quantitative parameterization.

A significant portion of the earth's surface is covered by rock and soil devoid of vegetation. Soils are formed by the weathering of crustal material and forms particles that have diameters mostly larger than 0.1 μm. The loose silicate material, usually with considerable amounts of organic matter attached is then transported upward by air motions. It is well known that adjacent to a smooth surface a laminar boundary layer exists even if the airflow is otherwise turbulent. On the other hand, if the surface is rough due to the presence of irregular soil and sand particles, turbulent motion may prevail right down to the surface. Such turbulent flow can cause a rough surface to be eroded either by direct aerodynamic pickup of particles, or as a result of sand blasting. The latter, called *saltation*, is the bombardment by particles performing a jumping motion.

To initiate the motion of soil particles requires wind velocities in excess of a threshold value. Gellette and others have presented measured threshold friction velocities, v_*, for a number of dry soils of different types. Once the threshold velocity is surpassed, the flux of saltating particles increases rapidly with wind speed. Measurements show that the horizontal flux of particles through a plane perpendicular to both ground surface and wind direction increases with $v_*^2(v_* - v_{*0})$, where v_{*0} is the threshold friction velocity. This functional form derives from the assumption that all the momentum in the wind is transferred to the ground surface by saltating sand grains. The concentration of particles at some distance from the ground is fed from the friction layer by eddy diffusion with flux that for neutral stability conditions, when a logarithmic wind profile exist, is approximately proportional to v_*^2. Accordingly, the vertical flux of aerosol particles is expected to grow with the fifth power of the friction velocity, at least for sufficiently loose, sandy soils. Existing data indicate that a fifth-power law is observed in several cases.

The transfer of aerosol particles to earthen surfaces is a process very similar to that of transfer to a water surface. Both wet and dry deposition occur here also. The wet flux is quantified by Eq. 3.3-30, by the use of washout ratios. The dry flux is quantified by Eq. 3.3-31, where the deposition velocity is used. In the case of earthen surfaces it represents the gas-phase resistance to particle and gas transfer. It includes resistance by turbulent transport as well as that due to molecular diffusion in the laminar layer adjacent to the surface. An additional resistance occurs at the surface itself in the uptake of trace gases. The deposition velocity, v_d, is usually determined from concentrations measured at a height of 1 m above the ground, or over a forest at a similar height above the canopy. Near the surface in the 1- to 10-m height region the flux varies slowly with height so that v_d is not strongly dependent on the choice of the reference height.

For particles with radii greater than 1 μm, the deposition rate is given essentially by the sedimentation velocity. At higher wind speeds, however, the deposition rate is enhanced due to the increase in atmospheric turbulence. For submicrometer particles the deposition velocity attains a relative minimum of some E-4 m/s in the 0.1- to 1.0-μm size range, whereas for fine particles the deposition velocity rises again with decreasing particle size. The behavior described above is for surfaces with small roughness length, such as grass and water. High-growing vegetation is apparently more effective in collecting aerosol particles, in that stems and leaves of bushes and trees act as filters, intercepting particles by impaction. Some data indicate that the deposition velocities in forest canopies for submicrometer-sized particles may be as high as 1E-2 m/s. (The material above was derived primarily from Warneck,[37] Slinn,[39] Pruppacker and Klett.[38])

Sediment–Water. Waterborne particles, termed *suspended solids* consist primarily of sand, silt, clays, natural organic matter, and wastewater-derived organic particles. Once the particles enter the aquatic environment they are called *sediment*. In surface waters they are normally classified as wash load, suspended load, and bedload. Depending on water flow rate, a given particle may from time to time be any one of these, but usually the wash load is comprised of silt- and clay-sized particles (0.001 to 0.3 mm). The suspended load is comprised of the wash load and larger particles (fine sand) moving just above the bed. The bedload is comprised of sands and gravels moving in contact with the bed.

Sediments are placed into suspension by water runoff from the land, direct discharge from municipalities and industries, from resuspended bed material, and released by waterborne activities such as sediment dredging and disposal operations. The objective of formulating deposition and erosion rate expressions is to characterize the mobility of the fine-grained particles. Typically, it is these, which are less than 74 μm (silts and clays) and greater than 0.45 μm (colloids), which are hydraulically transported almost entirely in suspension rather than as bedload. Due to the capacity of these fines in absorbing chemicals from solution, they play important roles in the chemodynamics of aquatic environments. See Section 2.1 under the topic "Earthen Solid–Water Equilibrium Occurrences."

Silt- and clay-sized particles are also classified as cohesive sediment. Most particles in water are charged. They may coalesce and form cohesive aggregates, depending on their concentration and the nature of the electrostatic forces acting on them. The forces of attraction or repulsion between particles are conditioned by the electrical potential developing on the particle surfaces in the presence of the hydrogen and metal ions in solution. Cohesive aggregates may sink faster or slower than individual particles, depending on their bulk densities and sizes (see Stumm[40]). In contrast, sand particles are noncohesive. In a bed form the lifting forces can overcome the gravity force, and this is sufficient to move these noncohesive particles. Cohesive sediments occur predominantly in the form of mud and sludge covering the bottom.

Qualitatively, the sedimentation processes in a stream or estuary follow certain phases that vary in space and time and constitute the sedimentation cycle. This cycle includes erosion, advection, diffusion–dispersion, aggregation–flocculation, deposition, consolidation, and resuspension. More specifically, the cycle occurs as follows: Under the action of excessive hydrodynamic shear stresses, solid particles leave the bottom and are carried in suspension by the currents. While in suspension, the material is diffused and dispersed. The sediment remains in suspension until favorable conditions help create flocs and aggregates that settle under gravitational forces. Once deposited, the sediment, left undisturbed for a sufficient length of time, develops stronger bonds during consolidation and stays on the bed until the bonds are broken again and the material is resuspended into the water column. Otherwise, a hyperconcentration fluid mud is formed at the bottom which can easily be resuspended. Biological processes can significantly influence the bonds and consolidation processes, resulting in a sediment drastically different from that which existed in the as-deposited state. The following contains a brief introduction to the subject of particle deposition and erosion from the bed.

The basic equations used to describe particle behavior are mass continuity and the momentum balance. Most models are time dependent and two-dimensional, using the diffusion–advection equation (Eq. 3.2-11) for particles. A complete cohesive sediment transport model should account for erosion, deposition, and bed consolidation. At the bed interface, the material that is convected and diffusing is considered as part of the net depositional flux. Therefore,

$$(1 - P)v_s\rho_{32} + \mathscr{D}^{(t)}_{32y}\frac{\partial \rho_{32}}{\partial y} = j_{3y^-} - j_{3y^+} \tag{3.3-32}$$

where P is the probability that particle settling to the bed is reentrained, v_s the settling velocity, and $\mathscr{D}^{(t)}_{32y}$ the turbulent particle diffusion coefficient in the vertical direction. The two sediment flux terms on the right-hand side are the rates of particle deposition and erosion, respectively. Details on these two terms and appropriate rate expressions follow. (The material presented here has been abstracted primarily from Ariathurai[41] and Mehta and Partheniades.[42])

Rate of Sediment Deposition. A frequently utilized function expressing sediment deposition rate is composed of a number of sediment fractions with different physical characteristics. Clay and fine silt fractions form relatively uniform settling aggregates. Medium and coarse silt fractions settle at higher rates. The total rate of deposition can be expressed as the summation of all individual fractions following the linear deposition formula. For the ith fraction the deposition rate is

$$j_{3y^-} = \frac{k_a v_3 \rho_{32}(1 - \tau_0/\tau_d)}{h} \tag{3.3-33}$$

where h is depth of flow, k_a is an empirical coefficient accounting for aggregation, v_3 is the settling velocity, τ_d is the critical shear stress under which deposition occurs, and ρ_{32} is the concentration just above the bed. The term in parentheses is the probability that an aggregate reaching the bed will remain there. It varies linearly from 0 at a critical shear stress for deposition to 1 at zero bed shear stress, $\tau_0 = 0$. This functional form is from Krone.[43] All terms except τ_0, the bed shear stress, and h are for the ith fraction. Obviously, j_{3y^-} cannot be negative.

Rate of Sediment Erosion. Like deposition, erosion is related directly to the magnitude of the shear stress. At τ_0 above a critical value, particles are individually dislodged from the sediment bed as interaggregate bounds are broken. Particle resuspension, j_{3y^+}, is related to the shear stress in excess of a critical value, τ_θ, and to an erosion rate constant k_θ; thus

$$j_{3y^+} = k_\theta \left(\frac{\tau_0}{\tau_\theta} - 1 \right) \tag{3.3-34}$$

with $\tau_0 > \tau_\theta$. Deposition and resuspension tests must be performed to determine various constants, velocities, and critical shear stresses in both Eqs. 3.3-33 and 3.3-34.

Suspension concentrations above experimental eroding beds often reach equilibrium values that depend on the bed shear stress. Equilibrium concentrations of particles in suspension are achieved as erosion rates decrease with time to zero, while the flow remains constant. These particle concentrations have been found not to be related to the transport capacity of the flow (as for sand) but have been related to vertical differences or inhomogeneity in the bed (either particle characteristics or bed density) or to armoring by selective erosion of the bed surface.

Bioturbation. By a completely different mechanism, particles, porewater, and the associated chemical constituents within the upper layers of bed sediment are transported about. This mixing is subject to the activities of benthic organisms. Bioturbation occurs in several different ways. Some organisms, such as crabs and snails, mix surface sediment simply by crawling or plowing through it. Figure 3.2-9*d* contains a photograph documenting evidence of bioturbation. More important, other organisms, especially polychaete worms and bivalves, burrow into sediment and ingest the sediment particles. Such burrowing can extend to several tens of centimeters. Once their burrows are constructed, some organisms remain in them and flush the burrows with overlying seawater. This process is referred to as irrigation; it involves only the pore water and not the enclosing particles. Mathematical modeling of the process is presented in Chapter 5.

PROBLEMS

3.3A. Estimating Heat and Mass Transfer Flux Rates from Micrometeorological Data

Brooks and Pruitt[25] measured wind velocity, air temperature, and moisture near the ground. The data listed in Table 3.3A are a part of that work.

Assume that the fluid properties at the 90 cm elevation represent the bulk; do the following:

1. Obtain the shear stress at the ground, τ_0, in g/cm·s^2.
2. Estimate the gas-phase mass transfer coefficient for water vapor, ${}^3k'_{A1}$ in cm/h, from (a) Reynolds' analogy, (b) von Kármán's analogy, and (c) the Chilton–Colburn analogy.

Table 3.3A

Height (cm)	Air Velocity (cm/s)	Additional Data
16	185.4	
39	252.6	Temperature at 90 cm was 14.96°C
90	319.5	Temperature at 0 cm was 13.72°C
139	356.4	Humidity at 90 cm was 8.98 g/m^3
189	383.7	Humidity at 0 cm was 9.12 g/m^3
390	437.6	Volume was measured at STP
590	469.6	

3. Based on the Chilton–Colburn analogy value, estimate the water flux rate, n_A, in g/m$^2 \cdot$h, and the direction of transport.
4. Based on the Chilton–Coburn analogy, estimate the sensible heat flux rate, q in J/m$^2 \cdot$h, and the direction of transport.

3.3B. Failure of Reynolds Analogy for Liquids

Determine the error (%) in a particular mass transfer coefficient calculation (use $v_* = 0.1$ cm/s and $v_b = 5$ cm/s) if the Reynolds analogy is used. Sc = 3500 for the particular substance in aqueous solution.

1. Compare to the von Kármán equation.
2. Compare to the *j*-factor equation.

3.3C. Simple Relationships Between Sublayer Thicknesses

As momentum, heat, and mass are transferred across a fluid–solid interface, profiles of velocity, temperature, and concentration develop in the boundary layer next to the solid surface. Figure 3.3C shows the case for momentum transfer in the negative y direction, and for heat and mass transfer in the negative y direction. The sublayer thicknesses of velocity, temperature, and concentration are usually different, as shown.

1. Using the result of the film theory for mass transfer (Eq. 3.1-18) and Eq. 3.2-18, show that the Chilton–Colburn *j*-factor analogy (Eq. 3.3-26) predicts that the diffusive sublayer thickness δ_{A2} is related to the laminar sublayer δ_v by

$$\delta_{A2} = \frac{\delta_v}{\mathrm{Sc}_{A2}^{1/3}} \tag{3.3C-1}$$

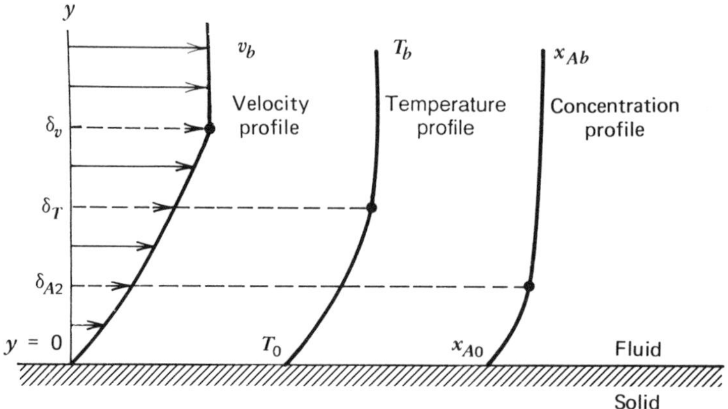

Figure 3.3C. Sublayer thickness and profiles.

2. Using the film theory for heat transfer,

$$q_0 = \frac{k_2}{\delta_T}(T_0 - T_b) \tag{3.3C-2}$$

and Eq. 3.2-18, show that the Chilton–Colburn j-factor analogy (Eq. 3.3-25) predicts that the thermal sublayer thickness δ_T is related to the laminar sublayer thickness δ_v by

$$\delta_T = \frac{\delta_v}{\mathrm{Pr}^{1/3}} \tag{3.3C-3}$$

3.3D. Concentration Profile in Turbulent Zone

Develop Eq. 3.3-17 by begining with Eq. 3.2H-4. Follow the procedure that led to Eq. 3.3-16.

3.3E. Field Measuring Technique for Chemical Flux Rates

Time-averaged concentration and velocities are measured at two elevations, y_1 and y_2, above a solid environmental interface but within the turbulent flow zone. Show the steps in developing the following equation:

$$n_A = \frac{\kappa_1^2(\rho_{Aj2} - \rho_{Aj1})(\bar{v}_{x2} - \bar{v}_{x1})}{[\ln(y_2/y_1)]^2} \tag{3.3E}$$

where $y_2 > y_1$ and the numerical subscripts denote the elevation.

3.3F. Chemical Sampling Height Above Air–Soil Interface

Deposition velocities are usually determined from concentrations measured at a height of approximately 1 m above the ground. Demonstrate that near the surface in the 1.0- to 10-m height region the flux varies slowly with height so that v_d, the deposition velocity, is not strongly dependent on the choice of sampling height.

REFERENCES

1. R. C. Reid and T. K. Sherwood, *The Properties of Gases and Liquids*, 2nd ed., McGraw-Hill, New York, 1966.
2. A. Lerman, "Time to Chemical Steady-States in Lakes and Oceans," in J. D. Hem, Ed., *Nonequilibrium Systems in Natural Water Chemistry*, Advances in Chemistry Series 106, American Chemical Society, Washington, D.C., 1971.
3. R. E. Bird, W. E. Stewart, and E. N. Lightfoot, *Transport Phenomena*, Wiley, New York, 1960, p. 656f.
4. W. Nernst, *Z. Phys. Chem.*, **47**, 52 (1904).
5. W. G. Whitman, *Chem. Met. Eng.*, **29** 146 (1923).
6. R. Higbie, "The Rate of Absorption of a Pure Gas into a Still Liquid During Short Periods of Exposure," *Trans. Am. Inst. Chem. Eng.*, **31**, 365 (1935).
7. P. V. Dankwerts, "Significance of Liquid-Film Coefficients in Gas Absorption," *Ind. Eng. Chem.*, **43**(6), 1469 (1951).
8. R. E. Treybal, *Mass-Transfer Operations*, 2nd ed., McGraw-Hill, New York, 1968, pp. 46–60.
9. T. K. Sherwood, R. L. Pigford, and C. R. Wilke, *Mass Transfer*, McGraw-Hill, New York, 1975, pp. 150–159.
10. R. E. Bird, W. E. Stewart, and E. N. Lightfoot, *Transport Phenomena*, Wiley, New York, 1960, p. 617.
11. R. E. Treybal, *Mass-Transfer Operations*, 2nd ed., McGraw-Hill, New York, 1968, p. 63.
12. C. R. Wilke and P. Chang, *Am. Inst. Chem. Eng. J.*, **1**, 264–270 (1955).
13. "Japanese Oil Spill Has Wide Repercussions," *Chem. Eng. News*, June 2, 1975, p. 13.
14. J. O. Hirschfelder, C. E. Curtis, and R. B. Bird, *Molecular Theory of Gases and Liquids*, Wiley, New York, 1954.
15. J. T. Davies, *Turbulence Phenomena*, Academic Press, New York, 1972.
16. D. H. Slade, Ed., *Meteorology and Atomic Energy*, U.S. Atomic Energy Commission Technical Information Center, Oak Ridge, Tenn., 1969, pp. 66–73.
17. M. Wimbush, "The Physics of the Benthic Boundary Layer," in I. N. McCave, Ed., *The Benthic Boundary Layer*, Plenum Press, New York, 1976, p. 8.
18. Committee on Sedimentation, "Nomenclature for Bed Forms in Alluvial Channels." *Proc. Am. Soc. Civ. Eng.*, **92**(HY3) (1966).
19. G. Newmann and W. J. Pierson, Jr., *Principles of Physical Oceanography*, Prentice Hall, Englewood Cliffs, N.J., 1966, pp. 29, 30, 327.

20. R. C. Y. Koh and L. N. Fan, "Mathematical Models for the Prediction of Temperature Distributions Resulting from the Discharge of Heated Water in Large Bodies of Water," 16130DW010/70, U.S. EPA, Water Quality Office, Washington, D.C., Oct. 1970, p. 127.
21. F. L. Parker and P. A. Krenkel, *Thermal Pollution: Status of the Art*, Report 3, School of Engineering, Vanderbilt University, Nashville, Tenn., 1969.
22. A. Defant, *Physical Oceanography*, Vol. 1, Macmillan, New York, 1961, p. 104.
23. C. O. Bennett and J. E. Myers, *Momentum, Heat and Mass Transfer*, 2nd ed., McGraw-Hill, New York, 1974, pp. 351–352.
24. T. H. Chilton and A. P. Colburn, *Ind. Eng. Chem.*, **26**, 1183 (1934).
25. F. A. Brooks and W. O. Pruitt, *Investigation of Energy, Momentum and Mass Transfer near the Ground*, Final Report, U.S. Army Electronics Command, Atmospheric Science Laboratory, Research Division, Fort Hauchuca, Ariz., 1965.
26. J. C. Slattery, *Momentum, Energy and Mass Transfer in Continua*, R. E. Krieger, Huntington, N.Y. 1981.
27. H. Schlichting, *Boundary-Layer Theory*, 7th ed., McGraw-Hill, New York, 1979.
28. R. E. Bird, W. E. Stewart, and E. N. Lightfoot, *Transport Phenomena*, Wiley, New York, 1960, p. 607.
29. T. R. Marrero and E. A. Mason, AIChEJ., **19**, 498 (1973).
30. E. N. Fuller, P. D. Schettler, and J. C. Giddings, *Ind. Eng. Chem.*, **58**(5), 18 (1966).
31. E. L. Cussler, *Diffusion-Mass Transfer in Fluid Systems*, Cambridge University Press, New York, 1984, p. 118.
32. A. Lerman, *Geochemical Processes: Water and Sediment Environments*, Wiley, New York, 1979, pp. 76–101.
33. E. C. Thornton and W. E. Seyfried, Jr., *Science*, **220**, June 10, 1983, pp. 1156–1158.
34. F. A. L. Dullien, *Porous Media: Fluid Transport and Pore Structure*, Academic Press, New York, 1979.
35. J. Bear, *Dynamics of Fluids in Porous Media*, American Elsevier, New York, 1972.
36. J. M. Smith, *Chemical Engineering Kinetics*, 3rd ed., McGraw-Hill, New York, 1981, p. 458.
37. P. Warneck, *Chemistry of the Natural Atmosphere*. Academic Press, San Diego, Calif., 1988, Chapter 7.
38. H. R. Pruppacher and J. O. Klett, *Microphysics of Clouds and Precipitation*, D. Reidel, Dordrecht, The Netherlands, 1978, Chapter 12.
39. W. G. N. Slinn, Ch. 4, *The Tropospheric Transport of Pollutants and Other Substances to the Oceans*, National Academy of Sciences, Washington, D.C., 1975.
40. W. Stumm, Ed., *Aquatic Surface Chemistry*, Wiley, New York, 1987.
41. R. Ariathurai, Ph.D. thesis, Department of Civil Engineering, University of California–Davis, 1974.
42. A. J. Mehta and E. Partheniades, Depositional Behavior of Cohesive Sediments, *Technical Report*, **16**, Coastal and Oceanographic Laboratory, University of Florida, Gainesville, Fl., 1973.

43. K. B. Krone, "Flume Studies of the Transport of Sediment in Estuarial Shoaling Processes," Final Report, Hydraulics Engineering Laboratory and Sanitary Engineering Research Laboratory, University of California–Berkeley, 1962.
44. E. McLaughlin, *Chem. Eng. Sci.*, **16**, pp. 76–81, (1961).
45. F. L. Holmes, "Antoine Lavoisier: The Conservation of Matter," *Chem. Eng. News*, Sept. 12, 1994, pp. 38–45.
46. D. R. Caldwell and T. M. Chris, *Science*, **205**, Sept. 14, 1979, pp. 1131–1133.

4

CHEMICAL EXCHANGE RATES BETWEEN AIR AND WATER

There are many times when a knowledge of the flux of an organic or inorganic chemical between water and air is valuable. For example, in estimating the movement of halocarbons in the biosphere it is necessary to know the rate at which the material moves between the atmosphere and the hydrosphere. In a different situation such as a spill of a chemical into a river, a knowledge of the rate of evaporation through the water–air interface is fundamental in estimating the resulting concentration as it moves downstream. The ability to predict rates of chemical transfer between various regions of the ocean surface and atmosphere is necessary to an understanding of the cycles of a number of trace atmospheric chemicals. These chemicals include naturally occurring gases such as carbon dioxide and synthetic trace chemicals such as chloroform and carbon tetrachloride. For some of these chemicals the ocean is a potential sink; for others the ocean has been proposed as a natural source. If we are to develop the capability to make long-range forecasts of the effect of pollution on concentrations of the trace chemicals, we must be able to establish their air–sea transfer rates. To do this it is necessary to measure not only the concentration differences across the air–sea interface but also the mass transfer coefficients governing the exchange.

The processes that control chemical movements in the real world are extremely complex when taken all together. A system of this degree of complexity cannot be described exactly by a theoretical development. Some information about the behavior of the system can, however, be obtained by analysis of highly simplified models. The reader may feel that the applications are too simple to be of interest; it is certainly true that they represent highly idealized situations, but the results find considerable use in an understanding of numerous topics in environmental chemistry and engineering.

In general, the information in this chapter is built around a very simple flux expression. The most convenient form for steady-state evaporation of A from a waterbody is the mass flux version of Eq. 3.1-35:

$$n_{Ay} = {}^1K'_{A2}(\rho_{A2} - \rho^*_{A2}) \tag{4.0-1}$$

where ${}^1K'_{A2}$ in L/t is the overall liquid-phase coefficient. The concentration of A in air is ρ_{A1} so that $\rho^*_{A2} = \rho_{A1}/H_\rho$ where H_ρ is Henrys' constant (see Eq. 2.1F). The concentration ρ_{A2} is for the molecular species of A in solution. It does not include A associated with suspended particles. For the absorption of vapor or gaseous species of A from the air into water, Eq. 4.0-1 applies; however, $\rho^*_{A2} > \rho_{A2}$ and the flux is negative. The concentration ρ_{A1} does not include A associated with aerosols, raindrops, fog droplets, snowflakes, and so on. The same transport coefficient is used and the phrase *dry deposition of vapors* is commonly used to describe the process. The flux of particle-bound A from air to water is quantified by a somewhat similar expression (see Eq. 3.3-31) and the process is also one of dry deposition. Wet deposition refers to the delivery of A to the surface of the water body by hyrometeors (i.e., rainfall). Delivery involves scavenging A from the vapor and particle phases (i.e., washout) followed by downward transport through the atmosphere on the descending hydrometeors. This flux is quantified differently (see Eq. 3.3-30). Only evaporation and dry deposition processes are considered in this chapter.

Various aspects of the chemical transport at the air–water interface are presented in the context of specific problems and applications of theory. The first section considers the problem of desorption (or evaporation) of gases and liquids from aerated basins. In this case the process is aided by mechanical means and involves air–water interfaces of the size of small ponds and lakes. The topics progress through the evaporation of pesticides from the surface of lakes, the flux of gases in both directions across the air–sea interface, the weathering of hydrocarbon components from sea-surface slicks, and the dry depositional processes from air to water.

4.1. DESORPTION OF GASES AND LIQUIDS FROM AERATED BASINS AND RIVERS

Natural and artificial basins (ponds, lagoons) are often used as aerobic biochemical oxidation reactors for neutralizing organic wastewater. Carbonaceous compounds in the wastewater are converted to carbon dioxide and water by a mixed culture of microorganisms. The organisms use the carbon compounds as an energy source. Oxygen must be available and mechanical means are frequently employed to increase the rate of transfer and supplement the natural absorption processes.

A common mechanical device employed is a surface aerator. Impeller blades connected by a shaft to a motor beat the water surface into an agitated state. Oxygen absorption in water is liquid-phase controlled, so that the agitation increases the absorption rate. Figure 4.1-1 shows a typical installation.

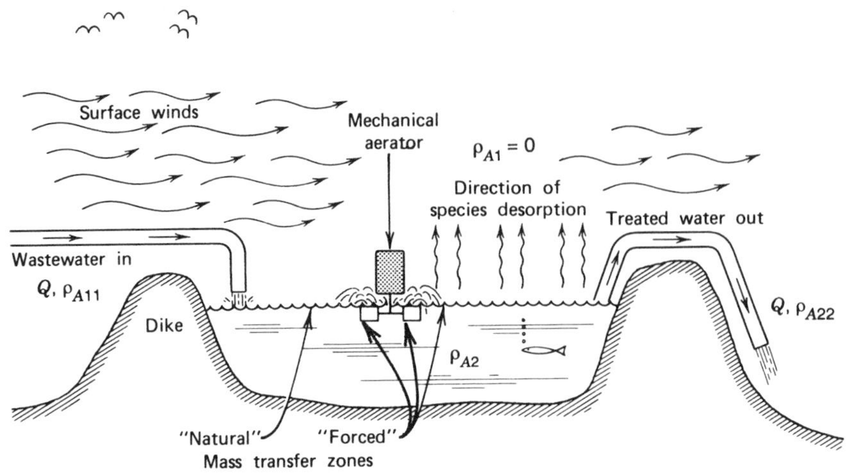

Figure 4.1-1. Idealized desorption wastewater treatment system.

Industrial wastewater is varied in nature and contains a collection of chemicals used in the manufacturing process. These waste chemicals, in small amounts, originate from raw material handling, intermediates, final products, and so on. Aerobic biochemical oxidation is a common means of removing these chemicals prior to discharging the water to a receiving stream or lake. It is not uncommon for the wastewater entering the basin to contain components that are volatile. These volatile components, both gases and liquids, can desorb from the water directly into the air. The combination of a large interfacial area between water and air, mechanical agitation, and high relative volatility can result in significant quantities being desorbed.

A necessary condition for the transfer of a volatile species from a liquid phase to a gas phase is a favorable chemical potential. This occurs when $\rho_{A2} > \rho^*_{A1})/H_\rho$ for dilute solutions in Eq. 4.0-1. Table 2.1-5 contains values of Henry's constant in mole fraction form, $H_{xA}(=H_\rho c_2/c_1)$, for common industrial components. In this development it is assumed that the air above the basin contains no A, so that $\rho^*_{A2} = 0$. It is also assumed that biochemical reaction is absent within the basin water, that there is no settling or absorption of species A by solids, and so on, only desorption through the air–water interface so that Eq. 4.0-1 is applicable.

Figure 4.1-1 illustrates the idealized, steady-state desorption system. Two idealized mixing models need to be considered. The completely mixed basin model reasonably depicts basins of small volume in which the mechanical device stirs the liquid so that a uniform concentration of A is maintained at all points. The plug flow basin model assumes no mixing and is approached by real-world basins made in the shape of long, narrow channels with flow in one end and out the other. Both models are highly idealized and represent the extreme in mixing and the extreme in unmixing, respectively. Real-world basins rarely approach either case but usually involve some intermediate mixing regime.

Case A: Completely Mixed Basin

This idealized mixing concept assumes that the liquid portion of the basin is in a state of mixing such that the concentration of A is the same everywhere. The inlet water ρ_{A21} is immediately mixed with the water in the basin to yield a uniform concentration ρ_{A22}, which is also the concentration in the exit water. A component balance for A over the entire basin volume V and surface area A according to Eq. 1.1-6 yields

$$Q\rho_{A21} = Q\rho_{A22} + {}^{1}K'_{A2}A(\rho_{A22} - 0) + 0 + 0 \tag{4.1-1}$$

where Q, in L^3/t is the volumetric flow rate of water through the basin. Solving Eq. 4.1-1 yields the exit concentration of A:

$$\rho_{A22} = \frac{\rho_{A21}}{1 + {}^{1}K'_{A2}A/Q} \tag{4.1-2}$$

The relative quantity of species A lost from the basin by desorption is of interest. The fraction desorbed in the completely mixed basin is $F_M \equiv (\rho_{A21} - \rho_{A22})/\rho_{A21}$, assuming no water evaporation or seepage from the basin. This definition, along with Eq. 4.1-2, yields

$$F_M = 1 - \frac{1}{1 + {}^{1}K'_{A2}A/Q} \tag{4.1-3}$$

which is independent of concentration. Fractional removal, decay, treatment, and so on, in any first-order process such as mass transfer is concentration independent. The desorption rate is concentration dependent, however, and can be expressed in terms of fractional removal by

$$w_A = Q\rho_{A21}F_M \tag{4.1-4}$$

These final two equations are useful in estimating the maximum fraction and quantity of component A desorbed from a completely mixed basin.

Case B: Plug Flow Basin

A *plug flow basin* is a completely unmixed basin. In this idealized mixing model the liquid portion of the basin flows through as a slug of fluid. The elements of fluid hold their position and do not mix with elements in front or back. The concentration of A falls continuously from the high of ρ_{A21} at the inlet to the low of ρ_{A22} at the outlet.

A differential equation describes the material balance on component A (see Problem 4.1A) over an elemental volume of basin. The integrated form of this equation is

$$\rho_{A22} = \rho_{A21} \exp\left(- \frac{{}^1\mathrm{K}'_{A2}A}{Q} \right) \tag{4.1-5}$$

The relative quantity desorbed from the basin is

$$F_p = 1 - \exp\left(- \frac{{}^1K'_{A2}A}{Q} \right) \tag{4.1-6}$$

and the desorption rate is given by Eq. 4.1-4 with F_p substituted for F_M.

These two equations are useful to estimate the maximum fraction and quantity of component A desorbing from a treatment basin. Any real-world basin should lie somewhere between a completely mixed state and a completely unmixed state, so that use of the preceding expressions will give the upper and lower bounds of chemical desorption without biochemical reaction and other competing mechanisms. Both of the preceding mixing models assume that the mass transfer coefficient is uniformly distributed throughout the basin volume. The following section is devoted to obtaining realistic estimates of the overall mass transfer coefficient (i.e., ${}^1K'_{A2}$) for use in the preceding equations.

It is convenient to divide the gas–liquid interface of a typical basin into two zones. One zone, located near the surface agitator, is called the *zone of forced convection*. In this zone the mechanically driven impeller forces the liquid and gas phases to move about and come into intimate contact. This forced convection zone extends to a radius of 3 m or more from the impeller shaft. The other zone is located beyond the immediate influences of the agitator and assumes characteristics of the natural environment. In this zone water flows outward from the aerator, not unlike a natural flowing stream. Surface winds dominate the movement of air over the water surface in this zone (see Fig. 4.1-1). The creation of these two zones reflects the physical state of the basin and guides the search for appropriate mass transfer coefficient correlations.

The two-resistance theory is a sufficient quantitative tool for describing the interphase desorption of volatile species in an aerated stabilization basin (ASB). Desorption is occurring in parallel from each of the zones so that the total rate is the sum of the "natural" (n) and "forced" (f) zones: $w_{A0} = w_{A0}^{(n)} + w_{A0}^{(f)}$. Applying Eq. 4.0-1 to each zone yields

$$w_A = ({}^1K'_{A2}A)^{(n)}(\rho_{A2} - 0) + ({}^1K'_{A2}A)^{(f)}(\rho_{A2} - 0) \tag{4.1-7}$$

where $({}^1K_{A2}A)^{(n)}$ is the natural overall liquid-phase desorption mass transfer coefficient area product (L^3/t) that accounts for the basin surface stream flow (impeller-induced) behavior of the liquid phase and the wind (air) surface behavior of the gas phase. The product $({}^1K_{A2}A)^{(f)}$ accounts for the pumping and phase-dispersing action occurring with the gas and liquid immediately adjacent to the surface agitator. As noted earlier, the forced convection zone occupies a region near each agitator and extends outward several meters from the shaft. In reality, even the natural zone is influenced by the aerator. Both $A^{(n)}$ and $A^{(f)}$ in Eq. 4.1-7 are plane areas and the sum equals that for the total basin (L^2).

The overall coefficients in Eq. 4.1-7 must be related to individual coefficients in each phase. The two-resistance theory developed in Chapter 3 is used (see Eq. 3.1-37) to give

$$\frac{1}{({}^1K_{A2})^{(n)}} = \frac{1}{({}^1k_{A2})^{(n)}} + \frac{1}{H_{xA}({}^2k_{A1})^{(n)}} \tag{4.1-8}$$

for the natural coefficient and

$$\frac{1}{({}^1K_{A2})^{(f)}} = \frac{1}{({}^1k_{A2})^{(f)}} + \frac{1}{H_{xA}({}^2k_{A1})^{(f)}} \tag{4.1-9}$$

for the forced coefficient. Therefore, a quantitative description of the rate of desorption of chemical A requires information on four individual phase mass transfer coefficients, the natural zone interfacial area and the forced zone interfacial area, plus Henry's constant. In the following section techniques used to obtain the individual coefficients and interfacial areas are presented in detail. Each coefficient area product will be developed as a product or individually as a coefficient and an area.

Estimating Mass Transfer Coefficients for Aerated Stabilization Basins

Mechanical Surface Agitator Liquid–Phase Coefficient. The rate of absorption of oxygen from air into water by mechanical agitation of the water surface has been studied extensively. Since oxygen absorption into water is liquid-phase controlling, this coefficient provides a very good measure of the forced desorption coefficient. Many experimental observations on scale-model and prototype units have resulted in a reliable correlating equation.[2] The reported oxygen coefficient transformed to yield a liquid-phase mass transfer coefficient applicable to any chemical species absorbing or desorbing in the basin area under the immediate influence of the agitator is

$$({}^1k_{A2}a_v)^{(f)} = 6.06\text{E}3\sqrt{\frac{\mathscr{D}_{A2}}{\mathscr{D}_{B2}}}\,\frac{n^1_{BO}EP\alpha}{V}\,(1.024)^{(T-20^\circ\text{C})} \tag{4.1-10}$$

where the subscript A denotes species A, B denotes molecular oxygen, and 2 denotes water. The coefficient area product has dimensions of lb mol/hr·ft^3, and the dimensions of the remaining terms may be found in Problem 4.1C. a_v (ft^2/ft^3) is the plane surface area A per unit mixed volume V, and both are a function of aerator size.[4] Both A and V are needed in (4.1-10) for obtaining a_v. The square-root dependence of the coefficient on liquid diffusivity assumes a penetration theory mechanism, that is short liquid exposure times for the liquid-phase transfer.

Mechanical Surface Agitator Gas Phase Coefficient. Reinhardt[5] performed experiments on the rate of absorption of ammonia at 30,000 ppm (volume) from air into acidified water stirred by a flat-bladed surface impeller. The experimental apparatus was designed to simulate surface aerator operations and to isolate the gas-phase resistance. He obtained the following relationship for the gas-phase Sherwood number:

$$\mathrm{Sh}_{A1} = 3.9\mathrm{E}\text{-}4\mathrm{Re}_1^{1.42} \cdot \mathrm{Sc}_{A1}^{0.5} \cdot \mathrm{Po}^{0.4} \cdot \mathrm{Fr}^{-0.21} \tag{4.1-11}$$

where $\mathrm{Sh}_{A1} = {}^2k'_{A1}d/\mathscr{D}_{A1}$, $\mathrm{Re}_1 = d^2s/\nu_1$, Po (power number) $= \omega E/\rho_1 d^5 s^3$, and Fr (Froude number) $= ds^2/g$. The impeller diameter is d, s is rotational speed (radians/time), ω is nameplate power, and E is efficiency. This coefficient, $({}^2k_{A1})^{(f)}$, is important for volatile chemicals that are miscible or very soluble in water.

Natural Surface Liquid-Phase Coefficient. Wind-induced momentum transport across the air–water interface creates turbulence in the surface water and deeper. Molecules from within the water column are brought to the interface, where vaporization occurs. Lunney et al.[6] reviewed the literature for data and correlations on this coefficient and found that the wind speed was the primary independent variable. Desorption experiments were performed with ethyl ether as the volatile substance and it was found that the water fetch/water depth ratio, l/h, was an important independent variable. The fetch is defined as the distance the wind travels over the water surface. However, for most surface impoundments and lakes $l/h > 50$ and the following correlation for ethyl ether was obtained:

$$({}^1k'_{A2})^{(n)} = 0.094v_x^2 \tag{4.1-12}$$

for velocities, v_x (in m/s) measured at 10 m above the surface. The coefficient is in cm/h and is from wind data of $v_x = 5$ to 16 m/s. Equation 3.1-25 with $n = \frac{1}{2}$ should be used to convert to other volatile substances.

The liquid-phase coefficient appears to be independent of wind velocity in the $v_x = 0$ to 5 m/s range. Here ${}^1k'_{A2}$ varies from 1 to 3 cm/h (see Fig. 4.2-3). The coefficient does not go to zero during periods of no wind. Other factors, not all of them known, but including gentle residual fluid motions and thermal gradients, create a low level of turbulence to maintain some degree of chemical transport on the liquid side of the interface.

Natural Surface Gas-Phase Coefficient. In the zone of natural convection the gas-phase resistance is related to the movement of air across the plane surface of the basin. Water evaporation is gas-phase controlling and gives a measure of this coefficient. Experimental field measurements on the evaporation of water from reservoirs and a resulting correlation developed by Harbeck[7] results in

$$({}^2k'_{A1})^{(n)} = 0.26v_8A_r^{-0.05} \tag{4.1-13}$$

where ${}^2k'_{A1}$ is in cm/h, v_8 is the wind velocity at 8 m above the water surface (in mi/hr), and A_r is the surface area of the reservoir (in acres). Equation 3.1-25 with $n = 1/2$ should be used to convert from water vapor to other volatiles. This coefficient quantifies the degree of air-side fluid turbulence processes that sweep molecules desorbing from the interface upward and away into the depths of the air boundary layer. As the wind velocity tends toward zero, weak air-side processes, including thermal gradients and water evaporation, maintain the coefficient at some finite value (see Section entitled "Desorption Under Quiescent Conditions."

With sufficient operating details on aerated basins the individual mass transfer coefficients of species A can be estimated by Eqs. 4.1-10 to 4.1-13. Using the two-resistance theory, Eqs. 4.1-8 and 4.1-9, and assuming parallel desorption, we obtain the overall coefficient

$${}^1K'_{A2}A = ({}^1K'_{A2}A)^{(n)} + ({}^1K'_{A2}A)^{(f)} \tag{4.1-14}$$

The completely mixed model equations or the plug flow model equations allow computation of desorption rates and percent removals (see Problem 4.1F).

Desorption Study of Aqueous Waste Impoundments

A study of 14 existing aerated basins[1] employed in the pulp and paper industry is summarized in Tables 4.1-1 and 4.1-2. Methanol and acetone were the chemicals chosen for the study. A summary of the individual mass transfer coefficients computed in the study is shown in Table 4.1-1. The variations reflect primarily the number of surface aerators, aerator horsepower, aerator

Table 4.1-1. Methanol Mass Transfer Coefficients for Aerated Basins (cm/h)

Coefficient	Average	Range
$({}^1k'_{A2})^{(f)}$	10,200	3,000–16,100
$({}^2k'_{A1})^{(f)}$	10,500	5,220–14,200
$({}^1k'_{A2})^{(n)}$	10.9	3.06–21.1
$({}^2k'_{A1})^{(n)}$	1,840	477–2,950

Source: Reprinted by permission from Ref. 1.

Table 4.1-2. Interfacial Mass Transfer Area in Aerated Basins

Model Input Generated Area[b] (ft^2/hp)	$a_v^{(f)}$ (ft^{-1}) Range	Average	Relative Quantity (%) of Natural Area[a] (Based on Average)
1.2	0.0000086–0.00044	0.00011	99+
56.2	0.00040–0.021	0.00495	95
217.0	0.0016–0.079	0.0191	84

Source: reprinted by permission from Ref. 1.
[a]The natural area average is 0.0980 with a range of 0.056 to 0.14 ft^{-1}.
[b]*Source:* Ref. 3.

blade diameter, and wind speed. Table 4.1-2 gives the interfacial areas in the basins and compares the size of the natural and generated surfaces. The surface agitators do not increase the interfacial area significantly; however, $({}^1k'_{A2})^{(f)}$ is approximately 1000 times $({}^1k'_{A2})^{(n)}$. These aerators create turbulence in the near-surface waters surrounding each unit denoted by the plane area, $A^{(f)}$ (not to be confused with the interfacial area, $a_v^{(f)}$, in Table 4.1-2). Roughly, $A^{(f)} = 21\ cm^2/W$ ($17\ ft^2/hp$) of delivered power for units of 10 kW and greater.

The following example illustrates the volatilization rate calculation procedure.

Example 4.1-1. Dioxin Vaporization from Wastewater Impoundment. 2,3,7,8-Tetrachlorodibenzo-*p*-dioxin (TCDD) is reputed to be the most toxic substance of anthropogenic origin. With a molecular weight of 322 g/mol it has a pure component vapor pressure of 0.74E-9 mmHg and water solubility of 7.9 ng/L at 25°C. At this same temperature its molecular diffusivity in air and water is 0.0525 and 5.6E-6 cm^2/s, respectively. Assuming none in the background air, estimate the maximum vaporization rate in g/day from an aqueous impoundment of 4 ha surface area containing four 50-kW aerators.

SOLUTION Without specific information, the coefficients in Table 4.1-1 will be used. Equation 4.0-1 gives the flux estimate. The surface areas are: $A^{(f)} = (21)(4)(50{,}000) = 4.2E6\ cm^2$. $A^{(n)} = 40{,}000 - 420 = 39{,}580\ m^2$. From Eq. 2.1F, Henry's constant $H_\rho = 16.04\ (0.74E\text{-}9)(322)/(298)(7.9E\text{-}6) = 1.62E\text{-}3$ L H_2O/L air. Convert the coefficients in Table 4.1-1 to TCDD values using Eq. 3.1-25 with n = 2/3:

$$({}^1k'_{A2})^{(f)} = 10{,}200 \left(\frac{5.6\text{E-}6}{1.32\text{E-}5}\right)^{2/3} = 5760 \text{ cm/h}$$

Similarly, $({}^2k'_{A1})^{(f)} = 5020$, $({}^1k'_{A2})^{(n)} = 6.2$ and $({}^2k'_{A1})^{(n)} = 880$ cm/h. The two-resistance equation for the overall coefficient is Eq. 4.1-8; however, see Problem

3.1I for concentration form.

$$({}^1K'_{A2})^{(f)} = \frac{1}{(1/5760) + 1/(1.62\text{E-}3)(5020)} = 8.12 \text{ cm/h}$$

Similarly, $({}^1K'_{A2})^{(n)} = 1.16$ cm/h. From Eq. 4.1-7, $w_A = [(0.0116)(39{,}580) + (0.0812)(420)]$ m^3/h (7.9E-3 mg/m^3) = 3.90 mg/h = 0.094 g/day. This is a maximum value because the concentration in water is the solubility.

Field observations for desorption rates have been made using the concentration profile technique[10] presented in Section 6.2. These measured rates have been compared to those computed from the desorption model above. EPTC (*S*-ethyl *N*,*N*-dipropylthiocarbamate) was measured vaporizing from a flooded field.[8] Benzene, toluene, ethylbenzene, and 1,1-dichloroethane were measured from two aqueous impoundments.[8] Methylene chloride and toluene rates were measured from similar impoundments.[9] Methanol and acetone were measured above four pulp and paper industry wastewater treatment impoundments.[10] Chemical emissions to air occurred at all locations. Compared to field measurements made under ideal meteorological conditions, the desorption model values were within a factor of ± 2. For example, the benzene field value was 0.095 for the model predicted value of 0.047 $ng/cm^2 \cdot s$.

Laboratory Simulations of Desorption (or Absorption) of Chemicals

Wachs et al.[11] report on the use of surface stirrers to promote the transfer of ammonia to the atmosphere. Ammonium chloride solution was introduced in the tank; the initial ammonia concentration was 80 mg/L. A series of experiments were carried out in a wind tunnel where the wind velocity could be regulated. Results of runs with and without surface stirring are shown in Fig. 4.1-2. There is a significant improvement in desorption resulting from the use of surface stirrers. Even when surface stirrers were used, wind velocity has an added appreciable effect on the desorption rates. Ammonia desorption rate is controlled by significant resistance in both phases.

It is common practice to perform laboratory- and pilot-scale experiments with finite batches of water. Many environmental simulation experiments such as ammonia desorption and oxygen absorption are performed much easier in batch experiments than in continuous-flow experiments. Time becomes the independent variable; however, with the proper mathematical description the important environmental coefficient can be extracted. It can then be used in flow situations with a degree of confidence.

Mass transfer involving absorption or desorption across a water–air interface is a first-order kinetic process (see Eqs. 3.1-34 and 3.1-35). The chemical of study (e.g., ammonia) is placed in the batch of water in a concentration corresponding to typical environmental situations. In this case the process is

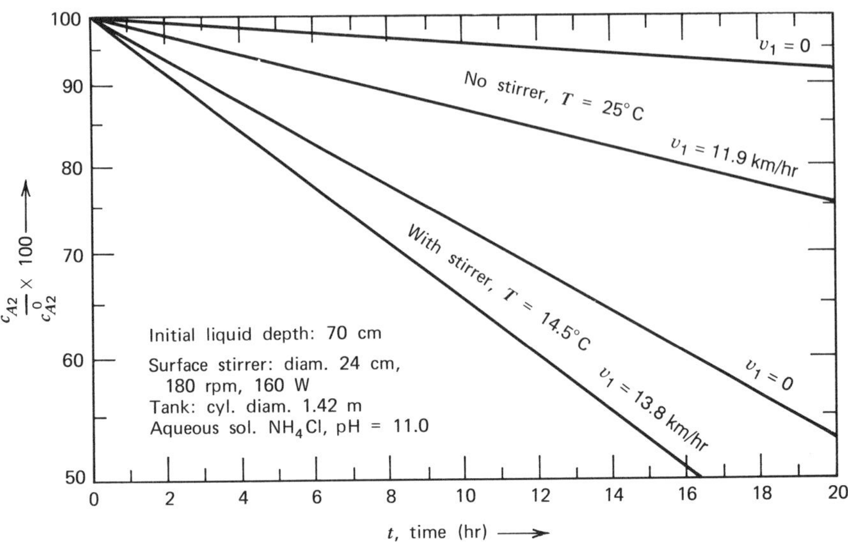

Figure 4.1-2. Ammonia desorption using surface stirrers. (Reprinted with permission from Ref. 11.)

unsteady state and the concentration of ammonia decreases with time just as does the desorption rate. The proper expression describing this transient behavior is

$$\ln \frac{c_{A2}}{c^0_{A2}} = -K_{\text{des}} t \tag{4.1-15}$$

where K_{des} is the experimentally observed desorption rate constant (t^1). Figure 4.1-3 shows a typical apparatus arrangement.

For the case of the absorption of oxygen or other chemicals into water from air, the proper expression is

$$\ln \frac{c^*_{A2} - c_{A2}}{c^*_{A2} - c^0_{A2}} = -K_{\text{abs}} t \tag{4.1-16}$$

where K_{abs} is the experimentally observed absorption rate constant. A denotes oxygen or any chemical being absorbed into the water.

Experimental investigations performed in the laboratory yield concentration–time data that can be used in Eqs. 4.1-15 and 4.1-16 to obtain the specific rate constants. Since water readily desorbs into air, corrections for water losses may need to be made if significant evaporation occurs. Once the proper constant is obtained, an additional transformation yields the overall mass transfer coefficient compatible with conventional rate equations. The experimental desorption coefficient is related to the overall liquid phase mass transfer

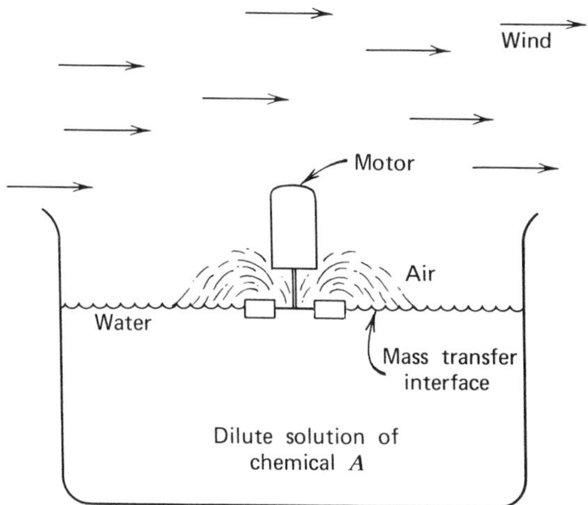

Figure 4.1-3. Laboratory simulation of batch desorption or absorption.

coefficient by

$$^{1}K_{A2} = \frac{m_2}{A} K_{\text{des}} = \rho_2 h K_{\text{des}} \tag{4.1-17}$$

where m_2 is the total mass of mixture in the batch, A the plane area of the gas–liquid interface, and h the depth of the liquid mixture in the uniform cross-sectioned vessel. Usually, the chemical of study is present as a very dilute solution so that ρ_2 is essentially the mass density of water.

Occasionally, the product $^{1}k'_{A2}a_v$ is useful, where a_v is the interfacial area per unit volume (see Eq. 4.1-10). In the case of a flat interface in a uniform cross-sectioned vessel, $a_v = 1/h$. For use in the rate equation

$$N_{A0} = {}^{1}K'_{A2}\,\Delta c_{A2} \tag{4.1-18}$$

where $^{1}K'_{A2}$ is the overall coefficient (in L/t), the following relationship is applicable:

$$^{1}K'_{A2} = hK_{\text{des}} \tag{4.1-19}$$

Equation 4.1-18 is a commonly used rate equation that employs a concentration driving force. K_{abs} can be substituted for K_{des} in the preceding equations for the absorption case.

Coefficients obtained from laboratory apparatus assembled to simulate environmental effects must be used with caution. In general, observed coefficients are overall mass transfer coefficients and measure the combination $^{1}k_{A2}$,

${}^2k_{A1}$, and H_{xA} (see Eqs. 4.1-8 and 4.1-9). Therefore, coefficients measured for one chemical do not necessarily apply to another, even though identical environmental conditions of temperature, wind, and liquid motion are known to exist. However, when the proper chemical is chosen, it is possible that the coefficient observed is one of the individual phase coefficients. For example, measurement of the oxygen absorption rate into water (i.e., aeration) is in fact a measure of the individual liquid-phase mass transfer coefficient.

Chemical Vaporization Rates Between River Surfaces and the Atmosphere

Mass transfer flux at the gas–liquid interface of rivers and streams are in general more rapid than in large bodies such as lakes and oceans, due primarily to enhanced liquid-phase coefficients. Vigorous liquid-side turbulence is present and is generated by the flowing water. For natural streamflow, a form of open-channel flow, the Reynolds number is

$$\mathrm{Re} \equiv \frac{4r_h v}{\nu_2} \tag{4.1-20}$$

where r_h is the hydraulic radius, defined as the area of the stream cross section divided by the wetted perimeter. Values of the hydraulic radius for some common cross sections are given in Table 4.1-3. The transition from laminar to turbulent flow in open channels occurs at Reynolds numbers between 2000 and 4000. Most natural streams are turbulent.

Table 4.1-3. Values of Hydraulic Radius and Area per Unit Volume for Various Stream Cross Sections

Shape and/or Cross Section	r_h[a]	a_v[b]
Rectangle, depth h, width b	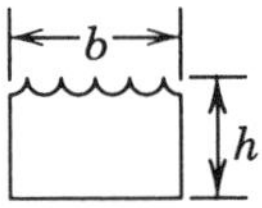 $\dfrac{bh}{b+2h}$	$\dfrac{1}{h}$
Semicircle, free surface on a diameter d	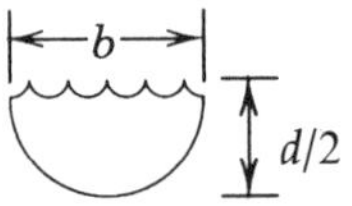 $\dfrac{d}{4}$	$\dfrac{8}{\pi d}$
Triangle trough, angle = 90°	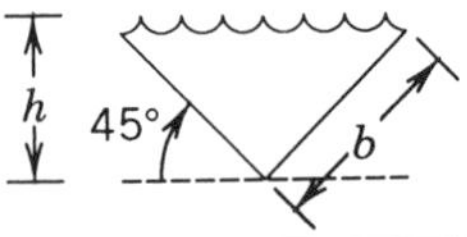 $\dfrac{b}{4} = \dfrac{h}{2\sqrt{2}}$	$\dfrac{2\sqrt{2}}{b} = \dfrac{2}{h}$

[a] $r_h \equiv$ cross-sectional area ÷ wetted perimeter.
[b] $a_v \equiv$ air–water interfacial area ÷ volume of water.

Ammonia Desorption from Rivers and Streams. The desorption of ammonia from natural streams is a specific case of some importance; however, the concepts developed may be applied to the desorption of any component. Free ammonia in water behaves as a dissolved gas and therefore exerts a vapor pressure in the surrounding air. Ammonia is frequently present in streams, its source being domestic wastewater, industrial wastewater, or agriculture runoff.

The equilibrium vapor pressure of free ammonia above an aqueous solution is a strong function of pH and temperature. Substances such as H_2S and NH_3 in water are partially dissolved as gases and partially ionized (dissociate) in water. Other important factors affecting reactive gas evolution (or solubility) are:

1. The solubilities of the undissociated gases themselves
2. The pH of the solution
3. The dissociation constant of the reactants
4. The concentration of other dissolved substances in solution

Substances that react in aqueous solution to accept or donate electrons (and consequently, donate or accept protons) dissolve to a degree dependent on solution pH and their dissociation constant K. The pH affects gas evolution through the interaction of hydrogen ions with gases that dissociate. The equilibrium reaction between ammonium ion and gaseous ammonia is

$$NH_4{}^+ \rightleftharpoons NH_3 + H^+ \tag{4.1-21}$$

Ionization removes the reactive gas from solubility considerations. The equilibrium constant for the ammonium ion dissociation reaction is

$$K \equiv \frac{c_{NH_3} c_{H^+}}{c_{NH_4{}^+}} \tag{4.1-22}$$

Equation (4.1-22) shows the relationship between pH (i.e., c_{H^+}) and free ammonia concentration, c_{NH_3}. Temperature affects both solubility and the dissociation constant. Click and Reed[12] have developed tables of equilibrium concentration of hydrogen sulfide and ammonia over aqueous solutions as a function of temperature and pH. Table 4.1-4 contains Henry's constant (mole fraction form) data for ammonia between water and air.

Bulk water motion moves dissolved ammonia from the stream depths and makes it available at the air–water interface. Desorption occurs at the interface and surface winds carry the ammonia molecules away from the air space near the interface. The flow of small, narrow streams approaches the plug flow model closely, while wide stretches of large streams and pools may more nearly follow the completely mixed flow model. Mixing in the vertical and longitudinal directions is usually incomplete. See Section 7.1, pages 468–479, for a

Table 4.1-4. Henry's Law Constant (Mole Fraction Form: $y_A = H_{xA} x_A$) for Ammonia Gas in Water

	Temperature (°F)			
pH	40	60	80	100
6	0.000266	0.000754	0.00198	0.00486
7	0.00266	0.00753	0.0197	0.0480
8	0.0263	0.0734	0.186	0.428
9	0.238	0.586	1.20	2.05
10	1.20	1.94	2.65	3.31

Source: Ref. 12.

mixing model applicable to streams that cannot be accommodated adequately by either of the ideal extremes.

Relations developed earlier for aerated basins may be applied directly to describe the desorption of ammonia from natural streams or any volatile chemical from a flowing body of water. Equation 4.1-6 in a slightly altered form, $A/Q = a_v \tau$, relates the fraction desorbed in plug flow:

$$F_p = 1 - \exp(-{}^1K'_{A2} a_v \tau) \qquad (4.1\text{-}23))$$

where ${}^1K'_{A2}$ is the overall liquid-phase mass transfer coefficient (L/t), a_v the stream air–water interfacial area per unit volume (L^2/L^3), and τ the mean water residence time (t). The completely mixed stream model is essentially Eq. 4.1-3:

$$F_M = 1 - \frac{1}{{}^1K'_{A2} a_v \tau + 1} \qquad (4.1\text{-}24)$$

Equation 4.1-4 gives the desorption rates of ammonia from the stream for the plug or mixed flow model. Equation 4.1-8 gives the overall coefficient in terms of the individual phase coefficients.

Equations appearing previously may be used to estimate the individual phase coefficients. The liquid-phase coefficient can be determined from those in Table 1.2-1 properly modified. The gas-phase coefficient can be estimated from Eq. 4.1-13; however, a specific study on the gas-phase coefficient for streams has been made.[13] Based on water evaporation from the San Diego Aqueduct in southern California the coefficient was found to be a linear function of wind speed:

$${}^2k'_{A1} = \exp\left(\frac{T - 26.1}{107}\right)(1730 + 650 v_{10}) \qquad (4.1\text{-}25)$$

The coefficient is in cm/h, wind speed, v_{10}, in m/s, and the temperature correction applies for T in °C between 18 and 48°C.

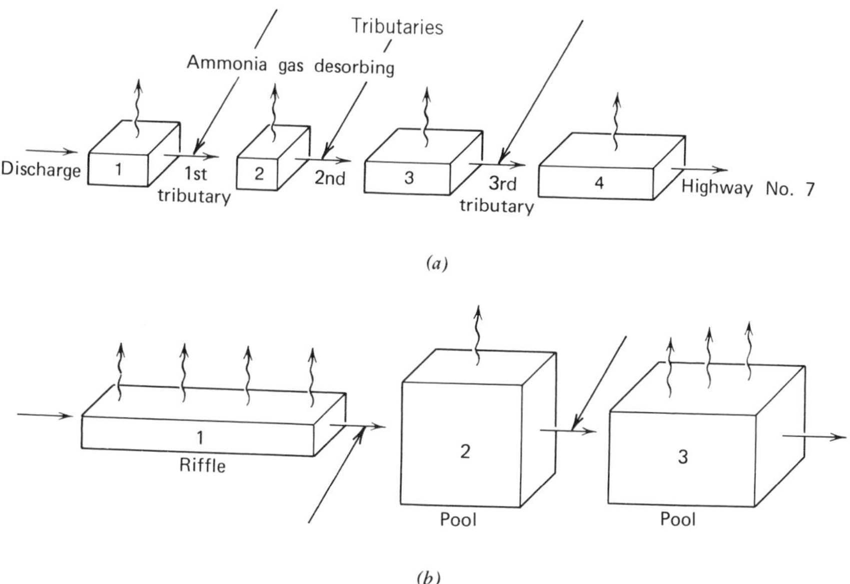

Figure 4.1-4. Stream subsectioning.

It is often desirable to subdivide a long stream into a number of subsections. Figure 4.1-4 shows two such subsectioning schemes. This is usually necessary because the flow characteristics of velocity, v_x, and depth, h, may change significantly from section to section. These changes can affect ${}^1K'_{A2}$, as is seen in Eq. 1.2-19, for example. Since the fraction desorbed, F, in either the plug or mixed-flow regimes is concentration independent, it is unaffected by tributaries free of the volatile species of concern (see Problem 4.1E). Even though F is concentration independent, the mass rate of desorption, which is given by Eq. 4.1-4, is a function of the concentration entering the first section.

Considering the subsectioned stream shown in Fig. 4.1-4a which has three tributaries, if each section is assumed to behave in a plug flow fashion, the fraction desorbed through the fourth is

$$F(4) = 1 - \exp[-({}^1K'_{A2}a_v\tau)_1 - ({}^1K_{A2}a_v\tau)_2 - ({}^1K'_{A2}a_v\tau)_3 - ({}^1K'_{A2}a_v\tau)_4] \tag{4.1-26}$$

where $({}^1K'_{A2}a_v\tau)_i$ reflects the flow conditions in the respective sections. If the three sections in the stream shown in Fig. 4.1-4b are modeled as a plug flow section followed by two mixed-flow sections, the fraction desorbed through the third section is

$$F(3) = 1 - \frac{\exp[(-{}^1K'_{A2}a_v\tau)_1]}{[1 + ({}^1K'_{A2}a_v\tau)_2][(1 + ({}^1K'_{A2}a_v\tau)_3]} \tag{4.1-27}$$

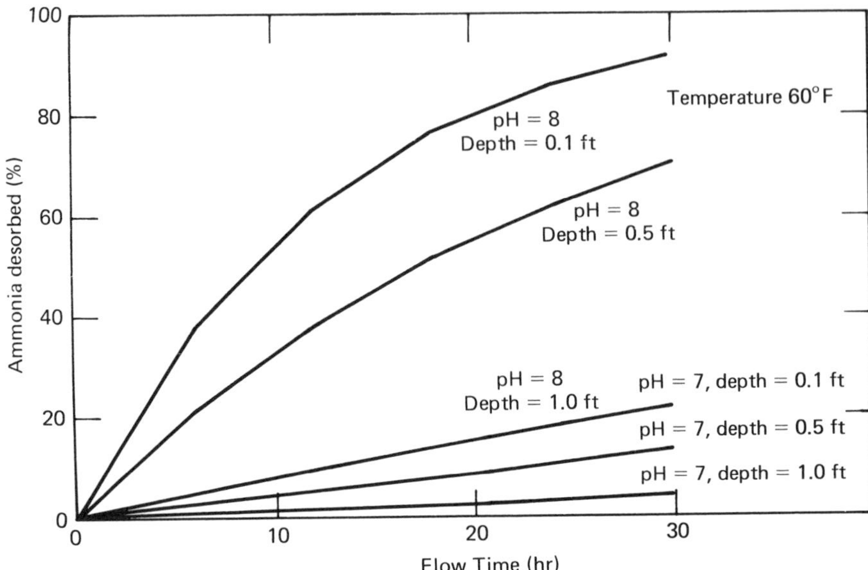

Figure 4.1-5. Predicted ammonia desorption.

Terrain resulting in pool-and-riffle streams may require this type of subsectioning. Because of the nonlinearity of the relations for F (i.e., Eqs. 4.1-23 and 4.1-24) the use of average values ${}^1K'_{A2}a_v\tau$ can produce significant errors, usually yielding lower desorptions estimates than what is actually occurring.

The stream depth is reflected for the most part by the a_v term. Table 4.1-3 gives algorithms for estimating a_v values for various stream cross sections. For the case of wide streams a_v is essentially the reciprocal of the stream depth. Accurate data on stream depth, which are particularly difficult to obtain for small streams, are essential if realistic desorption predictions are to be made. Figure 4.1-5 shows the sensitivity of the fraction of ammonia desorbed with stream depth for a small stream in southern Arkansas. Desorption in deep rivers is significantly less than in small streams, owing primarily to depth.

If the background air contains a significant quantity of the volatile species, ρ_{A1}, the minimum concentration in water at equilibrium is $\rho^*_{A2} = \rho_{A1}/H_\rho$. In this case the fraction desorbed is defined as

$$F \equiv \frac{\rho_{A2} - \rho^*_{A2}}{\rho_{A21} - \rho^*_{A2}} \tag{4.1-28}$$

Rather than representing the volatile fraction of the total quantity in the stream, F now represents the fraction in the stream that can be volatilized. With this definition all the relationships developed above for the mixed- and plug flow cases remain valid.

Suspended Solids in the Water Column. Flowing streams have the ability to carry solid particles in suspension. This is due to fluid turbulence whereby the instantaneous fluctuating vertical component of the water velocity (see page 23) is greater than the particle settling velocity. The origins of aquatic particulates vary with water type. In rivers, the great majority of the material is terrestrial in origin, coming largely from bank erosion, with the amount being strongly flow dependent. In some cases, such as the Rio Grande in the United States or the Yellow River of China, each liter of water may contain 10 g of suspended matter, while at the other extreme a clear mountain stream will have less than 1 mg/L. The world average value is 200 mg/L.[14] Theoretically the mass of solid particles in suspension has a buffering capacity with respect to the volatiles and retards the desorption rate and time.[16]

The total volatile species concentration is the sum of that in solution, ρ_{A2}, and that sorbed onto particles, $\omega_A \rho_{32}$, where ρ_{32} is the suspended solids concentration in water (mg/L). If equilibrium partitioning is assumed between the aqueous and solid phases,

$$\rho_{A2} + \omega_A \rho_{32} = \rho_{A2}(1 + \mathrm{K}^*_{A32}\rho_{32}) \tag{4.1-29}$$

The term in parentheses accounts for the additional capacity the suspended solids have for the volatiles. All relationships in this chapter that contain time, t, or residence time, τ, need to be modified accordingly. For example, Eq. 4.1-23 is now

$$F_p = 1 - \exp\left(-\frac{{}^1K'_{A2} a_v \tau}{1 + K^*_{A32}\rho_{32}}\right) \tag{4.1-30}$$

In reality it is the quantity of A in the volume V in $a_v(\equiv A/V)$ that is being modified (see Problem 4.1H).

Example 4.1-2. Dioxin Vaporization Fate in a River. A paper mill located on the Tennessee River (see Table 1.2C) is suspected of discharging traces of 2,3,7,8-TCDD with its water and a fate study is needed. The water is estimated to have 1 ng/L TCDD in solution, contains 500 mg/L suspended solids, and is at 25°C. Use 0.88 cm/h as to overall liquid-phase coefficient and assume plug flow conditions for the river and no TCDD in the air. The partition coefficient for TCDD is estimated to be 4680 L/kg for solids with 1% organic matter.

(a) Compute the TCDD residence time half-life in hours.

(b) Obtain the TCDD vaporization rate in ng/m$^2\cdot$h as a function of residence time.

SOLUTION (a) The half-life is given by Eq. 4.1H-2 and is $\tau_{1/2} = 0.693(120\text{ cm})(1 + 4680\cdot 500\cdot \text{E-6})/(0.88\text{ cm/h}) = 316\text{ h}$.

(b) Equations 4.0-1 and 4.1-5 are combined to yield the time-varying flux rate:

$$n_A = {}^1K'_{A2}\rho_{A20}\exp\left(-\frac{{}^1K'_{A2}a_v\tau}{1 + K^*_{A32}\rho_{32}}\right)$$

$$= (0.88\ \text{cm/h})(0.00\ 1\ \text{ng/cm}^3)\exp\left[-\frac{\tau(0.88)}{120(3.34)}\right]$$

$$= 8.8\exp(-0.00313\tau)$$

with τ in hours. Vaporization is a pathway for TCDD from the river, but the half-life is long and the rate is low.

There are some laboratory data on organic chemical volatilization from water in the presence of suspended solids. Dilling et al.[19] observed that peatmoss at $\rho_{32} \simeq 500$ mg/L increased the evaporation half-life of five chlorinated hydrocarbons. The effective value of $1 + K^*_{A32}\rho_{32}$ extracted from the data is 2.3. Singmaster and Crosby[43] observed that the evaporation half-lives of DDT, dieldrin, and aldrin in water containing suspended solids was two to three times larger than in solids-free water. San Francisco Bay, American River, and Sacramento River water was used; the latter contained settleable solids. Theoretically, for DDT in water with $\rho_{32} = 200$ mg/L and K^*_{A32} of 4600 L/kg, the term $1 + K^*_{A32}\rho_{32} = 1.92$. This corresponds to an approximate doubling of the evaporation half-life.

The equilibrium distribution of highly hydrophobic organics between the particles and the aqueous phase is probably a good assumption for those particles approaching 1 μm in size and smaller. For those particles approaching 100 μm in size, intraparticle diffusion effects become important and chemical transport within is important.[17] In the case of these larger particles, the vaporization rates are less than predicted by the models above.

Desorption Under Quiescent Conditions

Thermal gradients frequently exist across the air–water interface. Typically, the wastewater that is discharged from a manufacturing facility is warmer than the surrounding air. As it moves about in open flow channels, through holding basins or treatment vessels, and finally, to the receiving streams, it loses thermal energy and volatile chemicals to the atmosphere. In the absence of wind the transport processes are controlled by natural convection. Some fundamental concepts of the process appear in Section 3.2 on page 140.

Vertical profiles of temperature, volatile chemical concentrations, and air humidity are depicted hypothetically in Fig. 4.1-6 across an air–water interface where $T_2 > T_i > T_1$. Due to the temperature gradients, the fluids are unstable on either side of the interface. This occurs because warm air near the surface at T_i is less dense than that above at T_1 (see Problem 4.1M). Similarly, the

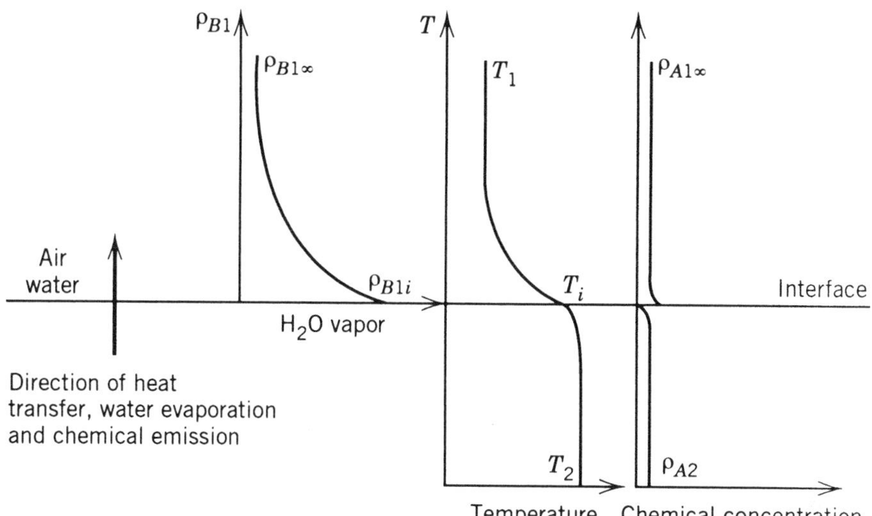

Figure 4.1-6. Gradients across the air–water interface.

water at T_2 is less dense than that near the surface at T_i. In addition, the air near the surface is less dense because it is more humid than the surrounding air. Water vapor has a molecular weight of 18 g/mol and that of dry air is 29 g/mol. Any density gradients due to volatile chemical gradients are usually negligible because the concentration is low.

Quantifying the effect of the air-side instabilities begins with Eq. 3.2-47, which is an equation of state for air density as a function of temperature and humidity. For the gas side the density gradient is

$$\frac{d\rho_1}{dy} = -\beta_1\rho_1\frac{dT_1}{dy} - \zeta_{B1}\rho_1\frac{d\rho_{B1}}{dy} \tag{4.1-31}$$

where β_1 is the coefficient of thermal expansion of air (T^{-1}), and ζ_{B1} is the mass coefficient of volume expansion for water vapor in air L^3/M (see Problem 4.1M). Gravity acting upon this density gradient will induce an upward fluid velocity, v_y, in the air immediately above the interface; the appropriate solution of Eq. 3.2-49 can yield an estimate of it. This flow transports the volatile constituent upward and away from the interface. A transport coefficient correlation based on laboratory data has been developed.[18] It is

$${}^2k'_{A1} = 0.5\mathscr{D}_{A1}^{2/3}\left(\frac{g}{v_1}\right)^{1/3}\left[\left(\frac{T_i - T_1}{T_i}\right)^{1/3} + 0.724\left(\frac{p_{Bi} - p_{B1}}{p_T}\right)^{1/3}\right] \tag{4.1-32}$$

where T_i(K) and p_{Bi}(atm) are the temperature and vapor pressure of water at the interface.

Quantifying the effects of the water-side instabilities follows a similar development. A horizontal, heated flat plate facing upward[15] provides a close simulation to cold water above warm. Substituting the Sherwood number for the Nusselt number and the Schmidt number for the Prandtl number (see Table No. 3.3-2) in the original correlation yields[18]

$$^{1}k'_{A2} = 0.14\left[\frac{g\beta_2(T_2 - T_i)\mathscr{D}^2_{A2}}{\nu_2}\right]^{1/3} \quad (4.1\text{-}33)$$

where β_2 has dimensions of T^{-1}. See Example 7.1-1 for information and numerical values of β_2.

In the case of vaporization under no-wind conditions the flux is quantified by Eq. 4.0-1 and the two-resistance theory applies. Typically, the individual transport coefficients under unstable, no-wind conditions are smaller than those under windy conditions. The air-side coefficient is approximately one hundredth and the water-side coefficient is approximately one-tenth of the windy values. For stable or neutral conditions of air and water under no-wind conditions the estimated values of both coefficients are even smaller.[41]

PROBLEMS

4.1A. Model for Chemical Desorption from Plug Flow Basin

1. Derive the differential equation that describes the desorption in a plug flow basin. (*Hint:* Using Eq. 1.1-6 and Fig. 4.1-1 as a guide, write a mass balance on component A for an elemental volume ΔV.)
2. Integrate the result of part 1 with the proper boundary conditions and use the definition of F to obtain Eq. 4.1-6.

4.1B. Chemical Vaporization from Water Is Vapor-Pressure Independent

1. Compute the fraction (F_M and F_p) of benzene and *o*-xylene desorbed at 25°C from a basin containing 50E6 gal with a flow of 10E gal/day. Use the average coefficients given in Table 4.1-1. Report the fraction desorbed as percentages. Note the difference in chemical vapor pressures.
2. Compute the desorption rate of each chemical for each basin type in g/day The inlet concentrations are 100 mg/L each. Are the rates comparable?
3. In the cases of benzene and *o*-xylene, which phase resistance controls the rate of mass transfer?

4.1C. Liquid-Phase Mass Transfer Coefficient for Mechanical Surface Agitators

As given in Section 3.1, individual phase mass transfer coefficients are scalars and are therefore independent of the direction of the concentration driving force. This means that a coefficient measured under absorption conditions can be transformed and used in desorption applications.

Experimental observations on the absorption of molecular oxygen into water by mechanical surface aerators has resulted in the correlation

$$\omega'_B = \frac{n'_{BO} E P \alpha}{9.17} (1.024)^{T - 20^\circ C} (\rho^*_{B2} - \rho_{B2}), \text{ lb } O_2/\text{hr} \tag{4.1C}$$

where $n'_{BO} = 2$ to 4 lb O_2/hr·hp, depending on the specific aerator; E is the specific aerator power delivery efficiency, 0.65 to 0.9; dimensionless; P the nameplate horsepower, hp; α the dirty water to clean water ratio, 0.8 to 0.85, dimensionless; T the water temperature, °C; ρ^*_{B2} the solubility of molecular oxygen in water at T(°C), mg O_2/L; and ρ_{B2} the actual concentration of oxygen in water, mg O_2/L. Transform Eq. 4.1C to Eq. 4.1-10. Use Eq. 3.1-25 to aid the transformation.

4.1D. EPTC Vaporization from a Flooded Field

EPTC vaporization from a flooded field was measured by Claith et al.[38] By a process known as herbigation it was applied at an average concentration of 2.2 ppm in the inlet water. The concentration in the exit water and wind speed was measured with time, as was the EPTC vaporization rate using the aerodynamic method (see Section 6.2). Table 4.1D contains the data. EPTC is *S*-ethyl *N*,*N*-dipropylthiocarbamate, with a molecular weight of 189.3, and $H_\rho = 7.29E - 3$ cm^3H_2O/cm^3 air at 30°C.

1. Using the desorption model, calculate the vaporization flux in g/ha·h at each time.
2. Compare the calculated and measured fluxes in time on a graphical plot.
3. Calculate the grams vaporized by each method over the 8-h period.

Table 4.1D. EPTC Field Data

Time (h)	Wind Speed at 1 m (m/s)	Average Concentration (ppm)	Flux (g/ha·h)
1600	1.0	2.0	35
1800	2.1	2.0	140
2000	2.7	1.8	260
2200	2.0	1.4	65
2400	1.8	1.4	50

4.1E. Volatilization from Rivers with Tributaries

When tributaries free of species A enter the main channel the concentration of the volatile constituent is diluted. Referring to Figure 4.1-4*a*, show that the fraction desorbed, F_M or F_p, at a point downstream of one or more freshwater tributaries is independent of the chemical concentration. (*Hint:* Use component mass balance at the mixing point between sections.)

4.1F. Desorption of Ammonia from Aerated Stabilization Basin

Compute the fractional removal and rate (kg/day) of ammonia by desorption using both the completely mixed and plug flow models. The following data are available ($A \equiv NH_3$, $B \equiv O_2$, $1 \equiv$ air, $2 \equiv H_2O$): $v_8 = 4$ mi/hr, $T1 = T2 =$ 25°C, $h = 7$ ft, $\alpha = 1.0$, $n'_{BO} = 3$ lb/hr·hp, E = 0.83, $A = 65.5$ acres, $Q_2 = 27E6$ gal/day, $V = 384E6$ gal, 13 low-speed (50 rpm) aerators of 100 NPHP each, impeller diameter = 3 ft, $\rho_{A21} = 15$ mg A/L, $Sc_{A1} = 0.78$ for A, $\mathscr{D}_{A1} = 0.28$ cm^2/s, $\mathscr{D}_{A2} = 2E\text{-}5$ cm^2/s, $\mu_2 = 1.2E\text{-}5$ lb$_m$/ft·s and $H_{Ax} = 0.018$.

4.1G. BTX Volatilization Fate in Lower Mississippi River

The fate of benzene, toluene and xylene (BTX) is of interest to the chemical industry located along the lower Mississippi River. Assuming Baton Rouge, Louisiana as the discharge point, compute and graph the fraction of each volatilized upon reaching the downriver locations represented by Plaquemine, LaPlace, New Orleans, and Pilot Town. Assume a flow of 500,000 ft^3/s with benzene mass transfer coefficients of ${}^1k'_{A2} = 4.1$ cm/h and ${}^2k'_{A2} = 1180$ cm/h. The respective BTX sediment–water partition coefficients are 41, 90, and 177 L/kg. Perform calculations for 1.0 and 10,000 mg/L suspended solids concentration.

4.1H. Vaporization with Suspended Solids in the Water Column

Assume equilibrium partitioning of the volatile species between the solid and water;

1. Redo the component balance for a completely mixed-flow basin with ρ_{32} suspended solids concentration containing ω_A weight fraction species A in addition to that in solution. How is Eq. 4.1-2 modified for suspended solids?
2. For the mixed-flow cases show that the residence-time half-life is

$$t_{1/2} = \frac{h(1 + K^*_{A32}\rho_{32})}{{}^1K'_{A2}} \tag{4.1H-1}$$

3. Repeat part 1 for the plug flow case. How is Eq. 4.1-5 modified?

Table 4.1I. Simultaneous Absorption and Desorption

Acetone Desorption Data		Oxygen Absorption Data	
Time (min)	Concentration (mg C/L)	Time (min)	Concentration (mg O_2/L)
0	3170	0.0	1.95
18	3140	1.5	2.50
35	2920	3.87	2.80
52	2480	5.83	3.20
67	2170	8.13	3.70
90	2050	10.9	4.10
127	1780	14.3	4.60
157	1490	20.1	5.20
363	660	26.6	5.80

4. For the plug flow case show that the residence time half life is

$$t_{1/2} = \frac{0.693h(1 + K^*_{A32}\rho_{32})}{{}^1K'_{A2}} \tag{4.1H-2}$$

where K^*_{A32} is the partition coefficient.

4.1I. Simultaneous Acetone Desorption and Oxygen Absorption

Acetone desorption and oxygen absorption were studied simultaneously in a pilot-scale surface-agitated vessel. Experimental conditions: 0.075-hp electric motor; single-blade 3.0-in. diameter impeller; vessel 50-in. diameter; water depth 8.75 in.; floor fan to simulate wind at 3 to 4 mph; temperature 18 to 24.5°C; impeller speed 465 to 545 rpm; 35.3 g of Na_2SO_3 and 1.98 g of CoO used to create an oxygen deficit in the vessel. The experimental results are shown in Table 4.1I.

1. Determine the experimental desorption coefficient for acetone (hr^{-1}).
2. Determine the experimental absorption coefficient for oxygen (hr^{-1}).
3. Explain the difference, if any, in the coefficients for absorption and desorption.

4.1J. Comprehensive Ammonia Desorption Problem

A small creek receives ammonia from the discharge of an ammonia nitrate manufacturing facility. Compute the fraction desorbed and the quantity desorbed after each subsection (see Table 4.1J). For mass transfer coefficients

Table 4.1J. Haynes Creek, South-Central Arkansas

Parameter	Section 1	2	3	4
Velocity (ft/sec)	0.28	0.31	0.34	0.86
Depth (ft)	0.35	0.44	0.32	0.52
pH	8.4	8.0	8.0	7.5
Temperature (°C)	20	15	15	15
Detention time, τ (hr)	3.27	1.68	3.55	2.5
Flow (ft^3/sec)	1.78	2.00	2.22	5.58

use ${}^2k_{A1} = 0.4$ lb mol/hr$\cdot$ft^2 and ${}^1k_{A2} = 0.43$ lb mol/hr$\cdot$ft^2. The ammonia concentration into subsection 1 is 50 mg/L.

4.1K. Evaporation Rates of Low-Molecular-Weight Chlorinated Hydrocarbon in Laboratory Studies

Solutions that contain 1.0 ppm (weight) of each of five chlorinated compounds were prepared. The solution of the chlorinated compounds in water (200 mL, solution depth 65 mm before stirring) was placed into a beaker and after starting the stirrer (200-rpm stainless steel shallow-pitch propeller in a 250-mL Pyrex beaker) samples were withdrawn and mass spectra scanned after 1 min and periodically thereafter. Peak heights were assumed to be proportional to concentration. Figure 4.1K shows typical evaporation results. Table 4.1K contains solubility and pure component vapor pressure data for the five chemicals.

1. Which phase resistance controls the evaporation?
2. Will stirrer speed affect the evaporation rate?
3. Compute the ${}^1K'_{A2}$ (cm/h) average for the five compounds based on the data in Fig. 4.1K. Evaporation temperature was 25°C.

Table 4.1K. Chlorinated Hydrocarbon Data

Compound	Solubility (ppm)	Vapor Pressure (mmHg)
CH_2Cl_2	19,800	426
$CHCl_3$	7,950	200
CH_3CCl_3	1,300	123
$CHCl{=}CCl_2$	1,100	74
$CCl_2{=}CCl_2$	400	19

Source: Ref. 19.

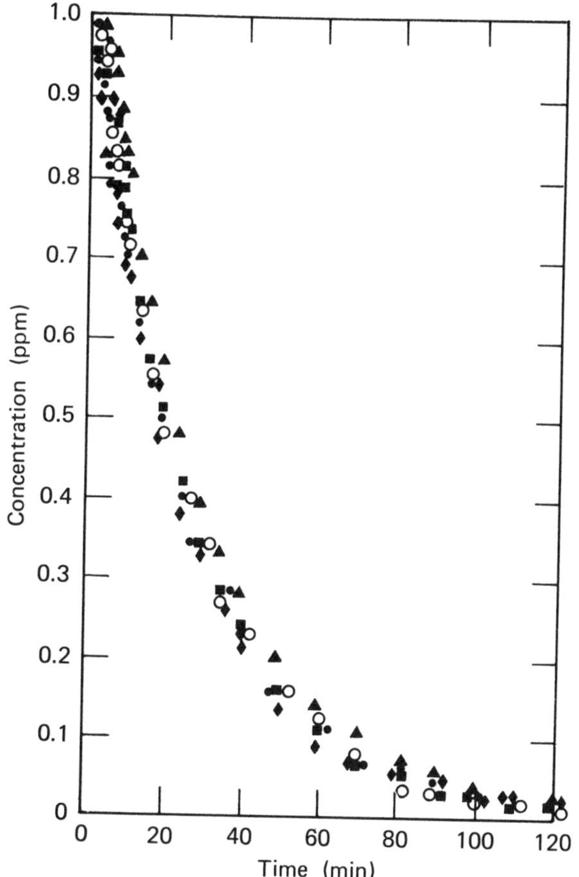

Figure 4.1K. Evaporation of rates of CH_2Cl_2 (◆), $CHCl_3$ (●), CH_3CCl_3 (○), $CHCl{=}CCl_2$ (■), and $CCl_2{=}CCl_2$ (▲) from water. (Reprinted with permission from Ref. 19. Copyright by the American Chemical Society.)

4.1L. Evaporation Rates of Chloroethanes and Propylenes from Dilute Aqueous Solutions

Laboratory studies of the kinetics of evaporation of three compounds shown in Table 4.1L appear in Fig. 4.1L.

1. Make a visual comparison of the data in Fig. 4.1L with those in Fig. 4.1K. Ignoring the data scatter in the region of 0 to 5 min, explain the wide variation of evaporation rates observed in Fig. 4.1L. Note that the experimental technique for obtaining the data is identical.
2. Compute the overall liquid-phase mass transfer coefficient (cm/min) for $CHCl_2CHCl_2$ from the experimental data in Fig. 4.1L.

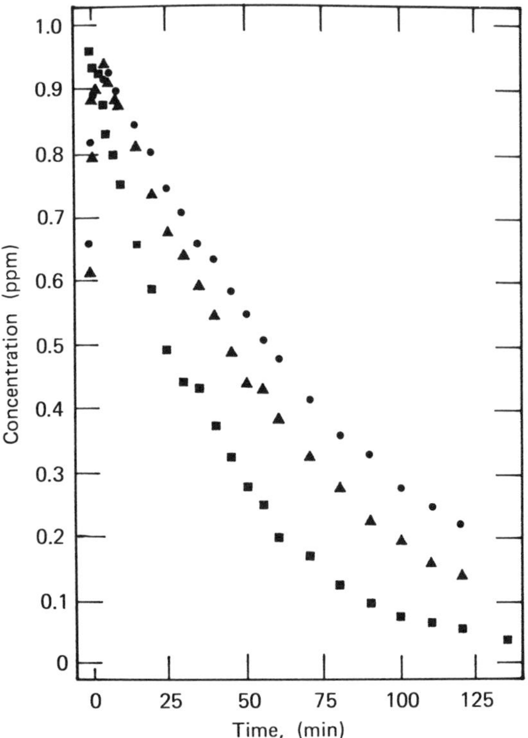

Figure 4.1L. Evaporation rates of $CH_2{=}CHCH_2Cl$ (■), CH_2ClCCl_3 (▲), and $CHCl_2CHCl_2$ (●) from water. (Reprinted with permission from Ref. 20. Copyright by the American Chemical Society.)

3. Compute the overall liquid-phase mass transfer coefficient (cm/min) for $CHCl_2CHCl_2$ from the two-resistance theory, given

$$k'_{A1} = 50\left(\frac{\mathrm{M}_{\mathrm{H_2O}}}{M_A}\right)^{1/2} \quad \text{cm/min}$$

$$\frac{1}{{}^1K'_{A2}} = \frac{1}{{}^1k'_{A2}} + \frac{1}{H_\rho^2 k'_{A1}} \tag{4.1L}$$

$${}^1k'_{A2} = 0.33\left(\frac{M_{\mathrm{CO_2}}}{M_A}\right)^{1/2} \quad \text{cm/min}$$

4.1M. Air Density Instability Parameters near Warm Water

Warm air residing below cooler air is an unstable microclimate. For dry air:

1. Show that density decreases with increasing T at constant p_T and composition.

Table 4.1L. Evaporation Parameters of Chlorohydrocarbons in Water

Compound	Solubility (ppm)	Vapor Pressure (mmHg at 20°C)	Partition Coefficient, H_ρ^a
$CH_2{=}CHCH_2Cl$	3370	361	0.44
CH_2ClCCl_3	1100	13.9	0.11
$CHCl_2CHCl_2$	3000	6.5	0.019

Source: Ref. 20.

[a] $H_\rho = c_1 H_{xA}/c_2$.

2. Develop the relationship $\beta_1 = T^{-1}$, T in K.

Humid air residing below drier air is also an unstable microclimate.

3. Show that the density of a gas is proportional to its molecular weight at constant T.

4. For a binary gas mixture consisting of water vapor ($\equiv B$) and dry air ($\equiv A$) develop the following relationship for ζ_{B1}. At the earth's surface air density is primarily a function of temperature and the concentration of water vapor; $\rho_1 = f(T, \rho_{B1})$.

$$\zeta_{B1} = -\left[y_B + \frac{M_A}{M_B - M_A}\right]^{-1} \tag{4.1M}$$

5. Show the development of Eq. 4.1-31. (*Hint:* Assume the ideal gas equation of state applies).

4.1N. Measurement of Volatile Chemical Emissions from Wastewater Basins[4]

The working equation for the *concentration profile method* or *aerodynamic method* of determining emission rates in the field is

$$n_A = -\left(\frac{\mathscr{D}_{A2}}{\mathscr{D}_{B1}}\right)^{2/3} \frac{S_v S_\rho \kappa_1^2}{\phi_m^2 \mathrm{Sc}_{A1}^{(t)}} \tag{4.1N}$$

S_v and S_ρ are the slopes of wind velocity, v_x, and volatile species concentration in air, ρ_{A1}, versus. $\ln y$ data, respectively, where y is elevation taken in a turbulent boundary layer above the basin that is undisturbed by upwind obstructions. (See Section 6.2 for the development of Eq. 4.1N and details for its application). Table 4.1N contains data from a surface impoundment at a hazardous waste disposal facility. It represents an approximate 30-min sample time in the afternoon of August 26, 1981. Calculate the benzene emission rate in $ng/cm^2 \cdot s$.

Table 4.1N. Field Data in Boundary Layer Above a Wastewater Basin

Wind Velocity		Air Temperature		Benzene Concentration	
Elevation, y (cm)	Speed, v_x (cm/s)	Elevation, y (cm)	T_1 (°C)	Elevation, y (cm)	ρ_{A1} (μg/L)
58.1	274	47.0	19.65	6.35	0.138
78.1	282	59.7	19.6	15.2	0.123
118	288	77.5	19.6	24.3	0.123
198	307	102.0	19.55	55.2	0.089
278	318	150.0	19.55	104.0	0.124
358	333	220.0	19.5	232.0	0.058

4.1O. Transport Coefficients Without Wind[42]

An existing surface impoundment receiving aqueous waste containing several volatile chemicals, benzene being a major component, is suspected of emitting these to air.

1. Estimate the air-side coefficient under no-wind conditions for $T_i = 31$ and $T_1 = 25°C$ at 0% relative humidity.
2. Estimate the water-side coefficient for $T_2 = 35°C$.

The atmospheric pressure is 29.96 in. Hg. β_2 at 35°C is 3.6E-4 K^{-1}. Answers in cm/h.

4.2. EXCHANGE OF CHEMICALS ACROSS THE AIR–WATER INTERFACE OF LAKES AND OCEANS

Lakes and oceans differ from rivers and small basins in more ways than just physical size. Chemical exchanges at the air–water interface are dominated by the effect of the wind on the water. The flow of water through the system, resulting in a detention time, is not pertinent for large lakes and oceans; however, density stratification is important. An idealized picture of a stratified body of water is a well-mixed layer at the surface, a layer with a more or less well developed density gradient (pycnocline) below it, and a well-mixed layer below the pycnocline. In many freshwater lakes the density stratification is thermal in origin (see Fig. 7.1-1 for an example.)

Rates of Loss of Low-Solubility Contaminants from Water Bodies Such as Lakes and Oceans

Chlorinated hydrocarbons such as pesticides and polychlorinated biphenyls (PCBs) have been transported widely throughout the global environment, even to remote arctic and antarctic regions. The major route by which these contaminants are transported is apparently through the atmosphere. Analysis of rainwater in England has shown concentrations of total pesticide residues of 104 to 229 ppt DDT; concentrations of 40 ppt have been reported in meltwater from antarctic ice. These residues presumably originate as vapors or absorbed on dust particles and may be carried many thousands of miles from the original source. Some of these source materials are applied by spraying techniques, which allow the possibility of direct evaporation; however, most are used as solids, liquids, or wettable powders in which transport to the atmosphere can take place only by natural evaporative processes when exposed to the atmosphere.

Phase Equilibrium of Low-Solubility Contaminants. As presented in Chapter 2, the partial pressure of chemical A and its concentration in the aqueous solution may be linked by equating the fugacities in both phases (e.g., Eq. 2.1-2). For the case of low-solubility chemicals it is then possible to obtain a phase equilibrium relationship from the accessible properties p_A^0 and x_A^* or ρ_{A2}^* (mg/L):

$$p_A\,(\equiv y_A p_T) = \frac{p_A^0}{x_A^*}\,x_A \qquad (4.2\text{-}1)$$

This is another form of Henry's law; see Eq. 2.1-12. Some pure-component vapor pressure and solubility data for important environmental chemicals appear in Table 4.2-1.

Quantification of Rates of Evaporation from Aqueous Solution or Suspension. The following is a modification of the original source materials[21,22] that includes suspended solids. In deriving the rate equations it is convenient to consider a column of water 1 m^2 in cross section of depth h m containing h m^3 of water, as shown in Fig. 4.2-1. The concentration of the evaporating compound is given by Eq. 4.1-29 so that the quantity in the column is $\rho_{A2}(1 + K_{A32}^*\rho_{32})h$ grams. Equation 4.2-1 is suitable for phase equilibrium. The mass flux across the phase boundary can be expressed by Eq. 4.0-1:

$$n_A = {}^1K'_{A2}(\rho_{A2} - \rho_{A2}^*) \qquad (4.2\text{-}2)$$

assuming that there is a nonnegligible atmospheric level of the contaminant.

Unsteady-State Model. A mass balance on A for the volume of water shown in Fig. 4.2-1, assuming no other methods of loss, yields

$$\frac{d\rho_{A2}}{dt} = -\frac{{}^1K'_{A2}(\rho_{A2} - \rho_{A2}^*)}{(1 + K_{A32}^*\rho_{32})h} \qquad (4.2\text{-}3)$$

Table 4.2-1. Evaporation Parameters for Various Compounds at 25°C

Compound	Molecular Weight, M_A	Solubility, ρ^*_{A2} (mg/L)	Vapor Pressure p^0_A (mmHg)	${}^1K'_{A2}$ (m/h)
Alkanes				
n-Octane	114	0.66	14.1	0.124
2,2,4-Trimethylpentane	114	2.44	49.3	0.124
Aromatics				
Benzene	78	1780	95.2	0.144
Toluene	92	515	28.4	0.133
o-Xylene	106	175	6.6	0.123
Cumene	120	50	4.6	0.119
Naphthalene	128	33	0.23	0.096
Biphenyl	154	7.48	0.057	0.092
Pesticides				
DDT ($C_{14}H_9Cl_5$)	355	0.0012	1E-7	9.34E-3
Lindane	291	7.3	9.4E-6	1.5E-4
Dieldrin	381	0.25	1E-7	5.33E-5
Aldrin	365	0.2	6E-6	3.72E-3
Polychlorinated biphenyls (PCBs)				
Aroclor 1242	258	0.24	4.06E-4	0.057
Aroclor 1248 ($C_{12}H_6Cl_4$)		5.4E-2	4.94E-4	0.072
Aroclor 1254 ($C_{12}H_5Cl_5$)		1.2E-2	7.71E-5	0.067
Aroclor 1260 ($C_{12}H_4Cl_6$)		2.7E-3	4.05E-5	0.067
Other				
Mercury	201	3E-2	1.3E-3	0.092

Source: Reprinted with permission from Ref. 22. Copyright by the American Chemical Society.

Integrating this equation with ρ^0_{A2}, the concentration at zero time, results in

$$\rho_{A2} = \rho^*_{A2} + (\rho^0_{A2} - \rho^*_{A2}) \exp\left[-\frac{{}^1K'_{A2}t}{(1 + K^*_{A32})h}\right] \tag{4.2-4}$$

If the atmospheric level of the contaminant is low and a half-life $\tau_{A1/2}$ is defined as the time required for the concentration to drop to half its original value, Eq. 4.2-4 becomes

$$\tau_{A1/2} = \frac{0.693(1 + K^*_{A32}\rho_{32})h}{{}^1K'_{A2}} \tag{4.2-5}$$

The unsteady-state model represents a situation in which a water body becomes depleted of a compound introduced at a point in time (e.g., from an accidental spill). It also represents a situation of agricultural runoff containing

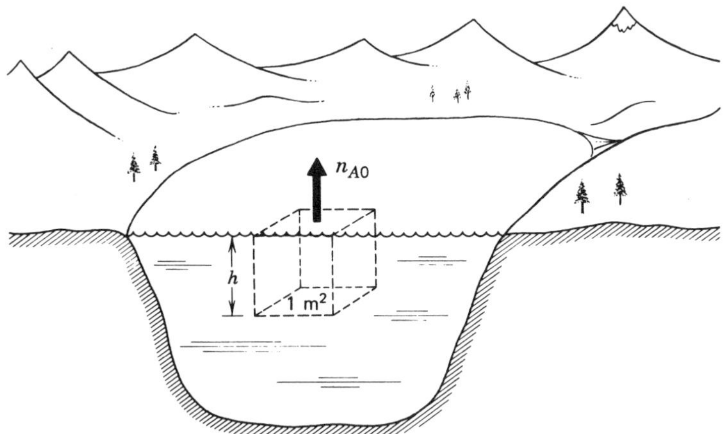

Figure 4.2-1. Chemical evaporation from lake ecosystem.

a pesticide in which the warmer river water "floats" on the surface of the receiving lake to a depth h and becomes depleted of the compound. The $K^*_{A32}\rho_{32}$ group represents the relative quantity of A sorbed reversible onto suspended sediment. Equilibrium desorption of A back into the aqueous phase must occur prior to evaporation, and this process increases the evaporation half-life as shown in Eq. 4.2-5.

Steady-State Model. Situations arise where there is a fairly constant influx from other sources and the evaporation rate of the contaminant is exactly balanced by this influx. The concentration ρ_{A2} adjusts to a value such that these rates are equal. If the influx rate is i_A (g/m$^2\cdot$h), equating this to n_A in Eq. 4.2-2 yields

$$\rho_{A2} = \frac{i_A}{{}^1K_{A2}} + \rho^*_{A2} \tag{4.2-6}$$

If the mean residence of time of A in the volume is defined

$$\tau_A \equiv \frac{(\rho_{A2} + \omega_A\rho_{32})h}{i_A} \tag{4.2-7}$$

it can then be calculated from Eq. 4.2-6 that

$$\tau_A = (1 + K^*_{A32}\rho_{32})h\left(\frac{1}{{}^1K'_{A2}} + \frac{\rho^*_{A2}}{i_A}\right) \tag{4.2-8}$$

Equations 4.2-7 and 4.2-8 are valid for $i_A > 0$. If ρ^*_{A2} is negligible, the mean residence time differs from the unsteady-state model half-life only by the factor

0.69. Table 4.2-1 gives the overall liquid-phase mass transfer coefficient for various compounds at 25°C. For low suspended solids concentration, $\rho_{32} \rightarrow 0$, the half-lives vary from 4.81 hr for benzene to 12,940 hr for dieldrin. Most of the resistance to mass transfer resides in the liquid phase. Calculations of half-life and residence time are left as an exercise for the student (see Problem 4.2C).

The half-lives for a water depth of 1 m show that most of these compounds evaporate rapidly from solution. In situations where the water body is turbulent, with frequent exchange between the surface water layer and the bulk (e.g., in a fast-flowing shallow river or during whitecapping on a lake or ocean), the liquid-phase mass transfer coefficient may be increased considerably and the evaporation rate increased accordingly. For depth greater than 1 m, the half-life is increased correspondingly.

The model equations developed above are valid for any body of water that is well mixed through the depth *h*. Clearly, this is not the depth of the water column from the sediment–water interface to the air–water interface in many lakes and the ocean. For lakes, a reasonable value of h is the depth of the epilimnion. The epilimnion is mainly surface waters above the thermocline characterized by nearly uniform mixing and near isothermal conditions. A similar region exists in ocean waters. Other rate-limiting transport processes at depths in water bodies, for example, through the thermoclines, were not considered in the preceding treatment. The topic of intraphase diffusional processes is presented in Section 7.1. A brief introduction and some consequences of intraphase diffusion follow.

Times to Chemical Steady States in Lakes and Oceans

In natural systems of large dimensions many chemical processes are controlled by the transport of the species through the system. The distribution of chemical species in natural systems is only too often not homogeneous; concentration gradients and more or less abrupt changes in abundance from one part of an environment to another are commonplace. With reference to vertical migration of chemical species through the water column of lakes and oceans, the relevant transport process is eddy diffusion.

The difference of several orders of magnitude between the molecular and eddy diffusion coefficients reflects the much more rapid dispersal by turbulent eddies in natural bodies of water. The much higher values of the eddy diffusivities in surface waters are owing to the greater effect of the wind-generated turbulence compared with the deeper parts of the basin (see Table 3.1-4). Chemical mixing times and depths in surface waters place conditions on the appropriateness of using mass transfer coefficient–based flux expressions such as Eq. 4.0-1. The use is appropriate in aquatic impoundments, rivers, and shallow lakes, where a well-mixed water column (i.e., large $\mathscr{D}_{A2}^{(t)}$) exists; however, this is not the case in most large aquatic systems where $\mathscr{D}_{A2}^{(t)}$ is finite and relatively small in the vertical dimension. It is therefore instructive to determine the time period and the depth for the surface layer in which such flux expressions can be used.

According to Eq. 4.2-4, a well mixed water column changes exponentially with time to a sudden step change in the interface concentration. For the water side the time constant is $(1 + K^*_{A32}\rho_{32})h/{}^1k'_{A2}$. With ${}^1k'_{A2} = 20\,\text{cm/h}$, water free of suspended solids 1 m in depth has a time constant of 5 hr. For a hydrophobic with $K^*_{A32} = 1000$ L/kg in water with 1000 mg/L suspended sediment concentration the time constant is 10 hr. The interface concentration must remain uniform for a period of time equal to three time constants in order for steady-state conditions to be 95% achieved. It is clear that this time period is 15 to 30 hr in the cases above and that the water column responds very slowly to the air column changes. In a similar fashion for an air mass of height 10 m with air-side coefficient, ${}^2k'_{A1} = 30\,\text{m/h}$, the time constant is $\frac{1}{3}$ hr. For steady state to be achieved, the air time period needed is therefore 1 hr. These periods set the time limits for use of Eq. 4.0-1 in estimating the steady-state flux. Without uniform concentration for at least these time periods before sampling, the air and water steady-state flux conditions cannot be assured.

The appropriate depth-scaling parameter is contained in Eq. 7.1-44. Here if ${}^1k'_{A2}h/\mathscr{D}^{(t)}_{A2} \sim 0.1$, a very shallow concentration gradient (i.e., good mixing) exists in the surface waters. Under normal wind–wave conditions (see Fig. 7.1-9), where $\mathscr{D}^{(t)}_{A2} = 100\,\text{cm}^2/\text{s}$, the top 10 to 100 cm is virtually gradientless. Using Eq. 4.0-1 for the flux, based on a water sample obtained from this zone, is very appropriate.

Deeper within lakes and oceans it is possible to choose reasonable lower and upper limits of the diffusion coefficients and thereby to bracket the model in short and long time estimates. Lerman[24] considers an idealized three-layer water column with a mixed upper layer, a less mixed pycnocline, and a well-mixed layer below the pycnocline. When a three-layer system remains closed and the dimensions of the water layers do not change, a conservative chemical species in one of the mixed layers redistributes itself between the two layers because of the diffusional flux down the concentration gradient from one mixed layer into the other.

For the case of transport from the lower into the upper mixed layer, Lerman calculated the change in concentration in the upper layer as a function of time. Calculations were made for a 60-m-deep water column with lower layer 25 m, pycnocline 10 m, and upper layer 25 m for three different eddy diffusion coefficients in the pycnocline ($\mathscr{D}^{(t)}_{A2} = $ 5E-3, 1E-2, and 5E-2 cm^2/s). These values of the eddy diffusion coefficients are in the range reported for pycnoclines in stratified lakes. The calculations show that the concentration of a chemical species in the two mixed layers would equalize in a period of 10 to 40 years.

The time required to attain equal concentrations in any given lake depends on the eddy diffusivity in the pycnocline and on the vertical dimensions of the individual layers characteristic of the particular lake. The characteristics in the preceding example reasonably represent the time to chemical steady state in many lakes. In light of the period of time for intraphase movement it appears that the evaporation half-lives of 4.81 to 12,940 hr reported by Mackay et al. are very short. It appears that if a chemical species is uniformly distributed in a water column but calculations are made using a 1-m mixing depth, the result will be a minimum estimate of the half-life.

Flux Rate of Gases Across the Air–Sea Interface

Liss and Slater[25] describe the use of the two-resistance model of interphase mass transfer to estimate the flux of eight common gases across the air–sea interface. Using reasonable estimates of the individual phase mass transfer coefficients plus a knowledge of the Henry's law constant in conjunction with Eqs. 4.2-2 and 3.1-37, together with observed concentrations differences of each gas across the interface, the flux of the gases can be calculated. Mean concentrations of these gases in oceanic air and seawater are required. The concentration measures for many chemicals have been made by sampling from research vessels at sea and are available in the literature.

Factors Affecting Interface Resistance. A review of the published literature established that reasonable values of the gas- and liquid-phase mass transfer coefficients appropriate to the sea surface for unreactive gases are ${}^2k'_{A1} = 3000$ cm/h and ${}^1k'_{B2} = 20$ cm/h, where $A \equiv H_2O$ and $B \equiv CO_2$. To obtain the gas-phase coefficient for gases other than water vapor Graham's law, Eq. 3.1-30, may be used. The value of the liquid-phase coefficient is based largely on measurements of CO_2 exchange and is probably valid for gases of molecular weight 40 ± 25. For substances outside this range, Eq. 3.1-31 is applicable.

In chemically reactive gases, transport in the liquid phase may be more complex than by straightforward diffusion processes. For example, with CO_2 as well as the usual concentration gradient for gas molecules in physical solution at a suitable pH, there is also a similar gradient for HCO_3^- and CO_3^{2-} ions. It has been shown in the laboratory that for pH > 5 and under moderately calm conditions, the ionic species gradient can contribute significantly to the exchange of CO_2 across the interface. The chemical enhancement of oceanic CO_2 gas exchange results from the chemical reaction

$$CO_2 + H_2O + CO_3^{2-} \rightleftharpoons 2HCO_3^- \tag{4.2-9}$$

Hoover and Berkshire[26] have used a one-layer (liquid) film model to derive the following equation, which successfully predicts the exchange enhancement of CO_2 found in laboratory experiments:

$$\alpha_B = \frac{K^*_{B2} K\delta_{B2}}{[(K^*_{B2} - 1)(K\delta_{B2}) + \tanh(K\delta_{B2})]} \tag{4.2-10}$$

where $K \equiv \sqrt{K_{B2}K^*_{B2}/\mathscr{D}_{B2}}$. Here α_B is the fractional increase in the liquid-phase mass transfer coefficient due to chemical reaction Eq. 4.2-9, K^*_{B2} is the

Table 4.2-2. Exchange Constants for Selected Gases Across the Air–Sea Interface

Gas	${}^2k'_{A1}$ (cm/h)	α_B	${}^1k'_{A2}$ (cm/h)	H_{AX}	${}^1K'_{A2}$ (cm/h)
SO_2	1,600	1,721	34,420	47	1,600
N_2O	1,900	1.0	20	2,000	20
CO	2,400	1.0	20	62,000	20
CH_4	3,180	1.0	20	52,000	20
CCl_4	1,030	1.0	10.7	1,300	10.7
CCl_3F	1,085	1.0	11.3	6,200	11.3
MeI	1,070	1.0	11.1	300	10.6
$(Me)_2S$	1,620	1.0	20.0	370	19.2

Source: Ref. 25; reprinted by permission, Macmillan Journals Ltd.

ratio of total to ionic forms of inorganic carbon, K_{B2} is the hydration reaction rate constant for CO_2 in water, and δ_{B2} is the thickness of liquid film. Therefore, for a chemically reactive gas it is appropriate that the liquid-phase mass transfer coefficient by corrected for chemical enhancement by multiplying it by the value of α_B calculated from Eq. 4.2-10.

Table 4.2-2 contains exchange constants and Henry's law constants for a number of gases crossing the air–sea interface. For all gases considered other than SO_2, liquid-phase resistance controls the exchange. For SO_2, the gas-phase resistance is all-important. This is in part due to the high solubility of SO_2, but also because of its extremely rapid reaction rate constant in seawater. Exchange of other gases, with high solubilities and for rapid aqueous-phase reaction, such as NH_3, SO_3, and HCl, is probably also controlled by the gas-phase resistance.

Based on the exchange constants in Table 4.2-2, Liss and Slater computed the flux directions and rates across the air–sea interface for the gases listed. The results of these calculations appear in Table 4.2-3. Further details appear in Problem 4.2B. The authors were able to relate all or a fraction of the air–sea interface flux rates to other models or mass balance methodologies that quantify terrestrial or oceanic sources plus other fates for these substances. The

Table 4.2-3. Calculated Fluxes of Selected Gases Crossing the Air–Sea Interface (g/yr)

SO_2	N_2O	CO	CH_4	CCl_4	CCl_3F	MeI	$(Me)_2S$
1.5E14 (+)	1.2E14 (−)	4.5E13 (−)	3.2E12 (−)	1.3E10 (+)	5.3E9 (+)	2.8E11 (−)	7.2E12 (−)

Source: Ref. 25; printed by permission Macmillan Journals Ltd.

[a]The sign (+) denotes into the sea and (−) out of the sea.

other methodologies included the burning of fossil fuels for SO_2, N_2O from nitrogen inflow to oceans not removed by sediments, CO injected into the atmosphere due to human activities, CCl_4 and CCl_3F from industrial production figures, iodomethane, and dimethyl sulfide from global mass balances for these substances.

The preceding account makes readily apparent the role that interphase flux calculations can play in assessing the fate of natural and synthetic chemicals within the global environment. Obviously, the results are only as good as the data used in the calculations. There is much uncertainty in choosing mean values for the gas and liquid exchange constants. It is not known from measurements at the sea surface how ${}^1k'_{A2}$ varies with wind speed. Even greater uncertainties arise involving the air and water concentrations of some of these gases. As further data become available it will be a relatively simple matter to recalculate the fluxes and to extend the approach to other gases.

Estimating Mass Transfer Coefficients of Air–Sea Interface. Liss and Slater employed the reasonable values of the gas and liquid phase of 3000 and 20 cm/h, respectively, for calculations of the flux of selected chemicals. Several investigators have undertaken the study of mass transfer coefficients in the ocean environment. In general, these investigations begin with laboratory experimentations employing wind tunnel–water tank apparatus (see Fig. 4.2-2) to simulate conditions in the ocean. The exchange of simple chemicals such as oxygen, carbon dioxide, and water vapor is studied to gain a basic understanding of the importance of wind speed, fetch, waves, salinity, surface contamination, and so on, on the individual phase coefficients. From observation aboard

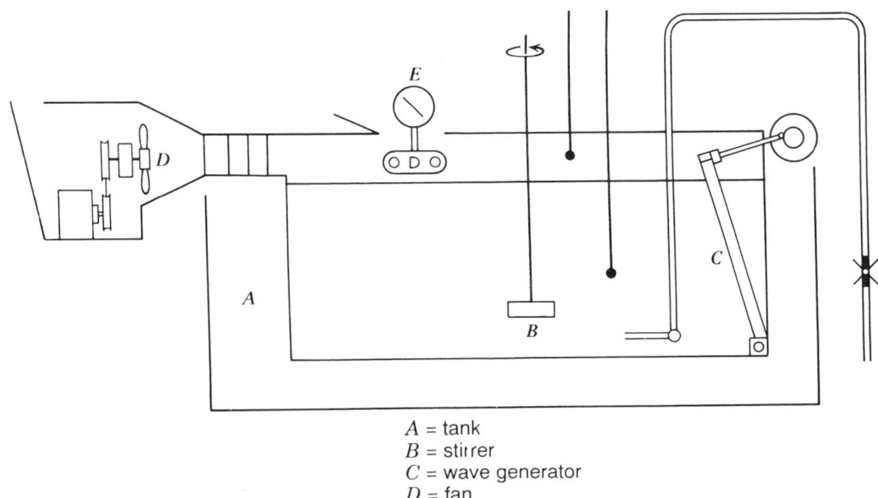

Figure 4.2-2. Apparatus for studying the effects of surface agitation on rate of solution of oxygen in water. (Reprinted with permission from Ref. 27.)

research vessels at sea it is possible to obtain better estimates of gas exchange rates. The concentrations of radon gas (half-life 3.85 days) and radiocarbon data (carbon 14) measured at sea provide this information. The following is a brief review of laboratory and oceanic studies.

Effect of Wind. The effect of wind–water interaction on gas absorption and desorption in laboratory tanks and in situ (lakes and oceans) has been studied by several researchers. Cohen[44] reviewed many of these data. Investigators generally report the experimental mass transfer coefficients as a function of wind velocity, v_x, measured at a height of 10 m (for in situ studies) or 10 cm (laboratory tank studies) above the water surface. The results of seven investigations are shown in Fig. 4.2-3. The noticeable feature is the substantial increase in the liquid-side mass transfer coefficient, ${}^1k'_{A2}$, at about 3 m/s wind velocity, reaching an order of magnitude higher at 10 m/s. In this region there is an appreciable wave growth and the air flow is aerodynamically rough. The corresponding increase in surface area is much less than 50%; hence this alone cannot account for the dramatic mass transfer enhancement. Figure 4.2-3 contains significant scatter, suggesting that average wind velocity in the wind tunnel is not a sufficient correlating parameter.

Downing and Truesdale[27] studied the effects of a number of factors on the rate of solution of oxygen in fresh and saline water to provide information about reaeration in the polluted Thames Estuary. In an estuary the water is

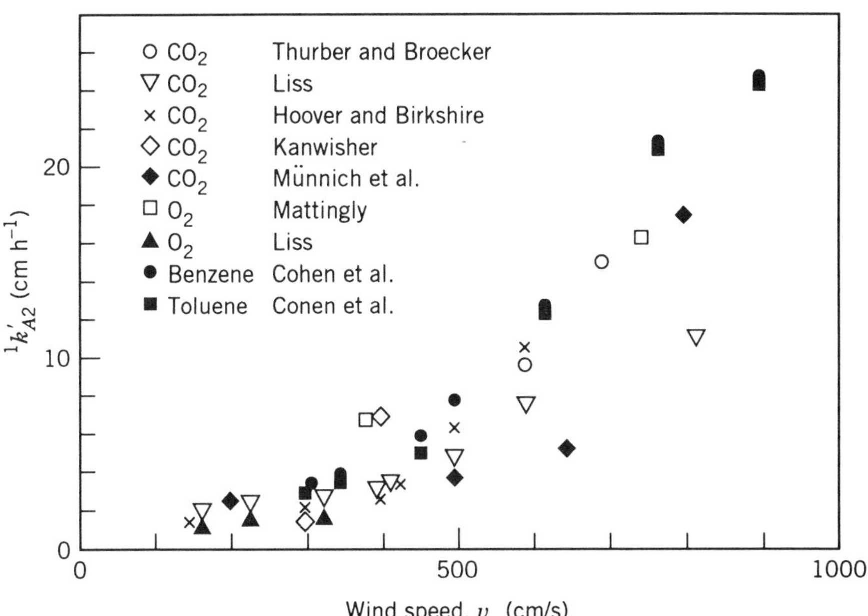

Figure 4.2-3. Liquid-phase mass transfer coefficient, ${}^1k'_{A2}$ versus wind velocity v_x (measured 10 cm above the water surface). (From Ref. 44.)

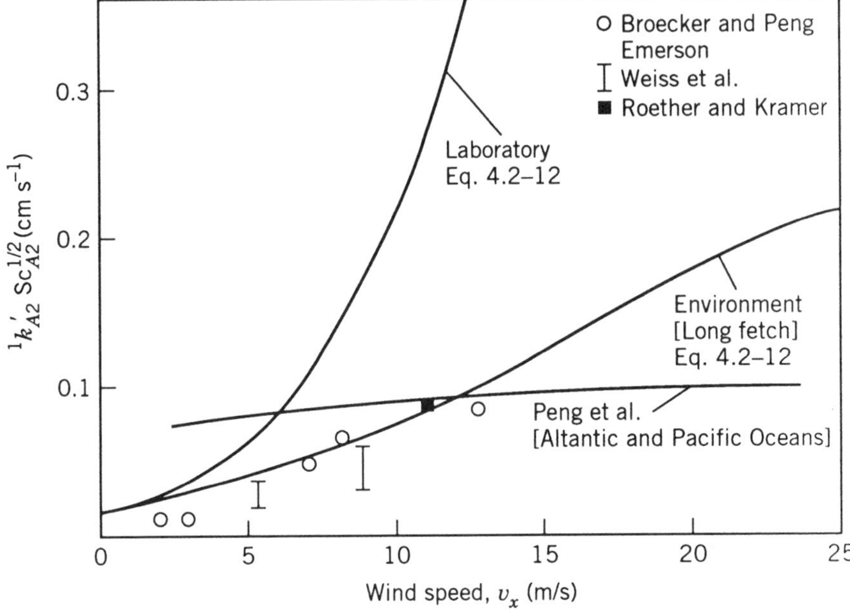

Figure 4.2-4. Prediction of $^1k'_{A2}$ in environmental water bodies. (From Ref. 44.)

subjected to considerable agitation both at and below the surface, owing to the motion of currents caused by tides and the flow of fresh water, but the chief agent causing disturbances at the surface is the wind. To simulate the former effect, the water in the laboratory experiments was stirred at varying speeds with impellers. To simulate the effects of wind directly was impossible since a very long fetch is required to work up waves of the magnitude of those observed in the estuary. The local action of wind in producing ripples and wavelets on the surface was studied in a small experimental tank, and larger waves were generated mechanically in the wave tank shown in Fig. 4.2-2.

The authors report that the laboratory results are not in good agreement with estimates of the effect of wind obtained from the results of direct measurement of the rate of reaeration in an estuary. These indicate a gradual steady increase in rate of solution with increasing wind velocity up to values on the order of 25 cm/h at a velocity of 10 m/s. Such behavior is contrasted in laboratory and in situ measurements in Figure 4.2-4.

It appears from these laboratory experiments that the liquid-phase coefficient at the ocean surface at any place will be proportional to the average of the quantity (wind velocity)^2. To obtain the average of (wind)^2 it is necessary to know the degree of variability of the wind. In the trades, where it is fairly constant, the two are not very different. In midlatitudes, however, where much of the gas exchange rate may be due to short periods of high wind velocities (storms), the use of average wind speed may cause large error. The result is that the liquid-phase coefficient at sea can vary considerably with latitude.[27]

Effect of Waves. To measure this effect, progressive waves of different heights and the same wavelength of 99.3 cm were generated at a constant frequency of 75 per minute. The rate of solution of oxygen increased almost linearly from 9.6 to 37 cm/h as the height of the waves increased from 2.8 to 10.8 cm. Varying the frequency of waves 8 cm in height from 36 to 75 waves per minute increased the rate of solution from 14.6 to 27.2 cm/h. The results are to be compared with those obtained from direct determinations of the rate of entry in the Thames Estuary, which indicate a roughly linear increase in rate of solution with increasing wave height from 5 cm/h for calm conditions to 25 to 30 cm/h for wave heights of 50 cm. For waves 10 to 13 cm high the rates of solution were 10 to 12 cm/h, less than half those in the laboratory tank.

The role of breaking waves and bubbles in enhancing the transport of atmospheric gases across the air–sea interface has been discussed by Kanwisher.[28] He concludes that when conditions are sufficiently rough for bubbles to be produced this may contribute significantly to the gas exchange.

Effect of Temperature, Oil Films, and Surfactants. Downing and Truesdale also report results of the effect of temperature, oil films, and soluble surface-active agents on the oxygen mass transfer coefficient. In most cases the coefficient increase is approximately linear with increasing temperatures in the range 0 to 35°C. In terms of the change in the coefficient at 20°, the values are correlated by

$$^1k_{A2}(\text{at } T,{}^\circ\text{C}) = {}^1\text{k}_{A2}(\text{at } 20^\circ\text{C})\alpha^{T-20^\circ\text{C}} \tag{4.2-11}$$

with α between 1.015 and 1.035 ($A \equiv O_2$).

Oil films had little effect until the thickness was greater than 1 μm; films of greater thickness tended to reduce the coefficient. No similar experiments appear to have been carried out at sea. It would seem reasonable to assume that except in a region of accidental oil spillage, the amount of oil likely to be found at the sea surface would be insufficient to have any measurable effect on the transfer coefficient.

There is some evidence that monomolecular surface layers composed of organic molecules can decrease the rate of the exchange of dissolved gases, such as oxygen and carbon dioxide. However, the film has to be in the close-packed condition to offer appreciable resistance to the passage of carbon dioxide. The same limitation almost certainly applies to transfer of oxygen and other dissolved gases. It seems unlikely that such a close-packed monomolecular film could exist over more than a very small part of the ocean surface.

Laboratory and Field Coefficient Measurement Summary. Liss[29] measured the transfer coefficients of O_2, CO_2, and water vapor across an air–water interface using both a laboratory tank filled with water and a wind–water tunnel. When the water pH is less than 5, CO_2 exists in water only as the physically dissolved species. The measured coefficient for the desorption of

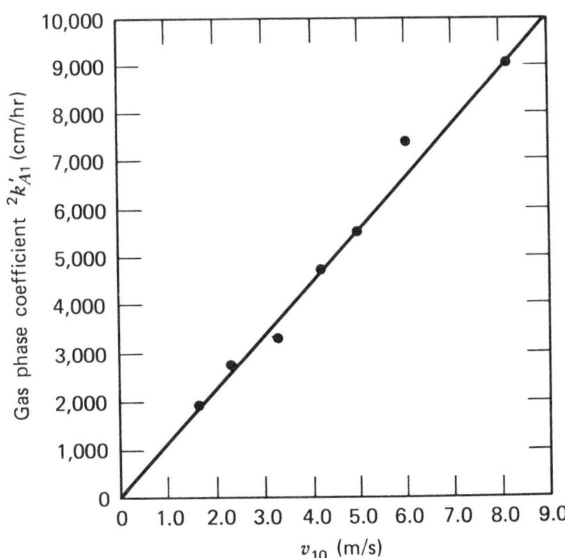

Figure 4.2-5. Variation of the exchange constant for water vapor ($^2k'_{A1}$), with the wind velocity measured at a height of 10 cm above the water surface (v_{10}). (Reprinted with permision from Ref. 29. Copyright 1963, Pergamon Press Ltd.)

CO_2 was the same as for the absorption of oxygen. Results indicate that the exchange of water vapor is controlled by resistance in the air. The wind-tunnel experiments demonstrate that the mass transfer coefficients for both CO_2 and O_2 increase approximately as the square of the wind velocity, whereas the coefficient for water vapor increases linearly with wind velocity (see Fig. 4.2-5).

Liss has compiled the results of liquid-phase mass transfer coefficients measured at sea. The range of values reflects the difficulty of making these measurements. The rather large spread may be due to different degrees of turbulence at the interface or to experimental error. The compiled values appear in Table 4.2-4.

Cohen et al.[30] have studied the volatilization of toluene and benzene in an effort to measure the rate of air–water mass transfer of nonionizing low-solubility pollutants. Measurements of the liquid-phase mass transfer coefficients in a laboratory wind–water tank are presented and the effects of waves on mass transfer rate is discussed.

Later Cohen revisited this subject and develops a theoretical model for the effects of wind shear on the mass transfer rate.[44] He begins with the surface renewal model of mass transfer, Eqs. 1.2-14 and 1.2-15, for building a model of the combined effect of wind-generated waves and water drift on $^1k'_{A2}$. Relations for the root-mean-squared velocity fluctuations perpendicular to the interface region, $\sqrt{\overline{(v'_y)^2}}$, and the mixing length, l, are approximated by the Kolmogorov microscales to the rates of energy dissipation per unit mass. This is further

Table 4.2-4. Liquid-Phase Mass Transfer Coefficients Measured in Situ at Sea Surface

Gas	Transfer Process	${}^1k'_{A2}$ (cm/h)	Comments
Rn	Desorption	7	Average wind speed 8 m/s
Rn	Desorption	25	Speed < 8 m/s
O_2	Absorption/desorption	22	Whole year average
CO_2	Absorption	29	Speed < 4.2 m/s, waves < 30 cm
CO_2	Absorption/desorption	36	Three experiments
CO_2	Desorption	4–14	Method unknown
CO_2	Absorption	11	Seven-year average

Source: Reprinted with permission from Ref. 29. Copyright 1973, Pergamon Press, Ltd.

related to the mean-squared vorticity of the liquid. Vorticity is principally due to the rate of strain, the total of which is approximated to be a linear combination of the wave induced and the wind stress on the liquid surface. The stress term is then related to the drift current and the wave term is related to amplitude, radian number, and frequency of the waves. The final expression is in terms of the water-side friction velocity, v_{*2} (see Eq. 3.2-37 for the definition of friction velocity). Based on laboratory data the resulting correlation is

$$ {}^1k'_{A2}\mathrm{Sc}_{A2}^{1/2} = 0.0029 + 0.048v_{*2}^{1.015} \tag{4.2-12} $$

where both coefficient and friction velocity are in cm/s. The correlation fits the benzene and toluene data with average error <10% up to $v_{*2} \sim 4.5$ cm/s.

Both the wave-field contribution and the drift current correlate theoretically with the friction velocity; however, the wave field contribution is most important at low wind speeds when the drift current is weak. A wind velocity of about 4 m/s is the breakpoint for laboratory measurements. The wave-field contribution is negligible at high wind speeds, where turbulent energy dissipation is associated primarily with the drift current. A linear dependence of the coefficient on friction velocity results (Eq. 4.2-12) as the wind speed increases. To use Eq. 4.2-12 a relationship for v_{*2} as a function of wind speed is needed.

The wind velocity profile for fully developed turbulent flow over a rough water surface is well represented by the logarithmic velocity distribution, Eq. 3.2-41. The stress exerted upon the water surface is τ_0, defined as $\rho_1 v_{*1}^2$, where v_{*1} is the air-side friction velocity. A common correlation for measured sea surface wind stress is[45]

$$ \tau_0 = C_D \rho_1 v_x^2 \tag{4.2-13} $$

where v_x is the average wind speed at a fixed height above the water surface. C_D, a resistance coefficient commonly termed the wind-stress coefficient or drag coefficient, depends on the hydrodynamic character of the underlying surface and varies with the roughness of this surface and varies with wind speed. Assuming shear stress continuity at the interface yields

$$v_{*2} = \sqrt{\frac{C_D \rho_1}{\rho_2}}\, v_x \tag{4.2-14}$$

This relation is used with Eq. 4.2-12 to estimate environmental (in situ) mass transfer coefficients. For a fully developed open sea, Cohen[44] offers the multiformula representation of the wind-stress coefficient, C_D, developed by J. Wu:

$$C_D = 8.5\text{E-4}, \quad v_x < 5 \text{ m/s} \tag{4.2-15a}$$

$$= [0.85 + 0.11(v_x - 5)]\text{E-3}, \quad 5 \text{ m/s} < v_x < 20 \text{ m/s} \tag{4.2-15b}$$

$$= 2.5\text{E-3}, \quad v_x > 20 \text{ m/s} \tag{4.2-15c}$$

Figure 4.2-4 summarizes the available ${}^1k'_{A2}$ values measured in environmental bodies. There appears to be fair agreement among the reported in situ values. The comparison of the predicted and the measured values are shown, and until more definitive in situ data become available, Eq. 4.2-12 should be adequate for estimating ${}^1k'_{A2}$ in the marine environment.

In the work above the role of breaking waves in chemical transport is not considered since significant wave breaking was not encountered at the range of wind speed considered.

Measuring Mass Transfer Coefficients at the Air–Sea Interface. The foregoing laboratory results are important to the task of estimating mass transfer coefficients at the air–sea interface. The question of how applicable these results are to the actual interface is open. Table 4.2-4 and Figure 4.2-4 give some values of the liquid-phase coefficient ${}^1k'_{A2}$ obtained at sea. The following is a demonstration of how such numerical results are obtained.

Measurements of the vertical distribution of radon (${}^{222}Rn$) in surface seawater offer a means of determining the gas exchange rate between the ocean and the atmosphere.[31] Radon (half-life 3.8 days) is generated within the sea by the decay of dissolved radium:

$${}^{226}Ra \rightarrow {}^{222}\text{Rn} + \alpha$$

The partial pressure of radon in the seawater produced in this manner greatly exceeds the radon pressure in oceanic air. Radon therefore desorbs, leaving the surface waters with a lower concentration than that accounted for by the decay of the parent radium. By measuring the difference between the radium and radon activity as a function of depth, it is possible to determine the rate of radon loss, and in turn the liquid-phase mass transfer coefficient.

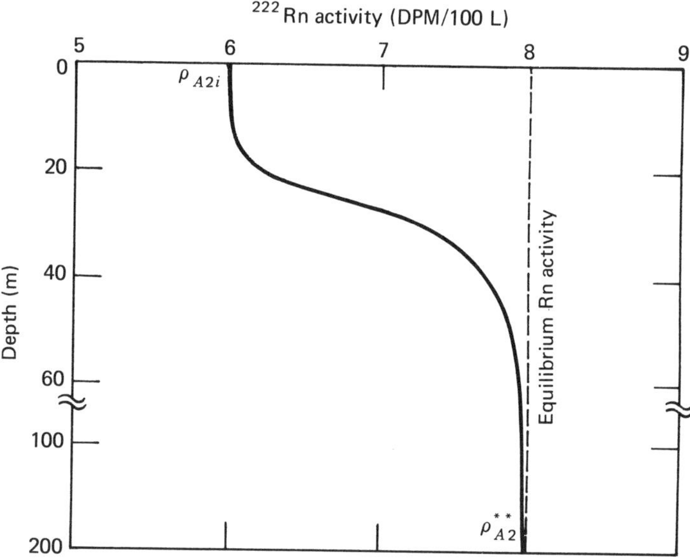

Figure 4.2-6. Radon content below the sea surface.

The assumption is made that the main barrier to radon desorption is the resistance on the sea side of the oceanic interface. The net loss of radon atoms through each unit area of the sea surface must exactly balance the integrated deficiency of radon activity in the water column. Figure 4.2-6 shows a typical profile of radon content for various depths below the surface. The integrated deficiency of radon content m_A where species ($A \equiv {}^{222}\mathrm{Rn}$) is

$$m_A = A \int_0^\infty (\rho_{A2}^{**} - \rho_{A2})\, dy \tag{4.2-16}$$

where ρ_A^{**} is the equilibrium mass concentration of radon supported by the radium present in the water, y the distance below the sea surface, and A the surface area.

For convenience, the value of the integral can be expressed in terms of ρ_{A2i}, the radon concentration at the surface, and h, the depth to which the surface anomaly would have to be carried, such that

$$A \int_0^\infty (\rho_{A2}^{**} - \rho_{A2})\, dy = A(\rho_{A2}^{**} - \rho_{A2i})h \tag{4.2-17}$$

A steady-state mass balance, Eq. 1.1-6, on radon for a volume Ah of seawater of uniform concentration ρ_{A2i} near the surface yields that the production rate of radon must equal the decay rate plus the rate of loss at the air–water interface:

$$(-r_B)h = \rho_{A2i} h k_A''' + {}^1k_{A2}'(\rho_{A2i} - \rho_{A2}^*) \tag{4.2-18}$$

where $-r_B$ is the rate of disappearance of radium ($B \equiv {}^{226}\text{Ra}$), k_A''' the fraction of radon atoms decaying per unit time, and ρ_{A2}^* the equilibrium solubility of atmospheric radon. A similar mass balance on a volume V of water located well below the surface where no radon gradients exist yields that $(-r_B)$ $V = \rho_{A2}^{**} V k_A'''$, from which $-r_B$ can be obtained:

$$(-r_B) = \rho_{A2}^{**} k_A''' \tag{4.2-19}$$

Equations 4.2-18 and 4.2-19 can be combined and rearranged to yield an expression for the liquid-phase coefficient. If, as is almost always the case for radon, $\rho_{A2i} \gg \rho_{A2}^*$,

$${}^1k_{A2}' = k_A''' h \left(\frac{\rho_{A2}^{**}}{\rho_{A2i}} - 1 \right) \tag{4.2-20}$$

This result is useful in estimating ${}^1k_{A2}'$ values at sea from measurements of the vertical distribution of radon. Equation 4.2-17 is used to compute h. Implied assumptions necessary for the development of this simple expression are that the radium content of seawater is constant and there is no turbulent diffusion of radon in the lateral and vertical directions. (See Problem 4.2E for an application of Eq. 4.2-20 to data obtained from the vessel R. V. Rockaway in the Bomex area.)

Evaporation from Sea Surface. Defant[32] presents H. U. Sverdrup's simple mass transfer model for the evaporation of water from the surface of the sea. He assumed that the air space above the sea surface was composed of two zones. The lower zone is immediately above the water surface and consists of a thin boundary sublayer through which water vapor transport proceeds only by molecular diffusion. Above this boundary sublayer, the water vapor transport proceeds through turbulent exchange. The mass fraction water vapor ($A \equiv H_2O$) in air is ψ_A. The two zones are illustrated in Fig. 4.2-7 and are identical to the model used for the deposition of gases and particles onto water.

The thickness of the boundary sublayer immediately above the water surface depends on the wind velocity. The sublayer itself can hardly be regarded as invariably composed of the same air particles. Since the turbulent eddies penetrate down to and into the boundary sublayer, it must be clearly understood that this layer occasionally disappears completely; however, after some time it is always reformed, so that a mean thickness, δ_{A2}, of this sublayer can be introduced (see Problem 3.3C).

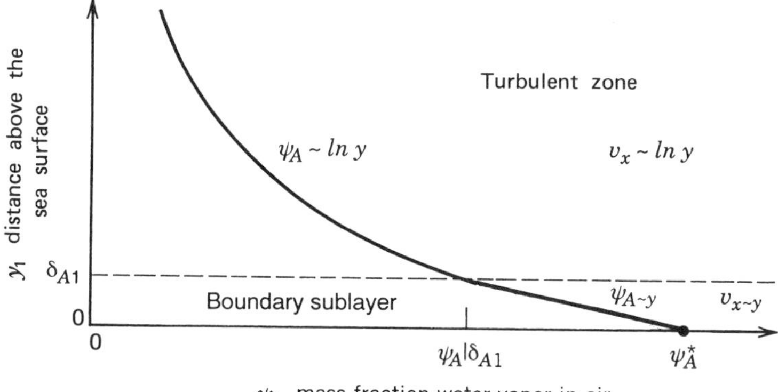

Figure 4.2-7. Two-layer surface at the air–sea interface.

The turbulent diffusivity is a linear function of the height above the water surface and depends on the roughness of the surface, as shown in Eq. 3.2H-5:

$$\mathscr{D}_{A1}^{(t)} = \kappa_1(y + y_0)v_* \tag{4.2-21}$$

The flux of water vapor in the turbulent zone is

$$j_A = -\kappa_1(y + y_0)v_*\rho_1 \frac{d\psi_A}{dy} \tag{4.2-22}$$

Since j_A is constant for steady-state evaporation from the sea surface, Eq. 4.2-22 may be integrated from the upper edge of the boundary sublayer $\psi_A|_{\delta_{A1}}$ at $y = \delta_{A1}$ to an arbitrary distance y at which ψ_A is $\psi_A|_y$, to give

$$j_A = \frac{\rho_1\kappa_1 v_*}{\ln[(y + y_o)/(\delta_{A1} + y_o)]} (\psi_A|_{\delta_{A1}} - \psi_A|_y) \tag{4.2-23}$$

Within the boundary sublayer, mass transfer occurs by molecular diffusion, for which Eq. 3.1-18 can be used in the form

$$j_A = \frac{\mathscr{D}_{A1}\rho_1}{\delta_{A1}} (\psi_A^* - \psi_A|_{\delta_{A1}}) \tag{4.2-24}$$

where ψ_A^* is the mass fraction of water in equilibrium with the salinity and temperature of the sea surface. By combining Eqs. 4.2-23 and 4.2-24, the unknown vapor concentration at the junction between the boundary sublayer and the turbulent zone, $\psi_A|_{\delta_{A1}}$, can be obtained and then used in Eq. 4.2-24 to yield an expression for the vapor flux:

$$j_A = \frac{v_*}{\dfrac{\ln[(y + y_0)/(\delta_{A1} + y_0)]}{\rho_1 \kappa_1} + \dfrac{v_* \delta_{A1}}{\rho_1 \mathscr{D}_{A1}}} (\psi_A^* - \psi_A|_y) \qquad (4.2\text{-}26)^*$$

Now using Eq. 3.2-42 to replace v_* yields the final desired vapor flux expression obtained by Sverdrup:

$$j_A = \frac{\kappa_1/\ln[y + y_0)/y_0]}{\ln[(y + y_0)/(\delta_{A1} + y_0)]/\rho_1\kappa_1 + (v_*\delta_{A1}/\rho_1\mathscr{D}_{A1})} (\psi_A^* - \psi_A|_y) v_x|_y \qquad (4.2\text{-}27)$$

Observations at sea were necessary to establish y_0 and δ_{A1}. From data gathered aboard the research vessel *Atlantis*, a value of y_0 of 0.6 cm was obtained. From values of wind velocity above the water surface, temperature, humidity, salinity, and evaporation rates δ_{A1} can be determined from Eq. 4.2-27. It was found that the mass transfer boundary sublayer thickness could be roughly correlated with v_* from which the simple equation $\delta_{A1} = 4.12/v_*$ was obtained. The term δ_{A2} is in cm, and v_* is in cm/s.

Employing the relation between δ_{A2} and v_* and a reasonable range of v_* values observed at sea ($13.2 \leqslant v_* \leqslant 36.3$ cm/s), it is possible to simplify Eq. 4.2-27. If one chooses to measure ψ_A and v_x at $y = 10$ m, with $\psi_A = \rho_{A1}/\rho_1$, Eq. 4.2-27 becomes

$$j_A = \frac{k_{10} v_x}{\rho_1} (\rho_{A1}^* - \rho_{A1}) \qquad (4.2\text{-}28)$$

and k_{10} lies between 1.86E-6 and 1.89E-6 g/cm^3. From inspection of this equation, the relationship for the air-side coefficient for water vapor is $k_{10}v_x/\rho_1$.

Equation 4.2-27 is quite remarkable since it is based solely on theoretical considerations. The theory of evaporation discussed earlier involves a hydrodynamically smooth surface with a laminar boundary sublayer and a turbulent layer of air above it. The preceding development is essentially the Prandtl–Taylor analysis. In the Prandtl–Taylor analysis, molecular transport of mass is assumed to be the only mechanism of importance in the laminar sublayer. In the turbulent core, the Reynolds analogy is applied. The linear dependence of velocity of the gas-phase mass transfer coefficient is supported by field observations. See, for example, Harbeck's result for evaporation from the surface of lakes, Eq. 4.1-13, and Schooley's data, Figs. 4.2D and 4.2-5.

*Equation number 4.2-25 has been intentionally omitted.

Dry Deposition of Gases and Particles

The process of transferring gases and particles from the atmosphere to water at the surface of lakes, oceans, and other aquatic bodies is termed *dry deposition*. A brief introduction to aerosols and the dry deposition process for these particles is given in Section 3.3 along with the flux expression, Eq. 3.3-31, which defines the deposition velocity, v_{dA}. This section is designed to extend the foregoing material to include the transfer of gaseous species in addition to the particle-bound species. It begins with a brief review of equilibrium partitioning of volatile chemical species between the gas and the particle–solid phases in air.

Gas and Particle Equilibrium Distribution of Volatile Organic Chemicals. The basic principles governing the equilibrium distribution of volatile chemical species between the solid particle phase and the gas phase of the atmosphere are presented in Chapter 2. The concentration of particulate matter in air is the total suspended solids, ρ_{31}. A typical value for urban air is 100 μg/m^3 with surface area $a_m = 11$ m^2/g. The solid surface area per unit volume air, a_v, is then 1.1E-5 cm^3/cm^3. These values are from Bidleman.[46] Based on the single-component BET absorption isotherm, Junge[47] developed a theoretical model for the exchangeable fraction in air adsorbed to the aerosols Φ_A.

Beginning with Eq. 2.1-30 and assuming that $p_A \ll p_A^0$ as Junge did, yields

$$\omega_A = \frac{\omega_A^* B_1 p_A}{p_A^0} \tag{4.2-29}$$

for the mass of A sorbed onto the solid phase in equilibrium with the partial pressure of A, p_A, in air. Using the ideal gas law to convert the result to the equilibrium concentration of A in the gas phase, ρ_{A1}, yields

$$\omega_A = \frac{\omega_A^* B_1 \rho_{A1} RT}{M_A\, p_A^0} \tag{4.2-30}$$

For air with total suspended solids concentration ρ_{31}, the fraction A on the solid phase is

$$\Phi_A = \frac{\omega_A \rho_{31}}{\omega_A \rho_{31} + \rho_{A1}} \tag{4.2-31}$$

where Φ_A is the ratio of mass A on solid to total mass A on solid plus gas. Using Eq. 4.2-30 for ω_A and $\rho_{31} = a_v/a_m$ yields

$$\Phi_A = \frac{c a_v}{c a_v + p_A^0} \tag{4.2-32}$$

which is Junge's model, where the constant $c \equiv \omega_A^* B_1 RT/M_A a_m$.

Volatile and semivolatile organic chemicals bound to atmospheric particles appear to consist of both nonexchangeable fraction, which is strongly adsorbed to active sites or embedded within the particle matrix and is not in equilibrium with the vapor phase, and an exchangeable fraction adsorbed to solid surfaces. The characteristic time for the concentration profile in the gas phase close to the surface of the particle to relax to the steady-state value, or in this case the equilibrium value, is proportional to $d^2/4\mathscr{D}_{A1}$, where d is the particle diameter. This is on the order of E-3 seconds or smaller for all particles of atmospheric interest.[49] We can conclude that for a situation of little or no relative motion between the aerosols and air which constitutes the atmosphere, the exchangeable fraction of a volatile species is at equilibrium between the solid and gas phases. For the exchangeable fraction Junge's model predicts that those substances with small pure component vapor pressures can be expected to be associated with the particles. As a first approximation, Junge assumed that c in Eq. 4.2-32 was constant and estimated it to be 1.7E-4 atm·cm. Figure 4.2-8 shows the exchangeable particle-bound fraction, Φ_A, as a function of the vapor pressure for four cases of air quality: clean background, average background, background plus local, and urban air, for which the respective a_v values are 4.2E-7, 1.5E-6, 3.5E-6, and 1.5E-5 cm^2/cm^3. Bidleman[46] is the source of the values. Clearly, the equilibrium partitioning is a strong function of the solids concentration in air, ρ_{31}, and the effective external surface area of the solids, a_m, for adsorbing or desorbing the volatile species since $a_v = \rho_{31}/a_m$.

The available data seem to support such a theoretical equilibrium distribution in a qualitative sense.[46] The sizable differences in Φ_A between model predictions and data are probably due to a number of uncertainties. These

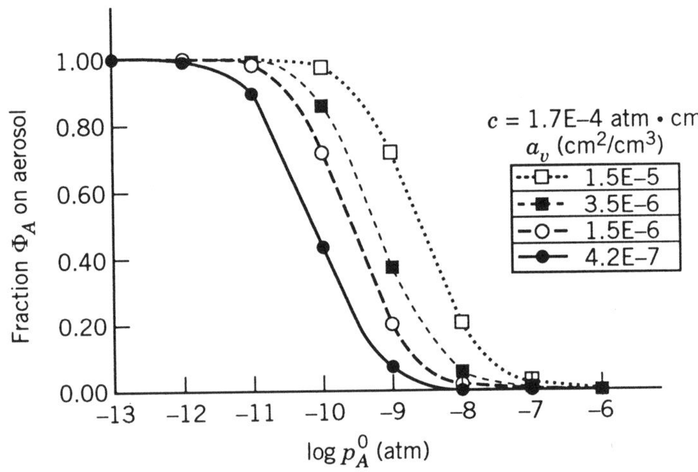

Figure 4.2-8. Theoretical fraction semivolatile organic chemical on particles versus pure component liquid vapor pressure.

uncertainties include using p_A^0 of the subcooled liquid or the crystalline solid at the air temperature T, measurements that reflect the nonexchangeable portion within the particles, and differences in adsorption sites for particles at various geographic locations. Theoretically, the coadsorption of other species, particularly water vapor, cannot be ignored.[48] Bidleman[46] suggests that until we have a fuller understanding of the magnitude of such factors on the equilibrium partitioning of semivolatile chemicals between the particle and gas phase in the atmosphere, Eq. 4.2-32 provides a method for reasonable quantitative prediction.

Multilayer Atmosphere. A multilayer view of the atmosphere is realistic and necessary in order to construct a comprehensive model for the deposition of particles and gaseous species to the ocean surface. Slinn et al.[50] and Slinn[56] considers four atmospheric layers. The topmost, the aloft layer, is above the atmospheric boundary layer and extends upward from $y = 1$ km. The next is the atmospheric boundary layer, occupying the region 1 km $\geqslant y \geqslant$ 100 m. The next is the constant flux layer or mechanical turbulence layer within 100 m $> y \geqslant \delta$, where δ_v is the depth of the viscous sublayer. Adjacent to the water surface is the deposition layer of depth conceptually proportional to the viscous sublayer in thickness but influenced by surface roughness, wave height and period, wind speed, and other factors. Since the focus here is in the region near the air–water interface, attention will be directed onto the two bottom layers (see Fig. 4.2-9).

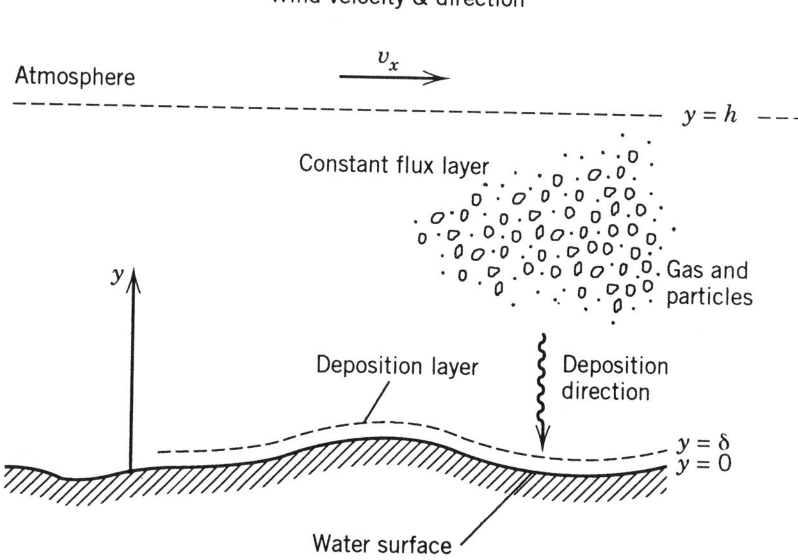

Figure 4.2-9. Schematic of simple two-layer atmospheric model.

Continuity Equation for Gases and Particles

The analysis begins with Eq. 3.1J, the continuity equation for species A in the atmosphere

$$\frac{\partial \rho_{A1t}}{\partial t} + \nabla \cdot \mathbf{n}_A = r_A \tag{4.2-33}$$

where ρ_{A1t} is the concentration of A, particle bound and gas phase. Besides species A the atmospheric mixture contains water vapor (species B), air (species C), and particles (phase 3). The particulate phase includes solid or liquid aerosols. Here it should be noted that the definition of the mass average velocity, $\mathbf{v}$, includes the sum of all constituents, so that

$$\mathbf{v} = \frac{\mathbf{n}_A}{\rho_1} + \frac{\mathbf{n}_B}{\rho_1} + \mathbf{v}_c + \mathbf{v}_3 \tag{4.2-34}$$

where $\mathbf{n}_B$ is the flux of water vapor, $\mathbf{v}_c$ the air velocity, $\mathbf{v}_3$ the particle velocity, and ρ_1 the mass density of the atmosphere.

For the steady-state transport of A from the atmosphere downward to the water surface without reaction, only the y component of Eq. 4.2-33 is required:

$$n_{Ay} = v_y \rho_{A1t} - \mathscr{D}_{A1} \frac{d\rho_{A1t}}{dy} \tag{4.2-35}$$

This result will be applied to the atmospheric conditions in each layer.

Constant-Flux Layer. Figure 4.2-9 illustrates the deposition process from the two layers for a moving air mass confined within a height h. This height can be specified arbitrarily or it can represent the thermal inversion height for the current atmospheric conditions of the day (see Fig. 6.1-2). In normal wind conditions, $v_x = 0.1$ to $10\ \mathrm{m/s}$, transport in the constant-flux layer is dominated by mechanical turbulence and is influenced secondarily by thermal turbulence.

Because of the presence of turbulence in the constant-flux layer, average values of concentration and velocity are the only measurements typically available. This is due to the relatively long response time of most instruments that measure concentrations. It is therefore necessary to average the y-component flux for a time period. These time averages are denoted by tildes (e.g., $\tilde{\rho}_{A1}$) and the rapid fluctuations about the average by primes (e.g., ρ'_{A1}). In the constant-flux layer the velocities of the individual gaseous species (i.e., A, B, and C) will be combined and denoted v_{1y} since they will move as a unit with the turbulent eddies. Retaining the particle velocity separately, Eq. 4.2-34 becomes, for the y component,

$$v_y = v_{1y} + v_{3y} \tag{4.2-36}$$

and in terms of average and fluctuating components it becomes

$$v_y = \tilde{v}_{1y} + v'_{1y} + \tilde{v}_{3y} + v'_{3y} \tag{4.2-37}$$

Substituting this into Eq. 4.2-35 with $\rho_{A1t} = \tilde{\rho}_{A1t} + \rho'_{A1t}$ in similar components and time averaging (see Problem 3.2C) yields

$$n_{Ay} = \tilde{v}_{1y}\tilde{\rho}_{A1t} + \overline{v'_{1y}\rho'_{A1t}} + \tilde{v}_{3y}\tilde{\rho}_{A1t} + \overline{v'_{3y}\rho'_{A1t}} - \mathscr{D}_{A1}\partial\tilde{\rho}_{A1t}/\partial y \tag{4.2-38}$$

The diffusion term is negligible compared to the terms that account for the turbulent mass transfer flux. Also, there is no significant downward air velocity, so $\tilde{v}_{1y} \simeq 0$. Eq. 4.2-38 simplifies to

$$n_{Ay} = \overline{v'_{1y}\rho'_{A1t}} + \tilde{v}_{3y}\tilde{\rho}_{A1t} + \overline{v'_{3y}\rho'_{A1t}} \tag{4.2-39}$$

A differentiation between the fluctuating gas and the fluctuating particle turbulent mass-transport terms is not made and they are typically combined into a single term, $\overline{v'_{1y}\rho'_{A1t}}$, for conditions in the constant-flux layer. The result is a flux equation for species A that contains a turbulent mass transport term and a gravitational setting term.

Similarity theory is used to parameterize the turbulent term (see the section "Similar Turbulent Profiles" in Chapter 3). For some height y within the constant-flux layer and above the deposition layer (i.e., $y \geqslant \delta > 0$), Eqs. 3.3-15 and 3.3-17 can be combined to yield the steady-state flux of A as

$$\overline{v'_{1y}\rho'_{A1t}} = \frac{\kappa_1 v_*^2}{v_x}\left[\rho_{A1t}(y) - \rho_{A1t}(\delta)\right] \tag{4.2-40}$$

The tildes have been omitted; however, the quantities are average values. This result assumes that the wind speed within the deposition layer is much smaller than at elevations within the constant-flux layer. This equation contains the characteristic transfer coefficient for the constant-flux layer $\kappa_1 v_*^2/v_x$. Here v_x is the average wind speed above the water surface at 10 m height.

The friction velocity v_* is a convenient and useful measure of the turbulence intensity. In cases for which the wind speed is measured directly, the simple proportionality

$$v_* = 0.037 v_x \tag{4.2-41}$$

is compatible with the drag coefficient expression given by Eq. 4.2-13 for typical sea conditions. Slinn[50] gives the error in v_* introduced by this relatively simple relationship as $\pm 20\%$ and notes that similar errors can arise from the neglect of atmospheric stability. Section 6.1 contains a means of correcting the flux for stability (see Eq. 6.2-17). Also, the friction velocity can be obtained

Table 4.2-5. Cunningham Correction Factor C_c for Spherical Particles in Air at 20°C, 1 atm

d_p (μm)	C_c	d_p (μm)	C_c	d_p (μm)	C_c
0.001	216	0.05	4.95	2.0	1.08
0.002	108	0.1	2.85	5.0	1.03
0.005	43.6	0.2	1.87	10.0	1.02
0.01	22.2	0.5	1.33	20.0	1.01
0.02	11.4	1.0	1.16	50.0	1.00

Source: Modified from a table given by Seinfeld.[49]

more precisely from velocity profile measurements (see Eq. 3.2-41), so more accurate flux estimates are possible.

The gravitational settling velocity v_{3y} can be estimated from the modified Stokes velocity equation

$$v_{3y} = \frac{d^2 \rho_3 g C_c}{18 \rho_1 v_1} \tag{4.2-42}$$

where C_c is the Cunningham correction factor.[49] It accounts for the noncontinuum effects, which become important as d, the particle diameter, approaches the size of the gas molecules. Table 4.2-5 contains C_c values for spherical particles.

Deposition Layer. This layer is illustrated in Fig. 4.2-9 and occupies the very thin region from the air–water interface upward to height δ. Within the layer the airflow is laminar in the x direction, and transport is dominated by viscous forces, so that turbulence is negligible. It is necessary to return to Eq. 4.2-35 to begin the analysis of the deposition layer.

The y component of velocity in the layer needs to be resolved. From Eq. 4.2-34,

$$v_y = \frac{n_{By}}{\rho_1} + v_{3y} \tag{4.2-43}$$

since the velocity contribution due to the downward flux of A is exceedingly small and the downward air velocity is zero at the interface, so it must also be through the sublayer. Substituting into Eq. 4.2-35 yields

$$n_{Ay} = \frac{n_{By}}{\rho_1} \rho_{A1t} + v_{3y} \rho_{A1t} - \mathscr{D}_{A1} \frac{d\rho_{A1t}}{dy} \tag{4.2-44}$$

Here the remaining flux terms account for the mean slip velocity, caused by the diffusiophoresis of water evaporating from the surface, gravitational particle settling, and diffusion. The equation applies for both the particulate and gaseous forms of species A. Its further use requires the total concentration expressed into the two fractions followed by some simplifying assumptions to yield the final algorithms.

Equation 4.2-38 contains the particle flux term $\overline{v'_{3y}\rho'_{31}\omega_A}$. This is a turbulence-generated flux and theoretically cannot exist in the viscous sublayer because no fluctuating velocities are present here. In reality, however, residual momentum possessed by particles near and above the sublayer, energized by the eddies in the constant flux layer, causes particle penetration into the sublayer and the result is a particle flux. Slinn[52] has made provisions for this flux. He modeled the convection of particles across the viscous sublayer using the viscous "jet" at the leading edge of Joukowski's caterpillar tread that describes eddies at a solid channel wall. "The vortex flow curls itself up into separate elliptical eddies.... The centers of these elliptical eddies move horizontally at velocities which increase with the distance from the channel bottom.... The elliptic filaments advance like a caterpillar tread." Based on this physical picture, Slinn developed an expression for the deposition velocity that includes particle impaction due to momentum. Using data from the literature on jet collection efficiencies, he arrived at an empirical expression for this efficiency of $10^{-3/\mathrm{St}}$, where St is the Stokes number $\equiv (v_{3y}/g)v_*^2/\nu_1$ and v_{3y}/g is the particle relaxation time. His approximate expression for this deposition velocity, including the Brownian diffusion term, is

$$v_d\ (\text{impact} + \text{Brownian}) = \frac{v_*^2}{\kappa_1 v_x(h)}\,(10^{-3/\mathrm{St}} + \mathrm{Sc}_{31}^{-1/2}) \qquad (4.2\text{-}44a)$$

where Sc_{31} is the Schmidt number of the particles in air.

The evaporation of water from the sea surface was considered in detail earlier. This process results in an upward gas velocity and Davis notes that aerosol particles suspended in a nonuniform but isothermal gas mixture move due to existing concentration gradients. This he calls *diffusiophoresis*. The water vapor flux also acts to retard the downward flux of gaseous species A as well. From Eq. 4.2-28 the upward water vapor–induced velocity, $v_{By} \equiv n_{By}/\rho_1$, is

$$v_{By} = \frac{k_{10} v_x(y)}{\rho_1^2}\,[\rho_{B1}^* - \rho_{B1}(y)] \qquad (4.2\text{-}45)$$

where wind speed, v_x, and humidity, ρ_{B1}, is determined at height $y = 10$ m, for example.

The particle settling velocity in the middle term of Eq. 4.2-44 is given by Eq. 4.2-42. This term accounts for the transport of species A through the sublayer as a result of this gravitational settling velocity. The final flux term in Eq. 4.2-44 need be resolved into its gaseous and particulate fractions to account for the total molecular diffusion through the deposition layer. The Stokes–

Einstein equation for the Brownian diffusion of aqueous-phase particles is given by Eq. 3.1-27. In the case of aerosols the preferred form is

$$\mathscr{D}_{31} = \frac{k_B T C_c}{3\pi\mu_1 d} \tag{4.2-46}$$

where C_c is the Cunningham correction factor in Table 4.2-5.

In addition to the transport mechanisms presented in relation to Eq. 4.2-44, there are a few others that have been identified as possibly significant in the deposition layer. In the steady state, under the influence of a temperature gradient in the gas and of friction, aerosol particles move with constant velocity toward the low temperature.[54] This phenomenon, called *thermophoresis* is not accounted for. The effect of breaking waves and spray formation in high winds and particle growth in the humid region near the air–water interface has been incorporated into the flux relationships by Williams.[55]

Two-Layer Model for Dry Deposition to Water. The mechanisms of transport leading to dry deposition are complex, and thus in describing the rate of removal of a species by dry deposition one does not attempt to represent in detail the processes at their most fundamental level. The simplified theoretical phenomena represented by Eqs. 4.2-39 and 4.2-44 for the constant flux and the deposition layers, respectively, although susceptible to many improvements, can provide a good working, first-order approximation. The model that follows is a modification of the particle dry deposition model presented by Slinn and Slinn.[52] Difference equations, rather than formal integration over y, will be used for the steady-state flux through the two layers. Also, both fractions, particles plus gaseous, of species A in the atmosphere will be considered. In this way the final result will apply to chemical species associated entirely with the aerosol phase. This includes both solid particles and liquid droplets, of which soot and acid mists are examples. It includes semivolatile chemical species that can be partially associated with the aerosol fraction or the gaseous fraction of the atmosphere. The PNAs and some pesticides are examples. It includes species that exist entirely in the gaseous state, such as dimethyl sulfide. See Seinfeld[49] for a summary of these types of air pollutants. However, because of mechanism differences in the layers, the flux of each fraction will be developed separately and then recombined to give the total.

In the constant-flux layer the downward flux of particles carrying species A is due to the combined processes of turbulent diffusion and gravitational settling of particles. Equation 4.2-39 with 4.2-40 become

$$j_{A3} = \frac{v_*^2 \omega_A}{v_x(h)} [\rho_{31}(h) - \rho_{31}(\delta)] + v_{3y}(d_d)\omega_A \rho_{31}(h) \tag{4.2-47}$$

where $\rho_{31}(h)$ and $\rho_{31}(\delta)$ are the particulate concentrations at height $y = h$ and at the top of the sublayer, $y = \delta$. The gravitational settling term is based on the dry particle diameter, d_d. The gaseous fraction forced downward by the particles in the $v_{3y}\rho_{A1t}$ term is assumed to be negligible.

The portions of Eq. 4.2-44 that apply to particles in the deposition layer are selected and put into difference equation form to yield

$$j_{A3} = \left[\frac{n_{By}}{\rho_1} + v_{3y}(d_w)\right]\omega_A\rho_{31}(\delta) + \frac{v_*^2}{v_x(h)\kappa_1}\cdot\omega_A(10^{-3/\mathrm{St}} + \mathrm{Sc}_{A3}^{-1/2})[\rho_{31}(\delta) - \rho_{31}(0)] \tag{4.2-48}$$

where $\rho_{31}(0)$ is the particulate concentration at the air–water interface. In the sublayer the particles are assumed to possess their wet diameter, d_w, the equilibrium size for the high value of the humidity that exists here. The relative humidity may approach 100% over fresh water and is limited to about 98.3% over salt water, due to Raoult's law. Dramatic increases in particle sizes occur between 0%, 99%, and 100% relative humidity (RH) and this effects v_{3y} as well.

The term for the overall transfer velocity, called the particle deposition velocity, v_{dp} is defined by

$$j_{A3} = v_{dp}\omega_A[\rho_{31}(h) - \rho_{31}(0)] \tag{4.2-49}$$

The following steps are taken: (1) let $\rho_{31}(0) = 0$ (i.e., ignore particle resuspension), (2) assume steady-state conditions; (3) use Eqs. 4.2-47 and 4.2-48 to eliminate $\rho_{31}(\delta)$ in Eq. 4.2-47, and (4) set the resulting expression equal to Eq. 4.2-49. The expression for the particle deposition velocity is then

$$v_{dp} = \frac{C(A + B)}{A + C} \tag{4.2-50}$$

where

$$A \equiv \frac{v_*^2}{v_x(h)} \qquad B \equiv v_{3y}(d_d) \qquad C \equiv v_{3y}(d_w) + \frac{n_{By}}{\rho_1} + \frac{A(10^{-3/\mathrm{St}} + \mathrm{Sc}_{31}^{-1/2})}{\kappa_1}$$

Figure 4.2-10 shows plots of Eq. 4.2-50 for three wind speeds, $n_{By} = 0$, and for three cases of particle growth.

There are several sets of water–wind tunnel measurements of particle deposition velocities that follow the general pattern shown in Fig. 4.2-10.[50] The measured minimum in v_{dp} is near 0.01 cm/s and occurs at a particle diameter of 0.1 to 1 μm. Brownian diffusion through the deposition layer controls the deposition process for the smaller particles, with v_{dp} increasing as diameter decreases, approaching molecular dimensions. Gravitational settling dominates the process for the large particles, with v_{dp} increasing with d^2 according to the settling velocity law.

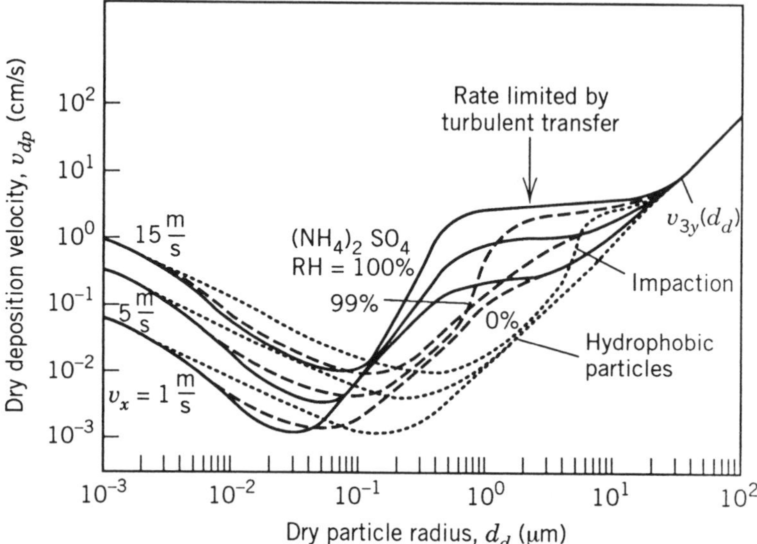

Figure 4.2-10. Plots of v_{dp} in Eq. 4.2-50 for three wind speeds and three types of particles: dotted curves, hydrophobic (dry) particles; dashed curves, for particles that grow to the equilibrium size for $(NH_4)_2SO_4$ particles exposed to a relative humidity of 99%; solid curves, expected behavior of $(NH_4)_2SO_4$ particles for the case of deposition to lakes with relative humidity near 100% in the deposition layer. (From Ref. 52.)

For aged, atmospheric aerosol particles with mass mean radius near 1 μm, Slinn and Slinn[52] recommend that $v_{dp} = A$ instead of Eq. 4.2-50 is an acceptable approximation consistent with reported data. This assumes that the deposition process is rate limited by turbulent transfer through the constant-flux layer (see Fig. 4.2-10). Giorgi[53] observes that in comparison with experimental data, even if the data refer to conditions that do not reproduce the characteristics of natural water surfaces, the results are encouraging, and the simplified models, although susceptible to many improvements, can provide a good working first-order approximation.

In the constant-flux layer the downward flux of gaseous species A is due to turbulent diffusion only. From Eqs. 4.2-39 and 4.2-40

$$j_{A1} = \frac{\kappa_1 v_*^2}{v_x(h)} [\rho_{A1}(h) - \rho_{A1}(\delta)] \tag{4.2-51}$$

where $\rho_{A1}(h)$ and $\rho_{A1}(\delta)$ are the concentrations of species A in the gaseous state at height $y = h$ and at the top of the sublayer $y = \delta$, respectively.

There are two terms in Eq. 4.2-44 that apply to the gaseous fraction of species A in the deposition layer:

$$j_{A1} = \frac{n_{By}}{\rho_1} \rho_{A1}(\delta) + {}^2k'_{A1}[\rho_{A1}(\delta) - \rho_{A1}(0)] \tag{4.2-52}$$

where the molecular diffusion term is expressed in terms of a gas-phase mass transfer coefficient flux equation. The other term is the diffusiophoretic effect of water vapor transport on the gaseous fraction.

In an earlier section on the evaporation of water from the sea surface, a simple relationship was obtained for the air-side coefficient. This relation is ${}^2k'_{A1} = k_{10}v_x/\rho_1$ and is from Eq. 4.2-28. Being a scalar it is also applicable for deposition onto the sea surface, but it is limited to water vapor. A somewhat parallel development, which led to Eq. 4.2-28, based on fluid velocity measurements within the laminar sublayer and the buffer layer of turbulent flow past a flat plate[51] yields a more general result as follows.

Equation 3.2-43 is the accepted correlation for the thickness of the laminar sublayer. The buffer layer thickness extends to a height approximately five times larger. There is also some momentum resistance in the turbulent zone. It and the buffer layer resistance can be approximated by using a pseudothickness of $\delta_v = 30\nu_1/v_*$. The film theory result, Eq. 3.1-18, for the gas-side coefficient is

$$ {}^2k'_{A1} = \frac{\mathscr{D}_{A1}}{\delta_{A1}} \tag{4.2-53} $$

where δ_{A1} is the diffusive sublayer thickness. Equation 3.3C-1 relates this to the laminar sublayer thickness. Combining the three equations above yields

$$ {}^2k'_{A1} = \frac{v_*}{30\mathrm{Sc}_{A1}^{2/3}} \tag{4.2-54} $$

This general result is consistent with the coefficient from Eq. 4.2-28. Using Eq. 4.2-41 for v_* with $\mathrm{Sc} = 0.6$ for water vapor at 20°C in the above yields $k_{10} = 2.06\text{E-6 g/cm}^3$. This is the approximate numerical value observed for water evaporation at sea. Slinn et al.[50] suggest that the numerical constant in Eq. 4.2-54 is the inverse of von Kármán's constant but note that this is a tentative estimate until further research is performed. The number 1.7 as the constant in Eq. 4.2-54 has been reported for water surfaces in the laboratory.

To obtain an expression for the overall transfer or dry deposition velocity of the gaseous fraction v_{dg}, defined here via:

$$ j_{A1} = v_{dg}[\rho_{A1}(h) - \rho_{A1}(0)] \tag{4.2-55} $$

similar steps as for the particle fraction development are taken except that $\rho_{A1}(0) \neq 0$. The final expression for the gas fraction deposition velocity is

$$ v_{dg} = \frac{A(D + E)}{A + D + E} \tag{4.2-56} $$

where $D \equiv n_{By}/\rho_1$ and $E = v_*/30\mathrm{Sc}_{A1}^{2/3}$. Unlike particles, there are very few observations, field or laboratory, of the dry deposition velocity of gaseous species.[46]

The total flux of species A to the water is the sum of the particle and gaseous fractions. Adding Eqs. 4.2-49 and 4.2-55 yields

$$j_A = j_{A3} + j_{A1} = v_{dp}\omega_A \rho_{31}(h) + v_{dg}[\rho_{A1}(h) - \rho_{A1i}] \tag{4.2-57}$$

where ρ_{A1i} is used to denote the gaseous concentration at the air–water interface. The particle flux of A to the water is not effected by the concentration of A (in solution) in the mixed surface water layer of concentration ρ_{A2}. The gaseous flux is intimately connected to the process on the water side and is also affected directly by the concentration ρ_{A2}. Anticipating model applications that require connecting the atmosphere to the water column, ρ_{A1i} is eliminated from Eq. 4.2-57, yielding

$$j_A = v_{dp}\omega_A \rho_{31}(h) + v_{dg}\frac{\rho_{A1}(h)\,{}^1k'_{A2}/H_\rho + {}^1k'_{A2}\rho_{A2} - v_{dp}\omega_A \rho_{31}(h)}{v_{dg} + {}^1k'_{A2}/H_\rho} \tag{4.2-58}$$

where H_ρ is Henry's constant for the volatile species. The quantity of A delivered by the particle fraction will affect the quantity delivered by the gaseous fraction through the net accumulation of A in water as reflected in ρ_{A2}. The final equation can be used to obtain the steady-state flux of species A from the atmosphere to a water body. If species A is entirely in particulate form, $v_{dg} = 0$ and, in addition, if particle deposition alone is desired, $\omega_A = 1$. If species A is entirely gaseous, $v_{dp} = 0$.

Closure. The equilibrium distribution of semivolatile organic chemicals between particle and gaseous phases will remain constant, according to Eq. 4.2-32, only if the deposition rates of particles and gaseous fractions remain constant. This is a very improbable occurrence. A cursory inspection of Eq. 4.2-58 will show that it is highly unlikely that the flux rates of A by particles and gaseous phases behave so. Furthermore, since the particles are continually lost from the air mass to the water surface, there must be the shunting of molecules between phases within the lower portion of the constant flux layer and within the deposition layer if any semblance of equilibrium distribution is to be maintained. If water vapor plays a major role in the equilibrium distribution process between phases, this factor may be important because of the large changes in air humidity that occur with proximity of the water surface. Clearly, dry deposition is a complex mass transfer phenomena in need of much further study.

Example 4.2-1. Particle Deposition into Lake Pontchartrain. Southerly winds move aerosols generated in the New Orleans urban area over the waters of Lake Pontchartrain, a nearly circular lagoon 45 km in diameter. For an atmospheric mixing height of 100 m determine: (a) which transport mechanisms dominate the deposition process, (b) the particle deposition velocity, and

(c) the fraction of incoming particles deposited. Meteorological data at New Orleans in July from tables in Appendix E are to be used. Assume that the dry particle diameter, d_d, is 0.39 μm, and use $d_w = 4.5 d_d^{1.04}$ at RH = 99% for the wet particle diameter[55]. Use an incoming suspended solids concentration of 136 μm/m^3 and 1.4 g/cm^3 dry particle density. Neglect the diffusophoretics of water vapor.

SOLUTION (a) The mechanisms are represented by the terms in Eq. 4.2-50:

(1) Turbulent mass transport in the constant-flux layer (CFL); use Eq. 4.2-41; $v_*^2/v_x = 1390$ cm/h.
(2) Gravity settling in the CFL; use Eq. 4.2-42 $v_{dp}(d_d) = 3.53$ cm/h.
(3) Gravity settling in the deposition layer (DL); $v_{dp}(d_w) = 47.3$ cm/h.
(4) Impact through the DL; see page 225 for St, $(v_*^2/v_x\kappa_1)(10^{-3/\mathrm{St}}) = 0.0$ cm/h
(5) Particle Brownian diffusion in the DL; Eq. 4.2-46 for $\mathscr{D}_{31}$, $(v_*^2/v_x\kappa_1)$ $[\mathrm{Sc}_{31}^{-1/2}] = 3.59$ cm/h

In the CFL processes, (1) and (2) are in parallel, with the turbulent mass transport dominating. In the DL, (3), (4), and (5) are in parallel where gravity settling dominates.

(b) The overall transport process between layers is in series and represented by Eq. 4.2-50; $v_{dp} = (47.3 + 0 + 0.0 + 3.53)(1390 + 3.53)/(1390 + 50) = 49.2$ cm/h.

Clearly, the slow process of gravity settling in the DL controls the process. In this case diffusiophoresis of water vapor is an important process; see Exercise problem 4.2L.

(c) For deposition from a uniform flow with average velocity v_x within an atmospheric mixing height h, the particle concentration in air above the water is $\rho_{31} = \rho_{31}^0 \exp(-v_{dp} l / v_x h)$, where l is the fetch. For the lake, $l = 4.5\mathrm{E}6$ cm, $v_x = 1.35\mathrm{E}6$ cm/h, $\rho_{31} = 136(0.983) = 115$ ng/m, and there is a 1.6% reduction in particle concentration.

Fate and Transport Processes for Oil Spilled on Water

When oil is spilled in an aqueous environment (see Fig. 4.2-11), it is altered from its original composition by a series of processes termed *weathering*. This fate and transport process of oil constituents on surface waters is a subject that has received much study. Renewed interest emerges after each major oil spill.[57] Two reports containing detailed literature reviews, commissioned by the U.S. Department of Interior Minerals Management Service, taken together provide a comprehensive coverage of the physical, chemical, and biological processes of oil and its constituents spilled at sea.[58,59] Such a detailed presentation is beyond the scope of this book. The following is a brief narrative description of

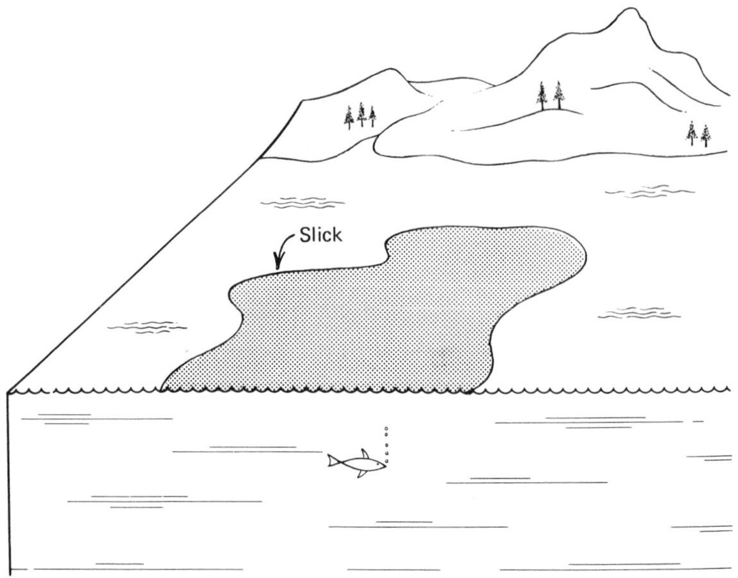

Figure 4.2-11. Oil slick at sea.

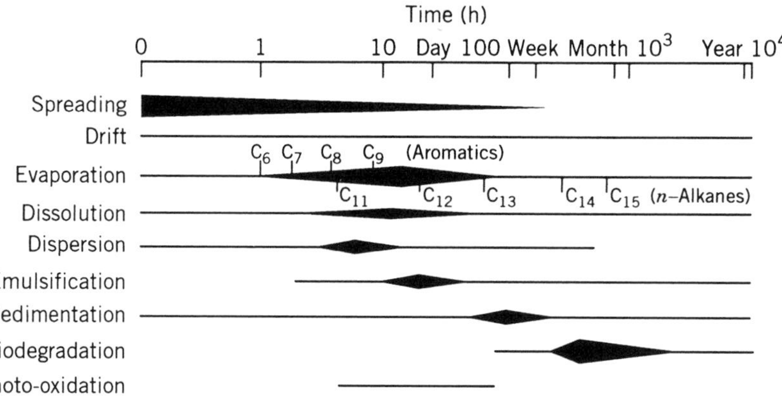

Figure 4.2-12. Processes versus time elapsed since the spill. (Modified from Ref. 58.)

the processes listed in Fig. 4.2-12 based on the sources above. In the figure line length indicates the probable time span of any process, and line width indicates the relative magnitude of the process through time. Following a brief description of all processes, evaporation and dissolution from a sea surface slick will be presented in detail.

Spreading. With little wind or wave action to effect the movement of the oil the major physical processes causing spreading are gravity and interfacial tension forces. Retarding the spread are the inertial and viscous forces.

Spreading is non-uniform. Patches of thicker oil are typically surrounded by a larger area of sheen. The wind, through drag and sea surface waves, enhances the spreading. It ruptures the slick into bands and streaks, moving it over significantly larger areas (see Problem 3.1F).

Drift. This is simply the horizontal movement of the center of mass of the slick on the water surface. Water current, wind driven or otherwise, is the motive force.

Evaporation. The loss of volatile species to air by the mechanisms of diffusion and equilibrium partitioning. This can result in 15 to 20% of the total oil being lost.

Dissolution. The movement of soluble species from the oil phase, either slick or droplet form, to the seawater by equilibrium partitioning and diffusion. It is expected to account for around 1% of the mass in the original slick.

Dispersion. In addition to dissolution, oil may enter the water column as colloidal or suspended droplets in amounts greater than from dissolution. These droplets, less than 1 μm to more than 1000 μm in diameter, are created by water turbulence which strips away globules from the slick and entrains them. Spontaneous dispersion can occur in addition to this mechanical dispersion. Settling and turbulent diffusion move the droplets down the water column. The presence of and oil attachment to suspended particulate matter in the water column aids the dispersion process.

Emulsification. A water-in-oil emulsion (mousse) results when the slick is vigorously churned by a storm. Emulsification can occur within a few hours due to the large wave size and high energy levels. The water content can be 70 to 85%, tripling the original oil volume. In this form, spreading, dispersion, evaporation, and dissolution almost completely disappear. The mousse has increased viscosity and adhesion capabilities with a solid surface such as beach material.

Sedimentation. The sinking of oil and its pollution of the sea-bottom sediments have been observed and presents the single greatest long-term threat to the environment from oil spill accidents. Fortunately, it appears to involve a small fraction of the quantity of oil spilled. Being lighter than seawater, oil droplets sink when they adhere to suspended particulate matter, which is heavier than seawater. Bioturbation can deliver oil to depths of 5 to 7 cm and occasionally deeper into the sediment bed.

Biodegradation. This is a process by which microorganisms transform and eventually mineralize the hydrocarbon components.

Photooxidation. This is a process by which oil undergoes oxidation in the presence of sunlight. Oxygenated products are generated and the process becomes significant a week or more after the spill.

In a system where all these processes are occurring, quantification and modeling are extremely difficult, compounded by the multitude of components in the oil. The following is a simple transport model study of the disappearance of aromatic and aliphatic components from small sea-surface slicks.

Simultaneous Evaporation and Dissolution. Water-soluble aromatic and aliphatic hydrocarbons may have sublethal effects on marine organisms at concentrations of 10 to 100 ppb, lethal toxicity at 0.1 to 1.0 ppm for most larval stages, and lethal effects at 1 to 100 ppm for most adult organisms. It is thus of interest to investigate the rates of disappearance of specific aromatic and aliphatic components from crude oil slicks under conditions of essentially constant water temperature and nearly constant air temperature. Studies of the fate of slick components during the early stages of slick aging are crucial because the lower-boiling fractions contain most of the lethal components of the slick.[33] These components evaporate or dissolve within a few hours of slick initiation. Although little is known of the relative percentages of loss of these slick components due to evaporation and dissolution, it is assumed that they are mainly lost by evaporation, at least under conditions of low sea-surface roughness.

Lower-boiling components leave the slick by two routes: mass transfer to the air and to the seawater. Figure 4.2-13 shows the directions of chemical movements and concentration of a typical low-boiling component within the slick and in the adjoining phases. The total rate of mass transfer for component A is

$$N_{A0} = {}^4k'_{A1}(c_{A1i} - c_{A1}) + {}^4k'_{A2}(c_{A2i} - c_{A2}) \tag{4.2-59}$$

where ${}^4k'_{A1}$ is the gas-phase mass transfer coefficient above the slick surface, c_{A1i} is the air concentration of component A at the slick interface in equilibrium with the concentration of A in the slick, c_{A4}. The slick is denoted as phase 4. On the water side of the slick, ${}^4k'_{A2}$ is the mass transfer coefficient, and c_{A2i} is the seawater concentration of chemical A at the slick interface in equilibrium with c_{A4}. Far removed from the interfaces, the concentrations in air and water are c_{A1} and c_{A2}, respectively. The liquid-phase mass transfer coefficients in the oil (i.e., ${}^1k'_{A4}$ and ${}^2k'_{A4}$) are assumed to be large and not rate limiting. This assumption is probably invalid for an oil layer thickness greater than 0.1 mm, within which mixing is slow due to little or no wave action on the sea surface.

Consider a slick containing a component mole fraction of chemical A, x_{A4}. This chemical has an air-phase partial pressure for component A immediately above the interface given by

$$p_{A1i} \equiv p_T y_{Ai} = x_{A4}\gamma_{A4}p_A^* \tag{2.1-9}$$

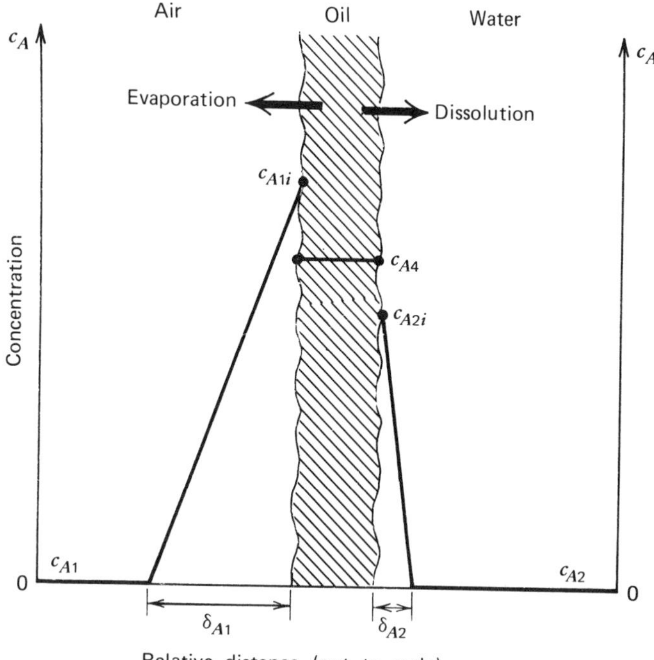

Figure 4.-13. Mass transfer from oil slick, film theory interpretation

Partial pressure in air can be converted to concentrations by using the ideal gas law relationship $c_{A1} = p_{A1}/RT_1$. If component A has an aqueous-phase solubility of c^*_{A2}, the water–slick interface concentration can be approximated by

$$c_{A2i} = c^*_{A2} x_{A4} \gamma_{A4} \tag{4.2-60}$$

See Problem 2.1L for details on developing Eq. 4.2-60.

Now if it is assumed that the oil phase is well mixed, the background concentration of chemical A in air and seawater is zero, a differential mass balance on a square meter of slick with a "thickness" corresponding to $\mathscr{M}/A$, mol/m^2, and containing component A yields

$$\frac{d(x_{A4}\mathscr{M}/A)}{dt} = -{}^4k_{A1}c_{A1i} - {}^4k_{A2}c_{A2i} \tag{4.2-61}$$

where $\mathscr{M}$ is the total moles of oil in the slick and A is the area of the sea surface covered. The oil thickness is $h = \mathscr{M}/Ac_4$ where c_4 is the molar density of the oil. Employing Eq. 2.1-9, the ideal gas law, and Eq. 4.2-60 in the preceding mass balance yields a simple differential equation with x_{A4} as the dependent variable:

$$\frac{d(x_{A4}\mathcal{M}/A)}{dt} = -\frac{{}^4k'_{A1}x_{A4}\gamma_{A4}p^*_A}{RT} - {}^4k'_{A2}c^*_{A2}x_{A4}\gamma_{A4} \tag{4.2-62}$$

Rearranging and assuming that $\mathcal{M}/A$ is constant and integrating between limits of $x_{A4} = x^0_{A4}$ at $t = 0$ to x_{A4} at t gives

$$x_{A4} = x^0_{A4}\exp(-Kt) \tag{4.2-63a}$$

where

$$K = \frac{{}^4k'_{A1}\gamma_{A4}p^*_A/RT + {}^4k'_{A2}\gamma_{A4}c^*_{A2}}{\mathcal{M}/A} \tag{4.2-63b}$$

The preceding description follows closely that of Harrison et al.[33]

The rate at which chemical A leaves the slick and enters the water decreases as the slick ages and the concentration decreases. The rate at which chemical A enters the air behaves similarly. In both cases the rate is proportional to the individual phase mass transfer coefficients and concentration of the chemical. The rate of evaporation is

$$N_{A0}(\text{evap}) = {}^4k'_{A1}\frac{\gamma_{A4}p^*_A}{RT}x^0_{A4}\exp(-Kt) \tag{4.2-64}$$

and the dissolution rate is

$$N_{A0}(\text{diss}) = {}^4k'_{A2}c^*_{A2}\gamma_{A4}x^0_{A4}\exp(-Kt) \tag{4.2-65}$$

In the interest of assessing the fate of chemical A, it is important to know the quantity entering each phase. The quantity of A that has moved into the air per unit of sea surface covered by the slick of age t is

$$\mathcal{M}_A(\text{evap}) \equiv \int_0^t N_{A0}(\text{evap})\,dt,$$

$$= {}^4k'_{A1}\frac{\gamma_{A4}p^*_A}{RT}\frac{x^0_{A4}}{K}[1 - \exp(-Kt)] \tag{4.2-66}$$

The quantity that transferred to the water during a similar period is

$$\mathcal{M}_A(\text{diss}) \equiv \int_0^t N_{A0}(\text{diss})\,dt \tag{4.2-67}$$

$$= {}^4k'_{A2}\gamma_{A4}c^*_{A2}\frac{x^0_{A4}}{K}[1 - \exp(-Kt)] \tag{4.2-68}$$

The last four expressions show that the movement of chemical A into the adjoining phases occurs rapidly immediately after the slick is formed but that the rate decreases in an exponential manner.

Harrison et al.[33] report on the dynamic behavior of two components in small experimental sea-surface slicks formed of South Louisiana crude. The experiments were performed off the south shore of Grand Bahama Island. The crude oil was spiked with cumene (isopropyl benzene) to yield a solution of 4.2% cumene by weight. The concentrations of cumene, nonane, and other hydrocarbons within the oil phase were determined by gas chromatograph for the first few hours of slick aging.

Cumene and all lower-boiling aromatics disappeared within the first 90 min. In general, cumene disappeared faster than nonane. The time to achieve 63% loss in spill 1 was 13 min for cumene and 28 min for nonane; 25 and 35 min, respectively, in spill 2; and 20 and 27 min, respectively, for spill 3. It was concluded from seawater samples that the dissolution rate was considerably slower than the evaporation rate and that only organisms in water that are or have been in close proximity to a spill for an extended time are likely to suffer toxic effects (see Problem 4.2F for further details).

The analysis above is a good first-order estimate of the vaporization and dissolution fate of slick constituents. It falls short of being a perfect model, for several reasons. In reality the slick spreads so that $\mathscr{M}/A$ is a function of time (see Problem 3.1F). With the loss of a portion of the volatile and soluble fractions the oil becomes more viscous, thereby decreasing internal mixing so that constituent diffusion within the slick begins to control the release rate. These factors can be included in a more complete theoretical model.

Multicomponent Evaporation. Crude oil and its distilled fuel products consist of a host of individual hydrocarbon compounds. Regnier and Scott[34] performed evaporation rate studies of n-alkane components of Arctic diesel 40, a No. 2 fuel, to clarify the multicomponent behavior of slicks. Gram size (i.e., 12-g) samples were placed in 90-mm-diameter dishes, yielding a 3-mm oil depth. Temperature was controlled in the environmental chamber and a 21-km/h wind speed maintained. The air movement over the samples produced small wavelets that resulted in a stirring action. Samples were taken for gas chromatographic analysis.

A detailed study of the 10 n-alkanes, c_9 through c_{18} revealed that a first-order model was sufficient to quantify the evaporation rate of each. Since Eq. 4.2-63b reflects first-order kinetics of evaporation, the rate constants must be related to the gas-phase mass transfer coefficient, vapor pressure, temperature, evaporation surface area, and total moles by

$$K = \frac{{}^4k'_{A1}\gamma_{A4}p_A^* A}{RT\mathscr{M}} \qquad (4.2\text{-}69)$$

The dissolution term is omitted. It was found that a 1:1 correspondence existed between K and p_A^*. A least squares fit of those data with $p_A^* > 1\text{E-}6$ atm was performed. The following equation with p_A^* in atm and K in min^{-1}

$$\log p_A^* = 1.25 \log K + 0.160 \tag{4.2-70}$$

makes it possible to calculate the evaporation rate constant of an n-alkane if its vapor pressure is known. Although the result is for an oil layer 3 mm in thickness, Eq. 4.2-69 suggests that $K \sim 1/h$, so that corrections to other thicknesses are possible.

Example 4.2-2. Half-Life of Normal Hexane in Oil Slick. Estimate the half-life of n-hexane in 1-mm-thick oil slick of temperature 75°F. The vapor pressure of n-hexane is 3.0 psia.

SOLUTION Use Eq. 4.2-70 to estimate the evaporation rate constant, and use Eq. 4.2-63a for the time. This procedure assumes that evaporation is the dominant mechanism of n-hexane disappearance from the slick. From Eq. 4.2-70

$$\log K = \frac{\log(3/14.7) - 0.160}{1.25}$$

$$K = 0.209 \text{ min}^{-1}$$

For a 1-mm layer, $K = 0.209\ (3\text{ mm}/1\text{ mm}) = 0.627\ \text{min}^{-1}$. From Eq. 4.2-63a with $x_{A4} = x_{A4}^0/2$,

$$t_{1/2} = \frac{-\ln(1/2)}{K} = 1.11 \text{ min}$$

Closure. The simple models presented above contain several essential aspects of transport processes associated with oil slicks or similar chemical slicks on the surface of water. It appears that evaporation is a dominant mechanism, and from the viewpoint of interphase chemical movement, the air receives the bulk of the transported contaminants. The evaporation process occupies only a brief period during the life of a crude oil slick. Figure 4.2-14 contains evaporation rate data of an arctic diesel. It appears that the evaporation rate can decrease by as much as two orders of magnitude in 8 days.

During the early period of evaporation, 0 to 50,000 sec in Figure 4.2-14, the data appear linear as $\log n_{A0}$ versus t^{-1}. The theoretical flux equation for evaporation, Eq. 4.2-64, displays this same functionality, and the slope of the curves at a particular time is equal to $-K$. The data show that as time increases the magnitude of K decreases and the rate falls more slowly. During the period where the curve is nonlinear, the basic assumptions behind the transport model are violated. As the more volatile components evaporate, the mixture becomes more viscous, and diffusion processes within the oil phase progressively dominate the rate of evaporation. It has been observed that a

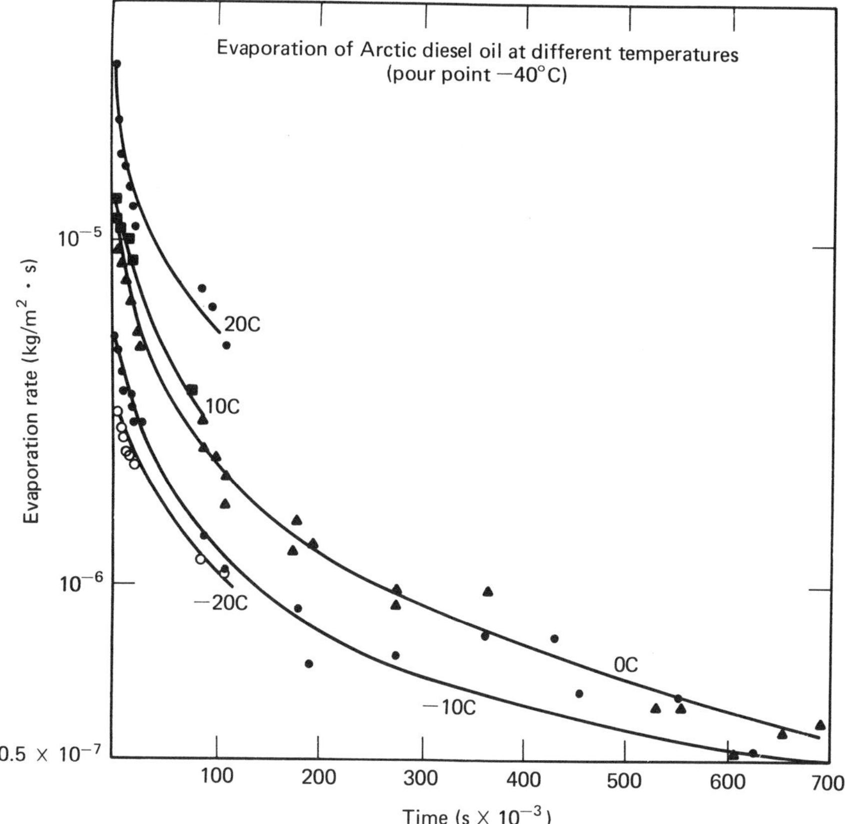

Figure 4.2-14. Evaporation rates of arctic diesel as a function of time and temperature. (From Ref. 35.)

type of film forms on the surface of unmixed crude oil that essentially stops evaporation of the lighter fractions.

PROBLEMS

4.2A. Tritium Loss from Water Exposed to the Atmosphere[36]

Laboratory and field observations were made on the fate of small quantities of tritium as HTO found in process water in a certain nuclear facility. The water is usually released into streams or open basins excavated in the ground.

In the field study, an impermeable basin (13,000 m^2 in area) containing tritium was isolated from further waste discharges and studied for a 3-year period. Monthly, water was taken from a depth of 1 ft at one location and the tritium content determined. The field data are shown in Fig. 4.2A. A slow mixing of surface water with underlying liquid was observed. This occurred because of density stratification; the waste underneath the fresh water layer on

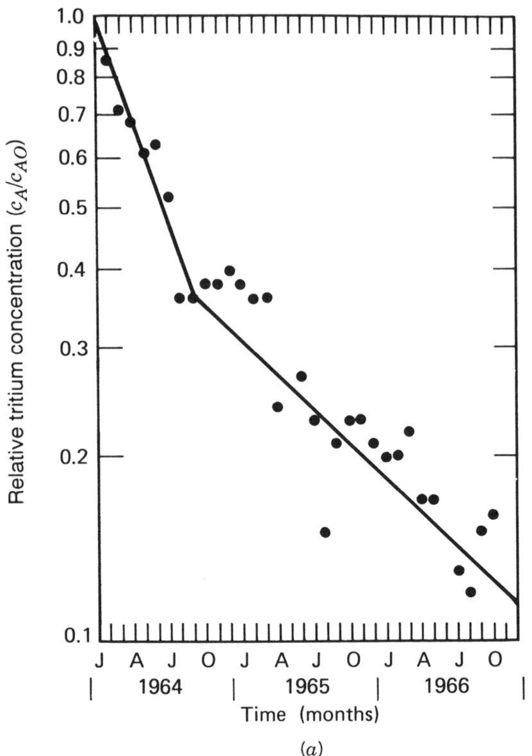

(a)

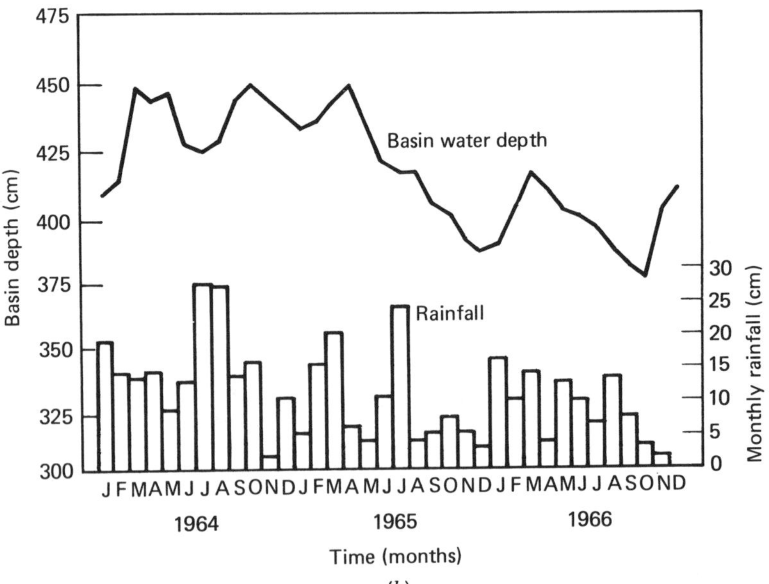

(b)

Figure 4.2A. (*a*) Tritium concentrations in an isolated impermeable basin from January 1964 through December 1966. (*b*) Depth of water in an isolated impermeable basin containing tritium and the rainfall received from January 1964 through December 1966. (Reprinted with permission from Ref. 36. Copyright by the American Chemical Society.)

the surface contained dissolved salts. The value of the overall mass transfer coefficient depends on the rate of mixing.

1. Compute an average overall liquid-phase mass transfer coefficient for the basin using the field data. Answer in cm/s. Does the value seem low? Opine how this can be.

2. Estimate the Henry's constant, H_ρ, cm^3 H_2O/cm^3 air, for HTO at 25°C. At 25°C the vapor pressure of HTO is 0.92 that of H_2O.

3. Confirm by calculation that tritium loss is liquid-phase controlled, owing apparently to slow water mixing. Use 750 cm/s as a reasonable gas-phase coefficient for earthen basins. Assume that the basin is located in South Carolina.

4.2B. Air–Sea Interface Flux Calculations[25]

Calculate the flux rate in g/yr and direction of transfer for the gases given in Table 4.2B at the air–sea interface. Use 3.6E 18 cm^2 as the air–water surface area of the oceans. Compare your answers with the values given in Table 4.2-3.

4.2C. Evaporation of Low-Solubility Contaminants from Water Surfaces[21,22]

Evaporation parameters for several chemicals appear in Table 4.2-1.

1. Calculate the vaporization half-life and mean residence time for *n*-octane, benzene, DDT, and Aroclor 1242. Assume a water depth of 1 m free of suspended solids.

2. Repeat the calculation for Aroclor 1242 using Hickory Hills Pond Sediment (Table 2.1-6) with suspended solid concentrations of 1, 10, 100, 1000, and 10,000 mg/L.

4.2D. Evaporation in a Short-Fetch Water–Wind Tunnel[37]

In a comparison of evaporation in the laboratory and at sea, Schooley employed a short-fetch water–wind tunnel to measure evaporation for wind velocities between 200 and 800 cm/s. The average wind speed was measured with a small commercial wind-velocity probe. *A* spring-balance arrangement weighed the tunnel before and after each experiment to determine the volume evaporated. The experimental data are shown in Fig. 4.2D.

Table 4.2B. Concentration of Gases in the Marine Environment

Gas	In Air	In Water	Gas	In Air	In Water
SO_2	3 μg/m	0	CCl_4	71.2E-6[a]	60E-12[b]
N_2O	0.25[a]	0.4 ppm[c]	CCl_3F	50E-6[a]	7.6E-12[b]
CO	0.13[a]	6E-8[b]	MeI	1.2E-6[a]	135E-12[b]
CH_4	1.4[a]	4E-8[b]	$(Me)_2S$	~0	1.2E-11 g/cm^3

Source: Ref. 25.

[a]Units of concentration are parts per million (by volume).

[b]Units of concentration are cm^3 gas at STP ÷ cm^3 liquid water.

[c]This is the N_2O concentration in air in equilibrium with the ocean surface water N_2O

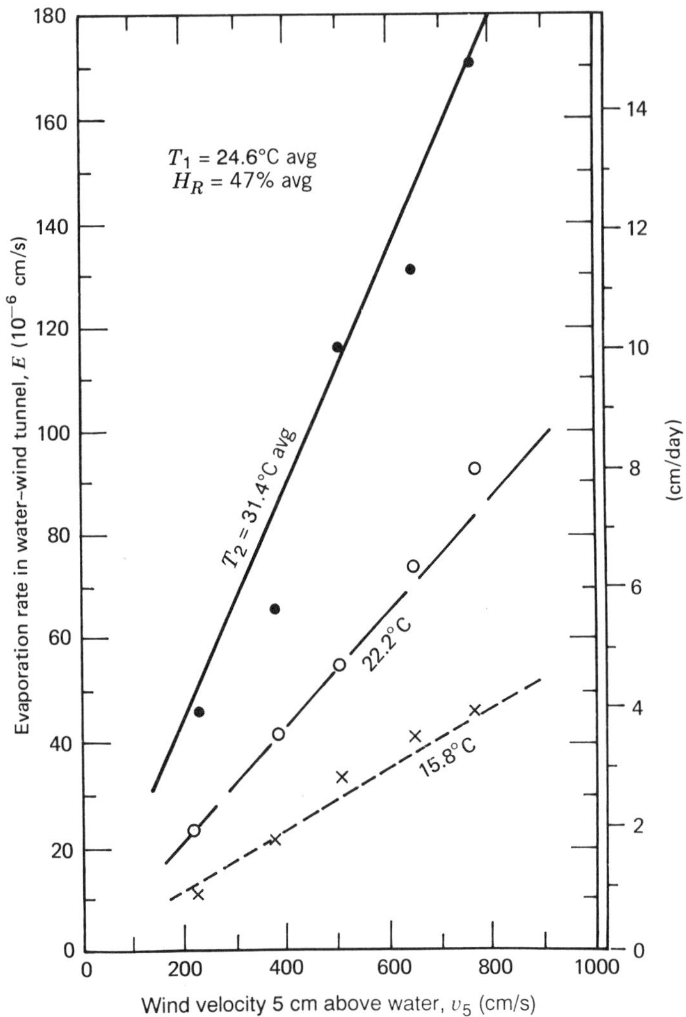

Figure 4.2D. Evaporation rate versus average wind speed at 5 cm above water for three different water temperatures in a short-fetch water–wind tunnel. Air temperature and relative humidity held constant. (Reprinted with permission from Ref. 37.)

1. Show that Eq. 4.2-28 contains the correct parameterization for the mass transfer coefficient by displaying the data as in Figure 4.2-5.

2. Determine k_{10} for each data point obtaining the average and standard deviation. Assume that wind speed at 5 cm is 40% that at 10 m.

4.2E. Sea-Surface Liquid-Phase Coefficient from Measurements of Radon[31]

Subsurface water samples (10 L) were collected using a Niskin sampler while surface samples were drawn through tubing lowered from the fantail of the

Table 4.2E. Radon Content at Various Depths Below the Sea Surface

Depth (m)	^{222}Rn (dpm[a]/100 L)	Number of Observations
1	6.1	16
10	5.9	8
20	6.2	4
25	7.1	5
35	7.4	6
50	7.8	6
75	7.5	2
100	8.2	6
200	8.0	4

Source: Ref. 31.

[a]Disintegrations per minute.

ship. The samples reported here were collected from the Coast Guard vessel *Rockaway* during the period May 2 to June 9, 1969, from a position 15°N, 46°W. The radon results are given in Table 4.2E corrected for salinity. Using the data in this table, calculate ${}^{1}k'_{A2}$ in cm/h and compare your results with values reported in Table 4.2-4.

4.2F. Evaporation and Dissolution of Cumene and Nonane from Small Sea-Surface Slicks[33]

A small sea-surface slick was created by placing 275 gal (1.04 m^3) of crude oil spiked with cumene ($c_{A4}^0 = 0.34$ mol/L). Water temperature was essentially constant at 23.6°C, air temperature was 20.5 to 24.1°C, relative humidity was 60 to 79%, wind at 3 m was calm to 18 mph with gusts to 22 mph (9.8 m/s), and sea-surface conditions ranged from calm with gentle swell to extensively whitecap covered.

1. Compute the time (min) for 63% of the cumene to disappear from a slick 1 mm in thickness. Use the following data: $\mathscr{M}/A = 5$ mol/m^2 (1 m^3 spread over 10^3 m^2); ${}^{4}k'_{A1} = 1\text{E-}2$ m/s, ${}^{4}k'_{A2} = 5.5\text{E-}5$ m/s; $p_A^* = 5.3\text{E-}3$ atm (4.1 mmHg); $c_{A2}^* = 50$ mg/L; $\gamma_{A4} = 1.0$. Compare the computed time with the observed times reported in the text.

2. Compute the time (min) for 63% of the original nonane to disappear. Use the following data: $p_A^* = 3.0$ mmHg; $c_{A2}^* = 0.22$ mg/L.

3. Using the model results, show that for $t \to \infty$ the percent A evaporated is $100\ {}^{4}k'_{A1}H_p/({}^{4}k'_{A1}H_p + {}^{4}k'_{A2})$, where is H_p, Henry's constant, in concentration units. Compute the percent evaporated for both cumene and nonane. For these two constituents, which mechanism dominates the disappearance from the slick?

4. Constituent disappearance includes both evaporation and dissolution. Separate the two and compute $t_{63\%}$ for cumene evaporation and cumene dissolution. Repeat for nonane. For these constituents, which process is more rapid?

4.2G. Slick Evaporation: Theory Versus Data

Equation 4.2-69 is a theoretical result for constituent evaporation, while Eq. 4.2-70 is the result of laboratory experiments.

1. Show that Eq. 4.2-69 can be put in the general form of Eq. 4.2-70.
2. Show that Eq. 4.2-70 can be transformed to

$$K \simeq \frac{9(p_A^*)^{4/5}}{4h} \tag{4.2G}$$

where K is in min^{-1}, p_A^* is atm, and h is slick thickness in mm.

3. Using this result, repeat the $t_{63\%}$ calculation for cumene and nonane evaporation in part 4 of Problem 4.2F.

4.2H. Steady-State Radon Mass Balance

Show the procedure and assumptions for applying Eq. 3.0-1 to obtain Eq. 4.2-18.

4.2I. Particle-Bound Fractions for Selected Air Contaminants

Using the data in Table 4.2-1, determine which chemical species are likely to be predominately particle bound when at equilibrium with the gas phase of urban air at 25°C. Which are predominately in the gas phase, and which are somewhat evenly distributed?

4.2J. Deposition Flux Algorithms for Particles and Gases

1. Complete the transformation steps of Eq. 4.2-39 for the constant flux layer to Eqs. 4.2-47 and 4.2-51.
2. Complete the transformation steps of Eq. 4.2-35 for the deposition layer to Eqs. 4.2-48 and 4.2-52. Give and justify the assumptions in the transformations.

4.2K. Diffusiophoretic Flux of Water in Deposition Layer

Estimate n_{By}/ρ_1 in cm/h for water evaporating from a lake surface at 28°C into air at 65, 75, 85, 95, and 99% relative humidity. Use $v_* = 375$ m/h.

4.2L. PCB Deposition into Lake Pontchartrain

Continue Example 4.2-1; the A-1254 total concentration in the atmosphere is 1.0 $\mu g/m^3$.

1. Assume that the particles have a sorptive surface area of 11 m^2/g and estimate the equilibrium fraction on the particle phase.
2. Estimate the quantity deposited in the lake for zero concentration in water (g/h).
3. Determine the fraction PCB deposited by particles.

4.2M. Chemical Evaporation from Open Pit

Frequently, a used organic chemical is placed in an open holding pit as a means of disposal. Open pits collect rainwater, and immiscible chemicals lighter than water will float (see Fig. 4.2M).

1. Develop an equation to predict pool lifetime. Perform a transient material balance on a pool of pure chemical A of surface area A and depth h. Assume in the model development that evaporation occurs from the top of the pool and peels away successive layers (so to speak) of constant area A. *Answer:*

$$t = \frac{\rho_A h}{{}^4k'_{A1}\rho^*_{A1}} \tag{4.2M}$$

   where t is lifetime, ρ_A the density of chemical A, ${}^4k'_{A1}$ the gas-phase mass transfer coefficient, and ρ^*_{A1} is the air concentration of chemical A in equilibrium with the liquid pool ($4 \equiv$ volatile liquid phase).

2. Estimate the time (h) required to evaporate 112,200 gal of styrene (C_8H_8) from an earthen pit of surface dimensions 300 by 300 ft. Neglect the dissolution of styrene into the water, and consider only the evaporation into air. Assume that this pit is located in an industrial park near Ft. Smith, Arkansas. Use typical winter-day conditions of 42°F and 8.2 mi/hr wind speed.

   Styrene data: molecular weight 104.14, density 0.903, boiling point 145°C, melting point -31°C, colorless liquid, $\mathscr{D}_{A2} \simeq 0.07$ cm^2/s, vapor pressure

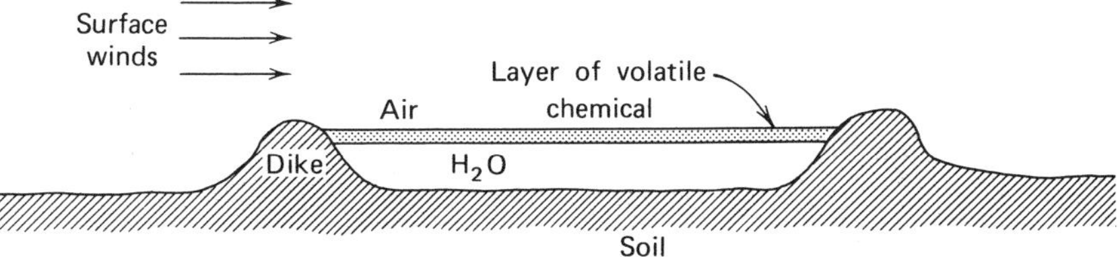

Figure 4.2M. Open-pit evaporation of volatile liquids.

(mmHg) versus temperature (°C): (1, −7.0), (5, 18.0), (10, 30.8), (20, 44.6), (40, 59.8).
(This problem was suggested by Albert Hood.)

4.3. HEAT TRANSFER ACROSS THE AIR–WATER INTERFACE

Temperature is an important variable in assessing chemical movement rates across the air–water interface. The temperature at the water surface significantly affects the vapor pressure and solubility of the chemical species and also affects the magnitude of the transfer coefficients and reaction rates. Fundamental concepts of heat transfer were presented in Section 3.3. In this section specific attention is devoted to the rate equations for heat transfer between water bodies and is generally applicable at the surfaces of streams, lakes, and oceans.

A study of heat transfer in water bodies begins with the equation of energy. If one considers a fixed volume of water V containing an air–water interface of area A_{12} and a water–sediment interface of area A_{23}, the equation of energy can be written simply as

$$\rho_2 \hat{C}_p \frac{dT}{dt} = \frac{1}{V}(q_{12}A_{12} + q_{23}A_{23}) \tag{4.3-1}$$

where gradients of temperatures, convective flow, and heat generation within the volume are neglected. Equation 4.3-1 states that the change in enthalpy of a fixed water body is the result of heat flow through the interfaces. The sign convention adopted is that energy additions to a surface are positive and energy losses are negative. All energy or heat flux relations are in $J/m^2 \cdot s$. ($1\ Btu/ft^2 \cdot hr = 3.15\ J/m^2 \cdot s$.)

If we now assume steady state and neglect heat losses through the bottom and sides of the lake, pond, or stream, the heat (or energy) flux through the air–water interface is seen to consist of four major sources:

$$q_{12} = q_{lw} + q_e + q_c + q_s \tag{4.3-2}$$

where q_{lw} is the longwave radiant energy flux, q_e the evaporative heat flux, q_c the sensible heat flux, and q_s the shortwave radiant energy. Figure 4.3-1 illustrates these transfer mechanisms. As shown, the magnitude of the fluxes depends on the water body, its geographic location, and time. All fluxes except q_s can be positive or negative. Heat flux across the sediment-water interface, q_{23}, is covered briefly in Example 5.4-1.

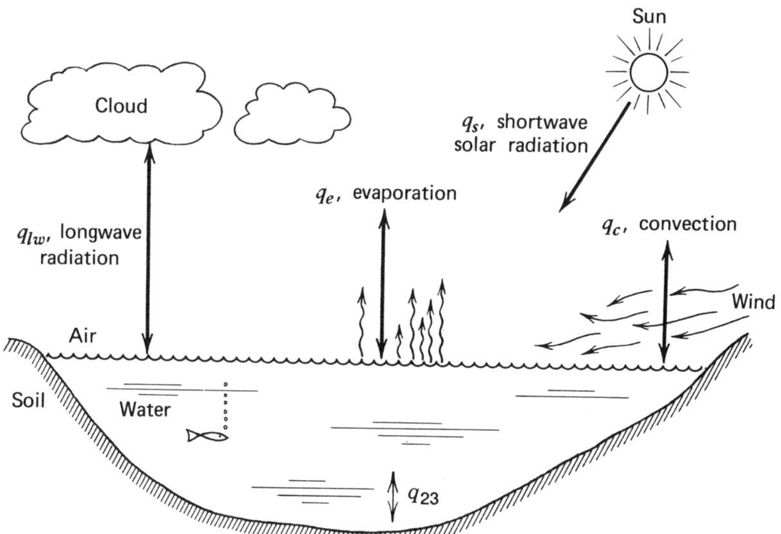

Figure 4.3-1. Heat and energy transfers at the air–water interface.

Components of Energy Balance Equation

Shortwave Radiation. Shortwave radiation originates directly from the sun. The amount of solar radiation incident on a horizontal surface varies, depending on the geographic location, elevation, season, and meteorological conditions. Solar radiation intensity observations are made at a number of U.S. Weather Bureau stations. Long-term meteorological data is published by the Environmental Data Service, National Oceanic and Atmospheric Administration, National Climate Data Center, U.S. Department of Commerce. The data available are solar radiation observed at ground stations, tabulated in langley. A langley/day = 4.1868 J/cm$^2 \cdot$day. Monthly daily average values are available at numerous locations in the United States.

An alternative source is the total daily solar radiation at the top of the atmosphere. Table 4.3-1 are the Smithsonian Meteorological Tables for this radiation flux. The entries are cal/cm$^2 \cdot$day (1 cal = 4.1868 J). Cloud cover decreases the incoming radiation. Stations recording cloud cover are more numerous than those recording ground-level radiation. Average percentage of possible sunshine in selected climatic regions is given in Table E.2e. If this type of cloud cover information is available, the following equation may be used to estimate the shortwave radiation[39]:

$$q_s = q_{s0}(0.803 - 0.340n - 0.458n^2) \tag{4.3-3}$$

where q_{s0} values are obtained from Table 4.3-1 and n is the fractional cloud cover which is 1 − fraction sunshine.

Table 4.3-1. Total Daily Solar Radiation at the Top of the Atmosphere[a]

	Longitude of the sun															
	0°	$22\frac{1}{2}$°	45°	$67\frac{1}{2}$°	90°	$112\frac{1}{2}$°	135°	$157\frac{1}{2}$°	180°	$202\frac{1}{2}$°	225°	$247\frac{1}{2}$°	270°	$229\frac{1}{2}$°	$315\frac{1}{2}$°	$337\frac{1}{2}$°
	Approximate Date															
Latitude	Mar. 21	Apr. 13	May 6	May 29	June 22	July 15	Aug. 8	Aug. 31	Sep. 23	Oct. 16	Nov. 8	Nov. 30	Dec. 22	Jan. 13	Feb. 4	Feb. 26
−90°		423	722	999	1077	994	765	418								
−80	155	423	760	984	1060	980	754	418	153	7						7
−70	307	525	749	939	1012	934	742	519	303	129	24				24	131
−60	447	635	809	934	979	929	801	629	442	273	146	72	49	73	146	276
−50	575	732	867	958	989	954	859	725	568	414	286	204	176	205	289	419
−40	686	807	910	972	991	967	901	798	677	545	429	348	317	350	434	553
−30	775	865	929	967	975	960	921	856	765	663	564	492	466	494	568	670
−20	841	894	923	935	935	930	916	884	831	760	685	627	605	630	691	769
−10	882	897	893	881	873	877	886	887	871	835	789	748	733	752	795	845
−0	895	873	837	804	790	800	830	863	885	886	870	851	843	855	878	896
−10	882	824	760	707	687	704	753	814	871	910	927	931	933	936	936	921
−20	841	750	660	593	567	590	654	741	831	907	959	988	999	993	968	918
−30	775	654	543	465	436	463	538	646	765	877	964	1020	1041	1025	973	888
−40	686	538	413	329	297	328	409	533	677	819	944	1027	1059	1032	953	828
−50	575	408	276	193	165	192	274	404	568	743	901	1014	1056	1018	909	752
−60	447	269	140	68	47	68	139	266	442	644	840	987	1046	992	847	652
−70	307	127	23				23	126	303	532	778	993	1081	998	785	539
−80	155	7						7	153	429	790	1041	1132	1046	796	434
−90										429	801	1056	1149	1062	809	434

Source: R. J. List, *Smithsonian Meteorological Tables*, 6th rev. ed., 1949.

[a]Values are in cal/cm^2 and apply to a horizontal surface.

Longwave Radiation. By virtue of a temperature difference, the water and its surrounding air canopy exchange radiant energy. The rate equation for the net exchange is

$$q_{lw} = \beta\sigma T_1^4 - e_2\sigma T_2^4 \tag{4.3-4}$$

where β is the cloud cover factor, σ the Stefan–Boltzmann constant (5.67E-12 $J/s \cdot cm^2 \cdot K^4$), T_1 the air temperature in kelvin, e_2 the emissivity of the water surface, and T_2 the water surface temperature in kelvin.

The first term on the right-hand side of Eq. 4.3-4 is the longwave atmospheric radiation. β is a function of the type of cloud cover, given by

$$\beta = a + bp_{A1} \tag{4.3-5}$$

where p_{A1} ($A \equiv H_2$ vapor) is the partial pressure of water vapor in inches of mercury (1 in. Hg = 3.377E3 Pa), and a and b are constants. For each value of cloud cover, these constants can be approximated as shown in Table 4.3-2. Over the range 0 to 50°C the vapor pressure of water in Pa can be calculated within $\pm 4\%$ error by

$$p_{A1} = 99.62\left[\exp\left(21.66 - \frac{5431.3}{T}\right)\right] H_R \tag{4.3-6}$$

where T is in kelvin and H_R is the fraction relative humidity, $0 < H_R < 1$.

The second term on the right-hand side of Eq. 4.3-4 is the longwave radiation originating at the water surface. A value of 0.97 is reasonable for the emissivity of water.

Table 4.3-2. Cloud Cover Constants

Cloud Cover (tenths)	β
0	$0.74 + 0.15p_{A1}$
1	$0.75 + 0.15p_{A1}$
2	$0.76 + 0.15p_{A1}$
3	$0.77 + 0.143p_{A1}$
4	$0.783 + 0.138p_{A1}$
5	$0.793 + 0.137p_{A1}$
6	$0.80 + 0.135p_{A1}$
7	$0.81 + 0.13p_{A1}$
8	$0.825 + 0.12p_{A1}$
9	$0.845 + 0.105p_{A1}$
10	$0.866 + 0.09p_{A1}$

Source: Industrial Pollution, edited by N. Irving Sax, © 1974 by Litton Educational Publishing, Inc. Reprinted by permission of Van Nostrand Reinhold Company.

Evaporative Energy Exchange. If the air above a body of water is less that 100% saturated with water vapor (i.e., $H_R < 1$), there is a potential for evaporation from the surface. The evaporated water requires energy, latent heat of evaporation, as it changes from a liquid to a vapor. Equation 3.3-5 relates the mass and energy flux rates. The Lake Hefner study (see Eq. 4.1-13) resulted in the following equation:

$$q_e = -30.5 v_x (p_{A1}^0 - p_{A1}) \tag{4.3-7}$$

where q_e is the evaporative heat loss in $J/m^2 \cdot s$, v_x the wind speed in km/h, p_{A1}^0 the vapor pressure of water in in. Hg at T_2 (the water surface temperature), and p_{A1} is the partial pressure of water vapor in the air far removed from the lake surface. Note that p_{A1}^0 can be obtained from Eq. 4.3-6 with $H_R = 1$ and $T = T_2$. Atmospheric conditions of temperature (T_1), relative humidity (H_R), and wind velocity (v_x) are usually measured at 8 to 10 m above the surface. These values can be used in Eq. 4.3-7.

It would be expected that coefficients for evaporation and heat lost would be much different for rivers and streams than for lakes and would be dependent on water velocity and turbulence. Rathbun and Tai[13] studied water evaporation from the San Diego Aquaduct. Their result is summarized in Eq. 4.1-25. Once the water flux is obtained, Eq. 3.3-5 is used to obtain the evaporative heat loss rate.

Sensible Heat Exchange. Heat enters or leaves water by conduction if the air temperature is greater or less than the water temperature. There is a transfer of sensible heat and the rate expression is of the form of Eq. 3.3-3. The rate expression is

$$q_c = 0.27 v_x (T_1 - T_2) \tag{4.3-8}$$

where q_c is the sensible heat transfer rate in $J/m^2 \cdot s$, and T_1 and T_2 are the air and water temperatures, respectively, in kelvin.

PROBLEMS

4.3A. Maximum Radiant Energy Loss from Lake Surface

Compute the rate of radiant energy lost from the surface of a lake at 20°C on a clear night. Assume that the sky is black with no clouds (i.e., a perfect adsorber). Report your answer in $J/m^2 \cdot s$.

4.3B. Typical Energy Transfer Rates Through Reservoir Surface

Compute the energy transfer rate through the air–water interface of Beaver Reservoir. Account for all energy and heat transfer mechanisms. Note the direction and magnitude. Compute the totals for each day. Use the following environmental conditions:

1. Julian day 123 (approximately May 3): wind speed 10.4 mi/hr, relative humidity 71%, air temperature 66°F, water temperature 16°C.

2. Julian day 280 (approximately October 7): wind speed 9.7 mi/hr, relative humidity 68%, air temperature 69°F, water temperature 21°C.

Report your answers in $J/m^2 \cdot s$.

REFERENCES

1. L. J. Thibodeaux and D. G. Parker, "Desorption Limits of Selected Industrial Gases and Liquids from Aerated Basins," in C. Rai and L. A. Spielman, Eds., *Air Pollution Control and Clean Energy*, AIChE Symposium Series 72, 1976, p. 156.
2. Technical Practice Committee, Subcommittee on Aeration in Wastewater Treatment, *Manual of Practice 5*, Water Pollution Control Federation, Washington, D.C., 1971.
3. J. S. Boundurant, S. Luce, and H. Townsend, Unpublished chemical engineering senior project report, Department of Chemical Engineering, University of Arkansas, Fayetteville, Ark., 1973.
4. L. J. Thibodeaux, D. G. Parker, and H. Heck, "Measurement of Volatile Chemical Emissions from Wastewater Basins," Final Report, *NTIS PB 83-135 632*, U.S. EPA, IREL, Cincinnati, Ohio. Jan. 1982, p. 79.
5. J. A. Reinhardt, "Gas-Side Mass-Transfer Coefficient and Interfacial Phenomena of Flat-Bladed Surface Agitators," unpublished doctoral dissertation, University of Arkansas, Fayetteville, Ark., 1977.
6. P. D. Lunney, C. Springer and L. J. Thibodeaux, *Environ. Prog.*, **4**(3), 203–211 (1985).
7. G. E. Harbeck, Jr., "A Practical Field Technique for Measuring Reservoir Evaporation Utilizing Mass-Transfer Theory," *Geological Survey Professional Paper 272-E*, U.S. Government Printing Office, Washington, D. C., 1962.
8. L. J. Thibodeaux, C. Springer, P. D. Lunney, S. James, and T. T. Shen, "Air Emission Monitoring of Hazardous Waste Sites," *Proceedings of the Conference on Management of Uncontrolled Hazardous Waste Sites*, Washington, D.C., Nov. 1982, pp. 70–75.
9. C. Springer, L. J. Thibodeaux, P. D. Lunney, R. S. Parker, and S. James, "Secondary Emissions from Hazardous Waste Disposal Lagoons: Field Measurements," *EPA-600/9-83-018,* EPA Research Symposium, Ft. Mitchell, Ky., Sept. 1983, pp. 58–69.
10. L. J. Thibodeaux, D. G. Parker, and H. H. Heck, "Chemical Emissions from Surface Impoundments," *Environ. Prog.*, **3**(2), 73–78 (May 1984).
11. A. M. Wachs, Y. Folkman, and D. Shemesh, "Use of Surface Stirrers for Ammonia Desorption from Ponds," Application of New Concepts of Physical–Chemical Wastewater Treatment, Sept. 18–22, 1972.

12. C. N. Click and J. C. Reed, "Atmospheric Release of Hydrogen Sulfide and Ammonia from Wet Sludges and Wastewater," in L. K. Cecil, Ed., *Proceedings of the 2nd National Conference on Complete Water Reuse*, American Institute of Chemical Engineers Symposium Series, New York, 1975, p. 426.

13. R. E. Rathbun and D. T. Tai, "Gas-Film Coefficient for Streams," *J. Environ. Eng.*, **109**(5), 1111–1127 (1983).

14. E. Tipping, "Colloids in the Aquatic Environment," *Chem. Ind.*, Aug. 1, 1988, pp. 485–490.

15. W. H. McAdams, *Heat Transmission*, 3rd ed., McGraw-Hill, New York, 1954.

16. D. Mackay and Y. K. Yuen,"Volatilization Rates of Organic Contaminants from Rivers," *Water Pollut. Res. J. Can.*, New Series, **5**(2), 83–98 (1980).

17. S.-C. Wu and P. M. Gschwend, "Numerical Modeling of Sorption Kinetics of Organic Compounds to Soil and Sediment Particles," *Water Resources Res.*, **24**(8), 1373–1383 (Aug. 1988).

18. T. Hedden, C. Springer, and L. J. Thibodeaux, "Pilot-Scale Studies of Volatile Chemical Emission to Air from Warm Water in the Absence of Wind," *Hazard. Waste Hazard. Mater.*, **7**(3), 223–237 (1990).

19. W. L. Dilling, N. B. Tefertiller, and G. J. Kallos, "Evaporation Rates and Reactivities of Methylene Chloride, Chloroform, 1,1,1–Trichloroethane, Trichloroethylene, Tetrachloroethylene and Other Chlorinated Compounds in Dilute Aqueous Solutions," *Environ. Sci. Technol.*, **9**(9), 833–838 (1975).

20. W. L. Dilling, "Interphase Transfer Processes, II. Evaporation Rates of Chloro Methanes, Ethanes, Ethylenes, Propanes and Propylenes from Dilute Aqueous Solutions: Comparison with Theoretical Predictions," Environ. Sci. Technol., **11**(4), 405 (1977).

21. D. Mackay and Q. W. Wolkoff, "Rate of Evaporation of Low Solubility Contaminants from Water Bodies to Atmosphere," *Environ. Sci. Technol.*, **7**(7), 611–614 (1973).

22. D. Mackay and P. J. Leinonen, "Rate of Evaporation of Low-Solubility Contaminants from Water Bodies to Atmosphere," *Environ. Sci. Technol.*, **9**(19), 1178–1180 (1975).

23. J. M. Prausnitz, *Molecular Thermodynamics and Fluid Phase Equilibria*, Prentice Hall, Englewood Cliffs, N.J., 1969.

24. A. Lerman, "Time to Chemical Steady-States in Lakes and Oceans," in J. D. Hern, Ed., *Nonequilibrium Systems in Natural Water Chemistry*, Advances in Chemistry Series 106, American Chemical Society, Washington, D.C., 1971, pp. 31–76.

25. P. S. Liss and P. G. Slater, "Flux of Gases Across the Air–Sea Interface," *Nature*, **247**, 181–184 (1974).

26. T. E. Hoover and D. C. Berkshire, "Effect of Hydration on Carbon Dioxide Exchange Across an Air–Water Interface," *J. Geophys. Res.*, **74**, 456 (1969).

27. A. L. Downing and G. A. Truesdale, "Some Factors Affecting the Rate of Solution of Oxygen in Water," *J. Appl. Chem.*, **5**, 570–581 (Oct. 1955).

28. J. Kanwisher, "On the Exchange of Gases between the Atmosphere and the Sea," *Deep-Sea Res.*, **10**, 195–207 (1963).

29. P. S. Liss, "Processes of Gas Exchange Across an Air–Water Interface," *Deep-Sea Res.*, **20**, 221–228 (1973).
30. Y. Cohen, W. Cocchio, and D. Mackay, "Laboratory Study of Liquid-Phase Controlled Volatilization Rates in Presence of Wind Waves," *Environ. Sci. Technol.*, **12**(5), 553–558 (1978).
31. W. S. Broecker and T.-H. Peng, "Gas Exchange Rates Between Air and Sea," *Tellus*, **26**, 21–35 (1974).
32. A. Defant, *Physical Oceanography*, Vol. 1, Macmillan, New York, 1961, pp. 226–231.
33. W. Harrison, M. A. Winnik, P. T. Y. Kwong, and D. Mackay, "Crude Oil Spills: Disappearance of Aromatic and Aliphatic Components from Small Sea-Surface Slicks," *Environ. Sci. Technol.*, **9**(3), 231–234 (1975).
34. Z. R. Regnier and B. F. Scott, "Evaporation Rates of Oil Components," *Environ. Sci. Technol.*, **9**(5), 469–472 (1975).
35. R. O. Ramseier, "Oil Pollution in Ice-Infested Waters," in I. Hoffman, Ed., *Proceedings of the International Symposium of the Identification and Measurement of Environmental Pollutants*, Ottawa, Ontario, Canada, 1971, pp. 273–276.
36. J. H. Horton, J. C. Corey, and R. M. Wallace, "Tritium Loss from Water Exposed to the Atmosphere," *Environ. Sci. Technol.*, **5**(4), 338–343 (1971).
37. A. H. Schooley, "Evaporation in the Laboratory and at Sea," *J. Mar. Res.*, **27**, 335 (1969).
38. M. M. Claith, et al., *J. Agric. Food Chem.*, **28**, 610–613 (1980).
39. D. M. Gates, "Radiant Energy, Its Receipt and Disposal," *Meteorol, Monogr.*, **6**, 28 (1965).
40. F. L. Parker, "Thermal Pollution and the Environment," in N. Irving Sax, Ed., *Industrial Pollution*, Van Nostrand Reinhold, New York, 1974, p. 160.
41. L. J. Thibodeaux, C. Springer, T. Hedden, and P. Lunney, "Chemical Volatilization Mechanisms from Surface Impoundments in the Absence of Wind," *Proceedings of the 8th Annual Research Symposium, EPA 600/9-82-002,* U.S. EPA, Ft. Mitchell, Ky., 1982, pp. 161–173.
42. C. Springer, P. D. Lunney, K. T. Valsaraj, and L. J. Thibodeaux, "Emission of Hazardous Chemicals from Surface and near Surface Impoundments to Air," *Final Draft Report: Project 808/60-02,* U.S. EPA, Cincinnati, Ohio, Dec. 1984.
43. J. A. Singmaster, and D. G. Crosby, "Volatilization of Hydrophobic Pesticides from Water," *Paper 6,* Division of Pesticide Chemistry, American Chemical Society Meeting, New Orleans, La., Mar. 1977.
44. Y. Cohen, "Mass Transfer Across a Sheared, Wavy Air–Water Interface," *Int. J. Heat Mass Transfer*, **26**(9), 1289–1297 (1983).
45. G. Neumann and W. J. Pierson, Jr., *Principles of Physical Oceanography*, Prentice Hall, Englewood Cliffs, N.J., 1966, p. 208–209.
46. T. F. Bidleman, "Atmospheric Processes," *Environ. Sci. Technol.*, **22**(4), 361–367 (1988).
47. C. Junge, "Basic Considerations About Trace Constituents in the Atmosphere as Related to the Fate of Global Pollutants," in I. H. Suffet, Ed., *Fate of Pollutants in the Air and Water Environments*, Part 1, Wiley, New York, 1977, pp. 7–25.

48. L. J. Thibodeaux, D. C. Nadler, K. T. Valsaraj, and D. D. Reible, "The Effect of Moisture on Volatile Organic Chemical Gas-to-Particle Partitioning with Atmospheric Aerosols: Competitive Adsorption Theory Predictions," *Atmos. Environ.*, **25A**(8), 1649–1656 (1991).
49. J. H. Seinfeld, *Atmospheric Chemistry and Physics of Air Pollution*, Wiley, New York, 1986, p. 253
50. W. G. N. Slinn, L. Hasse, B. B. Hicks, A. W. Hogan, D. Lal, P. S. Liss, K. O. Munnich, G. A. Sehmel and O. Vittori, "Some Aspects of the Transfer of Atmospheric Trace Constituents Past the Air–Sea Interface," *Atmos. Environ.*, **12**, 2055–2087 (1978).
51. V. G. Levich, *Physicochemical Hydrodynamics*, Prentice Hall, Englewood Cliffs, N.J., 1962, pp. 28–32.
52. S. A. Slinn and W. G. N. Slinn, "Modeling of Atmospheric Particulate Deposition to Natural Waters," Chapter 2 in S. J. Eisenreich, Ed., *Atmospheric Pollutants in Natural Waters*, Ann Arbor Science, Ann Arbor, Mich., 1981.
53. F. Giorgi, "A Particle Dry-Deposition Parameterization Scheme for Use in Tracer Transport Models," *J. Geophys. Res.*, **91**(D9), Aug. 20, 1986, pp. 9794–9806.
54. C. N. Davis, *Aerosol Science*, Academic Press, New York, 1966.
55. R. M. Williams, "A Model for the Dry Deposition of Particles to Natural Water Surfaces," *Atmos. Environ.* **16**(8), 1933–1938 (1982).
56. W. G. N. Slinn, "Dry Deposition and Resuspension of Aerosol Particles: A New Look at Some Old Problems," *Proceedings of the Atmospheric–Surface Exchange of Particulate and Gaseous Pollutants: 1974 Symposium* Richland, Wash., Sept. 4–6, 1974.
57. J. A. Galt, W. J. Lehr, and D. L. Payton, "Fate and Transport of the Exxon Valdez Oil Spill," *Environ. Sci. Technol.*, **25**(2), 202–209 (1991).
58. G. A. L. Delvigne, J. A. Roelvick, and C. E. Sweeny, "Research on Vertical Turbulent Dispersion of Oil Droplets and Oiled Particles, Interim Report," *OCS Study MMS 86-0029*, U.S. Department of the Interior, Minerals Management Services, Washington, D.C., June 1986.
59. T. W. Kana, et al., "Development of a Coastal Oil Spill Smear Model. Phase I: Analysis of Available and Proposed Models, Interim Report," *OCS Study MMS 85-0098*, U.S. Department of the Interior, Minerals Management Services, Mar. 1985.

5

CHEMICAL EXCHANGE BETWEEN WATER AND ADJOINING EARTHEN MATERIAL

The interfaces at the bottom of water bodies such as streams, lakes, estuaries, and the oceans are unfamiliar to most humans except for some oceanographers, professional divers, other underwater specialists, and recreation divers. We spend most of our time on the Earth at the other two interfaces and in general have little experience of the basic happenings at the water–earthen material interface. Although much can be derived from the analogy with the air–earthen material interface, this does not replace feeling the bottom currents, seeing and treading on the bottom geometric forms, and directly sampling either phase at the interface. In tracking chemical movements at this interface, we must by necessity draw on a body of knowledge that depends on remote sensing and remote sampling plus laboratory simulation for most of the data base.

There are a host of biogenic and anthropogenic chemicals whose movement at the water–earthen material interface is important from the viewpoint of both pollutants and ecosystem balance. In this chapter we investigate chemical movements at the interface of interest by focusing on specific applications. Just as in Chapter 4, two goals are accomplished. First, the principles presented in Chapters 2 and 3 are demonstrated, and then specific relevant problems are presented and studied in detail. This approch should demonstrate to the student the process of translation from basic principles to specific application. The student should be able to attack new and different chemodynamic problems.

In general, the natural processes associated with the sediment–water interface that control chemical transport and fate can be organized into sections reflecting the three major aquatic systems. The three are rivers and streams, lakes and reservoirs, and estuaries and oceans. Although this structure is used to introduce the various mechanisms and general characteristics of the individual aquatic system, the student should realize that one particular system may contain all or several mechanisms. However, at a specific locale only one or two mechanisms probably dominate chemical transport and fate. Identifying the dominant processes and quantifying the flux are key objectives of the chemodynamic analysis. The water-side processes are presented first, followed by the sediment-side processes. Example problems illustrate some key processes in each system. The exercise problems extend the text material.

5.1. CHEMICAL TRANSPORT AT BOTTOM OF FLOWING STREAMS

Forced-Convection Dissolution of Sinkers

This section begins with a description of the dissolution process for pure, immiscible, sinker chemicals on the bottom of a stream. Although it is a special case, the rate equation and transport coefficient that quantify the process are applicable to the general case of trace contaminant movement near the bottom.

The accidental spills of liquid materials into rivers, lakes, estuaries, and so on, occur with waterborne traffic and are caused by transportation accidents but can also result from inadvertent releases from production facilities located near a water body or from routine disposal procedures. Table 5.1-1 contains a list of "sinker" chemicals along with density, solubility, and interfacial tension. As will be seen, the latter parameter is important to the on-bottom geometric shapes.

Materials and chemicals, both solid and liquid, heavier than water ($\rho_4 > 1$) move toward the bottom immediately on being spilled or released. Natural flow, chemical processes, and physical processes operate to transport, disperse, cover, dissolve, adsorb, and transform the material. Translocation occurs because of the bulk flow of the aqueous body, and dispersion occurs by flow-induced fluid turbulence. While the material resides on the bottom, it can be covered by sediments, adsorbed by the natural bottom materials, and undergo microbial attack and dissolution.

Some of the natural processes that occur immediately after a spill of a sinker can be anticipated. Figure 5.1-1 depicts the process for a hypothetical spill in a river. A barge containing the material (i.e., chloroform) is involved in an accident resulting in the release of a quantity m_A. Assume that a fairly large hole ($\geqslant 10\,\text{cm}$) is formed in a river barge and that a dense liquid enters into a deep ($\gtrsim 4\,\text{m}$), slow-moving body of water. The following sequence of events, illustrated in Fig. 5.1-1, describes the spill process:

1. From the hole a liquid jet of equal diameter emerges.
2. Once in the water the jet quickly disintegrates into large globules of hole size.
3. These are, in turn, unstable, and globule breakup occurs until a maximum stable droplet size, d, is achieved.
4. A cloud consisting of very small droplets through size d is formed.

Table 5.1-1. Water-Soluble High-Density (ρ > 1) Immiscible Chemicals

Species	Density in Air (g/cm^3)	Solubility in Water (mg/L)	Interfacial Tension (dyn/cm)[a]		
			Air	Water	Vapor
1. Acetic acid	1.06	50,000	$68.0_{30°}$	—	$27.8_{20°}$
2. Aniline	1.022	34,000	44.0	—	$42.9_{20°}$
3. Benzaldehyde	1.04	1,000	40.04	$15.51_{20°}$	—
4. Benzyl alcohol	1.043	46,000	$39.0_{20°}$	$4.75_{22.5°}$	$39.0_{20°}$
5. Bromine	2.93	41,700	$41.5_{20°}$	—	$41.5_{20°}$
6. Carbon disulfide	1.26	2,200	—	$48.36_{20°}$	—
7. Carbon tetrachloride	1.595	500	—	$45.0_{20°}$	$26.95_{20°}$
8. Chloroform	1.5	5,000	$27.14_{20°}$	$32.8_{20°}$	—
9. Dichloroethane	1.256	9,000	$23.4_{35°}$	—	—
10. Ethyl bromide	1.431	10,600	—	$31.2_{20°}$	$24.15_{20°}$
11. Ethylene bromide	2.18	4,300	—	$36.54_{20°}$	$38.37_{20°}$
12. Furfural	1.159	83,100	$43.5_{20°}$	—	$43.5_{20°}$
13. Mercury[b]	13.54	0.0005	470	$375_{20°}$	—
14. Naphthalene	1.15	30	$28.8_{127°}$	—	$28.8_{127°}$
15. Nitrobenzene	1.205	1,900	$43.9_{20°}$	—	$43.9_{20°}$
16. Phenol	1.071	93,000	$40.9_{20°}$	—	$40.9_{20°}$
17. Trichloroethane	1.325	10	$22.0_{114°}$	—	—
18. *N*-Propylbromide	1.353	2,500	—	—	$19.65_{20°}$
19. Tetrachloroethane	1.60	3,000	$36.3_{22.5°}$	—	—
20. Water[b]	1.00	N.A.	$73.05_{18°}$	N.A.	72

[a]In air, water, and its own vapor. Temperature is °C.
[b]Mercury and water data included for reference.

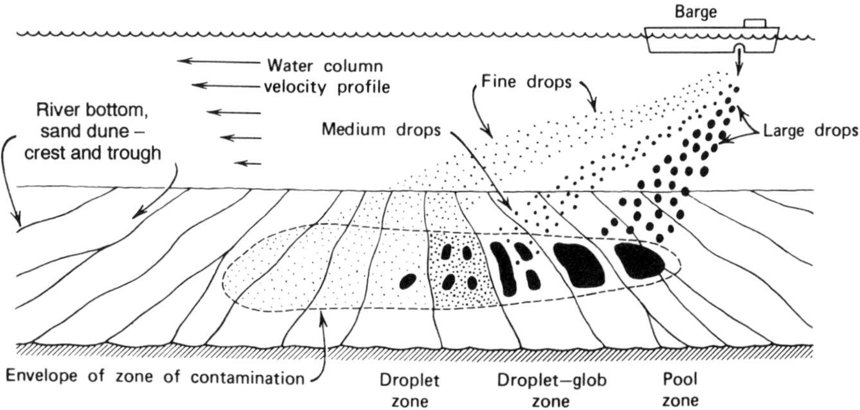

Figure 5.1-1. Hypothetical spill incident.

5. Due to stream flow a longitudinal size classification occurs. Large drops arrive at the bottom first. The smaller ones arrive later and are located farther downstream. Some very small droplets remain suspended in the turbulent eddies.
6. As the drops of liquid arrive at the bottom, they tend to accumulate at specific locations downstream from the point of release.
7. At points nearest the discharge droplet crowding occurs and coalescence creates pure liquid pools in the bottom depressions and sand wave troughs.
8. During massive spills when a depression becomes full, the dense liquid overflows downstream into the next depression.
9. The smaller downstream residing drops that are not crowded together enough to coalesce remains as isolated spheres on the bottom.
10. The flowing water easily move the spheres and causes "herding." This crowding results in coalescence, with several droplet spheres forming a glob.
11. The liquid chemical achieves its final quasi-steady state bottom form, and the on-bottom dissolution process begins. The liquid is in three geometric shapes: Spherical drops, globs, and pools. *Globs* are defined as pancake-shaped pools several centimeters in diameter and a few millimeters thick.

Most of the occurrences noted above have been observed in the laboratory.[1] There can and will be many variations on this idealized spill mechanism. If it occurs in shallow water, a droplet cloud may never form; the liquid may ooze down in large globules and cover the bottom. The spill of a small quantity of liquid in deep water may result in the bottom being splattered with individual drops only. Stream turbulence in a fast-moving body of water can produce significant changes. The high flow velocity may keep the liquid moving along the bottom. Movement is not unlike the sediment bedload phenomenon that causes sand and silt to move downstream. In this extreme case the spillage moves out as a slug and behaves more or less like the spill of a misicible material.

Dissolution begins immediately on water contact and occurs from globules and droplets in transit to the bottom. Normally, the duration of the in-transit time is short, typically seconds to minutes. With slightly soluble materials, the major part of the dissolution occurs while the material resides on the bottom. Of all the on-bottom processes that can occur to a spilled liquid chemical, dissolution is rapid and initiates other chemical and biochemical effects within the aquatic environment. These may include biological oxygenation of the solute, volatilization, sorption to suspended particles, transport within the bed, hydrolysis, and so on, in addition to toxic effects. We now consider the on-bottom interphase mass transfer aspects of the problem.[2]

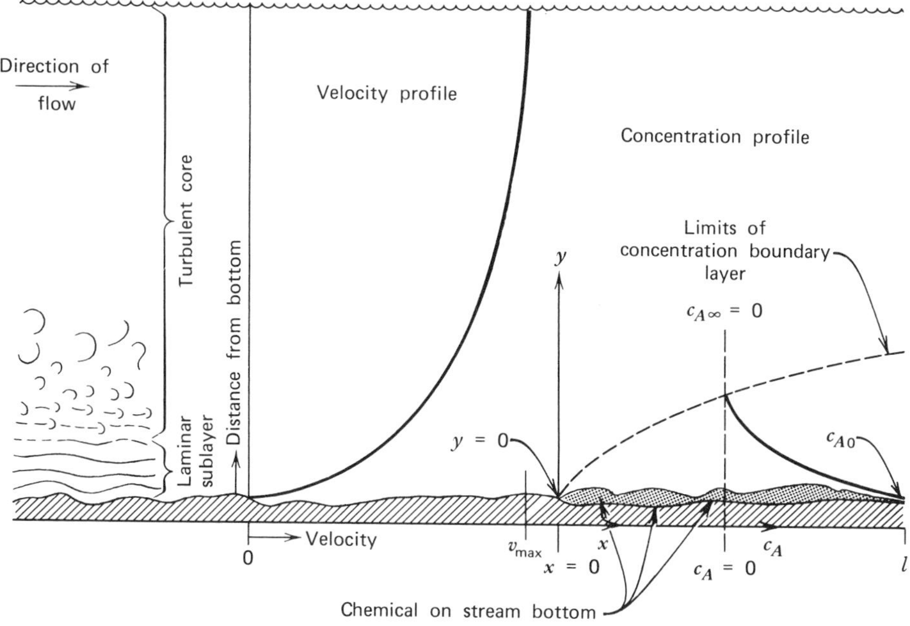

Figure 5.1-2. Chemical dissolution in flowing stream.

As shown in Fig. 5.1-2, the chemical occupies a portion of the stream bottom, length l and width w. In this case the area of the zone of contamination is simply lw. Since the chemicals of interest are slightly soluble, Eq. 3.1-16 is sufficient to describe the dissolution rate in the water:

$$w_A = {}^4k'_{A2}A(\rho^*_{A2} - \rho^0_{A2}) \tag{5.1-1}$$

where w_A is the mass rate, ${}^4k'_{A2}$ the water-phase mass transfer coefficient above the liquid (2 ≡ water phase, 4 ≡ chemical phase), A the interfacial area betewen the liquid chemical and water, ρ^*_{A2} the solubility of the chemical, and ρ^0_{A2} the background concentration of the chemical in water.

Consider a quantity of pure liquid chemical A mass of m_A in place on the bottom of a moving stream of flow rate Q. The chemical displays an interfacial area A (m_A) that is a function of the mass remaining. Some simplifying assumptions can be made as follows: $\rho^0_{A2} = 0$ is valid if the water approaching the spill site contains no species A in solution and ${}^4k'_{A2}$ is a constant independent of geometric shape. The following simple differential equation describes the dissolution process:

$$\frac{dm_A}{dt} = -A(m_A){}^4k'_{A2}\rho^*_{A2} \tag{5.1-2}$$

The downstream concentration of species A, ρ_{A2}, is the quotient of the mass dissolution rate and the stream volumetric flow rate:

$$\rho_{A2} = \frac{A(m_A)^4 k'_{A2} \rho^*_{A2}}{Q} \tag{5.1-3}$$

Here the chemical at dissolution rate w_A (g/s) is assumed to mix instantly with the water in the stream of flow rate Q (m^3/s) at a plane immediately downstream of the zone of contamination. The result is a completely mixed steady-state flow model that yields the chemical concentration at the plane of mixing. The spillage lifetime on bottom t_A due to the dissolution process is obtained by separating the t and m_A variables in Eq. 5.1-2 and integrating from $t = 0$, $m_A = m_{Ai}$ to $t = t_A$, $m_A = 0$:

$$t_A = \frac{1}{^4k'_{A2} \rho^*_{A2}} \int_0^{m_{Ai}} \frac{dm_A}{A(m_A)} \tag{5.1-4}$$

These expressions can be employed to obtain two important water quality predictions associated with the spill. The importance of $A(m_A)$ in predicting ρ_{A2} and t_A is readily apparent at this point. Specifying the three bottom geometric forms yields the function $A(m_A)$.

***Dissolution Kinetics of Geometric Forms*[2].** The liquid on the stream bottom is assumed to be present in three geometric forms: spherical drops, globs, and pools. The interfacial area displayed is a function of the geometric shape.

Drops. Liquid drops arriving at the bottom of a water body that do not coalesce into pools or globs remain as isolated spheres. The bottom will undoubtedly be splattered with drops of various diameters. An equation is available for estimating the maximum stable diameter d of a drop falling through water.[3] If these drops are assumed not to break up when they arrive on the bottom, the diameter can be calculated by

$$d = 3.79 \sqrt{\frac{\sigma_{A2}}{\rho_A - \rho_2}} \tag{5.1-5}$$

where d is diameter in cm, σ_{A2} the interfacial tension of the liquid in water in N/m, and $\rho_A - \rho_2$ the density difference in g/cm^3. The interfacial area of a mass m_{Ad} of uniform drops of diameter d is

$$A_d = \pi \left(\frac{6 m_{Ad}}{\pi \rho_A} \right)^{2/3} \tag{5.1-6}$$

If dissolution is assumed to proceed such that the drops remain spherical with decreasing diameter, Eq. 5.1-6 substituted into Eq. 5.1-4 yields a mass–time relationship from which lifetime can be obtained:

$$t_d = \frac{\rho_A d}{2\,{}^4k'_{A2}\rho^*_{A2}} \tag{5.1-7}$$

The downstream concentration of the chemical resulting from the drops is obtained from Eq. 5.1-3:

$$\rho_{Ad} = \frac{6\,{}^4k'_{A2}m_{Ad}\rho^*_{A2}}{\rho_A dQ}\left(1 - \frac{2\,{}^4k'_{A2}\rho^*_{A2}t}{\rho_A d}\right)^2 \tag{5.1-8}$$

Globs. Liquids residing on flat surfaces do so provided that interfacial forces result in "nonwetting" of the bottom material. The height of a glob h_g is controlled by the water–chemical interfacial tension σ_{A2} and the density difference $\rho_A - \rho_2$:

$$h_g = \sqrt{\frac{2\sigma_{A2}}{g(\rho_A - \rho_2)}} \tag{5.1-9}$$

This equation is similar in form and derivation to a relationship for an oil film on water.[4] The interfacial area of globs is

$$A_g = \frac{m_{Ag}}{\rho_A h_g} \tag{5.1-10}$$

where m_{Ag} is the mass of chemical in the shape of globs. As dissolution proceeds, the interfacial tension forces cause A_g to decrease proportionately to the mass remaining while h_g remains constant. The globs therefore shrink laterally.

It is assumed that globs are somewhat large so that the perimeter interfacial area is small compared to the top interfacial area. This assumption is invalid until glob diameter approaches h_g. At this point the glob becomes a drop. However, dissolution is assumed to occur from the top of the glob only so that interfacial area is proportional to mass remaining. Substituting Eq. 5.1-10 into Eq. 5.1-4, we get a mass–time relationship from which the lifetime can be established:

$$t_g = \frac{\rho_A h_g}{{}^4k'_{A2}\rho^*_{A2}} \ln \frac{1}{f} \tag{5.1-11}$$

where f is some small fraction of the liquid remaining, but for all practical purposes the dissolution process is completed. A reasonable f is 0.05 (i.e., 5% undissolved) and $\ln(1/f) \simeq 3$. The downstream concentration resulting from shrinking globs is

$$\rho_{Ag} = \frac{{}^4k'_{A2}m_{Ag}\rho^*_{A2}}{\rho_A h_g Q} \exp\left(-\frac{{}^4k'_{A2}\rho^*_{A2}t}{\rho_A h_g}\right) \tag{5.1-12}$$

Pools. The geometric makeup of the bottom of most large actively flowing streams, such as rivers, consists of sand waves as ripples and dunes. The bottom material is mainly sand, and spilled liquids accumulate in the troughs of sand waves.

Yalin[5] maintained that if flow is tranquil ($Fr < 1$), two kinds of sand waves can be present: ripples and dunes. Ripples and dunes are similar in their shapes; they both have an upstream surface with a gentle, gradual varying slope and an abrupt downstream face with a constant slope (which is approximately equal to the tangent of the angle of repose). The idealized nonsymmetrical shape of ripples or dunes is shown schematically in Fig. 5.1-3. Ripples and dunes are distinguished from each other by the difference in their sizes.

Simons et al.[6] have reported on the sedimentary structures generated by flow in alluvial channels. In the low-flow regime the bedform is either ripples or dunes or a combination of ripples and dunes, all of which are triangular-shaped elements of irregular shape. They also observed that in natural streams and rivers, dunes with ripples superimposed on dunes are the dominant bedforms in the low-flow regime. Ripples have a length Λ of about 30 cm or less from crest to crest and an amplitude Δ of 0.6 to 6.0 cm in height and have rather small width normal to the direction of flow. If the boundary shear stress is increased, a magnitude of velocity and a degree of turbulence are soon achieved that cause large sand waves called *dunes* to form. Viewed in elevation, the dunes are large triangular-shaped elements similar to ripples. Their lengths range from 60 cm to several meters, and their height from 6 cm to a few meters, depending on the scale of flow. In experiments with a large flume the dunes range from 60 cm to 3.0 m in length and from 6 to 30 cm in height. In the Mississippi River dune lengths of a hundred meters and heights as great as 10 m have been reported.

A sand wave trough partially filled to a height h_w with liquid chemical, as shown in Fig. 5.1-3, is considered to be unsaturated. If $h_w = \Delta$, wave depth, the bottom is saturated with the chemical. The interfacial area for mass transfer

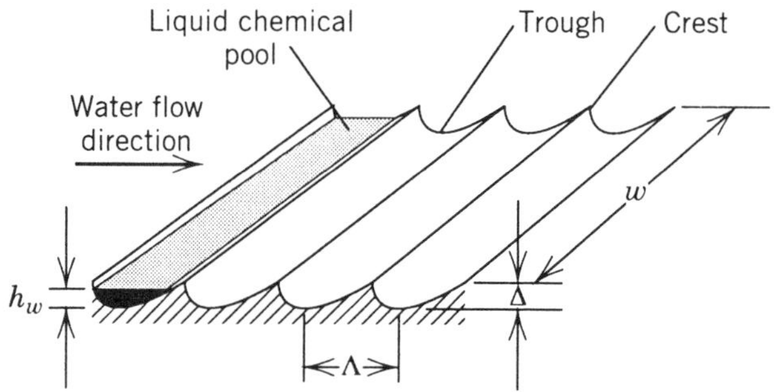

Figure 5.1-3. Nonsymmetrical sandwaves on river bottom.

for a mass of liquid m_{Aw} in the wave valleys shown in Fig. 5.1-3 is

$$A_w = \frac{2m_{Aw}}{h_w \rho_A} \tag{5.1-13}$$

The width can extend completely across the stream. Equation 5.1-14 is used to obtain the lifetime for the chemical in the waves, which is

$$t_w = \frac{\rho_A h_w}{{}^4k'_{A2} \rho^*_{A2}} \tag{5.1-14}$$

If the wave structure is saturated, Δ should replace h_w in Eq. 5.1-14. The downstream concentration is obtained from Eq. 5.1-3:

$$\rho_{Aw} = \frac{2\,{}^4k'_{A2} m_{Aw} \rho^*_{A2}}{\rho_A h_w Q} \left(1 - \frac{{}^4k'_{A2} \rho^*_{A2} t}{\rho_A h_w}\right) \tag{5.1-15}$$

where t is the time, which begins when the spillage arrives in place on the stream bottom.

Combined Forms. The total concentration-time history of the spilled chemical at mile 0.0 is the sum of the contribution from each source:

$$\rho_{A2} = \rho_{Ad} + \rho_{Ag} + \rho_{Aw} \tag{5.1-16}$$

Substituting the appropriate expressions from Eqs. 5.1-8, 5.1-12, and 5.1-15, we get the concentration–time history in the flowing stream:

$$\begin{aligned} \rho_{A2} = \frac{{}^4k'_{A2} \rho^*_{A2}}{\rho_A Q} \Bigg[&\frac{6m_{Ad}}{d} \left(1 - \frac{2\,{}^4k'_{A2} \rho^*_{A2} t}{\rho_A d}\right)^2 + \frac{m_{Ag}}{h_g} \exp\left(- \frac{{}^4k_{A2} \rho^*_{A2} t}{\rho_A h_g}\right) \\ &+ \frac{2m_{Aw}}{h_w} \left(1 - \frac{{}^4k'_{A2} \rho^*_{A2} t}{\rho_A h_w}\right) \Bigg] \end{aligned} \tag{5.1-17}$$

The terms in parentheses must be positive or zero; otherwise, they are not included. Also,

$$m_A = m_d + m_g + m_w \tag{5.1-18}$$

must be satisfied.

Equation 5.1-17 is useful for estimating in-stream concentration of spilled liquid chemicals. Many terms in this equation can be estimated a priori. Streamflow (Q) and the total quantity spilled (m_A) can be obtained from the spill site, actual or projected. The mass transfer coefficient (${}^4k'_{A2}$), globe height (h_g), and drop diameter (d) can be estimated from equations given in this book. Solubility (ρ^*_{A2}) and density (ρ_A) are available in handbooks. The remaining four variables (i.e., m_d, m_g, m_w, and h_w) can be reduced to three unknowns by use of Eq. 5.1-18. To obtain a realistic concentration–time prediction, the three remaining variables or unknowns must be specified from special knowledge of the spill scenario, stream bottom, water velocity, depth of water, and so on, and their effects. In the absence of this information the individual models taken separately can yield reasonable estimates of maximum concentration and minimum lifetime due to a projected spill (see Problem 5.1A).

Closure. Several spills of sinker chemicals into streams have been documented. These include chloroform in the Mississippi River,[34] PCBs in the Duwamish River,[16] and perchloroethylene in the St. Clair River,[35] all in North America, and a chemical mixture containing organophosphate pesticide and organic mercurial compounds in the Rhine River, Switzerland.[36] Behavioral observations in both pilot- and laboratory-scale spill simulations have been reported by Thibodeaux[1] and Ashworth.[37] Both observed some liquid being buried by light, mobile bed material, as did Lau and Marsalek.[38] Ashworth observed carbon tetrachloride penetrating into a coarse gravel bed and into a compact mud bed, although no penetration was observed into a sand bed. Considerable laboratory and field research is still needed to understand the many factors involved with the behavior of sinker liquids and solid flakes in natural stream environments.

Stream Bottom Mass Transfer Coefficients

A necessary piece of information for estimating on-bottom lifetimes and in-stream concentrations is the bottom mass transfer coefficient, ${}^4k'_{A2}$. Although there are several "standard" mass transfer correlations available, these may not be altogether applicable. Laminar boundary layer theory for tangential flow along a sharp-edged, semi-infinite flat plate with mass transfer is well developed. The general correlation is given by Eq. 3.1-23. Since a laminar-flowing, natural stream is the exception rather than the rule, this equation is of little use. Equation 3.1-24 is a similar correlation for turbulent flow parallel to flat plates. For this equation as well as Eq. 3.1-23, the hydrodynamic boundary layer begins at the same point (i.e., at the sharp edge) as the concentration boundary layer. For the case of a spilled liquid in a flowing stream, the hydrodynamic boundary layer is developed before the flow enters the zone of contamination.

Kramers and Kreyger[7] obtained experimental measurements of the rate of solution on rather short surfaces, length l, of benzoic acid in water in laminar and turbulent flow. The analysis is treated as the diffusion of a solute from a plane surface into a laminar flow with a constant velocity gradient. The final correlation is

$$^4k'_{A2} = 0.449\left[\frac{(g_x\Gamma_v)^{2/3}\mathscr{D}_{A2}^2}{\nu_2 l}\right]^{1/3} \tag{5.1-19}$$

where g_x is the acceleration of gravity in the direction of flow, and $Re \equiv 4\Gamma_v/\nu_2 > 2360$. Here Γ_v is the volumetric flow rate per unit channel bottom width, $L^3/t \cdot L$. This correlation gave a reasonable fit to the experimental data for the Re range 1500 to 7000. The soluble section was located 330 mm downstream from the water film inlet. A hydrodynamic boundary layer developed before the water encountered the soluble section (5 to 80 mm in length).

The gradient of the water surface of a flowing stream, s, is related to g_x by

$$g_x = g \sin\alpha \tag{5.1-20}$$

and is very nearly equal to the slope of the bottom. Here $s = \sin\alpha$, where α is the angle of the stream bottom from the horizontal. In open-channel flow it is possible to estimate the gradient from Manning's formula:

$$v_x = \frac{r_H^{2/3}s^{1/2}}{n} \tag{5.1-21}$$

This equation is dimensional, and v_x is the mean flow velocity in m/s, r_H the mean hydraulic radius of the wetted surface in m, s the slope of the water surface, and n a coefficient of roughness. Values of n are given in Appendix E.

There are other correlations in this book that may be used for estimating $^4k'_{A2}$. Equation 5.1-19 more nearly mimics the stream flow dissolution process; however, the experimental apparatus was small and the turbulence level low. The major drawback of Eq. 5.1-19 is its flat geometry since most streams have a wave-type bottom structure. A wavy bottom reduces the mass transfer coefficient from that observed on a flat geometry because of decreased turbulence in the troughs.[8]

Sinkers. Chang[8] performed a series of dissolution experiments in three laboratory-scale models of flowing streams. Furfural and chloroform were placed in shallow circular pans embedded into the sand bottom so that only the liquid surface was in contact with the water. The dissolution mass transfer coefficient was computed by

$$^4k'_{A2} = \frac{\Delta m_A}{A\Delta t\rho^*_{A2}} \tag{5.1-22}$$

where Δm_A is the mass dissolved in water, A the interfacial area, Δt the dissolution time, and ρ^*_{A2} the chemical solubility in water. Observations were performed with a flat sand bed and with the sand bed formed into a repeated wave structure, as illustrated in Fig. 5.1-3.

As noted earlier, natural stream bottoms consist of sand waves known as ripples and dunes. Various wave amplitudes and periods were used in the laboratory experiments. A fixed wave structure of amplitude $\Delta = 5.1$ cm and 15 cm was used, the sand wave length/depth ratio, Λ/Δ being 5 in both cases. The circular pans were placed in the valleys of the waves, this being the likely resting place of the liquid chemical. In general, higher coefficients were achieved at the flat bed.

Christy[12] used Chang's[8] sinker dissolution data and developed a correlation for a forced-convection bottom-water mass transfer coefficient that is applicable to both flat and wavy bedforms. The result is

$$^4k'_{A3} = \frac{0.114v_*}{\mathrm{Sc}_{A2}^{2/3}[1 + 9.6(\Delta - h_w)^{1/2}]} \tag{5.1-23}$$

where Δ (m), is the wave height and h_w (m), is the chemical depth in the wave trough. The coefficient assumes the units of v_*. In the case of a flat bed or transport from isolated drops and globs, set $\Delta = h_w$. The flat bed data are shown in Fig. 5.1-4. Although forced convection mass transfer dominates, some

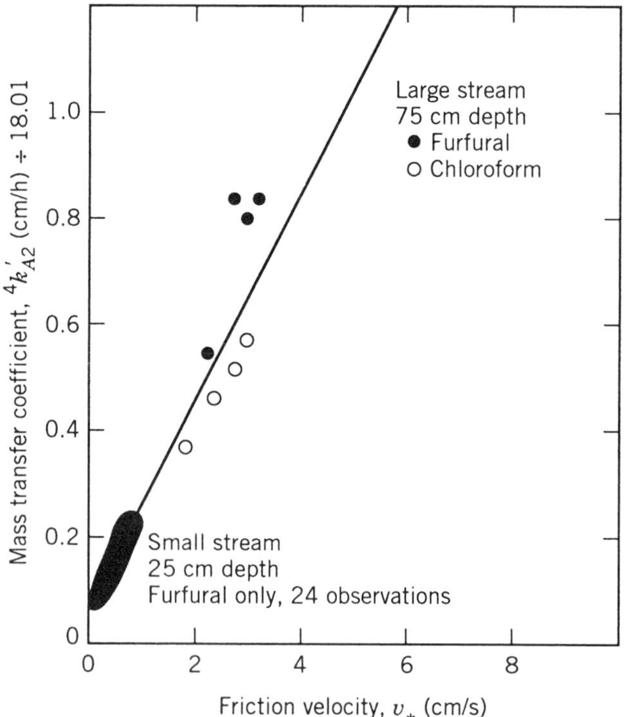

Figure 5.1-4. Stream-bottom mass transfer coefficient for a flat sand bed. The intercept value of 0.046 is due to natural convection. (Reprinted with permission from Ref. 8.)

Table 5.1-2. Stream Sinker Chemical Mass Transfer Coefficients

Chemical Stream	CH_3Cl, Miss. R.	Furfural, Lab. Flume	CCl_4, Kanawha R.	PCB, Duwamish R.	Ethylene Chloride, Earth Ditch
Depth (m)	16.5	0.425	2.44	13.7	0.315
Flow (m/s)	7590	0.0668	442	4300	9.4E-4
Coefficient (cm/h)	1.82	6.81	4.53	0.875	0.498

Source: Ref. 1.

natural convection mass transfer is present and the two contributions should be added when pure material is present. The subject of natural convection is covered in Section 5.2.

The friction velocity, v_*, is directly related to the bottom shear stress, τ_0, which can be approximated by

$$\tau_0 = \rho_2 shg \tag{5.1-24}$$

where s is the slope of the water surface, h is water depth, and g the gravitational acceleration. Equations 5.1-20, 5.1-21 and 5.1-24 can be combined to estimate v_* from common stream parameters.

$$v_* = \frac{v_x n}{r_H^{2/3}} \sqrt{hg} \tag{5.1-25}$$

The stream parameters within the fraction are dimensional; see Eq. 5.1-21 for the correct dimensions. Table 5.1-2 contains the stream mass transfer coefficients for five sinker chemicals computed with the SHICHEM model.[1] In some streams the water flow is very slow much of the time, as in the bayous of southern Louisiana. In this case water mixing and therefore bottom turbulence is usually caused by surface winds. Refer to Section 5.2 for information on this water-side coefficient.

Trace Contaminant Water-Side Coefficients. Recognizing that life-sustaining mass transfer processes take place between flowing water and the bottom and bank organisms of natural streams; Novotny[9] made a study to explain the process and develop an expression for the mass transfer coefficient. The bottom is a collection point for settleable waste organic matter of biogenic or anthropogenic origin. Organisms that reside on the bottom must draw on the oxygen resources of the flowing stream so that the bottom boundary layer resistance is an important aspect of the transfer of dissolved oxygen. Novotny points out that in some cases bottom organisms may be anaerobic even though the oxygen concentration in the flowing water is high and that this can be explained by the fact that the amount of oxygen diffusing through the boundary layer into the benthal layer is not sufficient for aerobic conditions.

Another aspect of the role of the resistance in the boundary layer is regulation of the mineral composition of the water. As a stream flows over various geologic formations, the water dissolves indigenous minerals, and this determines in part the final mineral content of the stream. An acidic stream can become neutralized as it flows over a section of limestone or dolomite.

The flow in a stream can be divided into two zones: (1) the free turbulent flow zone and (2) the laminar sublayer. When dealing with the diffusion phenomenon, the diffusive boundary layer, a third zone, should be considered. Figure 3.3C shows the three zones.

Novotny began with this basic idea. He and Levich[39] assumed that the bottom roughness projections, characterized by y_0, protrude through the diffusive sublayer allowing turbulent eddies, parameterized by the mixing length, l, to exist close to the interface. His final result expressed as a mass transfer coefficient is

$$^{3}k'_{A2} = \frac{\sqrt{\nu_2}\, v_*^{1/2}}{\mathrm{Sc}_{A2}^{2/3}\, y_0^{1/2}} \tag{5.1-26}$$

written in the form of Eq. 5.1-23. Both v_* and y_0 can be obtained from stream velocity profiles. This theoretical result is similar to Eq. 5.1-23 in some respects. The roughness height functionality generally reflects that found for the wave height, but the friction velocity functionality is different. Further experimental work may support the theoretical function of the friction velocity since the data set is too small to distinguish clearly between the $\frac{1}{2}$ and 1 power.[8]

Equation 5.1-26 was developed for the transport of oxygen, calcium, and other geochemicals across the aquatic boundary layer of flowing streams. It can also be used for trace contaminants in the same place. Although it appears to give reasonable numerical values for $^{3}k'_{A2}$ when used with consistent values of v_* and y_0, Eq. 5.2-23 is recommended because it is backed by supportative data.

Example 5.1-1. Phenol Coefficient for Mississipi River. For phenol as a trace contaminant moving through the sediment–water interface of the Mississippi River at 20°C, compute the mass transfer coefficient in cm/h. Assume that the river is at low flow and that the sand wave height is 50 cm.

SOLUTION Equation 5.1-23 is used for the coefficient and Eq. 5.1-25 is for the friction velocity. From Fig. E.1, $v_x = 4$ ft/s, $w = 1300$ ft, and $h = 25$ ft; from Table E.1, n = 0.025; from Table C.7, $\mathrm{Sc}_{A2} = 1200$.

The fraction in Eq. 5.1-25 is 0.00807 with $r_H = 24$ ft; therefore,

$$v_* = 0.00807\sqrt{980.7\ \text{cm/s}^2(25)30.48\ \text{cm}} = 6.98\ \text{cm/s}$$

For a flat bed Eq. 5.1-23 gives

$${}^3k'_{A2} = 0.114(6.98\ \text{cm/s})1200^{-2/3}(3600\ \text{s/h}) = 25.4\ \text{cm/h}$$

For $\Delta = 50$ cm, ${}^3k'_{A2} = 3.26$ cm/h. For a triangular-shaped wave the arithmetic average is 14.3 cm/h. No correction for natural convection is required for a trace quantity.

Particle Processes at Sediment–Water Interface

The dissolution of a hydrophobic pollutant from a contaminated bed is usually accompanied by the parallel transport from the bed in a particle-bound state. Stream-flow turbulence is capable of eroding sediment particles from the bed surface and sustaining their presence in the water column. The concentration of particles in the water column, usually referred to as *suspended sediment*, is the result of a dynamic equilibrium process between particle erosion from and deposition onto the bed. As shown in Fig. 5.1-5, the concentration of suspended sediment typically displays a U-shaped relationship with stream flow, increasing exponentially at high flows. At low flows, suspended sediments are dominated by natural organic materials, but the contribution of the particulate organic carbon declines as the amount of clay, silt, and sand in suspension increases during high flows.[41]

Singh[40] and others have observed that the hydraulics of sediment transport in channels and estuaries is better understood than that in overland flow. Some information on channel particle dynamics was presented in Chapter 3; see Eqs.

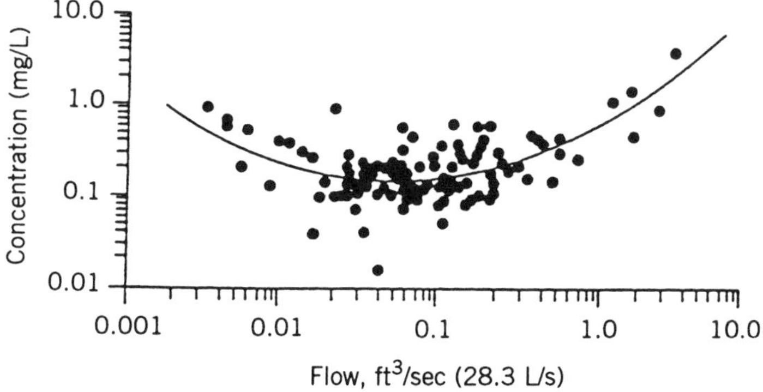

Figure 5.1-5. Concentration of particulate matter as a function of stream flow in the Hubbard Brook Experimental Forest of New Hampshire. (Reprinted with permission. From Ref. 41.)

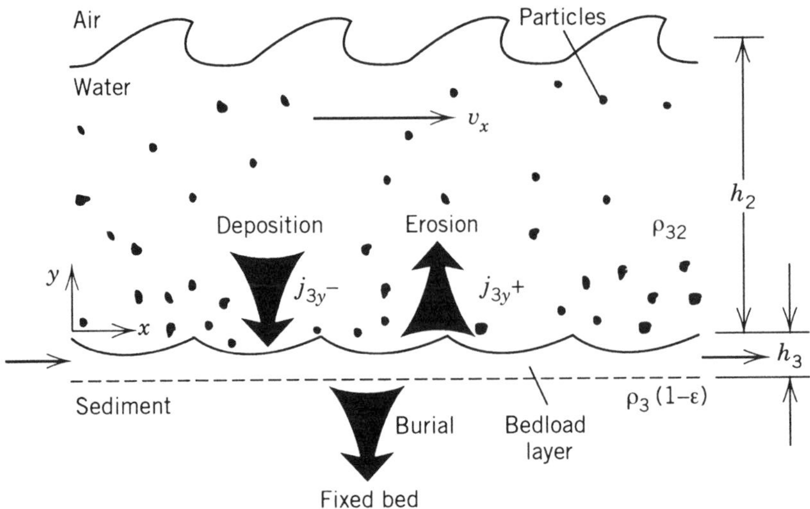

Figure 5.1-6. Major particle transport processes in stream.

3.3-33 and 3.3-34 where a brief introduction is given to the theory of particle deposition and erosion from the bed. Figure 5.1-6 illustrates the major processes occurring. The particle processes are depicted as occurring between three reservoirs: the water column, a moving layer of particles near the surface called the bedload, and the fixed bed below. The concentration of suspended sediment in the water column is ρ_{32} (m/L). In the bedload the particle concentration is $\rho_3(1-\varepsilon)$ (m/L), where ε is bed porosity. If deposition exceeds erosion, particle burial occurs to maintain a constant depth of the bedload layer, h_3. Disinterment of particles from the fixed bed occurs if erosion exceeds deposition. At dynamic equilibrium conditions, burial is zero.

Particle depositional flux is given by Eq. 3.3-33. It is convenient to reformulate this flux equation as the product of a transport coefficient, ${}^3k_2'$ (L/t), and the suspended sediment concentration. In this form Eq. 3.3-33 shows that as the bottom shear stress, τ_0, increases, ${}^3k_2'$ decreases. In other words, deposition is slower at higher stream flows. Equation 3.3-34 is the particle flux expression for erosion from the bed. It is also a linear function of τ_0 and for values above some critical value, erosion occurs. This flux expression does not contain the bed sediment concentration, $\rho_3(1-\varepsilon)$, so that one will be invoked to reformulate this flux expression as a product with the erosion transport coefficient, ${}^2k_3'$ in L/t. The idealized processes depicted in Fig. 5.1-6 and use of the terms defined above allow a simple demonstration of some key aspects of particle dynamics in streams. A mathematical analysis of an idealized steady-state particle regime in a stream as presented by Delos et al.[52] follows.

Assume that a tributary with high suspended sediment concentration enters the main stream and after mixing the river suspended sediment concentration is ρ_{32}^0. A steady-state particle mass balance yields the following expression for concentration, ρ_{32}, as a function of distance downstream, x, and stream parameters defined above:

$$\rho_{32} = \rho_{32}^{0} \exp\left(-\frac{{}^{3}k_2' x}{h_2 v_x}\right) + \left[1 - \exp\left(-\frac{{}^{3}k_2' x}{h_2 v_x}\right)\right] \frac{{}^{2}k_3' \rho_3 (1 - \varepsilon)}{{}^{3}k_2'} \quad (5.1\text{-}27)$$

This result shows that for large values of x the suspended sediment concentration approaches a constant value that is the dynamic equilibrium concentration for the particular steady-state flow regime of the stream. This concentration is denoted by ρ_{32}^{*}, and upon inspection of Eq. 5.1-27 it is determined to be

$$\rho_{32}^{*} \equiv \frac{{}^{2}k_3' \rho_3 (1 - \varepsilon)}{{}^{3}k_2'} \quad (5.1\text{-}27a)$$

so that if $\rho_{32}^{0} > \rho_{32}^{*}$, as is the case with the high-sediment-load tributary example, the suspended sediment concentration falls with increasing x until ρ_{32}^{*} is achieved. If the initial concentration is less than ρ_{32}^{*}, the concentration will rise with x until ρ_{32}^{*} is achieved.

The burial rate is the difference between deposition and erosion at a distance x. Burial flux, defined here as a positive quantity, is

$$j_{3y} \equiv {}^{3}k_2' \rho_{32} - {}^{2}k_3' \rho_3 (1 - \varepsilon) \quad (5.1\text{-}28)$$

Using Eq. 5.1-27 it transforms to

$$j_{3y} = [{}^{3}k_2' \rho_{32}^{0} - {}^{2}k_3' \rho_3 (1 - \varepsilon)] \exp\left(-\frac{{}^{3}k_2' x}{h_2 v_x}\right) \quad (5.1\text{-}29)$$

This result also applies to disinterment of particles from the fixed bed since if erosion flux is greater than deposition flux, Eq. 5.1-28 yields a negative quantity for j_{3y}. This completes the idealized model.

The suspended solids concentration versus stream flow data to the right of the minimum shown in Fig. 5.1-5 conforms to the foregoing analysis of stream-particle dynamics. If it is assumed that at each time the stream was sampled it was at a condition of dynamic equilibrium, Eq. 5.1-27a represents the general shape of the curve. As flow increases, erosion increases and deposition decreases, so that ${}^{2}k_3'$ is larger and ${}^{3}k_2'$ is smaller. According to the equation, this results in an increase in the ρ_{32}^{*}; the behavior expected. In the case of particulate organic carbon the data to the left of the minimum result from a different type of particle transport phenomenon from the bed.

There is evidence to indicate that colloidal organic particles are produced in streambeds as a consequence of biological activity therein. These originate from decaying plant and animal matter comprised primarily of aggregates of humic acids that are stable in low-salinity environments.[42] Transport to the water column probably occurs by several processes, Brownian diffusion being one known process.[43] At low-flow conditions a stable bed is conducive to the growth of microbial life forms as well as larger benthic species. This activity in the bed gives rise to colloid production and eventual release to the water column. Since the magnitude of the release rate is little influenced by the rate of water flow and assuming that it is constant, the in-stream concentration should decrease inversely with the water flow rate. This is the general shape of the curve to the left of the minimum in Fig. 5.1-5.

Hydrophobic organics chemicals (HOCs) sorb onto all the particles including the organic colloids.[44] See Chapter 2 for information on water solid equilibrium of HOCs. The Hudson River has been contaminated with PCBs from St. Edward, N.Y. to New York City.[45] The mass of this substance resides totally within the bed sediment. The primary mechanisms resulting in the depletion from the bed are transport to the water column, transport to the groundwater–subsoil, and biological or chemical degradation. The first mechanism is probably the dominant. Figure 5.1-7 shows the solution plus particle-bound PCB concentration in the water as a function of discharge. These data were obtained during 1977, when it was being demonstrated that PCBs were moving from the upper river basin to the lower river. The highest concentrations were associated with high flows of approximately 1E6 L/s, and the lowest concentrations were associated with the medium river discharges of 2 to 6E5 L/s . With even lower flows, <2E5 L/s , the concentration increased, indicating that a relatively constant load of PCBs was being released from the bed to the water column. Based on particle dynamics in streams, it is not

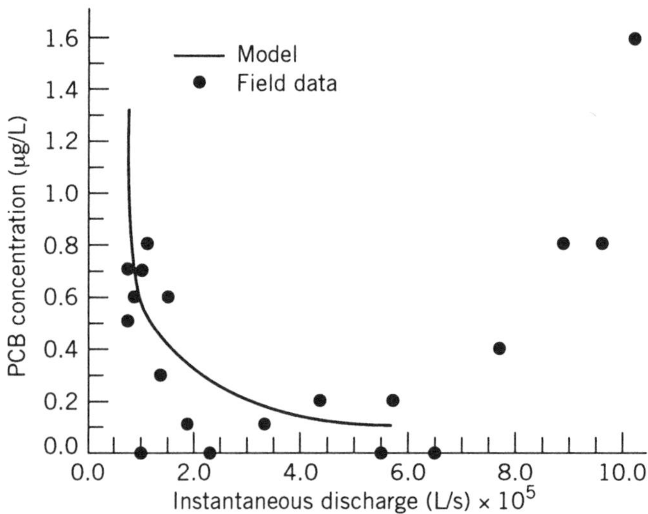

Figure 5.1-7. Hudson River at Schuylerville: PCB concentration versus river discharge, March 30 September 27, 1977. (Reprinted with permission. From Ref. 45.)

surprising that the streamflow behavior of PCBs in the Hudson is similar to the behavior of suspended sediment. See Problem 5.1B for additional information on transport of PCBs in the Hudson River.

Closure. At this time it appears that many aspects of particle transport mechanisms across the sediment–water interface of flowing streams are known. The movement of particles is one important factor. Medine and McCutcheon[47] give a presentation of the basic concepts on near-bed sediment transport and advective dispersion transport in the water column as it relates to tracking contaminated sediment. A vignette model based on a simplified version of the full three-dimensional particle continuity equation is given. The result is instructive for the student, in that it contains relatively simple equations of the vertical suspended particle concentration profile in streams, $\rho_{32}(y)$, and the mass rate of particles carried by the flow, $Q\rho_{32}$. Based on the bottom shear stress introduced by the water flow and employing the Shields parameters, Gschwend et al.[68] demonstrate, with example problems, the connection between bed erosion of noncohesive sediment and contaminant release kinetics from those suspended particles to the aqueous phase. With site-specific sediment data coupled with the appropriate erosional and depositional flux equations, the solid-phase chemodynamics between bed and water column can be estimated, but much research is still needed on the chemical movement associated with particle processes for both cohesive and noncohesive sediments. Transport within the full range of sediment movement regimes is needed as indicated in the following case.

An understanding of the role that particle residence time on the bed surface plays in regulating chemical transport is needed. This factor appears to be more important for contaminated transport from the bed. In the case of transport to the bed, the delivered particles are likely to be in chemical equilibrium with the overlying water, but in the case of bed-originating contaminated particles, their exposure time at the surface, prior to reburial, may be insufficient for chemical equilibrium to be achieved with the overhead water.

Once sediment particles are set in motion, three types of movement can be identified: rolling/sliding, saltation, and suspended particle motion. At low sediment transport rates, sediment particles move by rolling and sliding in a thin layer along the surface of the uneven sediment bedforms. This particle movement is referred to as bedload movement. Saltation occurs when the particles move by irregular jumps. Suspended particle movement occurs at very high sediment transport rates when the sediment particles erode from the surface and are washed away.

Savant[46] performed experimental investigations on particles during the rolling/sliding movement regime with the bottom surface in a waveform. After moving up the upstream face of the wave in a thin layer, the particles avalanched down the lee face of the wave. The moving layer was approximately two particle diameters thick. In this fashion, continuous downstream migration or translation of the bedform (i.e., dune) occurred as new layers of particles were constantly being exposed to the free-stream flow. Laboratory studies illustrated the layer *peeling and relocation process* and this led to the development of mathematical relationships for the exposure times of particles in the surface layer and of the dune. Savant[46] estimated that the characteristic bedload movement times for three rivers varied from 0.3 to 3200 days. Based on these, the particle residence times on the surface are very short, 3.4 min to 6.1 hr. A theoretical analysis was not performed to compare the dissolution chemodynamics of this bedload process to those of diffusion and advection. This analysis must await further developments on the intraparticle desorption process since the thin layer of particles on the surface essentially behave this way (see the discussion "Transient Sorption and Desorption from Particles" in Section 5.2).

Some Sediment-Side Chemodynamic Processes for Streams

Two transport processes are introduced in this section: molecular diffusion and advection in porous media. From studying riverbed, cross-section maps and sediment characteristics, silt deposits are typically near the banks in the lower-velocity areas, and sand and gravel deposits are generally in the center section of the channel. Other mechanisms, including sediment movement and advection, dominate chemical transport in the near-bank regions. Although presented here in the context of streambed sediment processes, the principles of molecular diffusion and advection apply to the sediment beds of lakes, estuaries, and oceans.

Molecular Diffusion in Beds. An analysis of the steady-state flux of chemicals in solution within porous media begins with Fick's first law as represented by Eq. 3.1-13. In mass concentration units it appears as

$$n_{Ay} = \frac{(n_{Ay} + n_{By})\rho_{A2}}{\rho_2} - \rho_2 D_{A3} \frac{d\phi_A}{dy} \tag{5.1-30}$$

where n_{Ay} denotes the upward flux of species A of mass fraction in solution, ϕ_A, and D_{A3} is the effective diffusivity in the porous media. The flux of water ($=B$) is n_{By} and $\rho_2 = \rho_{A2} + \rho_B$ for the binary solution within the pores of the media. For negligible water movement and a dilute solution of A, Eq. 5.1-30 takes on the more familiar form of Fick's first law, Eq. 5.1-31.

The molecular diffusivity alone is not sufficient to describe the diffusion within porous solids that have interconnected voids or pores in the solid. The diffusion is greatly affected by the size and type of the voids. Figure 5.1-8 shows a sketch of a cross section of such a porous solid. A reexamination of Fick's first law is in order.

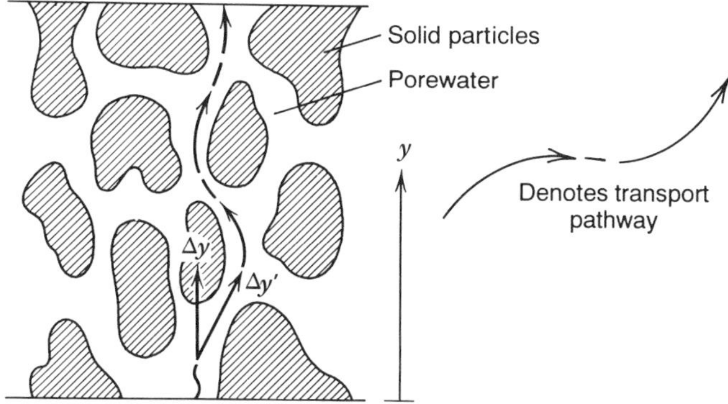

Figure 5.1-8. Pore diffusion in porous media.

Fick's law is replaced with an effective diffusion coefficient, D_{A3}, for the case of diffusion within porous solids and is normally written as follows:

$$j_{Ay} = -D_{A3} \frac{d\rho_{A2}}{dy} \tag{5.1-31}$$

Two factors operate to make the effective diffusion coefficient less than the molecular diffusivity. The interfacial area through which the chemical moves is reduced because the free or open cross section for diffusion is but a fraction of the total because of fill particles. The diffusivity is also effectively reduced because the diffusion distance along the tortuous path is greater, as shown in Fig. 5.1-8. The effective diffusion coefficient is then defined formally as

$$D_{A3} = \frac{\mathscr{D}_{A2}\varepsilon}{\tau_h} \tag{5.1-32}$$

where $\mathscr{D}_{A2}$ is the molecular diffusivity, ε the void fraction of the sand- or sediment-filled hole, and τ_h the *tortuosity factor.*

The factor τ_h is introduced to allow for the fact that the diffusion path is greater than the distance traveled normal to the face, and for varying cross section of the pores, which are not straight, round tubes. This correction factor must normally be obtained experimentally except for fill of exceedingly uniform structure and pore size. Values obtained from experimental data show that τ_h varies from unity to more than 6.[13] If we assume that on average, the pore makes an angle of 45° with the vertical y direction in a resultant two-dimensional path, $\tau_h = \sqrt{2}$.

Greskovich et al.[14] performed an experimental study of diffusivities through soils and calculations of hindrance factors to provide a basis for predicting rejuvenation rates of polluted muds. The ultimate goal of the investigation was to estimate the hindrance factor H and hence the rate of diffusion through different types of streambed media. The hindrance factor is defined as

$$D_{A3} = \frac{\mathscr{D}_{A2}}{H} \tag{5.1-33}$$

This factor relates to particle shape, tortuosity, and bed void fraction. Because the pore size in the bed is usually much larger than the molecular diameter of the diffusing species, a hindrance factor based on one diffusing species is also accurate for other species of similar molecular diameter that exhibit the same effects with the muds.

A Stefan cell was used to measure the hindered diffusivities. Potassium chloride was used as the diffusing species for the determination of the hindrance factors. Table 5.1-3 contains the experimental data for various test media. The results indicate that the diffusion rates differ for various bed types with different packing characteristics and different particle compositions. The hindrance factors shown vary from 3.04 for fine sand to 1.58 for loosely packed streambed silt.

A general correlation is needed that relates the diffusion coefficients of dissolved species to the porosities and tortuosities of sediments. This is usually done by a quantity known as the formation factor, f. Lerman[48] has reviewed much of the work on formation factors and recommends that Archie's law, $f = \varepsilon^{-2}$, may be used to estimate the effective diffusion coefficient in the pore waters of sediment for the lack of a better model. A theoretical model developed by Millington and Quirk[49] for surface soils is applicable to sediments. For saturated media the result is $f = \varepsilon^{-4/3}$, which yields

$$D_{A3} = \mathscr{D}_{A2}\varepsilon^{4/3} \tag{5.1-34}$$

for the effective diffusion coefficient. This equation has been used with success by the author and others.

Table 5.1-3. Experimental Hindrance Factors and Void Fractions for Test Media

Medium	Particle Diameter (cm)	Void Fraction	Hindrance Factor, H
Sand	0.0147–0.0208	0.491	2.82
Sand	0.0208–0.0295	0.507	3.04
Sand	0.0417–0.0589	0.510	2.70
Clayey-silt soil	~50% < 0.0035	—	2.24
Dried streambed silt	~70% < 0.002	—	2.46
Fresh streambed silt	~50% < 0.002	—	1.58

Source: Ref. 14.

Example 5.1-2. NO_3^- Flux into Seine River Sediment. The diffusivity of NO_3 into water at 25°C is 19E-6 cm²/s. The anion and its counterion both diffuse into water along the same concentration gradient as the condition of electrical neutrality must be maintained. The River water is at 20 mg/L. If the porewater at 10 cm depth is 0.0 mg/L and bed porosity is 65%, estimate the flux in g/m²·h using the Archie (A) and Millington–Quirk (MQ) formation factors.

SOLUTION The effective diffusion coefficients, D_{A3}, are 8.03E-6 cm²/s for (A) and 10.73E-6 cm²/s for (MQ). The steady-state flux equation is

$$n_A = \frac{D_{A3}(\rho_{A22} - \rho_{A21})}{z_2 - z_1} \tag{E5.1-2}$$

where the second numerical subscript denotes the position. Using the (A) diffusivity the flux is

$$\begin{aligned} n_A &= 8.03\text{E-}10 \text{ m}^2/\text{s } (20 - 0.0 \text{ g/m}^3) \text{ } 3600 \text{ s/h}/0.1 \text{ m} \\ &= 5.78\text{E-}4 \text{ g/m}^2 \cdot \text{h} \end{aligned}$$

and for the (MQ) value of D_{A3}, $n_A = 7.70\text{E-}4 \text{ g/m}^2 \cdot \text{h}$, which is 33% larger.

Advection in Beds. *Advection*, referred to as *convection* in most chemical engineering literature, is the chemical flux as a result of bulk media movement. It first appears in Eq. 3.0-3. In this case the media is water in the pores of the sediment bed. The advective term is the first on the right hand side of Eq. 5.1-30. The quotient $(n_{Ay} + n_{By})/\rho$ can be written as v_{2y}, the apparent velocity of water, that is, the volumetric flow divided by total cross-sectional area. The average water velocity within the pore spaces is v_{2y}/ε. It is often convenient to rewrite Eq. 5.1-30 as

$$n_{Ay} = v_{2y}\rho_{A2} - D_{A3}\frac{d\rho_{A2}}{dy} \tag{5.1-35}$$

where the product of velocity and concentration at a plane perpendicular to the direction of the velocity vector is the advective flux term.

The water flow through the bed results from regional hydraulic gradients as well as localized hydraulic gradients. The regional gradients are those induced by groundwater in the stream banks and the terrain beyond. The piezometric heads may be above or below those in the stream where the term *gaining water* or *losing water* is applied to the flow. For example, Seine River water near d'Aubergenville, France has been estimated to flow through the bed sediment and into an underground aquifer at 1.5 cm/h. D'Arcy's law relates the water velocity through the bed, its permeability and the hydraulic gradient; see the information in Chapter 7 on the subject as it relates to groundwater flow.

Localized water advection in the bed is the result of streamflow over large bed-surface roughness features.

Apparently, spawning fish realized the importance of advection as a chemodynamic process in bed sediment. Vaux[50] reports that eggs deposited in constructed burrows survived because of better irrigation by oxygenated water. Thibodeaux and Boyle[51] verified what the fish knew by demonstrating that sandwave-shaped bottom forms on the bed surface forced stream water deep into the bed and returned it to the sediment surface and through the interface. The water generally enters the bed from the high-pressure side of the wave upstream from the crest and leaves on the low-pressure side downstream from the crest near the trough. The in-bed currents penetrate downward to a distance of approximately five wave heights (i.e., 5Δ). The average water velocity circulating within this zone is v_0, determined by

$$v_0 = \left(\frac{C v_x^2}{\lambda} + gs \right) \frac{K}{\nu_2} \tag{5.1-36}$$

where K is the bed permeability (L^2), λ is the wavelength (L), Δ is the wave height (L), and the bed friction factor is determined by $C = 0.5(\Delta/h_2)^{3/8}$. For a flat, formless bed, such circular flow does not occur, but due to the water surface slope, s (L/L), a velocity is induced in the downstream direction parallel to the bed surface.

The relative contribution of advection and diffusion in chemical transport is typically characterized by the Peclet number. For bed sediment it is defined as

$$\mathrm{Pe}_{A3} \equiv \frac{v_0 l}{D_{A3}} \tag{5.1-37}$$

where l is a characteristic length such as depth within the bed. Savant et al.[66] determined that within the main channel of three large rivers Pe_{A3} was 80 to 1700. This indicates that molecular diffusive transport at this locale within these sediment is negligible compared to advection.

Important Coupled Processes. Molecular diffusive and advection are parallel processes in many bed sediments, and it is instructive to consider their combined effect on the transport of chemicals. Consider the NO_3^- transport in Example 5.1-2. It is diffusing from the high interface ($y = 0$) concentration ρ_{A22} to ρ_{A21} at $y = h_3$. The porewater velocity is positive when it is in the same direction as the gradient; otherwise, it is negative. Integration of Eq. 5.1-35 over the foregoing boundary conditions yields

$$n_{Ay} = \frac{D_{A3}}{h_3} \frac{[\rho_{A22} \exp(\mathrm{Pe}_{A3}) - \rho_{A21}]\mathrm{Pe}_{A3}}{\exp(\mathrm{Pe}_{A3}) - 1} \tag{5.1-38}$$

where h_3 is the length l in the Peclet Number.

The situation described for NO_3^- in the Seine River sediment in Example 5.1-2 is used to illustrate the effect advection has on the flux of this ion into the sediment. The following results are obtained from Eq. 5.1-38. At $\text{Pe}_{A3} = 1$, a condition where advection and diffusion are numerically equal and into the bed, the flux is enhanced by 58% over diffusion alone. At $\text{Pe}_{A3} = 3$ it is enhanced by a factor of 3.2. For $\text{Pe}_{A3} = -1$, advection and diffusion are equal but in opposite directions and the flux is reduced 42% below diffusion alone. For $\text{Pe}_{A3} = -3$ it is less than 16% and at $\text{Pe}_{A3} = -7$ it is less than 1%. So an outward porewater velocity of 8.6E-6 cm/s over a 10-cm distance is sufficient to virtually overcome diffusion into the bed.

The remainder of this section is devoted to chemical transport in series under quasi-steady-state conditions. Here the resistance-in-series concept, introduced in Chapter 3, will be applied across the sediment–water interface. First, a sediment-side mass transfer coefficient and quasi-steady state will be defined.

Although invented for the fluid side, because of the turbulence that exists there, the concept expressed by Eq. 3.1-8, which creates a first-order concentration flux expression, is applicable to the particle side of the interface as well. The definition of ${}^2k'_{A3}$, the sediment-side coefficient, is therefore

$$j_{Ay} \equiv {}^2k'_{A3}\,\Delta\rho_{A2} \tag{5.1-39}$$

where j_{Ay} is the mass flux of species A in the y-direction perpendicular to the interface ($\text{M/t} \cdot \text{L}^2$), and $\Delta\rho_{A2}$ is the concentration difference of species A within the porewater between two levels in the bed. The coefficient is related to the depth within the bed, h_3, from which the concentration difference is derived. If the mechanism of transport is molecular diffusion, the appropriate parameterization of the coefficient is

$${}^2k'_{A3} = \frac{D_{A3}}{h_3} \tag{5.1-40}$$

The coefficient defined in Eq. 5.1-39 is general. There are several other transport mechanisms on the sediment side, and these, along with the appropriate expressions for their quantification, will be introduced in the subsequent sections on lakes, estuaries, and oceans.

All processes in nature are transient. However, for short periods of time the steady-state flux assumption is valid. It is particularly valid when the chemodynamic system response time is large. A good example of this is chemical molecular diffusion in beds contaminated with hydrophobic organics. The solid particles that make up the bed have the capacity to absorb large quantities of these substances; therefore, the content is depleted slowly (i.e., long response time). Steady-state flux over short time periods is valid and the condition is termed the *quasi-steady-state assumption.* The following mechanistic scenario for the interphase chemical transport employs the quasi-steady-state concept.

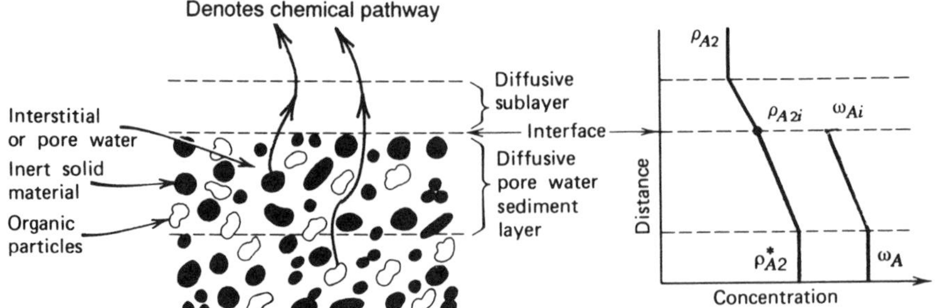

Figure 5.1-9. Sediment–water interface.

The sediment on the bottom of an aquatic environment consists of solid material that is a combination of both organic and inorganic matter. The organic matter may consist of decayed plant or animal bodies, whereas the inorganic is mostly silica in the form of sand and silt. Figure 5.1-9 illustrates what the sediment–water interface may look like upon close inspection of a thin section.

The mechanism for movement of chemical A from the sediment into the overlying water involves the following processes:

1. Release of the molecule from the cell matrix or the solid matrix
2. Diffusion through the cellular residue or the pores within the solid matrix
3. Desorption from the residue or solid surface into the interstitial water
4. Diffusion through the interstitial water
5. Diffusion through the sediment–water interface
6. Movement through the diffusive sublayer into the overlying turbulent water

In general, the individual rates of each step are unknown, but much research is under way aimed at understanding and quantifying the mechanisms. For example, the release of the molecule from the cell matrix mass involves a chemical reaction, whereas the release of the molecule from a solid particle matrix may involve a dissolution surface reaction or an ion-exchange reaction. Diffusion through the cellular residue or the pores within the solid matrix may be hindered or enhanced molecular diffusion. Desorption from the residue or solid particle interface into the interstitial water is likely to be a rapid process. Movement through the interstitial water is by molecular diffusion. Movement through the sediment–water interface plane is likely to be a rapid process also. There is likely some resistance within the diffusive sublayer.

The production of chemical A or its release from the substrate by chemical processes can be related by a first-order reaction as

$$r_A = k_A''' \omega_c \tag{5.1-41}$$

where ω_c is the concentration of some characteristic precursor molecule. Dissolution from the surface of a solid particle is also a first-order reaction (see Eq. 5.3-5).

The processes described above occur in series. Some are much faster than others and can be neglected. Little is known about some processes, so it is convenient to lump them together. This is the case of steps 1 through 3 above. Here it is common to assume that release, particle diffusion, and desorption can be characterized by some concentration manifest in the porewater, ρ^*_{A2} in equilibrium with solid concentration, ω_A. Step 4 is known to be very slow in the case of molecular diffusion, so that Eq. 5.1-40 applies. Theoretically, there should be no interface resistance since there is a continuous water pathway. It is unknown whether a bed particle surface charge will inhibit counterion or charged colloid movement here, however. Step 5 is usually neglected. The diffusive sublayer does represent a resistance, and Eq. 5.1-23 is applicable to streams.

The resistance-in-series concept yields a useful mathematical format for quantifying each step. The overall coefficient reflecting the entire six-step pathway is defined by

$$j_{Ay} \equiv {}^2K'_{A3}(\rho^*_{A2}\text{-}\rho_{A2}) \tag{5.1-42}$$

where ρ_{A2} is the chemical concentration in the stream. The resistance-in-series law for this scenario is simply

$$\frac{1}{{}^2K'_{A3}} = \frac{1}{{}^2k'_{A3}} + \frac{1}{{}^3k'_{A2}} \tag{5.1-43}$$

Here the total resistance is the sum of that on the sediment side and the water side. The flux of A from the bed is

$$n_{Ay} = v_{2y}\rho_{A2i} + {}^2K'_{A3}(\rho^*_{A2} - \rho_{A2}) \tag{5.1-44}$$

This equation accounts for the advective flux where the concentration at the interface, ρ_{A2i}, is required. It can be obtained from the relationship

$$\rho_{A2i} = \frac{{}^2k'_{A3}\rho^*_{A2} + {}^3k'_{A2}\rho_{A2}}{{}^2k'_{A3} + {}^3k'_{A2}} \tag{5.1-45}$$

The final three expressions provide a means of estimating the quasi-steady-state flux and the concentration profile from bed to stream. The algorithms above also work in reverse. This is the case when species A is moving from the stream through the interface and into the bed. The concentration difference in Eq. 5.1-42 is reversed. The advective term is positive when v_{2y} is in the same direction as the diffusive transport term or negative otherwise.

There is much evidence to indicate that the activities of benthic organisms, bottom currents, and gases evolving from decaying organic matter do mix the upper sediment layers. This transport activity below the sediment–water interface may be comparable to the water-side transport processes above the interface, thus ${}^{2}k'_{A3}$ is likely the order of ${}^{3}k'_{A2}$ or less. Realizing that Eqs. 5.1-43 to 5.1-45 are only order-of-magnitude models of the actual sediment layer processes, they do allow estimates of recovery times and water concentrations. A more analytically rigorous approach to the problem of the movement of chemicals in a sediment layer toward the interface is taken in Section 5.2.

Dissolution of Buried Sinkers and Other Contaminants

Earlier in this chapter, on the subject of sinkers spilled into streams, it was noted that quantities of the pure material ends up beneath the sediment water interface. This can occur by several natural processes and four are presented here.

1. In the case of the midchannel region the bedload, consisting of sand and silt, covers the liquid (or solid) as the *sandwaves* move downstream.
2. Near the stream bank where the sediment contains silt and clay it is very light in density and cannot support the weight of the more dense sinker. The sinker quickly settles through the less dense surface sediment, displacing it upward in bulk and therefore becoming buried.
3. The accumulation of pure material below the sediment surface also occurs as porewater is displaced when density differences and/or capillary pressure forces it in the sediment.
4. There are apparently some diagenic processes by which trace quantities of organics substances within the bed are concentrated to form a separate liquid phase. A diagenetic process is suggested by the oil film that typically forms on the water surface during mechanical dredging or other such disturbances of some contaminated bed sediment. Oil is a floater and the droplets can be observed rising to the air–water interface creating its usual iridescent sheen on the surface. Since it is less dense than water, it is unlikely that the substance could have been placed in the bed in droplet form. This suggests that some in-bed coalescence or other diagenic process has occurred which concentrates the substance from the dilute form that earlier settled from the water column attached to particles and/or was sorbed onto the bed material from solution.

By whatever process it is clear from the above that quantities of nonaqueous chemicals, either pure or as a mixture, accumulate at depth below the sediment–water interface. The theoretical dissolution process occurs in two steps as illustrated in Fig. 5.1-10. In part (*a*) the pure material is at the bottom of a depression. It exists at a distance h_3 below the surface, where the depth of

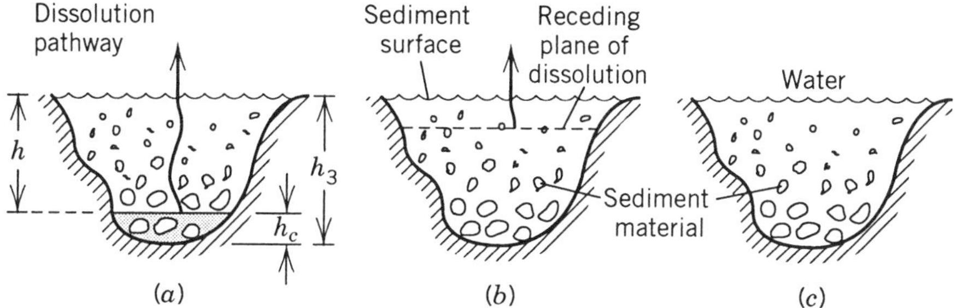

Figure 5.1-10. Dissolution of pure and sorbed chemical from within beds: (*a*) pure and sorbed; (*b*) sorbed; (*c*) clean.

the pool of pure material is h_c. As dissolution of this pool proceeds, h_c decreases with time. In the bed near the pool surface the concentration in porewater is the solubility, ρ^*_{A2}, and it is at some lower value at the interface, ρ_{A2i}. For dissolution from the upper plane surface of the pool only, the flux is obtained from Eqs. 5.1-43 and 5.1-44. If advection is neglected, the flux is

$$n_{Ay} = \frac{\rho^*_{A2} - \rho_{A2}}{1/{}^3k'_{A2} + h/D_{A3}} \tag{5.1-46}$$

The flux decreased with increasing h since as dissolution proceeds, the upper surface recedes further into the bed. A mass balance on the pool yields the following expression for the pool dissolution time, t_p:

$$t_p = \frac{\varepsilon\rho_A}{\rho^*_{A2} - \rho_{A2}}\left(\frac{h_c}{{}^3k'_{A2}} + \frac{h_3 h_c - h_c^2/2}{D_{A3}}\right) \tag{5.1-47}$$

where ρ_{A2} is the concentration of A in the stream water.

If local advection results in stream water penetrating the sediment and returning to the surface, the dissolution process is enhanced considerably. If the net effect of the advection is to sweep water of concentration ρ^*_{A2} from the upper surface of the pool and return it to the stream, the flux can be approximated as

$$n_{Ay} = \left(v_0 + \frac{1}{1/{}^3k'_{A2} + h_3/D_{A3}}\right)(\rho^*_{A2} - \rho_{A2}) \tag{5.1-48}$$

If Pe_{A3} is large, the flux is dominated by advection and the flux can be approximated by $v_0(\rho^*_{A2} - \rho_{A2})$. Typically, Pe_{A3} in sandwaves is 100 to 2000. In this case the pool dissolution time is

$$t_p = \frac{\varepsilon\rho_A}{\rho^*_{A2} - \rho_{A2}}\frac{h_c}{v_0} \tag{5.1-49}$$

Values of v_0 for flow in sandwaves can be estimated from Eq. 5.1-36. The dissolution time expression for the intermediate case where $10 > \mathrm{Pe}_{A3} > 1$ is left as an exercise.

In Fig. 5.1-10*b* the pool has completely dissolved and the sediment layers near the interface contain a reduced quantity of contaminant *A*. A significant quantity remains on the particles and in the porewater. This quantity is

$$m_A = Ah_3\varepsilon\rho_{A2} + Ah_3\rho_3(1-\varepsilon)\omega_A \tag{5.1-50}$$

If the porewater concentration is the chemical solubility in water and that on/in the particles is assumed to be in equilibrium with it, the mass is

$$m_A = Ah_3\rho^*_{A2}[\varepsilon + \rho_3(1-\varepsilon)K^*_{A32}] \tag{5.1-51}$$

As time proceeds during this period of dissolution, concentration gradients develop in space and time within the bed. The transport process is commonly described by Fick's second law modified for the chemical content on the particles, which in one-dimensional form is

$$[\varepsilon + \rho_3(1-\varepsilon)K^*_{A32}]\,\frac{\partial \rho_{A2}}{\partial t} = D_{A3}\,\frac{\partial^2 \rho_{A2}}{\partial y^2} \tag{5.1-52}$$

for the concentration in porewater. It is convenient to define D_{eff} as

$$D_{\text{eff}} \equiv \frac{D_{A3}}{\varepsilon + \rho_3(1-\varepsilon)K^*_{A32}} \tag{5.1-53}$$

so that the available analytical solutions to Fick's second law can be used. The particular solution for upward diffusion from a half-slab applies to a sediment bed if the bottom surface below the contamination is assumed to be impermeable to *A*. This solution, given in both graphical and analytical form, appears in Section 6.3, pages 408 to 412 for the analogous case of chemical volatilization from surface soils to air. To use this information to estimate concentrations in porewater, flux from the bed, and dissolution times, the student should transform the appropriate equations to bed sediment nomenclature.

Figure 5.1-10*c* illustrates the idealized clean conditions of the sediment bed. In theory and in practice this does not occur. The user of the relationships in Section 6.3 must decide what level of the final concentration of *A* represents "clean" before a definite time can be estimated. A zero value for the final concentration is not permissible with dissolution as the only chemical fate process in the bed sediment.

Closure. The student will find that the problems at the end of this section contain further useful information. Many problems relate to actual spill incidents and are constructed around the spill documentation. Problem 5.1A is concerned with a large spill of chloroform in the Mississippi River in 1973. Problem 5.1E concerns the spill of 250 gal of PCB into the Duwamish River in 1974, and Problem 5.1I is concerned with the spill of metallic mercury in the Shenandoah River between 1929 and 1950.

Calculations associated with spills that have already occurred are useful in estimating in-stream concentrations and on-bottom lifetime. Both of these calculations are relevant to exposure level and exposure time of the biota. Calculations are also relevant in deciding whether or not to proceed with cleanup of the chemical. Calculations associated with possible spills can be useful in transportation hazard assessment.

PROBLEMS

5.1A. Concentration–Time History of a Chloroform Spill in Mississippi River

Approximately 1.75E6 pounds of chloroform was released from a barge that sank near Baton Rouge, Louisiana, and the chemical began flowing down the Mississippi River toward the Gulf of Mexico. Although state health officials did not push the panic button, noting that they did not anticipate too much trouble from the accident, the U.S. Coast Guard warned downriver communities to keep a close surveillance on their water supply systems, particularly if intakes were close to the river bottom (chloroform is heavier than water).

Actually, the river was at a low flow state, and there is reason to believe that the heavy chemical remained in place on the bottom near the spill site. Using the dissolution model for each shape in turn:

1. Calculate in-stream concentration in mg/L.
2. Calculate the minimum on-bottom lifetimes in h.

Data: Bottom mass transfer coefficient, ${}^4k'_{A2} = 3.87\,\text{cm/h}$; river flow $Q = 7590\,\text{m}^3/\text{s}$; ripple (sandwave) amplitude $\Delta = 7.5\,\text{cm}$.

3. Compare the results of parts 1 and 2 to the data in Table 5.1A and arrive at some conclusions about the on-bottom chloroform mass distribution in drops, globs, and pools. At the time of the spill, the Dow Chemical Company obtained river water samples and measured concentrations.[15] A portion of the data record appears in Table 5.1A ($A \equiv$ chloroform.) Background concentration in the river is about 5 ppb. Assume that a total of 317,000 kg was accounted for in the river.
4. Using 54,150 kg as the mass of chemical in the shape of spheres, 230,730 kg as the mass of chemical in the valleys of sandwaves, and the remainder as globs, calculate the in-stream concentration (ppb) for each half-hour for 18 days.
5. Plot the calculated concentrations and the observed concentrations shown in Table 5.1A versus time on a single piece of graph paper.

Table 5.1A. Chloroform Spill in Mississippi River[a]

Date–Hour	ρ_{A2} (ppb)	Date–Hour	ρ_{A2} (ppb)	Date–Hour	ρ_{A2} (ppb)
8/19–2330	80	8/20–1230	121	8/24–1330	31
8/20–0030	220	1530	81	8/25–0400	24
0130	264	1730	59	2350	26
0230	352	8/21–0130	70	8/26–1600	20
0245	264	1430	70	8/27–1620	21
0330[b]	365	8/22–0600	53	8/28–0800	15
0430	326	2000	33	8/29–0800	6
0630	233	8/23–0400	25	8/30–0800	13
0730	202	1330	31	8/31–0800	12
0830	162	8/24–0400	25	9/4–0800	4
				9/6–0800	7

Source: Ref. 15.

[a]Sampled at mile 16.3.

[b]Chose 8/20 at 0330 as $t = 0$.

5.1B. PCB in Hudson River Bed Sediment

Horn et al.[45] theorize that the PCBs move by two mechanisms: dissolution from the bed and/or erosion of particles containing the substance. They observe that half of the PCB transport occurs at low and moderate flows and is not the result of riverbed scour. Perform an analysis of the transport process at these conditions for the Schuylerville station (see Fig. 5.1-7) and suggest one or more appropriate sediment-bed processes that support the data.

The equation for the solid curve shown in Fig. 5.1-7 is $\rho_{A2} = 5.5\text{E}4/Q$. It is a regression of the data for $Q \leqslant 5.6\text{E}5$ L/s, where ρ_{A2} is the total PCB concentration, particle plus soluble, in μg/L. The contaminated bottom areas above Schuylerville are total, 69E5m^2, and hot spots, 8.6E5 m^2. Concentration in the hot spots averaged 150 mg/kg; the other spots averaged 19 mg/kg. From sediment cores the maximum concentration was typically 15 to 30 cm below the surface. Assume no vaporization of PCBs through the air–water interface and use the properties of A-1242 at 25°C in your calculations.

5.1C. Buried Sinker Dissolution

Quantities of chloroform have become buried within the bed as a result of the spill described in Problem 5.1A. The puddles of pure material are estimated to be layers 0.5 cm in thickness located 5 cm below the surface.

1. For molecular diffusion as the assumed transport mechanism, estimate the pool dissolution time in days and the flux to water in ng/m$^2\cdot$h.
2. For local advection in the bed, $\text{Pe}_{A3} = 500$, estimate the pool dissolution time and the flux.

5.1D. Particle Dynamics in Streams

The process of particle transport between the bed and water column of flowing streams is illustrated in Fig. 5.1-6.

1. Show the details in deriving Eqs. 5.1-27, 5.1-27*a*, and 5.1-29.
2. Streams sampled frequently throughout the year at various locations x over the same reach yield ρ_{32} versus x data that display (a) stable or constant concentration, (b) net setting, or (c) net scour. Show how to use these data to evaluate the constants ${}^3k'_2$, ${}^2k'_3$, and ρ^*_{32} in Eqs. 5.1-27 and 5.1-27*a*. During sampling h_2 and v_x are constant. Typical values of $\rho_3(1-\in)$ range from 50,000 to 500,000 mg/L, according to Delos et al.[52]

5.1E. Spill of PCB in Duwamish River[16]

On September 13, 1974, an electrical transformer fell during loading operations and caused a spill of 250 gal of 100% PCB (Aroclor 1242) into the Duwamish River in Seattle, Washington. The lower Duwamish River is affected by tides up to 4 m and regularly flows at approximately 2 m/s. The spill site was a predominantly mud–silt bottom, with fresh water overlaying a saltwater wedge, approximately 14 m deep and 150 m wide. Environmental Protection Agency divers observed pools of free PCB (specific gravity 1.4) material on the bottom. There was evidence that the river current and tidal action had caused pockets of PCB to move about. Divers observed pools of PCB moving as much as 15 m with the tide from one day to the next. To investigate the fate of the PCB, perform the following dissolution calculations:

1. Estimate the dissolution lifetime.
2. Estimate the maximum water concentration in the Duwamish River (ppb).

Data needed for calculation: solubility in water = 2.4E-4 mg/L, $\sigma = 40$ dyn/cm, $\rho_A = 1.4\ \mathrm{g/cm^3}$, molecular weight 258, $\mathscr{D}_{A2} \simeq 0.8\text{E-}5\ \mathrm{cm^2/s}$, and ${}^4k'_{A2} = 4.54$ cm/h.

5.1F. Laboratory Measurements of Dissolution Rate of Heavy Chemicals in Aqueous Environments

Experimental measurements of the dissolution rate of heavy chemicals under laboratory conditions are helpful in attempting to predict dissolution rates in real-world flowing and nonflowing water environments. Open-top containers filled with pure liquids are placed inside laboratory flowing-stream simulators that reproduce river conditions. Dissolution is allowed to occur for a period of time, after which the quantity of liquid remaining in the container is redetermined. Similar experiments can be performed in quiescent tanks to simulate conditions in deep lakes and/or very slow-moving rivers.

Laboratory flowing-stream simulator experimental results: average velocity 0.28 ft/s, diameter of container 7.05 cm, temperature 18°C, initial weight

685.2 g, final weight 671.2 g, start time 10:25 A.M., and finish time 11:45 A.M.

Quiescent water body simulator experimental results: velocity 0.0 ft/s, diameter of container 7.05 cm, temperature 27°C, initial weight 321.6 g, final weight 303.1 g, start time 9:14 A.M., and finish time 4:25 P.M.

The chemical used in both experiments was furfural (C_4H_3OCHO).

1. Compute the observed mass flux rates (n_A) and the liquid-phase mass transfer coefficient (${}^4k'_{A2}$) from the data given for both experiments. Report the answer in the SI units of grams, centimeters, seconds, and kelvin.

2. Based on the coefficient in part 1, estimate the mass flux rate of chloroform for the flow condition only. Give your answer in SI units.

5.1G. Important Coupled Proceses

1. Show that when solute diffusion exists in the bed, the water-side resistance is insignificant (at the 5% level) for a diffusion path length of $h_3 \geqslant 20D_{A3}/{}^3k'_{A2}$.

2. Show the details in arriving at Eq. 5.1-38 for steady-state flux.

3. Verify the numeric details immediately below Eq. 5.1-38 using the Seine River data in Example 5.1-2. Give the flux values for the five scenarios, in $g/m^2 \cdot h$.

4. Show the details in developing the resistance-in-series concept result represented by Eqs. 5.1-42 to 5.1-45.

5. For the five scenarios in part 3, compute the flux values using the resistance-in-series equations. Use 7.9 cm/h for the water-side coefficient.

6. Are the flux values in parts 3 and 5 equal? Give reasons for the various results, if any.

5.1H. Dissolution Kinetics of Sinkers

During the dissolution of immiscible, slightly soluble, sinker chemicals in the shape of spheres, globs, and drops on a stream bottom, the area for mass transfer changes with time. Equations 5.1-2 to 5.1-4 are the general expressions for the dissolution kinetics. For each shape, derive the expression for:

1. The mass remaining on bottom as a function of time,

2. The concentration in the water as a function of time, and

3. The dissolution lifetime.

5.1I. Dissolution of Metallic Mercury from South Fork of Shenandoah River[17]

Because of its large density ($\rho_A = 13.6\ g/cm^3$) and liquid form, metallic mercury seems to find its way into sheltered nooks and crevices—of which the South River, with its irregular limestone bottom, has plenty—and is not easily dislodged. Major floods such as the one that followed tropical storm Agnes in 1972 have repeatedly scoured the river bottom, but although some of the mercury in the bottom sediments undoubtedly has been moved about, it has not been swept away.

It has been estimated that between 1929 and 1950, 35 L of liquid metallic mercury escaped to the river from DuPont's Waynesboro "old chemical building." High concentrations of mercury are still found in sediment samples taken from a natural trap formed by the remnants of a small dam. Estimate the dissolution lifetime for the metallic mercury in this stream.

Assume that potholes, nooks, and crevices account for 30% of the stream bottom and are filled with pebbles 1 cm in diameter. The zone of contamination is assumed to be a bottom area of 100 m^2. Assume that the potholes are 30 cm deep and 30 cm wide on the average. Assume that the mercury occupies the very bottom of the potholes. Report the lifetime in years.

5.1J. Riverbed Sediment Response to Massive Chemical Spill

A northern Louisiana newspaper reported in March 1981 that the Georgia-Pacific Corporation had dumped more than 23 tons of toxic chemical into the Mississippi River. The phenol spill, occurring over a 3-day period, was blamed for fouling downriver drinking water supplies for more than a week in February.

1. Assuming that the river was at average flow, calculate the steady-state phenol concentration in the water.
2. Assuming that the river behaves in a plug flow manner, compute the length of the spill in kilometers. Does it extend to the Gulf of Mexico?
3. Equation 5.1-52 characterizes the response of the sediment bed to this chemical insult if advection is neglected. Propose a model scenario and boundary conditions that provide a solution to Fick's second law that yields a reasonable simulation of phenol penetration into the bed. Assume that the bed surface is flat and very deep. Give the solution.
4. Using the solution to part 3, estimate the depth of phenol penetration into the bed. The phenol equilibrium partition coefficient is 0.25 L/kg.
5. Calculate the mass of phenol that penetrates the bed. What mass fraction of the quantity spill is this?
6. Nondimensionalize Eq. 5.1-52 to obtain a characteristic bed response time to chemical perturbations. Your answer should be

$$t_c = \frac{[\varepsilon + \rho_3(1-\varepsilon)K^*_{A32}]h_3^2}{D_{A3}} \tag{5.1J}$$

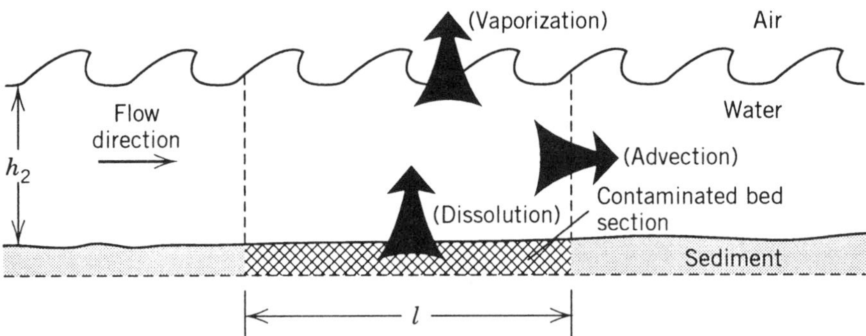

Figure 5.1K. Chemical pathways from the bed.

7. The characteristic bed response time is a strong function of chemical hydrophobicity. Use $K^*_{A32} = 0$, 1, 10, 100, 1E3, and 1E4 L/kg in calculations and compute the response time for a bed 20 cm in depth. Give your answers in years.

5.1K. Fate of Refractory Volatile and Semivolatile Bed Contaminants for Streams

For substances dissolving from the bed to the water column, vaporization through the air–water interface may be an important fate. Without degradation occurring in the water column, the pathways illustrated in Fig. 5.1K show the chemical being advected downstream and vaporizing.

1. Show that the fraction dissolved which eventually vaporizes is

$$F_v = \frac{1}{1 + (v_x/{}^1k'_{A2})(h_2/l)} \tag{5.1K}$$

   where l is the reach of streambed that is contaminated.

2. Using order-of-magnitude estimates of the parameters, calculate F_v and comment on chemical fate.

5.2. CHEMICAL MOVEMENT AT BOTTOMS OF PONDS, LAKES, AND QUIESCENT WATER BODIES

The term *quiescent water* is used here to contrast water bodies having a persistent flow direction and current such as those considered in Section 5.1; it does not imply stagnant water. Quiescent water bodies experience water column mixing through other environmental forces, including wind and thermal gradients. By this definition some streams, such as bayous in lowlands and sections of dammed rivers, are quiescent.

The section begins with the water-side processes for lakes, where thermal structure dominates chemical transport. It is followed by material on the sediment-side processes and ends with some important coupled chemodynamics processes in quiescent waterbodies.

Upsurge of Chemicals from Sediment–Water Interface of Lakes

Lakes that have been employed as waste dump sites can contain bottom sediments with a high content of specific chemicals. These chemical-laden bottom sediments can result from settleable solids and soluble substances that enter with streams. Chemical deposition from the atmosphere, both dry and wet, is also a contributor to bottom sediment contamination, as has been observed in alpine lakes removed from direct anthropogenic sources.

Methoxychlor [2,2-bis(*P*-methoxyphenyl)-1,1,1-trichloroethane] is a popular substitute for DDT [2,2-bis(*P*-chlorophenyl)-1,1,1 trichloroethane] used in the control of many insects. Methoxychlor reaches the aquatic environment by direct application, such as in the control of black fly larvae in streams or indirectly through terrestrial runoff. In a study of the Illinois waters of Lake Michigan and its tributaries, concentrations of methoxychlor in water averaged 15.1 ng/L in 1971 and 48.0 ng/L in 1972. Open-water sediments tested in 1970 and 1971 averaged 1.24 μg/L. Average DDT concentrations were somewhat higher than those for methoxychlor in both water and open-water sediments.

Organic and inorganic nitrogen (N) and phosphorus (P) compounds frequently find their way into lakes, in both particulate and soluble form. They are primarily responsible for the excessive growth of algae and other aquatic plants in lakes. These two chemicals are components of fertilizer and stimulate growth, in the spring, summer, and fall, of the lake vegetative matter. Lakes characterized by excessive yearly algal growths are termed *eutrophic*. At the end of the growing season, the algal cells settle to the bottom and become part of the sediment matter. On the bottom, chemical and biochemical processes occur to release the N and P from the cellular structure. Available as soluble compounds, these chemicals can work their way upward to the surface waters, where the photosynthesis process is active. Once again, these nutrients become part of the cellular mass of aquatic vegetation and the cycle has been completed. Dissolution and upsurge from the sediment–water interface play an important role in the nutrient-cycling processes.

If the dumping of N- and P-containing compounds is stopped, the rich bottom sediments retain and recycle these chemicals for many years, and the lake continues to display an eutrophic character. The rate at which these nutrient chemicals emerge from the bottom sediments determines in large part how soon a "'damaged" lake will return to a more normal state with respect to algae and aquatic plant productivity. Figure 5.2-1 shows the lake conditions of temperature and concentration during a period of upsurge of chemicals on a typical summer day. The model developed in this section uses the nutrient upsurge process in lakes; however, the results and the concepts presented apply

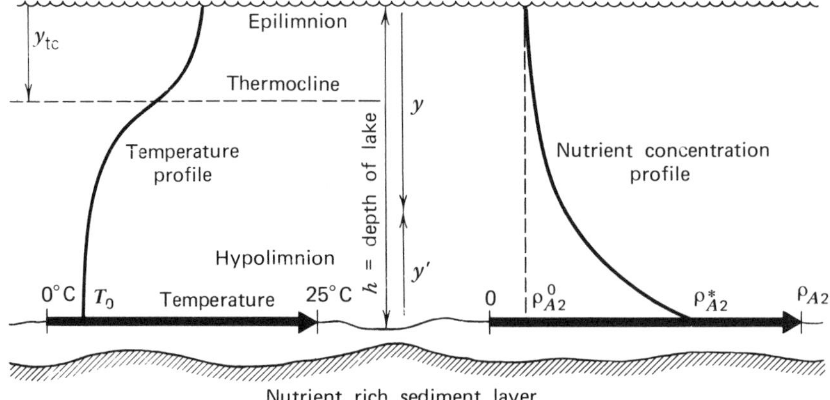

Figure 5.2-1. Typical lake-water column conditions.

to other chemicals emerging from or being transported to the sediment–water interface of lake or similar body of water.

Fickian Analysis of Upsurge of Chemical Nutrients.[18] The annual thermal history of a particular lake plays a dominant role in the movement of chemicals at the sediment–water interface. The thermal history of lakes as related to the seasons of the year is presented in some detail in Section 7.1. Also presented is a simple thermal model that describes the main features of the lake water column temperature and the thermocline position during the stratified period. Those not familiar with the thermal behavior of lakes are urged to review Section 7.1 prior to proceeding with the following treatment of chemical movement in the lake water column.

The period of time between the vernal equinox (March 21) and the autumnal equinox (September 23) is the dominant productive period for lakes in the temperate zone. During this period, lakes that contain nitrogen and phosphorus witness plentiful growth of algae due to high sunlight levels and warm water. This productive period is roughly from March 21 until September 23, Julian day 80 and Julian day 266, respectively, for a total of 186 days. This is also the heating period for the lake. As is pointed out in Chapter 7, during this period the net heat transfer into the lake is positive, and thermal stratification occurs. Increased air temperature and incoming radiation, coupled with a favorable water density–temperature relationship, cause a stable water column, consisting of warm water on top of colder water.

The calendar time of the vernal equinox is convenient for specifying a model time of zero (i.e., $t = 0$). Prior to this time the lake is at nearly uniform concentration with respect to temperature and chemical species. Ice may cover the surface but it begins to melt rapidly. A uniform temperature of 4 to 8 °C throughout is not uncommon. The time of the autumnal equinox is also important to the model structure.

The autumnal equinox marks the approximate end of the stratified thermal structure of the lake water column and marks the beginning of a more mixed state with respect to temperature and dissolved chemicals. At this time the net heat flux rate changes from positive to negative. Plunging cold surface water triggered by heat loss is the primary reason for the rapidly occurring mixed state and the subsequent fall turnover.

Turnover is a layman's term for the relatively rapid mixing process that can occur once the lake temperature becomes uniform. High wind drag at the air–water interface causes water motion and forces bottom water to the surface. If the bottom water was anaerobic, H_2S emission through the air–water interface makes turnover a noticeable undertaking. While the water column is at a uniform temperature, many turnovers can occur as the water mixing is rapid. This uniform-temperature water column condition can remain until the vernal equinox; however, if the water surface temperature falls below 4°C, reverse stratification can occur.

Reverse stratification occurs during the coldest part of winter. Ice or water having a temperature of less than 4°C is less dense and rides on the surface, and again a state of water column stability prevails. Another mixed period, sometimes called *spring turnover*, occurs during the period between the end of the reverse stratified period and the start of the normal stratified period, which occurs about March 21.

Lakes are extremely complex living systems. The chemical upsurge model presented here is a very important aspect of the life processes in a lake but is only a small part of the chemical and biochemical processes occurring in a given lake. Even if one considers only the transport of chemicals from the bottom sediments up the water column, it is necessary to consider at least two simple models. Arbitrarily, lakes are classified here as stratified and unstratified.

The stratified lake model accounts for the process of thermal stratification that occurs in most lakes to a greater or smaller degree. Stratification tends to retard the movement rate of chemicals originating at the sediment–water interface into the layers next to the air–water interface. The unstratified lake model is applicable to those lakes in which, for one reason or another, thermal stratification is not a dominant factor in chemical movement. Although both models are simple, they do reflect the major factors that influence chemical movement and serve as a starting point for the construction of more detailed transport models. The major factors include season of the year, time during the season, turbulent diffusivity, eddy thermal diffusivity, depth of the thermocline, initial chemical concentration in the lake water column, and chemical concentration at the sediment–water interface.

The mathematics of diffusion in a semi-infinite medium are employed as the quantitative tool to describe the chemical upsurge phenomena from the sediment–water interface. The medium in this case is water (see Fig. 5.2-1). The lake water is visualized as a medium occupying the space from $y' = 0$ (the bottom) to $y' = \infty$ (far removed from the bottom) with an initial concentration of ρ_{A2}^{0}. Nutrient A is present in the sediments at concentration ω_A and in equilibrium with interstitial water of nutrient concentration ρ_{A2}^{*}. (see Problem 2.1D for an example of the distribution of phosphorus between soil and water). At time $t = 0$ it is assumed that the bottom surface of the water is suddenly raised to concentration ρ_{A2}^{*} and maintained at that concentration for an extended period.

The initial condition above has some degree of realism for these nutrients.[56] In interstitial water, ammonium and phosphate ions can be adsorbed, more often reversibly, on various solid phases such as clay minerals, organic colloids, ferric oxides, and hydroxides. Values of K_{A32}^{*} in anoxic clay muds are generally within the range of 1 to 2.5 L/kg. Much lower values prevail, of course, in sands. The presence of ferric oxides in the upper oxidized layer of sediment has often been considered as a major factor controlling phosphorus recycling because of the strong adsorption properties of these phases for phosphates. Adsorption coefficients from 30 to over 3000 L/kg have been measured for oxic sediments. When it exists, the oxic layer can act as a trap for phosphates produced by benthic mineralization. When the oxic layer is eliminated because of oxygen depletion in the water column, as often occurs in lakes, the adsorption capacity of the sediment is destroyed and phosphates are released to the overlying water.

The stratified and unstratified lake models begin with a simplified form of the turbulent multicomponent equation of continuity Equation 3.2-11. In mass concentration units the equation can be simplified to

$$\frac{\partial \rho_{A2}}{\partial t} = \mathscr{D}_{A2}^{(t)} \frac{\partial^2 \rho_{A2}}{\partial y'^2} \tag{5.2-1}$$

where $\mathscr{D}_{A2}^{(t)}$ is the turbulent diffusivity and y' is the distance from the sediment–water interface in Fig. 5.2-1. The concentration ρ_{A2} is time smoothed. The initial and boundary conditions for Eq. 5.2-1 are

$$\text{IC:} \quad \text{at } t < 0, \quad \rho_{A2} = \rho_{A2}^{0} \text{ for all } y' \tag{5.2-1a}$$

$$\text{BC:} \quad \text{at } y' = 0, \quad \rho_{A2} = \rho_{A2}^{*} \text{ for all } t > 0 \tag{5.2-1b}$$

$$\text{BC:} \quad \text{at } y' = \infty, \quad \rho_{A2} = \rho_{A2}^{0} \text{ for all } t \tag{5.2-1c}$$

The solution to Eq. 5.2-1 is

$$\frac{\rho_{A2} - \rho_{A2}^{0}}{\rho_{A2}^{*} - \rho_{A2}^{0}} = 1 - \operatorname{erf}\left(\frac{y'}{\sqrt{4\mathscr{D}_{A2}^{(t)} t}}\right) \tag{5.2-2}$$

where the *error function*, erf, is defined as follows:

$$\operatorname{erf}\left(\frac{y'}{\sqrt{4\mathscr{D}_{A2}^{(t)} t}}\right) \equiv \frac{2}{\sqrt{\pi}} \int_0^{y'/\sqrt{4\mathscr{D}_{A2}^{(t)} t}} e^{-n^2}\, dn \tag{5.2-3}$$

Numerical values of the error function are presented in Appendix B.

Case I: Stratified Lakes. This case is applicable to lakes with a well-developed thermal profile. Nutrients entering the epilimnion from the sediment bed are capable of stimulating and supporting primary production. Nutrients enter the mixed epilimnion region because of the gradual deepening of the region and the turbulent diffusion through the thermocline. The flux rate of nutrient A through a horizontal plane located at the thermocline depth y_{tc} is

$$n_{Ay} = \left[(\rho_{A2} - \rho^0_{A2})v_y - \mathscr{D}^{(t)}_{A2}\frac{\partial \rho_{A2}}{\partial y'}\right]_{y_{tc}} \tag{5.2-4}$$

where v_y is the mass velocity of the thermocline downward. The distance from the bottom y' is related to the distance from the air–water interface y by $h = y' + y$, where h is the depth of the lake water. From Eq. 7.1-6 the thermocline velocity is

$$v_y \equiv \frac{dy_{tc}}{dt} = \sqrt{\frac{\alpha_2^{(t)}}{2t}} \tag{5.2-5}$$

and from Eq. 5.2-2,

$$\frac{\partial \rho_{A2}}{\partial y'} = \frac{-1}{\sqrt{\mathscr{D}^{(t)}_{A2} t\pi}} \exp\left(\frac{-y'^2}{4\mathscr{D}^{(t)}_{A2} t}\right)(\rho^*_{A2} - \rho^0_{A2}) \tag{5.2-6}$$

Substituting Eqs. 5.2-2, 5.2-5 and 5.2-6 into Eq. 5.2-4 results in

$$n_{Ay} = (\rho^*_{A2} - \rho^0_{A2})\left\{\left[1 - \operatorname{erf}\left(\frac{h - \sqrt{2\alpha_2^{(t)} t}}{\sqrt{4\mathscr{D}^{(t)}_{A2} t}}\right)\right]\sqrt{\frac{\alpha_2^{(t)}}{2t}} + \sqrt{\frac{\mathscr{D}^{(t)}_{A2}}{\pi t}}\exp\left[\frac{-(h - \sqrt{2\alpha_2^{(t)} t})^2}{4\mathscr{D}^{(t)}_{A2} t}\right]\right\} \tag{5.2-7}$$

This equation gives the flux rate of nutrient A into the epilimnion due to the upsurging components released from the sediment bed. In lake ecosystem modeling, this nutrient source term should be included along with the advective entering term, the advective exiting term, the bioconsumption term, the bioexcreta term, settling of particulates, and so on, to yield a complete material balance for the influx of nutrient A into the epilimnion.

Case II: Unstratified Lakes. This case is applicable to lakes with a poorly developed or nonexistent thermal profile. Sunlight penetrates and effectively stimulates primary production down to a depth h_e. This depth depends on the extinction coefficient of the water, self-shading by the algae, and so on. The flux rate of nutrient A into the sunlit surface region of the lake is

$$n_{Ay} = -\mathscr{D}^{(t)}_{A2}\left(\frac{\partial \rho_{A2}}{\partial y'}\right)\bigg|_{h_e} \tag{5.2-8}$$

Employing Eq. 5.2-6 for the derivative expression, we obtain

$$n_{Ay} = \sqrt{\frac{\mathscr{D}_{A2}^{(t)}}{\pi t}} \exp\left[\frac{-(h - h_e)^2}{4\mathscr{D}_{A2}^{(t)} t}\right] (\rho_{A2}^* - \rho_{A2}^0) \tag{5.2-9}$$

Both lake nutrient upsurge rate equations, Eqs. 5.2-7 and 5.2-9, require chemical concentrations and physical characteristics of the particular lakes. These parameters can be obtained from field observations. Equation 7.1-7 can be used to obtain $\alpha_2^{(t)}$ values from the thermal profile, and Eq. 5.2-2 is used to obtain $\mathscr{D}_{A2}^{(t)}$ values from chemical concentration profiles. Model time $t = 0$ can be obtained from inspection of graphs of the thermal and/or chemical concentration profiles for various times of the year (see Problem 5.2A). The sediment-bed equilibrium concentration of chemical A, ρ_{A2}^* and the initial concentration ρ_{A2}^0 can also be obtained from nutrient profile observations.

Generalization of Results of Fickian Analysis. The flux rate of nutrient A from the lake sediment bed is

$$n_{A0} = -\mathscr{D}_{A2}^{(t)} \left(\frac{\partial \rho_{A2}}{\partial y'}\right)\bigg|_{y'=0} \tag{5.2-10}$$

The concentration gradient at the bottom of the lake is obtained from Eq. 5.2-6 at $y' = 0$ and used in Eq. 5.2-10 to give

$$n_{A0} = \sqrt{\frac{\mathscr{D}_{A2}^{(t)}}{\pi t}} (\rho_{A2}^* - \rho_{A2}^0) \tag{5.2-11}$$

This equation gives the instantaneous flux rate of nutrient A into the water phase. A more useful result is the average flux rate for the time period t, obtained by integrating Eq. 5.2-11 to obtain

$$n_{A0} = 2\sqrt{\frac{\mathscr{D}_{A2}^{(t)}}{\pi t}} (\rho_{A2}^* - \rho_{A2}^0) \tag{5.2-12}$$

This equation is an expression for the gross rate of nutrient A entering the water through the sediment–water interface for the period t. The rate at which nutrient A arrives in the epilimnion and the photoactive zone of the lake is less. The latter rates are given by Eqs. 5.2-7 and 5.2-9, respectively.

The fraction of nutrient A leaving the sediments that then enter the epilimnion is significant because only this amount can stimulate algae growth. This fraction for case I is defined

$$\phi_{\text{I}} \equiv \frac{n_{Ay}}{n_{A0}} \tag{5.2-13}$$

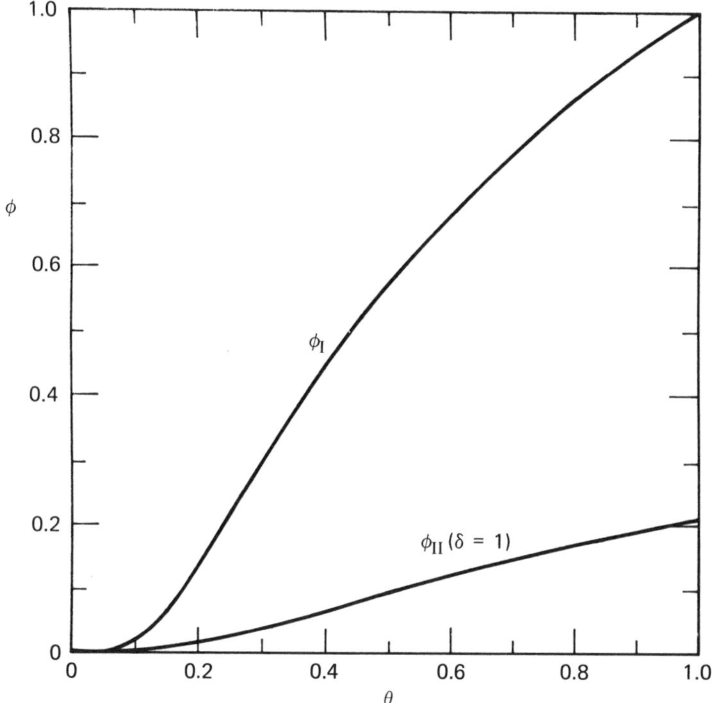

Figure 5.2-2. Fractions of upsurging nutrients in epilimnion (ϕ_I) and at surface (ϕ_{II}). (Reprinted by permission of American Water Resources Association, Minneapolis, Minn, from Ref. 18.)

Assuming that $\mathscr{D}_{A2}^{(t)} = \alpha_2^{(t)}$, Eq. 5.2-7 is time averaged to obtain n_{Ay}. The integration must be performed by numerical methods. It is necessary to define time $\theta = t/t_c$, where $t_c \equiv h^2/2\mathscr{D}_{A2}^{(t)}$ is a *characteristic time constant* of the particular lake. Figure 5.2-2 is a plot of ϕ_I versus θ. The upper curve, labeled ϕ_I, gives the fraction of upsurging chemical A that passes to the epilimnion as a function of dimensionless time θ.

In a similar manner the fraction of nutrient A leaving the sediments that enter the photoactive zone of depth h_e for case II can be defined

$$\phi_{II} \equiv \frac{n_{Ay}}{n_{A0}} \tag{5.2-14}$$

Equation 5.2-9 is time averaged to obtained $\bar{n}_{Ay}$. Numerical values of ϕ_{II} appear in Fig. 5.2-3. For this case it is necessary to define a dimensionless distance $\delta \equiv (h - h_e)/h$, which is the fractional distance from the photoactive zone to the lake bottom. A value of $\delta = 1$ is the lake surface and also appears in Fig. 5.2-2. The nonvolatile nutrients cannot pass through the air–water interface; thus $\delta = 1$ is a limiting value presented for completeness only. Figure 5.2-3 is useful for estimating the fraction crossing any plane δ where $\delta < 1$.

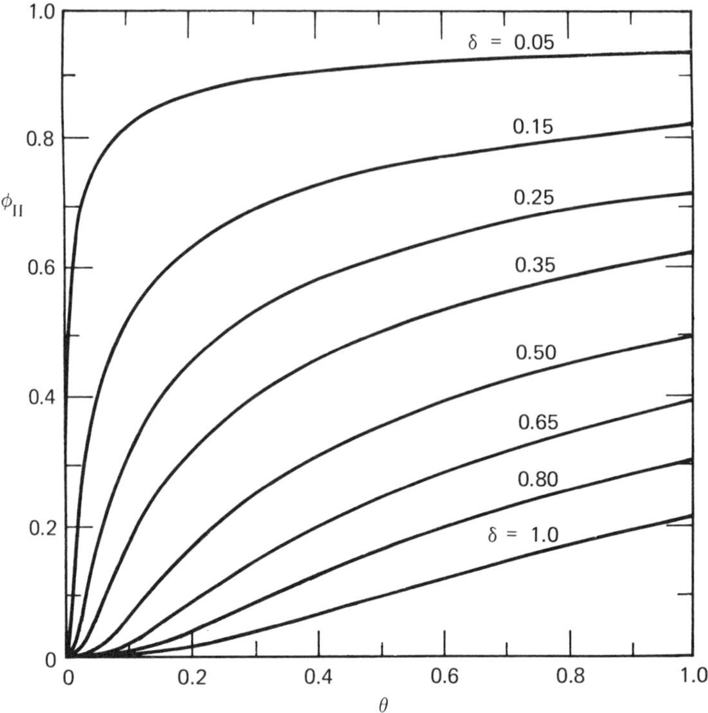

Figure 5.2-3. Fractions of upsurging nutrients in lake-water column (ϕ_{II}). (Reprinted by permission of American Water Resources Association, Minneapolis, Minn, from Ref. 18.)

Uses and Implications of Model. From the viewpoint of ecosystem modeling, the preceding development provides equations for estimating the rate at which bottom-originating chemical nutrients pass into upper regions of the lake water, such as the epilimnion. For this application Eqs. 5.2-7 and 5.2-9 should be used directly. In the interest of average flux rates for a period of time t, Eq. 5.2-12 modified by the fraction ϕ can be used:

$$n_{Ay} = 2\phi \sqrt{\frac{\mathscr{D}_{A2}^{(t)}}{\pi t}} \, (\rho_{A2}^{*} - \rho_{A2}^{0}) \tag{5.2-15}$$

Depending on the case, Fig. 5.2-2 or Fig. 5.2-3 is used to obtain ϕ.

The Fickian analysis results can also be used to help determine whether a fall bloom of algae is likely for a given lake based on the availability of bottom sediment–originating nutrients. The spring bloom of algae can deplete the lake surface water of a limiting nutrient, and an additional supply of this nutrient is needed if a fall bloom is to materialize. The depth of the lake h and the turbulent diffusivity $\mathscr{D}_{A2}^{(t)}$ are important lake characteristics regulating the time of travel of chemicals from the bottom sediments into the epilimnion. These two parameters are contained in the lake time constant, $t_c \equiv h^2/2\mathscr{D}_{A2}^{(t)}$.

If one assumes that 10% ($\phi_{\mathrm{I}} = 0.1$) of the limiting nutrient leaving the sediments is sufficient to stimulate increased algal growth in the epilimnion, then from Fig. 5.2-2, $\theta = 0.18$. According to this model, the time required for a significant quantity (i.e., 10%) of the bottom-originating limiting nutrient chemical to arrive at the epilimnion is

$$t_{\mathrm{I}} = \frac{0.18h^2}{2\mathscr{D}_{A2}^{(t)}} \tag{5.2-16}$$

In a similar fashion, if 10% ($\phi_{\mathrm{II}} = 0.1$) of the bottom-originating nutrient A travels 80% of the distance to the surface ($\delta = 0.8$), then from Fig. 5.2-3, $\theta = 0.35$. The model for case II gives the time of travel as

$$t_{\mathrm{II}} = \frac{0.35h^2}{2\mathscr{D}_{A2}^{(t)}} \tag{5.2-17}$$

It appears that less time is required for lakes with a well-developed thermal profile, but this is not the case. The effective turbulent diffusivity in stratified lakes is probably much smaller than in unstratified lakes. Also, stratified lakes are generally deeper.

The use of both Eqs. 5.2-16 and 5.2-17 gives a range of times after the vernal equinox in which the photoactive region of a lake begins to receive detectable quantities of nutrients from the sediment bed that may stimulate a fall bloom. If the time computed is less than or equal to 186 days, a fall bloom is likely. If the time is greater than 186 days, the declining length of the photoperiod and low water temperatures make a fall bloom unlikely.

Shagawa Lake: Comprehensive Example. Shagawa Lake is located in St. Louis County, Minnesota, near Ely. The lake is relatively small with a surface area of approximately 9.3 km^2, a mean depth of about 5.7 m, and a volume of about 53E6 m^3. There are three deep holes of about 13.7 m each, a gradually sloping bottom, and steep sides. A survey conducted in 1937 indicated that the bottom was 70% muck, 29% sand, and the remainder coarser soils.

The lake has probably been eutrophic for at least several decades, with local citizens commenting on excessive algae growths for over 40 years. The dominant source of phosphorus was the wastewater discharged by the city of Ely. An ideal opportunity existed to test the hypothesis that wastewater treated for phosphorus removal and allowed to enter a lake system would no longer exert its previous fertilizing influence, thus allowing the affected system to recover. The experiment began when the advanced wastewater plant went on stream in the spring of 1973. The plant was to remove 99% of the phosphorus entering. Only 19% of the phosphorus is entering the lake, compared to the period before the plant was built.

While the treatment plant was in operation, extensive field data were obtained to monitor the chemical, physical, and biota changes. With regard to phosphorus, the primary nutrient for algae growth, the bottom muds constitute the main source. The upsurge of this nutrient and others from the bottom muds may be observed by studying the field data.[19] (see Problem 5.2A). The lake thermal structure has an important influence on the nutrient upsurge concentration profiles. The following comments summarize a detailed study of the field data for Shagawa Lake:

Summary of the Temperature Profile Data

December, January, February, March. The lake is reverse stratified, with cold water above and warmer water below. Mixing should be mild because of the stability of the water column.

April, May. The lake is warming up. The density structure has been destroyed, and the lake is fairly isothermal. High winds in the spring keep the lake water fairly well mixed.

June, July, August. The lake continues to heat up. The lake is normally stratified with a warm water layer above colder water. Because of the stability of the water column, mixing below the thermocline should be at a low level.

September, October, November. The lake is cooling off. Plunging cold water created near the air–water interface increases mixing up and down the water column.

Summary of the total phosphorus concentration profile data

December, January, February, March. While the lake is reverse stratified, mixing of the water column is low. A "wave" of high concentration of phosphorus can be seen "upsurging" from the mud–water interface.

April, May. Mixing destroyed the phosphorus concentration profile of the preceding period. Concentration at all points along the water column is fairly constant.

June, July, August. The lake is normally stratified, and mixing has subsided. A "wave" of high concentration of phosphorus grows upward from the bottom again.

September, October, November. Plunging cold water from the surface causes mixing all along the water column. The total phosphorus concentration is once again fairly uniform.

Field data on orthophosphate, ammonia, nitrate, nitrite ions, and conductivity display the same development as total phosphorus. The preceding summary was for the year 1973. The data were almost identical for the year 1972. Not all lakes stratify twice during a year as Shagawa does. Lakes in southern regions of the United States may stratify only once. For these lakes the upsurge structure is simpler, usually consisting of two periods: one mixed and one unmixed.

Using Larsen's data, Thibodeaux[18] applied the Fickian upsurge model to Shagawa Lake. Equation 5.2-2 was used to determine $\mathscr{D}_{A2}^{(t)}$, turbulent diffusivity values. A total of 1463 observations of concentrations of ammonia, total phosphorus, and orthophosphate was used to calculate $\mathscr{D}_{A2}^{(t)}$ values. These data and other results from two other lakes are summarized in Table 5.2-1. Figure 5.2-4*a* contains typical profiles for ammonia concentration in Shagawa Lake. Figure 5.2-4*b* shows the profiles that result from application of the Fickian

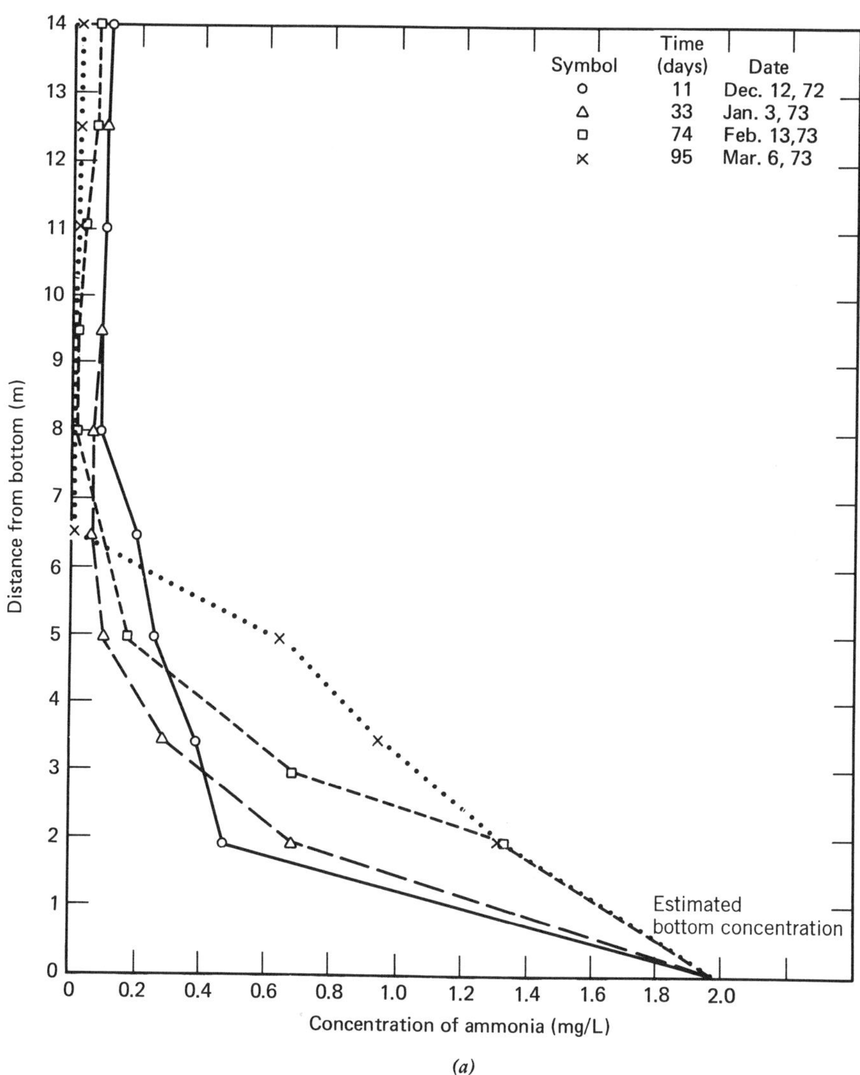

(*a*)

Figure 5.2-4. Shagawa Lake ammonia concentration station LBS winter 1972-1973: (*a*) field date; (*b*) calculated profiles (Reprinted by permission of American Water Resources Association, Minneapolis, Minn, from Ref. 18.)

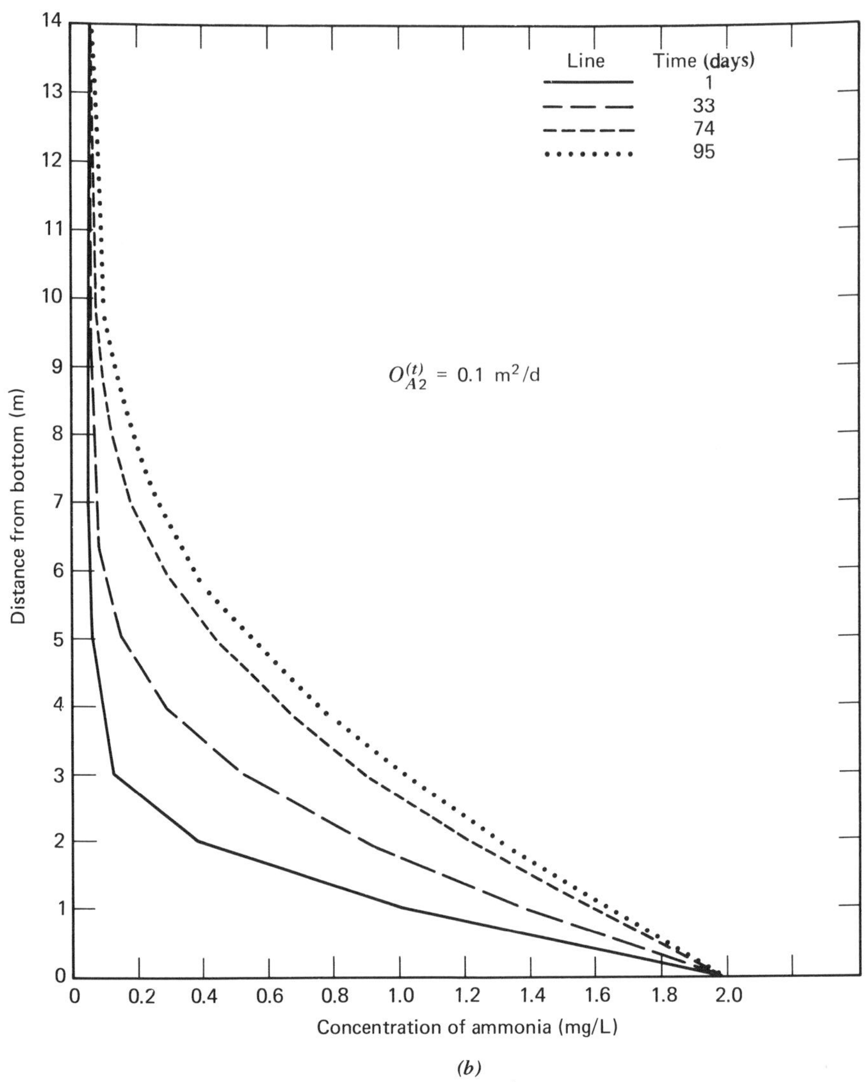

(b)

Figure 5.2-4. (*Continued*).

model to upsurging ammonia. See Problem 5.2A for an exercise in estimating $\mathscr{D}_{A2}^{(t)}$ from field data.

The Fickian model with its constant $\mathscr{D}_{A2}^{(t)}$ is a simplistic interpretation of a complex diffusion process. The mathematics are tractable, and the model serves as an easy introduction to and an instructional steppingstone to developing and understanding more sophisticated models. The next step in model development may be to use Eq. 3.2H-5 to replace the constant $\mathscr{D}_{A2}^{(t)}$ with

$$\mathscr{D}_{A2}^{(t)} = \kappa_1 (y + y_0) v_* \qquad (3.2H\text{-}5)$$

Table 5.2-1. General Lake Information

	Lake Shagawa	Lake Fayetteville	DeGray Reservoir
Location	Northern Minnesota	Northwest Arkansas	Southern Arkansas
Area (km^2)	9.3	4.2	54.3
Mean depth (m)	5.7	4.3	8.9
Maximum depth (m)	14.0	10.5	58.0
Volume (m^3)	53×10^6	3×10^6	794×10^6
Drainage basin (km^2)	110	—	1173
Eutrophic state	Eutrophic	Moderately eutrophic	Mesotrophic
$\mathscr{D}_{A2}^{(t)}$ (m^2/day)[a]	0.085	0.10	1.3
$O_{A2}^{(t)}$ range (m^2/day)[a]	0.022–0.17	0.014–0.453	$\sigma = 0.216$
$\alpha_2^{(t)}$ (m^2/day)	—	0.21	0.743
$\alpha_2^{(t)}$ range (m^2/day)	—	0.052–0.58	0.18–1.8
ρ_{A2}^* (mg/L)	0.52 total (P)	—	12.1 nitrate (N)
	0.36 Orthophosphate (P)	—	14.6 calcium (Ca)
	1.0 ammonia (N)	6.63–10.3 ammonia (N)	—
Julian day for $t = 0$	142 spring 335 fall	91	80

Source: Ref. 18.

[a]Turbulent diffusion coefficients are for the hypolimnion regions of the respective lakes.

or to assume that $\mathscr{D}_{A2}^{(t)}$ is one to two times $v_2^{(t)}$, as suggested by Bird et al.[53] However, if surface winds are assumed to be a primary reason for water movement, the preceding equation may be of little use. The surface winds change direction constantly; therefore, the bottom friction velocity v_* is also changing constantly. The change in direction creates turbulence itself. For this reason and other complexities that render lake turbulence a random process, a constant $\mathscr{D}_{A2}^{(t)}$ may be adequate for many purposes.

In the above concepts of chemical transport in the lake bottom-water region were developed using conventional pollutants, nitrogen and phosphorus. The theoretical approach is generally applicable to pesticides, chlorinated organics, hydrocarbons, metal species, and even nonsettleable colloids to and from the sediment–water interface. Because chemical transport here is driven by water turbulence, even in the quieter stratified periods, it is not usually necessary to attribute any specific molecular characteristics to $\mathscr{D}_{A2}^{(t)}$ Therefore, the values of $\mathscr{D}_{A2}^{(t)}$ given in Table 5.2-1, Section 7.1, and elsewhere[54] can be used for any and all species.

Bottom-Water Mass Transfer Coefficients

Average Annual Coefficient for Lakes. The average chemical flux rate for a period t from a surface of concentration ρ_{A2}^* to a water of background concentration ρ_{A2}^0 is given by Eq. 5.2-12. This equation is essentially the penetration theory result presented in Section 3.1. The turbulent diffusivity $\mathscr{D}_{A2}^{(t)}$ replaces the molecular diffusivity appearing in Eq. 3.1-21. In what follows, lake bottom-water characteristics will be used to develop an average, annual water-side mass transfer coefficient.

Beginning with the vernal equinox, a year in the life of a lake can be broken up into four time periods. Each period is either a stratified or an unstratified one of varying length. The sum of the four periods must be 365 days ($t_t = 365$ days):

$$t_s + t_f + t_w + t_v = t_t \tag{5.2-18}$$

where t_s is the summer stratified period, t_f the fall unstratified period, t_w the winter stratified period, and t_v the spring unstratified period. During the two unstratified periods the lake is assumed to turn over fairly rapidly, characterized by short time periods Δt on the order of days or weeks. Let Δt_f be the average time between turnovers for the fall period and Δt_v be the average time between turnovers for the spring period.

The average annual flux rate is determined by proportioning the flux rate of each period to the fraction of the year it occupies:

$$\bar{n}_{A0} = \sum_{i=1}^{4} n_{A0,i} \frac{\Delta t_i}{t_t} \tag{5.2-19}$$

Applying Eq. 5.2-12 to each time period, we obtain

$$\bar{n}_{A0} = \frac{2}{t_t\sqrt{\pi}} \left(\sqrt{\mathscr{D}_{A2,s}^{(t)} t_s} + \sqrt{\frac{\mathscr{D}_{A2,f}^{(t)} t_f^2}{\Delta t_f}} + \sqrt{\mathscr{D}_{A2,w}^{(t)} t_w} + \sqrt{\frac{\mathscr{D}_{A2,v}^{(t)} t_v^2}{\Delta t_v}} \right) (\rho_{A2}^* - \rho_{A2}^0) \tag{5.2-20}$$

Information about lake time periods and hypolimnion turbulent diffusivities is necessary to arrive at an average flux rate.

If it is assumed that a value of $\mathscr{D}_{A2}^{(t)}$ measured during the stratified periods is a fundamental lake parameter applicable to the mixed periods and that the high frequency of turnovers characterized by Δt is paramount in differentiating the mixed periods from the stratified periods, Eq. 5.2-20 can be simplified greatly and appear as

$$\bar{n}_{A0} = 2\sqrt{\frac{\mathscr{D}_{A2}^{(t)}}{\pi t_t}} \left(\sqrt{\frac{t_s}{t_t}} + \sqrt{\frac{t_f^2}{t_t \Delta t_f}} + \sqrt{\frac{t_w}{t_t}} + \sqrt{\frac{t_v^2}{t_t \Delta t_v}} \right) (\rho_{A2}^* - \rho_{A2}^0) \tag{5.2-21}$$

The average annual mass transfer coefficient is defined by

$$\bar{n}_{A0} = \overline{{}^{3}k'_{A2}}(\rho^{*}_{A2} - \rho^{0}_{A2}) \tag{5.2-22}$$

A further refinement of Eq. 5.2-21 occurs with the creation of the Nusselt number for mass transfer:

$$\text{Nu}_{A2} = 2\left(\sqrt{\frac{t_s}{t_t}} + \sqrt{\frac{t_f^2}{t_t \Delta t_f}} + \sqrt{\frac{t_w}{t_t}} + \sqrt{\frac{t_v^2}{t_t \Delta t_v}}\right) \tag{5.2-23}$$

where

$$\text{Nu}_{A2} \equiv \frac{\overline{{}^{3}k'_{A2}}}{\sqrt{\mathscr{D}^{(t)}_{A2}/\pi t_t}}$$

This equation shows that the average annual lake-bottom water-phase mass transfer coefficient is related to the duration of each stratified time period, each unstratified time period, and the respective Δt values within each unstratified period.

The relative importance of the various time periods may be studied by classifying lakes into two categories. The first category includes those lakes that have two stratified periods and two unstratified periods. For simplicity, the stratified periods are assumed to be of the same duration, $t_w = t_s$, and the unstratified periods are also assumed to be of the same duration, $t_v = t_f$, as are Δt_f and Δt_v, so that Eq. 5.2-18 becomes

$$2t_s + 2t_f = t_t \tag{5.2-24}$$

For category I lakes, Eq. 5.2-23 becomes

$$\text{Nu}_{A2,\text{I}} = 4\left[\sqrt{\frac{t_s}{t_t}} + \sqrt{\frac{(t_t - 2t_s)^2}{4t_t \Delta t}}\right] \tag{5.2-25}$$

The second category includes those lakes that have a single stratified period of duration t_s and a single unstratified period of duration t_w, for a total of

$$t_s + t_w = t_t \tag{5.2-26}$$

For category II lakes, Eq. 5.2-23 becomes

$$\text{Nu}_{A2,\text{II}} = 2\left[\sqrt{\frac{t_s}{t_t}} + \sqrt{\frac{(t_t - t_s)^2}{t_t \Delta t}}\right] \tag{5.2-27}$$

Table 5.2-2. Lake Bottom Nusselt Numbers

	$Nu_{A2,I}$		$Nu_{A2,II}$	
t_s/t_t	Δt = 7 days	Δt = 14 days	Δt = 7 days	Δt = 14 days
0	14.44	10.21	14.44	10.21
0.1	14.26	10.46	13.63	9.82
0.2	13.34	9.96	12.45	9.06
0.3	12.30	9.34	11.20	8.24
0.4	11.20	8.66	9.93	7.39
0.5	10.05	7.93	8.63	6.52
0.6	—	—	7.33	5.63
0.7	—	—	6.01	4.74
0.8	—	—	4.78	3.83
0.9	—	—	3.34	2.92
1.0	—	—	2.00	2.00

The Nusselt numbers contain only time ratios. It is now a simple task to choose reasonable values of the ratio t_s/t_t in order to study the effect of period duration and category variation on the annual average coefficient. Table 5.2-2 contains these calculations for Δt periods of 1 and 2 weeks.

Although Nusselt numbers are shown for a range $0 \leqslant t_s/t_t \leqslant 1$, it is unlikely that many lakes in the United States fall beyond the t_s/t_t range of (0.2, 0.5). Within this range the numbers are relatively constant, with a mean of 9.8 and a span of 13.34 to 6.52.

Example 5.2-1. Obtaining Estimates of Lake-Bottom Water Coefficient. Lorenz et al.[55] estimated the liquid-phase mass transfer coefficient for phosphorus ($A \equiv P$) at the bottom of Lake Washington to be ${}^3k'_{A2} = 36$ m/yr (see Example 3.3-1). Without any knowledge of the exact diffusivity or the seasonal periods for the lake, estimate the coefficient.

SOLUTION Without seasonal stratified and unstratified time periods, Eqs. 5.2-25 and 5.2-27 cannot be used, so the average Nusselt number of 9.8 is invoked. A turbulent diffusivity of 0.1 m^2/day is assumed. This value is reasonable and falls within the range of values reported in Table 3.1-4 for deep regions of lakes (0.001 to 1.0 cm^2/s).

$$\mathrm{Nu}_{A2} \equiv \frac{{}^3\bar{k}'_{A2}}{\sqrt{\mathscr{D}^{(t)}_{A2}/\pi t_t}} = 9.8$$

and solving for ${}^3\bar{k}'_{A2}$ yields

$${}^3\bar{k}'_{A2} = 9.8 \left(\frac{0.1\,\mathrm{m}^2}{\mathrm{day}} \left| \frac{1}{\pi} \right| \frac{1}{365\ \mathrm{days}} \right)^{1/2} \frac{365\ \mathrm{days}}{\mathrm{yr}} = 33.4\ \mathrm{m/yr}$$

The example above illustrates the use of the methodology in estimating bottom-water coefficients. Equation 5.1-20 is the more general result and it should be used when specific data are available to characterize the time periods and turbulent diffusivities.

In Chapter 7 of Lerman's book[48] he observes that the shape of a vertical concentration profile of a chemical species in water can yield information on its rate of transport and rates of chemical reactions producing or removing it. Also because of the complex nature of the chemical reactions involving nutrient species, they are not convenient tracers of fluxes from the sediments. He proposes the use of the natural radioactive isotope radon, ^{222}Rn, with half-life of 3.85 days, for obtaining the vertical eddy diffusivity, $\mathscr{D}^{(t)}_{A2}$, in bottom waters (see pages 215 and 406 for information on radon).

Radon forms in the sediment and diffuses into the water column, developing concentration profiles in freshwater lakes similar to those shown in Fig. 5.2-4. However, because of the decay of radon, a steady-state profile develops in the lower half of the water column. The one-dimensional transport model is

$$\mathscr{D}^{(t)}_{A2}\frac{d^2\rho_{A2}}{dy^2} - k'''_A\rho_{A2} = 0 \tag{5.2-28}$$

where k'''_A is the decay constant for radon. The solution to Eq. 5.2-28 is

$$\rho_{A2} = \rho_{A2i}\exp\left[-y\left(\frac{k'''_A}{\mathscr{D}^{(t)}_{A2}}\right)^{1/2}\right] \tag{5.2-29}$$

where ρ_{A2i} is a constant concentration at the lower boundary of the water column. This model can be used to measure the short-term variability of $\mathscr{D}^{(t)}_{A2}$ in water bodies and Lerman demonstrates the technique for a 1-month period. However, for water bodies that are mixed easily, the technique is not applicable because discernible concentration profiles do not have time to develop (see Problem 5.2I).

Wind-Enhanced Bottom-Water Coefficients for Unstratified Water Bodies.[57] In addition to the periods during the annual cycle of lakes, reservoirs, lagoons, and other natural water bodies, constructed wastewater impoundments are often unstratified. These contain hazardous substances that reside in the bottom sludge or sediment. When unstratified, the wind becomes a major factor in water mixing and hence regulating chemical transport across the water-side boundary layer.

In studying the impact of sediments as a nutrient source in shallow polluted lakes, Ryding and Forsberg[58] obtained evidence that the force and duration of the lengthwise wind in lakes were correlated with an increase in ammonia-nitrogen and total phosphorus in the overlying water. Lick[60] developed numerical models that are capable of realistically describing the currents throughout large lakes for the purpose of studying the dispersion of contaminants in near-shore areas. In modeling the wind-driven circulation in a lake, it was necessary to know the horizontal shear stress imposed as a boundary condition at the surface of the lake. Representative results of the calculations for Lake Ontario show the horizontal velocities at the surface due to wind of 5.2 m/s and velocities in a vertical and longitudinal plane consisting of the circulation of one large cell in the case of the one-layer model. Water velocities near the bottom were approximately 5 cm/s in a direction opposite to that of the wind.

Using a wind–water tank of the design of Cohen et al.,[59] laboratory measurements of ${}^3k'_{A2}$, the water-side coefficient defined by Eq. 5.2-22, were made.[57]. Water circulation pattern studies were performed to verify the general behavior reported by Lick[60] for the one-layer model. Realistic values of C_D, the aerodynamic drag coefficient defined by Eq. 4.2-13, were also measured in the tank. Benzoic acid in wafer form was the test chemical; it was placed on the sandy tank bottom during the experiment.

The mechanical energy balance for incompressible fluids was used to connect the wind stress at the air–water interface to the dissipation of energy as skin friction along the lakesides and bottom. The Fanning–Darcy equation was used for the friction loss in the hypothetical conduit where the equivalent length concept was used to relate depth, h and fetch, l. The circulating current was characterized by the hydraulic radius, which contains h and the lake width, w. This allowed the formulation of a simple relationship between the circulating water velocity, v, and the wind speed, v_1. Based in part on the experimental data and the well-known turbulent friction factor relationship for smooth pipe flow ${}^3k'_{A2}$ was assumed proportional to $v^{7/4}$. A semitheoretical correlation relating the coefficient to the key lake variables was obtained containing b, the one adjustable parameter:

$$
{}^3k'_{A2} = \frac{bC_D(\rho_1/\rho_2)v_1^2h^2}{lM_A^{1/2}} \tag{5.2-30}
$$

Three sets of experiments with v_1 (m/s), h (m), and l (m) held constant while the others were varied, yielded the same value of b. With $b = 18.9$ in Eq. 5.2-30 ${}^3k'_{A2}$ is in m/s. The drag coefficient $C_D = 0.00166$ for wind of 1 to 7 m/s and 0.00237 for wind 4 to 12 m/s. If land wind velocity is used multiply by 1.5 to get equivalent over-water wind velocity.

The experimental conditions were not far removed from lake conditions with respect to water velocity, which is a primary variable. Lick[60] reports unpublished water current measurements taken in Lake Erie during the unstratified period of 1964. During May and October, 24-hr resultant currents at 9.8 m below the surface ranged from 5 to 20 cm/s in a 10.1 m/s wind which was "fairly steady" for two or more days. At 14.9 m below the surface the currents ranged from 5 to 18 m/s. Most measurements were made in the central and eastern basins, which had average depths of 13.3 and 24.4 m, respectively.

Table 5.2-3. Lake Bottom-Water Coefficients, Unstratified

Lake	Wind (m/s)	Station	Depth (m)	Fetch (m)	${}^{3}k'_{A2}$ for P (cm/h)	
					Calculated	Observed
Erie	10.1	On site	18.3	161,000	0.083	—
Washington	4.1[a]	Seattle	38.0	11,300	0.81	0.41
Warner	5.8[a]	Boston	1.6	592	0.80	0.38

Source: Ref. 57.

[a] Average annual over land.

Experiments in the wind–water tank for wind 2 to 6 m/s yielded water velocities of 0.2 to 6.2 cm/s. Table 5.2-3 contains the mass transfer coefficient for Lake Erie, central basin, using Eq. 5.2-30. Few "field observations" of this coefficient are available. Those for Lorenz, et al.[55] for Lake Washington and DiGiano and Snow[61] for Lake Warner appear in Table 5.2-3 along with the estimated values using Eq. 5.2-30.

An interesting feature of Eq. 5.2-30 is the relationship of depth, h, and fetch, l, to the coefficient. Both the theoretical equation and the laboratory-scale simulation data display this unexpected behavior. It is a counterintuitive result; the coefficient increases with depth and decreases with fetch! Roughly, it appears that h/l, the depth/fetch ratio, reflects the ease of water circulation against friction on the bottom and sides. Increased circulation implies increased bottom turbulence and the coefficient increases. Additional work is needed to verify this behavior. Studies are also needed on the effect of bottom geometry, wind direction, wind shifts, and so on, on the coefficient.

Natural Convection Coefficients for On-Bottom Chemcals. It frequently occurs that liquid chemicals come to rest on the bottom of water bodies in which the water flow is low or stagnant. Because of the almost complete absence of shear, between the bottom material and the fluid, turbulence is absent. In the absence of flow and/or mechanical turbulence generated by flow, there can remain a sizable mass transfer rate.

Two important cases of natural convection dissolution are considered. These cases correspond to the microenvironment in which the liquid exists on the bottom. First, quantities of liquid chemicals can come to rest on the bottom and occupy perched positions, such as the globs mentioned previously. Interfacial tension in the range 0.03 to 0.04 N/m for the liquids of interest will cause the liquid glob surface to rise 1 to 2 cm above the sand bottom. In the second case the liquid may fill a shallow depression, pothole, or sandwave. In this case the liquid chemical–water interface plane is the same as the bottom plane.

Material in the form of globs and drops resting quietly on the bottom dissolves at fairly rapid rates. A thin layer of dense fluid mixture, generated by solubility and molecular diffusion processes, forms on the top of the liquid chemical surface. Since the chemical is heavier than water, this fluid mixture is more dense than the surrounding water and tumbles down the edge of the glob. The movement of the dense fluid near the interface induces local water movement, which speeds up the processes. At steady state a dense layer of liquid is moving radially from the center of the glob toward the edge. Fresh, lower-density water is pulled down from above, and the process continues in this fairly rapid fashion until the glob "“weathers” down and does not protrude above the bottom plane. Although the flow is laminar for the most part, a small amount of density-induced turbulence may be observed around the edge of the glob.

Dissolution experiments were performed with furfural in circular glob shapes.[8] Pan diameters of 5.0, 7.0, and 10.1 cm were used. Regression of the data in dimensionless number form yielded the following equation for the coefficient:

$$^{3}k'_{A2} = \frac{9.22(\mathrm{Gr}_{A2} \cdot \mathrm{Sc}_{A2})^{1/6} \mathscr{D}_{A2}}{d} \qquad (5.2\text{-}31)$$

where d, the characteristic length in Gr_{A2}, is the glob diameter. The dimensionless numbers are defined in Table 3.3-2 (see Problem 5.2D with regard to the Grashoft number).

If the liquid chemical fills the bottom depressions so that the chemical–water interface coincides with the bottom plane, there is no edge effect and the process of dissolution may be slower. Experimental observations have been made in the areas of heat transfer and mass transfer that are directly applicable to this microenvironment.

Fujii and Imura[10] observed and quantified a similar phenomena involving natural convection heat transfer from plates in water. In the case of a heated horizontal plate facing downward, they observed that the boundary layer was entirely laminar. This experiment corresponds very closely to one involving a heavy liquid in water. One major difference is that the liquid chemical can and will respond to the laminar flow stress at the interface. Although the liquid can flow, the solid heated plate cannot; however, this may be negligible. The heat transfer data were well correlated for both a 5- and a 30-cm heated plate by the equation

$$\mathrm{Nu} = 0.58(\mathrm{Gr} \cdot \mathrm{Pr})^{1/5} \qquad (5.2\text{-}32)$$

for $10^6 \leqslant \mathrm{Gr} \cdot \mathrm{Pr} \leqslant 10^{11}$. One may use the analogy theories to obtain a Nusselt (or Sherwood) number for mass transfer. Eq. 5.2D may be used to estimate the Grashoff number.)

Chang et al.[11] performed a series of natural convection dissolution experiments with salt samples approximately 1 m long in water and brine. The salt block was tested facing upward, vertically, and downward. The upward-facing position corresponds exactly to the dissolution process of a dense chemical on the bottom of a water body. Unfortunately, no upward-facing data beyond a 60° angle from the vertical were obtained. For this angle the correlating equation was

$$\mathrm{Sh}_{A2} = a(\mathrm{Gr}_{A2} \cdot \mathrm{Sc}_{A2})^{1/3} \tag{5.2-33}$$

and a ranged from 0.05 to 0.11. This equation will probably give high estimates of the natural convection mass transfer coefficient associated with a bottom-residing chemical, the interface of which coincides with the bottom plane.

Some Sediment-Side Chemodynamic Processes in Quiescent Water Bodies

The upper bed-sediment layers in quiescent waters enjoy a number of chemical transport mechanisms. They involve both particles and solution and are molecular diffusion, advection, bed accretion, colloid transport, and bioturbation. Molecular diffusion and solute advection have been covered in the earlier section on streams. Since the identical processes occur in bed sediment of lakes, ponds, and so on, that material applies directly. The others will be presented next, followed by some important coupled processes.

Bed Accretion. Lakes and similar quiescent bodies are usually net sediment traps in that they receive solid particles from streams and direct runoff from the surrounding lands. Most of this material enters as suspended particles or bedload. The term *bed accretion* is used to describe the general process by which layers of bottom sediment are built upward with time. It includes the individual processes of deposition, burial, and particle consolidation.

The process of particle settling from the water column and arriving at the sediment–water interface is termed *deposition*. Upon entering the lake proper the water velocity decreases and most particles cannot remain suspended in the milder turbulence that exist. Coagulation of small particles produces the light and fluffy material that settles much slower than the grainy pieces.

Concentration of a certain substance, grain size, faunal composition or water content, and so on, represents a property of a sediment.[63] At the time of deposition, contaminant content ω_A, for example, will vary laterally and vertically. Upon burial, vertical variations in concentration will also arise either by historical changes at the site of deposition or by diagenesis. *Diagenesis* refers to the sum total of processes that bring about changes in a sediment subsequent to deposition in water. The processes may be physical, chemical, and biological in nature and may occur at any time subsequent to the arrival of a particle at the sediment–water interface. Contaminant processes in sediment, according to Berner's[63] definition, is an early diagenesis process. Some other examples are the compactive dewatering of clay-muds, the destruction of lamination by burrowing benthic organisms, the diffusion of dissolved salts in lake beds, the bacterial decomposition of organic matter, and the formation of concretions.

The sediment–water interface is complex and not always a definite plane that separates sand particles and water. Formica et al.[62] noted during slicing (0.01-cm. increments) frozen sediment samples spiked with ^{14}C-labeled PCBs that the interface was a diffuse region, quite unlike the sharp demarcation existing at the air–water interface.

> Slicing commenced in frozen clear water and progressed to the true sediment underneath. Changes in color from translucent ice to opaque brown sediment, along with changes in the physical resistance encountered by the knife and visual graininess of the sliced material on the blade revealed three separate zones and a fairly precise demarcation of the true solid/particle interface.
>
> The three zones encountered were water, a flocculent layer, and consolidated solid particles. What appeared to be a thin flocculent layer was observed before true solid particles were encountered. It was obviously more porous and a different physical composition than the underlying sediment. Zero depth was defined upon encountering a slice that contained entirely solid particles. In all cases, this was also the point of maximum ^{14}C concentration except where physical evidence indicated it was deeper.
>
> The *flocculent layer* contained ^{14}C material in high concentration. It can be speculated that if this is an organic laden colloidal material, normally characterized by DOC, it is highly sorptive of the PCBs and should display such behavior. The origin of this material is either the result of fine particles settling out of the water column or the result of substances excreted from the sediment. It varied in thickness and was not present in all sample cups. This suggests sediment bioactivity may be the origin, since a few larvae type organisms appeared to be alive in the free sediment bed.

Here we deal with chemical concentration on particles, ω_A, as though it were a function of depth and time only. For sediments undergoing deposition we adopt the sediment–water interface as the origin; thus depth, y, will be measured positively downward from this plane. With this convention and using the total derivative to denote changes "following the motion" of the layer, we have

$$\frac{d\omega_A}{dt} = \frac{\partial \omega_A}{\partial t} + v_3 \frac{\partial \omega_A}{\partial y} \tag{5.2-34}$$

in which v_3 is visualized as the rate the sediment–water interface moves upward, which in the absence of complicating factors such as compaction, and non-steady-state deposition is simply the rate of deposition and of solids burial. In the case of lake sediment, if diagenesis is absent and total particle mass density replaces ω_A, Eq. 5.2-34 yields

$$v_3 = \frac{dy}{dt} \tag{5.2-35}$$

So that the increase in mass with time divided by the increase in mass over depth is the sedimentation rate, v_3

The phrase *rate of sedimentation* is often thought of in terms of a flux of sediment particles to the sediment surface and not as dy/dt as we have used it here. It is also the burial rate illustrated in Figure 5.1-6. The net flux of mass to the bed of rivers is given by Eq. 5.1-28; in the case where v_3 is known and it is viewed as the volume of sediment added to each unit area of sediment–water interface per unit time,

$$j_{3y} = v_3 \rho_3 (1 - \varepsilon) \tag{5.2-36}$$

is the flux of solids. Note that this flux holds for each depth whether or not there is compaction.[63] Under ideal conditions clean particles deposited on the bed form a natural cap retarding contaminant transport to the water column.

Perhaps the most common technique used in sedimentation rate determination in the United States is based on the rapid rise of ragweed pollen (*Ambrosia*) following extensive clear-cutting and soil cultivation in the latter part of the last century.[64] The most recent techniques use the natural radioisotope ^{210}Pb and the atomic bomb fallout nuclide ^{137}Cs. The relatively short-lived isotopes (22.3 years for ^{210}Pb and 30 years for ^{137}Cs) are ideally suited for lake sediments whose rates of sedimentation are on the order of a few millimeters per year. These events serve as layer markers in the sediment horizons. The depth below the surface, dy, divided by the lapse time, dt, yields a measured value, v_3, assumed constant over the period.

Compaction is defined as the loss of water from a layer of sediment due to compression arising from the deposition of overlying sediment.[63] Without compaction there may be no water advection from the sediment. During early diagenesis this entails the closer packing together of solid particles, with consequent expulsion of porewater and exhibited mainly in fine, grained muds. Total porosity, ε, introduced earlier, represents all the void spaces since they are usually interconnected in bed sediment and water may flow freely.

Advection and compaction are related. For surficial sands compaction is negligible, and in this case porosity does not change with burial. The sediment is buried as a solid body at a rate at which new material is added at the sediment–water interface. The porewater is also buried at the same rate. So that for no compaction only with an externally impressed flow is there a porewater velocity with respect to a fixed coordinate system. An impressed flow would be the sublake discharge of water percolating upward through the bed, for example.

For surficial muds with a high clay content compaction is not negligible, and in this case porosity does change with burial. Here steady-state compaction is a useful concept.[63] This is the common situation where sedimentation rate is constant and porosity at a fixed depth is constant. Even without an impressed flow there is a net upward water velocity through the sediment–

water interface. Porosity tends to diminish continuously until great depths are reached. Roughly, in surficial sediment, sands have a porosity of 0.3 to 0.5, silts 0.5 to 0.65, and clays 0.65 to 0.9 (see Berner[63] for more details).

Colloid Transport. Macromolecules and particulate matter exist in natural waters. Table 5.2-4 lists types and approximate sizes. Most of the material exist as a dispersed, insoluble phase, classically referred to as *hydrophobic colloids*. In addition, there are macromolecules in true solution, the *hydrophilic colloids*. In some lakes the particulate matter is dominated by algae; in others, riverine particles and/or sediment resuspensions may be important.[65] The subject of colloidal organic matter was introduced on page 272 in relation to particle behavior in the water column of streams. It was noted that the bed was one source of these colloids.

The concentration of organic colloids ($\equiv$C) in water, ρ_{C2}, is measured as dissolved organic carbon (DOC). The particulate fraction, washed from a bed-sediment sample, is fractionated by a series of filters to obtain the DOC size fractions. A distribution is shown in Table 5.2-5. Since the size of these particles is much smaller than the pore openings in the bed ($\gtrsim 1\ \mu$m), they are subject to Brownian diffusion. Transport within the bed and through the sediment–water interface can occur provided that the particle surface chemistry is such that there is no attraction (i.e., electrical charge) with the fixed solid phase. It appears that most of these particles bear a pH-dependent negative charge. Being repulsed, because of like charges on the surface of the fixed solid material, colloids are free to move about in the porewater.

Table 5.2-4. Colloidal Matter Suspended in Natural Waters

Type	Approximate size range (μm)
Organic macromolecules—humic substances, polysaccharides (10,000 to 100,000 mol. wt.)	0.001–0.01
Viruses	0.01–0.1
Iron oxides, aluminum oxides; clay minerals (e.g., illite, kaolinite, montmorillonite)	0.01–1.0
Sewage particles	0.1–100
Biologically precipitated MnO_2, bacteria (living or dead)	1.0–10
Detrital SiO_2 (quartz)	1.0–100
Soil aggregates	1.0–1000
Algae (living or dead), SiO_2 biologically precipitated	10–100
$CaCO_3$, chemically and biologically precipitated	10–1000
Fecal pellets, zooplankton	100–1000

Source: Ref. 65.

Table 5.2-5. Colloid Size Fractions and Diffusivities at 25°C

Size Fraction (μm)	DOC Concentration (mg/L)	Diffusivity, $\mathscr{D}_{C2}$ (cm^2/s)
0.05–0.4	14.0	2.84E-8
0.015–0.05	4.3	1.97E-7
0.003–0.015	18.3	7.11E-7
0.0013–0.003	12.3	2.98E-6
0–0.0013	5.0	9.85E-6
0–0.4	53.9 (sum)	1.86E-6 (average)

Source: Ref. 43.

Colloid transport in porous media is modeled as a Fickian process. The net flux as carbon in the y-direction is

$$n_{Cy} = v_y \rho_{C2} - D_{C3} \frac{d\rho_{C2}}{dy} \tag{5.2-37}$$

where the first term on the right hand side is the advective transport component. D_{C3} is the effective colloid diffusion coefficient in the bed. It appears that colloids behave similarly to molecules in solution so that Eqs. 5.1-32 to 5.1-34 are appropriate for estimating D_{C3}.[43] The Stokes–Einstein, Eq. 3.1-27, should be used to estimate $\mathscr{D}_{C2}$ values. Table 5.2-5 contains Brownian diffusivities based on the average particle radius. There appears to be limited experimental data of effective diffusivities of colloids in bed-sediment contexts. Thoma et al.[43] made such measurements and observed that the smaller size particles dominate the diffusivity.

Natural organic colloids are one of the many forms of natural organic matter (NOM) to which hydrophobic organic chemicals and trace metals adsorb. The trace metals include Cu, Zn, Cd, and Pb, among others. In solution these exist as ions and complexes with low-molecular-weight organics. Oxides and NOM are the strongest adsorbents for both HOCs and metals. Since these substances are present at low concentrations, their adsorption has little effect on the particle surface chemistry.[65] Instead, the significance of the adsorption is the ability of colloids to enhance the transport of the species. In the case of HOCs, if the partitioning between phases is fast, the local equilibrium assumption is justified. We can write

$$\omega_{AC} = K^*_{AC2} \rho_{A2} \tag{5.2-38}$$

where ω_{AC} is the amount of A sorbed to the colloids in units of g A/g DOC. K^*_{AC2} is the partition coefficient based on NOM with units of cm^3/g. It is further assumed that the colloidal particles consist of the same NOM as in sediments and soils, so that those correlations can be used to estimate K^*_{AC2} values (see, e.g., Eqs. 2.1-24 and 2.1-26).

For the fraction of species A sorbed to the particles the net flux is $n_{Ay} = n_{Cy}\omega_{AC}$ and with Eq. 5.2-37 can be written as

$$n_{Ay} = v_y \rho_{C2} \omega_{AC} - D_{C3} \frac{d(\rho_{C2}\omega_{AC})}{dy} \tag{5.2-39}$$

For the total flux of species A, the solution fraction must be combined with the particle-bound fraction. For those substances with high K^*_{AC2} values, the particle-bound fraction will dominate contaminant transport[43,67]. As in all previous cases, information on the water flow velocity and direction, v_y, is needed to account for the advective colloid flux contribution within or through the bed. Brownawell[67] plus Gschwend et al.[68] employed the foregoing approach to model colloid-mediated contaminant transport in bed sediment.

Bioturbation. This term was introduced in the discussion "Particles and Porous Media" in Section 3.3. The word implies turbulence or mixing brought about by the activity of organisms, specifically particles and porewater in the upper layers (ca. 10 cm.) of the bed. A literature review performed by Lee and Swartz[69] summarizes the organism–sediment interactions and compile rates of various processes, emphasizing differences in the rates and mechanisms of sediment modification among groups of functionally similar species. The review contains a brief description of vertical sediment stratification chemistry within which bioturbation exist.

The sediment system can be divided into three layers: (1) the surface oxidized layer, (2) a deeper reduced layer, and (3) a transition layer. The surface layer contains free O_2, CO_2, $SO_4{}^{2-}$, $NO_3{}^-$, and ferric oxides, which impart a yellowish-brown color to the sediment. Generally, free sulfides (H_2S, HS^-) are rapidly oxidized in this zone. Redox potential (Eh) is positive and ranges from about +400 to +100 mV. Due to the lack of O_2 there is a significant reduction in bioturbation below this layer. Deeper sediments constitute an anaerobic environment termed the sulfide system. Anaerobic heterotrophs produce a variety of reduced end products of which hydrogen sulfide is the most important, but which also include NH_3, CH_4 and simple organic compounds. Ion sulfide gives the sediment its characteristic grey or black color. Redox potential ranges from about −100 to −250 mV. The interface between the oxidized and reduced habitats is the redox potential discontinuity layer. It is characterized by a rapid decrease in Eh with depth. Both O_2 and H_2S are present in this layer and Eh values are typically +100 to −100 mV.

Bioturbation can modify patterns of horizontal stratification and affect the transport of particles, their sorbed ions, organic molecules, and so on, and entrapped porewater in both the horizontal and vertical directions. The most important activities producing bioturbation are feeding, burrowing, excavating, tube construction, and irrigation. It is convenient to subclassify bioturbation into three types that generally reflect particle and fluid dynamics produced by the benthic organisms: (1) plowlike, (2) conveyor belt, and (3) tube irrigation.

In plowlike bioturbation the animals mix the sediment by dragging their bodies on the surface or burrowing in from the top. Individual reworking rates range from 0.5 mg to 100 g per day and the mixing depths range from 1 mm to more than about 20 cm. The activities result in replacing the surface bed grains with material from below. Lee and Swartz[69] have defined these reworking rates and depths to facilitate comparisons and have converted and compiled numerous original literature sources into useful tables of data. The definitions are: particle reworking rates for individual species = $w'_{3i}(M/t \cdot \text{individual})$, depth of the reworked layer $= h^{(t)}$ (L), and total reworking rates for the population $= \Sigma_i^n w'_{3i} N_i (M/L^2 \cdot t)$, where N_i is the population density per unit surface area (number of individuals/L^2) and n is the number of active species. The data are for both plowlike and conveyor-belt types. Therefore, once the species are identified and their population known, it becomes possible to estimate the mixing depth and working rates. Reworking rates are strongly temperature dependent, and therefore yearly values are usually overestimates; nevertheless, the values are sufficiently accurate for first-order approximations. Existing data indicate the biodiffusion coefficient increases by a factor of 2 for each 10° rise in temperature[63]; this yields the temperature-dependent function $f_T = 1\text{E}[(T - 20°)/33]$ with T in °C.

Conveyor-belt bioturbation is caused primarily by tubular worms ingesting sediment below the surface and ejecting fecal pellets onto the surface. For example, tubificid Oligochaetes ingest silt and clay particles at 6 to 10 cm and egest it as pellets. This changes the sediment physically, usually adding organic matter. This process of vertically oriented deposit feeders can cover a polluted surface layer with pellets and protect it from erosion. With time the polluted surface gradually sinks and eventually reaches the feeding depth, where it is once again ingested and returned to the surface. Over several cycles the pollutant is homogenized within the reworked layer of depth $h^{(t)}$. Particles of larger, unsuitable size ($\geqslant 1$ mm) are not moved, creating a lag deposit, and graded bedding appears where the types are abundant.[69]

Tube-irrigation bioturbation involves primarily porewater. Thalassinid shrimp, also called "mud" or "ghost" shrimp, are tube builders. The tubes are relatively fixed domiciles, the more permanent being secreted or constructed from sediment selected from the surface. In this the excavators resemble conveyor-belt types. While occupying the tube the organism enhances porewater movement, hence the term *irrigation.* The vertical cylindrical burrows are somewhat uniformly spaced and flushed so rapidly that the water has the same composition as that near the sediment–water interface. Between burrows dissolved or colloidal substances can migrate horizontally by molecular diffusion to the nearest tube. Significant quantities of sedimentary particles are not moved and tube-irrigation bioturbation is therefore different from the others in this respect.

A complete theoretical description of diagenesis in the zone of bioturbation is extremely difficult. Not only are solutes migrating and particles altered by organisms, but these change with time due to seasonal temperature fluctuations. In addition, sporadic stirring disturbances by wave and current action (in shallow water) may occur to complicate the picture further. Nevertheless, theoretical approaches are needed if only as first attempts at describing, quantitatively, the multifaceted aspects of early diagenesis within this zone. Dimensional analysis has been invoked to connect reworking rates and depths to a common theoretical transport parameter.[72,73] A Fickian-type turbulent particle diffusivity, $\mathscr{D}_3^{(t)}$ commonly called the *biodiffusion* or *biogenic mixing coefficient*, so created is

$$\mathscr{D}_3^{(t)} \equiv \frac{\Sigma_{i=1}^{n} w'_{3i} N_i h_i^{(t)}}{\rho_3} \tag{5.2-40}$$

for particles with units of L^2/t. In the case of porewater the term is $\mathscr{D}_2^{(t)}$ and is typically 5 to 10 times larger than $\mathscr{D}_3^{(t)}$. It is termed an *irrigation biodiffusion coefficient* or *fluid biodiffusivity*. Where both types are appropriate in quantifying plowlike and conveyor-belt bioturbation only the $\mathscr{D}_2^{(t)}$ is appropriate for the tube-irrigation type. Aller[70,71] developed a detailed radial molecular diffusion model for ammonia transport between completely flushed vertical cylindrical burrows. He estimated that $\mathscr{D}_2^{(t)}$ in the y-direction perpendicular to the sediment–water interface was four times the effective molecular diffusivity in the radial direction, the latter being quantified by Eq. 5.1-34.

Biological activity and ocean bottom currents mix the surface layer of deep-sea sediments to a depth of the order of 10 cm. Guinasso and Schink[27] describe the biological mixing in deep-sea sediments in terms of a time-dependent eddy diffusion model where mixing takes place to a depth h at a constant eddy diffusivity $\mathscr{D}_3^{(t)}$. Microtektite data indicate that abyssal sediments are mixed from the surface to a maximum mixing depth that ranges between 17 and 40 cm below the surface. Estimates of $\mathscr{D}_3^{(t)}$ based on dimensional analysis of sediment reworking rates for near shore organisms (1 to 1E3 cm^2/yr) are used to predict abyssal mixing rates between 1E-3 and 1 cm^2/yr by invoking the assumption that mixing is proportional to benthic biomass.

There is a depth dependence on these coefficients since biological activity decreases with depth; however, a constant value is often used. On occasion a higher value may be used in the layers very near the interface, thereby creating a two-layer model. Below the bioturbed depth, $h^{(t)}$, molecular diffusion is assumed to be a significant process of transport.

To this point the subject of bioturbation has been concerned with particles and porewater. Using the two turbulent diffusivities defined above, the flux of species of A in the bed is

$$n_{Ay} = v_y(\rho_{A2} + \rho_{C2}\omega_{AC}) - (1-\varepsilon)\mathscr{D}_3^{(t)} \frac{d(\omega_A \rho_3)}{dy} - \varepsilon \mathscr{D}_2^{(t)} d(\rho_{A2} + \rho_{C2}\omega_{AC}) \tag{5.2-41}$$

Table 5.2-6. Biodiffusion Coefficients and Depths

Location	Coefficient, $\mathscr{D}_3^{(t)}$ (cm^2/s)	Depth, $h^{(t)}$ (cm)	Method[a]
Solid Particle Mixing			
Deep sea, various sites	3.2E-11 to 3.2E-8	10 to 48	DA
Mid-Atlantic Ridge	6E-9	8	^{210}Pb pattern
Long Island Sound	1.2 to 3.5E-6	4	^{234}Th pattern
Chesapeake Bay	1E-6	10 to 15	DA
New York Bight	5E-7	?	^{234}Th pattern
Rhode Island, 0 to 1 cm	29 to 1.6E-5	1	DA
Rhode Island, 2 to 10 cm	83 to 4.3E-6	8	DA
La Jolla, CA	1.5E-5	30	DA
Barnstable Harbor, MA	7.6E-8	6	DA
Long Island Sound	3.2E-7	2	DA
Long Island Sound	2E-6	3	DA
Laboratory	1E-5	3	Porewater
Laboratory	20 to 4.5 E-5	11	Porewater
Freshwater lake, mud	4.4E-8	0 to 6	?
Lake Huron, mud	1.2E-7	0 to 3/6	?
Porewater Mixing			
Long Island Sound, mud	>2.8E-5	0 to 8	?
Coastal North Sea, mud	E-4	0 to 3.5	?
Coastal North Sea, sand	0.5 to 2E-4	0 to >15	?
Narragansett Bay, mud	4E-5	0 to 25	?

Source: Refs. 56 and 69.

[a] DA, Dimensional analysis; Eq. 5.2-40.

where ω_A is the concentration on particles, ρ_{A2} that for solute in porewater, and $\rho_{C2}\omega_{AC}$ that for the colloids in porewater, all at depth $y \leqslant h^{(t)}$. Various values of $\mathscr{D}_3^{(t)}$ obtained by Eq. 5.2-40 in the field and in the laboratory appear in Table 5.2-6 summarized from Lee and Swartz[69] and Billen.[56] The original literature sources should be consulted for the species type.

Example 5.2-2. Steady-State Flux Within Bed by Bioturbation.[74] At a site in New Bedford Harbor the Aroclor 1242 depth–concentration profile is $(y, \omega_A) = (0.5, 1300; 1.5, 1100; 2.5, 1800; 3.5, 2900; 5.0, 3300; 7.0, 5600; 9.0, 6300)$ with y in cm and ω_A in μg/g sediment. For a 50% bed porosity, 2.5 g/cm^3 particle density, $\mathscr{D}_3^{(t)} = 10\,\text{cm}^2/\text{yr}$ over a 10-cm. bioturbed depth, and $K^*_{A32} = 7920$ L/kg, estimate the in-bed flux in g/m$^2\cdot$yr. The solubility of A-1242 in seawater is estimated to be $\rho^*_{A2} = 0.088$ mg/L.

SOLUTION Equation 5.3-41 applies. If solid/solute equilibrium partitioning is assumed at all depths, the porewater is saturated with A-1242, so that without concentration gradients for solute and colloids in porewater, only the particle-bound flux can be computed. The approximate gradient is $(6000 - 1200)/(8 - 1) = 686\,\mu\text{g/g}\cdot\text{cm}$. The flux assuming no water advection is

$$n_A = \frac{10\ \text{cm}^2}{\text{y}}\left|\frac{1.25\ \text{g}}{\text{cm}^3}\right|\frac{686\ \mu\text{g}}{\text{g}\cdot\text{cm}} = 8575\ \mu\text{g/cm}^2\cdot\text{yr} = 86\ \text{g/m}^2\cdot\text{yr}$$

Comments: If the concentration profile is well characterized, the assumed value of $\mathscr{D}_3^{(t)}$ is critical to the flux estimate, and in this case its relative error is estimated to be $\pm 6/10$. In reality, since the mass of PCB is finite, the flux is transient in the long time period (e.g., decades) and therefore decreases with time. However, for a period of time, in which the gradient can be assumed constant (e.g., months), quasi-steady-state conditions exist and the flux is constant.

Important Coupled Processes

Diagenetic Equation and Special Cases. The continuity equation for species A, Eq. 3.0-3, where ρ_{Aj} is the concentration on the solid or liquid phase, $j = 2$ or 3, in mass per unit volume, relates the three major processes affecting the fate of A within the bed: diffusion, advection, and diagenetic reaction. *Diagenesis* refers to the sum total of processes that bring about changes in sediments subsequent to deposition in water. Berner[63] notes that it is the most general expression for diagenesis and is designated in the field as the general diagenetic equation. For the case of linear, reversible, equilibrium adsorption/desorption one dimension (i.e., y-component) biodiffusion of solids and porewater, molecular and Brownian diffusion, porewater advection, depositional burial of solids plus porewater and reactions producing A, the transient form is

$$\begin{aligned}
&\frac{\partial}{\partial t}[(1-\varepsilon)\rho_3\omega_A + \varepsilon\rho_{A2} + \varepsilon\rho_{C2}\omega_{AC}] \\
&\quad = \frac{\partial}{\partial y}\left\{\mathscr{D}_3^{(t)}\frac{\partial}{\partial y}[(1-\varepsilon)\rho_3\omega_A]\right\} + \frac{\partial}{\partial y}\left[\varepsilon\mathscr{D}_2^{(t)}\frac{\partial\rho_{A2}}{\partial y} + \varepsilon\mathscr{D}_2^{(t)}\frac{\partial(\rho_{C2}\omega_{AC})}{\partial y}\right] \\
&\quad + \frac{\partial}{\partial y}\left[\frac{\varepsilon\mathscr{D}_{A2}}{\tau_h}\frac{\partial\rho_{A2}}{\partial y} + \frac{\varepsilon\mathscr{D}_{C2}}{\tau_h}\frac{\partial(\rho_{C2}\omega_{AC})}{\partial y}\right] \\
&\quad - \frac{\partial}{\partial y}[(1-\varepsilon)\rho_3 v_3\omega_A] \\
&\quad - \frac{\partial}{\partial y}[\varepsilon v_y(\rho_{A2} + \rho_{C2}\omega_{AC}) + \varepsilon v_3(\rho_{A2} + \rho_{C2}\omega_{AC})] \\
&\quad + \sum \varepsilon r_{A2} + \sum (1-\varepsilon)\rho_3 r_{A3}
\end{aligned} \tag{5.2-42}$$

This equation can be simplified to obtain useful special cases. For the case of linear, reversible equilibrium adsorption or desorption of A between solids and porewater, constant diffusion coefficients, constant velocities, constant solid properties, and constant colloid concentration, Eq. 5.2-42 becomes

$$[K_3 + (1 + K_C)]\frac{\partial \rho_{A2}}{\partial t} = \begin{bmatrix} K_3\mathscr{D}_3^{(t)} \\ (1 + K_C)\mathscr{D}_2^{(t)} \\ (\mathscr{D}_{A2} + \mathscr{D}_{C2}K_C)/\tau_h \end{bmatrix}\frac{\partial^2 \rho_{A2}}{\partial y^2} - \begin{bmatrix} K_3 v_3 \\ (1 + K_C)v_3 \\ (1 + K_C)v_y \end{bmatrix}\frac{\partial \rho_{A2}}{\partial y} + \sum r_{A2} + \frac{(1 - \varepsilon)}{\varepsilon}\rho_3 \sum r_{A3} \tag{5.2-43}$$

where $K_3 \equiv (1 - \varepsilon)\rho_3 K^*_{A32}/\varepsilon$ and $K_C \equiv \rho_{C2}K^*_{AC2}$. Dividing each term by $K_3 + (1 + K_C)$ allows a relative test of the effect chemical partitioning onto solids has on each process that effects the fate of substance A in the bed. Doing so indicates that fluid biodiffusion, molecular and Brownian diffusion, porewater advection, and reactions producing substance are less effective in relieving the burden of A within the bed than do biodiffusion of solids and depositional burial. This occurs because K_3 is large in comparison to K_C and other parameters in Eq. 5.2-43.

Transient Transport Within Beds. In the absence of biodiffusion, colloids (i.e., $\rho_{C2} = 0$), reactions producing A, porewater advection, particle deposition, and burial Eq. 5.2-43 reduces to

$$\frac{\partial \rho_{A2}}{\partial t} = D\frac{\partial^2 \rho_{A2}}{\partial y^2} \tag{5.2-44}$$

where

$$D \equiv D_{A3}/\varepsilon R_{A3} \tag{5.2-45}$$

Note that $R_{A3} \equiv 1 + (1 - \varepsilon)\rho_3 K^*_{A32}/\varepsilon$ is the *retardation factor* and $D_{A3} \equiv \mathscr{D}_{A2}\varepsilon/\tau_h$ is the effective diffusion coefficient of A in the bed.

The case of transient chemical transfer to and from the sediment–water interface has many applications and the semi-infinite media solution to Fick's second law has some useful formulations for estimating the flux. It has been used numerous times in this book, but some degree of caution must be used with Eq. 5.2-44 when applying existing solutions, developed for pure diffusive processes. These solutions (see Crank[77] for such solutions) should be used cautiously for retarded diffusion processes, that is, diffusion accompanied by sorption or desorption from particles. It is tempting to use diffusion-only solutions such as Eqs. 5.2-2 and 5.2-11. In this case $D_{A3}/R_{A3}\varepsilon$ can be substituted for $\mathscr{D}_{A2}^{(t)}$ in the concentration profile equation, Eq. 5.2-2, and it correctly describes the movement of A movement within the bed. It is incorrect, however, to make the substitution into Eq. 5.2-11 for the flux. For A entering or leaving the bed the correct flux expression is[62]

$$n_{A0} = \sqrt{\frac{D_{A3}R_{A3}\varepsilon}{\pi t}}\,(\rho_{A2}^{*} - \rho_{A2}^{0}) \tag{5.2-46}$$

Formica et al.[62] have demonstrated that an equation analogous to Eq. 5.2-2 is appropriate for retarded molecular diffusion of PCBs into lake bed sediment. The student is challenged to resolve Fick's second law for the retarded-diffusion case and derive Eq. 5.2-46.

Due to the slow water exchange near the surface, the water side may provide a significant chemical transport resistance; here the term *film-diffusion control* is used. Crank[77] presents the pure-diffusion solution to Fick's second law for a semi-infinite medium. It cannot be used for retarded diffusion because a flux boundary condition contains the same D as in the differential equation. As presented above, this is inappropriate.

The student should note that there are theoretical problems associated with modeling interphase chemical transport with particle bioturbation as a diffusive process characterized by $\mathscr{D}_3^{(t)}$. In the case of transport from bed to water a phase change must occur as sorbed A arrives on a particle and emerges on the other side of the sediment–water interface as a species in solution. Interphase chemical transport of this sort using coefficients obtained from the movement of insoluble radioactive species on particles (i.e., ^{210}Pb, ^{234}Th, etc.) or organism mixing rates (i.e., Eq. 5.2-40) with gradients of concentration on solids yielding the flux expression

$$n_{A0} = -\mathscr{D}_3^{(t)}\rho_3(1-\varepsilon)\left.\frac{d\omega_A}{dy}\right|_{y=0} \tag{5.2-47}$$

has not been fully demonstrated, but it has been[74] and is being employed.

Mathematically, particle bioturbation chemical transport is like pure diffusion, as can be seen by neglecting all others in Eqs. 5.2-43. So for a semi-infinite sediment without surface resistance Eqs. 5.2-2 and 5.2-11 are appropriate for particle bioturbation, replacing $\mathscr{D}_{A2}^{(t)}$ with $\mathscr{D}_3^{(t)}$ and ρ_{A2} with equivalent $\omega_A\rho_3(1-\varepsilon)$. For the case with surface resistance Crank[77] gives a solution. If the concentration in the semi-infinite bed is initially ω_A^0 throughout and the bed-to-water surface exchange is determined by

$$-\mathscr{D}_3^{(t)}\rho_3(1-\varepsilon)\left.\frac{\partial\omega_A}{\partial y}\right|_{y=0} = \left(\frac{{}^3k'_{A2}}{K_{A32}^{*}}\right)(\omega_{Ai} - \omega_A^{*}) \tag{5.2-48}$$

where ω_A^* is the equivalent concentration for ρ_{A2} in the water column and ω_{Ai} equivalent to the interface at ρ_{A2i}, the solution is

$$\frac{\omega_A - \omega_A^0}{\omega_A^* - \omega_A^0} = \operatorname{erfc}\left(\frac{y}{\sqrt{4Dt}}\right) - \exp(hy + h^2Dt)\operatorname{erfc}\left(\frac{y}{\sqrt{4Dt}} + h\sqrt{Dt}\right) \tag{5.2-49}$$

where $D \equiv \mathscr{D}_3^{(t)}\rho_3(1 - \varepsilon)$ and $h \equiv {}^3k'_{A2}/K^*_{A32}\mathscr{D}_3^{(t)}\rho_3(1 - \varepsilon)$. The flux of A from the bed, assuming that $\omega_A^0 > \omega_A^*$, is

$$n_A = \frac{{}^3k'_{A2}}{K^*_{A32}}(\omega_{Ai} - \omega_A^*) \tag{5.2-50}$$

the value of ω_{Ai} is obtained from Eq. 5.2-49 by setting $y = 0$.

Steady-State Flux and Resistances in Series. Bed-residing contaminants encounter a series of resistances before emerging into the water column. Some of these were described in Section 5.1, but modification is needed because bioturbation chemical transport usually involves the movement of particles within the bed. These operate in parallel and are:

1. Particles and porewater are transported in bulk from depth to the sediment–water interface.
2. The delivered porewaters with their solute and colloid load pass directly through the interface while the arriving particles appear there, residing briefly in the surface desorption zone, releasing a fraction of sorbed constituent A before being subducted once again into the depths of the bed.[76]
3. From the interface solute and colloids move through the diffusive sublayer and into the overlying turbulent water.

The release process mentioned in step 2 is discussed further in relation to processes 1 to 3 on page 280. Also, the physical injection of particles into the water column by benthic organisms has been observed and this undoubtedly contributes to the transport process.[78]

Although the desorption process was used as the illustration, the mechanisms also work for chemical transport in reverse, where it describes overall adsorption of solute from the water column, thereby contaminating a once clean sediment. This occurs in addition to contaminated particle deposition onto the bed, the common explanation for bed-sediment contamination.

The processes above can be represented as resistances in series for obtaining a steady-state flux. This approach is useful for the case in which the concentration of A, ω_A, is known at a single depth h within the bioturbated layer, as opposed to the case where the concentration profile is well established (see Example 5.2-2). Since porewater concentration is limited by solubility, chemical concentration on particles is used in the flux expressions that follow. An expression for the overall coefficient ${}^2K'_{A3}$ defined by

$$n_A = {}^2K'_{A3}\rho_3(1 - \varepsilon)(\omega_A - \omega_A^*) \tag{5.2-51}$$

is desired where ω_A^* is the concentration on particles equivalent to ρ_{A2} in the water column. The turbulent diffusive particle flux described in step 1 above is obtained assuming $\mathscr{D}_3^{(t)}$ and ρ_3 constant in Eq. 5.2-41 by integrating $d\omega_A/dy$ from $y = h$ to Δ:

$$n_A = \frac{\mathscr{D}_3^{(t)}\rho_3(1 - \varepsilon)(\omega_A - \omega_{A\Delta})}{h - \Delta} \tag{5.2-52}$$

Here Δ is the thickness of the sediment layer adjacent to the interface where A desorbs from the particles. Similarly, the particle process in step 2 is active over Δ to the interface with the same or a slightly larger $\mathscr{D}_3^{(t)}$, and the rate is

$$n_A = \frac{F_A\mathscr{D}_3^{(t)}\rho_3(1 - \varepsilon)(\omega_{A\Delta} - \omega_{Ai})}{\Delta} \tag{5.2-53}$$

where F_A accounts for the fraction desorbed. It is

$$F_A \equiv \frac{\omega_{A\Delta} - \omega_{Af}}{\omega_{A\Delta} - \omega_{Ai}} \tag{5.2-54}$$

defined as the ratio of the difference between the arriving level $\omega_{A\Delta}$ and the level after desorption, ω_{Af}, to the difference between the arriving level and the ultimate level of desorption for an infinite residence time at the interface, ω_{Ai}. The residence time of particles in the interface desorption layer, τ_i, is the mass in the layer divided by the mass rate of particles moving through the layer. The result is

$$\tau_i \equiv \frac{\Delta^2}{\mathscr{D}_3^{(t)}} \tag{5.2-55}$$

Estimating Δ is very problematic at this time. It is probably equal to some number of particle diameters, possibly one and as high as three or four, depending on particle packing and how freely water from above can circulate effectively in this zone. A value equal to two diameters has been suggested.[68] For a range of particle sizes the geometric mean gives a reasonable estimate for the diameter. The theory and quantitative relationships for determining F_A

based on physicochemical processes of sorption/desorption from particles are presented in the next section.

The flux equation through the water-side film, step 3, is Eq. 5.2-50. The overall resistance is the sum of the three individual resistances; it is

$$\frac{1}{{}^{2}K'_{A3}} = \frac{(h - \Delta) + \Delta/F_A}{\mathscr{D}_3^{(t)}} + \frac{K^*_{A32}\rho_3(1 - \varepsilon)}{{}^{3}k'_{A2}} \tag{5.2-56}$$

The parallel process for bioturbated porewater transport is similar to molecular diffusion, the resistances being

$$\frac{1}{{}^{3}K'_{A2}} = \frac{h}{\mathscr{D}_2^{(t)}\varepsilon} + \frac{1}{{}^{3}k'_{A2}} \tag{5.2-57}$$

for

$$n_A = {}^{3}K'_{A2}(\rho^*_{A2} - \rho_{A2}) \tag{5.2-58}$$

where h is the depth within the bed at which the porewater is saturated, concentration ρ^*_{A2}, and this depth may be less than that for simultaneous particle process. Obtaining the porewater gradient is problematic without a concentration profile measurement. By reversing the overall concentration differences in Eqs. 5.2-51 and 5.2-58, algorithms for the steady-state adsorption result.

Transient Sorption and Desorption from Particles. In step 2 above a process involving chemical desorption from particles residing on or very near the sediment–water interface was developed somewhat in the fashion of Karickhoff[76] and others.[68] The process of sorption–desorption from particles is a generic one for aquatic systems in that it also occurs with particles suspended in the water column as well as those deeper within the bed. The following quantitative description of this chemodynamic process follows generally the work of Jackman and Ng[75] and is applicable to all three situations.

They investigated the influence of external film diffusion and internal pore diffusion in controlling the kinetics of ion exchange of Sr adsorption replacing Ca on natural streambed sediments. Sized sediments displaying equilibrium sorption coefficients for Sr of 153 to 730 L/kg were exposed to a dilute Ca solution in a packed bed under flow conditions similar to those encountered near the surface of a streambed. Three transport and reaction models for a finite liquid volume systems were developed to interpret the experimental data; this led to a criterion that can be used to decide on the controlling mechanism. These models are presented here for the infinite-volume case, which is more useful for particles in the aquatic environment.

In all cases the particles are assumed to be spherical with radius r_0 and undergoing adsorption. In the film-plus-particle diffusion model two serial steps are involved in the physical transport: the diffusion of adsorbate from bulk fluid to the surface of the particles (referred to as film diffusion) and the diffusion of the adsorbate in the pores of the particles to the internal where actual reaction, exchange or adsorption, occurs. The spherical coordinates version of the single-component continuity equation that describes the fate of A in the particle is

$$\varepsilon \frac{\partial \bar{\rho}_{A2}}{\partial t} = \frac{1}{r^2} \frac{\partial}{\partial r}\left(D_{A3} r^2 \frac{\partial \bar{\rho}_{A2}}{\partial r}\right) + r_A \tag{5.2-59}$$

where overbars denote concentration in the fluid of the pores and r_A is the volumetric rate of production of A here. Treating ion exchange or adsorption as a reaction on the solid phase, the rate is of opposite sign since it leaves the fluid phase:

$$r_A = -\rho_3(1-\varepsilon) \frac{\partial \omega_A}{\partial t} \tag{5.2-60}$$

If we now assume that D_{A3} is constant and at every point within the particle the pore fluid is in equilibrium with the solid phase and follows a linear isotherm, the equations above can be combined to yield

$$[\rho_3(1-\varepsilon)K^*_{A32} + \varepsilon] \frac{\partial \bar{\rho}_{A2}}{\partial t} = \frac{D_{A3}}{r^2} \frac{\partial}{\partial r}\left(r^2 \frac{\partial \bar{\rho}_{A2}}{\partial r}\right) \tag{5.2-61}$$

a very familiar result. The boundary and initial conditions are

$$\frac{\partial \bar{\rho}_{A2}}{\partial r} = 0 \qquad \text{at } r = 0, \quad t \geqslant 0$$

$${}^3k'_{A2}(\rho_{A2} - \rho_{A2i}) = D_{A3} \frac{\partial \bar{\rho}_{A2}}{\partial r} \qquad \text{at } r = r_0, \quad t \geqslant 0 \qquad \text{(5.2-62a, b, and c)}$$

$$\bar{\rho}_{A2} = 0 \qquad \text{at } t = 0, \quad \text{all } r$$

where ${}^3k'_{A2}$ is the film coefficient, ρ_{A2} the adsorbate concentration in the bulk solution, and ρ_{A2i} the value at the particle surface. By rendering the equations and boundary conditions dimensionless, the Biot number for mass transfer appears:

$$Bi_A \equiv \frac{{}^3k'_{A2} r_0}{D_{A3}} \tag{5.2-63}$$

For those processes without reaction within the solid phase, it is the ratio of internal resistance to external resistance and serves as a general guide for two special case models. Crank[77] presents a concentration profile with time solution, $\bar{\rho}_{A2}(r, t)$, for this case. An integrated equation for the concentration of the total fraction of species A, F_A, entering on leaving the sphere is given. This fraction is

$$F_A \equiv \frac{\omega_A - \omega_A^0}{\omega_{A\infty} - \omega_A^0} \tag{5.2-64}$$

where ω_A^0 is the initial average concentration in the particle, ω_A is that at time t, and $\omega_{A\infty}$ is that at infinity. For the external concentration ρ_{A2}, $\omega_{A\infty} = K^*_{A32}\rho_{A2}$, the equilibrium value. Figure 5.2-5 shows curves of F_A plotted as a function of $\{D_{A3}t/r_0^2[\rho_3(1-\varepsilon)K^*_{A32} + \varepsilon]\}^{1/2}$ for several values of Bi_A and an infinite reservoir volume.

For small values of Bi_A most of the concentration difference is across the external film that controls the rate of adsorption. Since the internal process is relatively rapid, it is appropriate to use an average concentrations on the particle, ω_A, and in the pores, $\bar{\rho}_{A2}$. The mass balance is therefore

$$3k'_{A2}(\rho_{A2} - \rho_{A2i}) = \frac{r}{3}\frac{d}{dt}[\rho_3(1-\varepsilon)\omega_A + \varepsilon\bar{\rho}_{A2}] \tag{5.2-65}$$

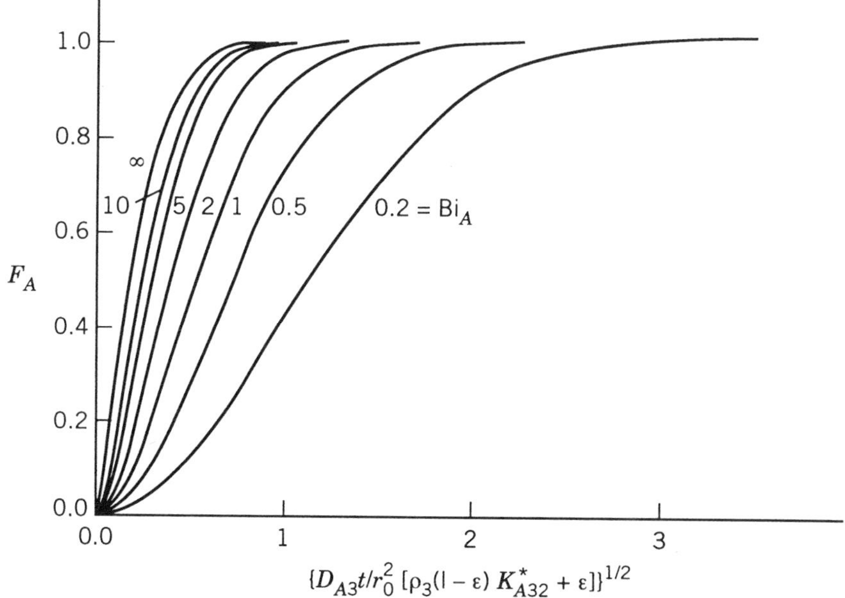

Figure 5.2-5. Sorption or desorption curves for the film-plus-particle diffusion model. (From Ref. 77.)

since concentrations are uniform within $\rho_{A2i} = \bar{\rho}_{A2}$. Converting all solution concentrations to these equivalent particle concentrations is done using $\omega_A = K^*_{A32}\rho_{A2}$. For a quantity ω^0_A initially in the particle the solution in terms of the fraction entering or leaving is

$$F_A = 1 - \exp\left\{\frac{-3^3 k'_{A2} t}{r_0[\rho_3(1-\varepsilon)K^*_{A32} + \varepsilon]}\right\} \tag{5.2-66}$$

This equation is called the *film-diffusion model*. It is also for a infinite volume of constant concentration ρ_{A2}.

For Bi_A values approaching infinity the concentration difference is within the particle and internal diffusion controls the rate of mass transfer. In this case we assume that the concentration within the particle porewater at the surface is equal to the concentration in the infinite fluid reservoir. Equation 5.2-62b is not required, being replaced with the boundary condition $\rho_{A2} = \bar{\rho}_{A2}$ at $r = r_0$. This case is termed the particle-diffusion model and Crank[77] gives a solution to this case also. In terms of the fraction A entering or leaving, the solution is

$$F_A = 1 - \frac{6}{\pi^2}\sum_{n=1}^{\infty}\frac{1}{n^2}\exp\left(\frac{-Dn^2\pi^2 t}{r_0^2}\right) \tag{5.2-67}$$

where $D \equiv D_{A3}/[\rho_3(1-\varepsilon)K^*_{A32} + \varepsilon]$. Figure 5.2-6 is a useful graphical solution in terms of the dimensionless parameter $\{D_{A3}t/r_0^2[\rho_3(1-\varepsilon)K^*_{A32} + \varepsilon]\}^{1/2}$ for an infinite volume of fluid.

For the case of particles in a finite volume of fluid, similar equations are needed coupled with a mass balance on A for the fluid reservoir. As noted above Jackman and Ng performed this type of experiment sorbing Sr to sediment. A numerical solution was required for the film-plus-particle diffusion model. Figure 5.2-7 shows an example of the correspondence between model and experimental data. It is seen that the film-diffusion model is valid for short periods of time while the particle-diffusion model is valid for long periods. The film-plus-particle diffusion model is valid throughout. Based on these and other data, they propose the following criteria for determining the appropriate model:

$$\gamma_{jn} \equiv \mathrm{Bi}_A[\rho_3(1-\varepsilon)K^*_{A32}]^{-1/3} \tag{5.2-68}$$

If $\gamma_{jn} < 0.5$, the film model is appropriate, and for values greater than 2 the particle model is appropriate, otherwise; the film-plus-particle model must be used. Wu and Gschwend[79] performed similar experiments with sediment suspensions using the more restrictive particle-diffusion model and obtained data for both adsorption and desorption of chlorinated benzenes.

An important parameter for estimating intraparticle diffusion coefficients is

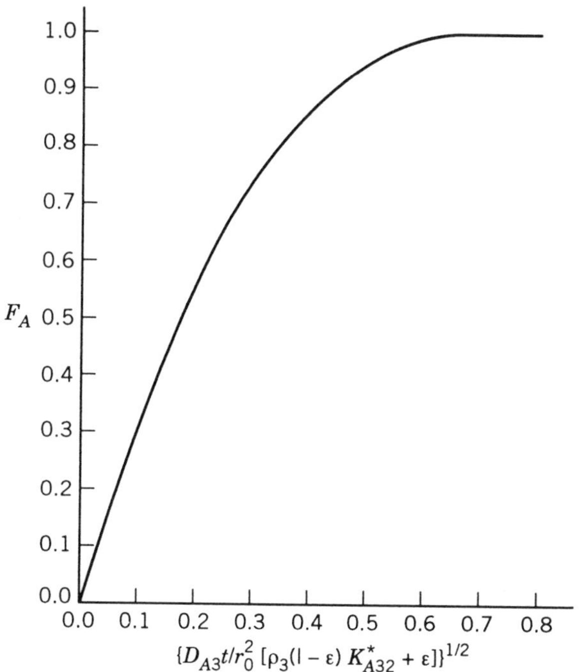

Figure 5.2-6. Sorption or desorption for the particle-diffusion model. (From Ref. 77.)

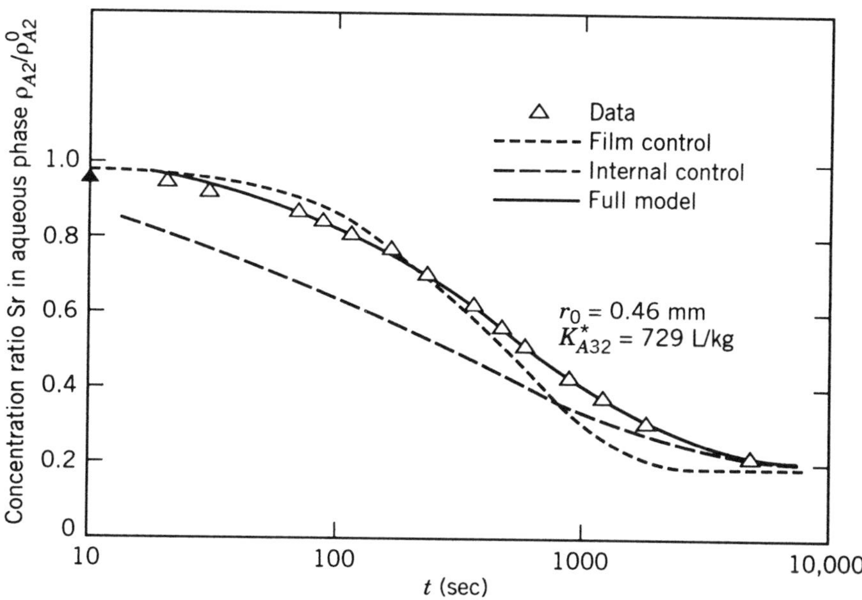

Figure 5.2-7. Data versus model for Sr adsorption on bed sediment. (Reprinted with permission. From, Ref. 75.)

Table 5.2-7. Measured Intraparticle Porosities

Material	Mean Particle Radius (mm)	ε	Comment
Cununurra clay soil	0.362	0.53	
	0.720	0.52	
Sepiolite	0.362	0.33	
	0.720	0.34	
Crushed porcelain	0.362	0.21	
	0.720	0.23	
Keweenaw 7 sandy loam	0.075	0.33	Micro. + macro. pores
Charles River sediment	0.088	0.18	Desorption expt.
		0.17	Sorption expt.
North River sediment	0.10	0.14	Desorption expt.
		0.08	Sorption expt.
Iowa soils	0.088	0.15	Sorption expt.
Range Point Marsh sediment	—	0.11	Sorption expt.
Little Lost Man Creek sediment	1.2	0.05	Porosimetry

Source: Refs. 47 and 75.

the internal porosity. Equation 5.1-34 is appropriate for porous media consisting of a bed of particles. The internal structure of individual particles is likely to be very different, but a similar form has been suggested

$$D_{A3} = \mathscr{D}_{A2}\varepsilon^2 \tag{5.2-69}$$

Medine and McCutcheon[47] have compiled the available measurements of intraparticle porosities. These appear in Table 5.2-7 along with the single value of Jackman and Ng[75], which is the last entry. Many additional data are needed on this important parameter plus effective diffusion coefficients for all types of suspended and bed particles.

Lake Water Column Coupled to Bed Sediment

As an illustration of these processes the following analysis is keyed to the release of phosphorus from the bed and its impact on lake water quality. Figure 5.2-8 is an exaggerated vertical scale schematic of a lake showing the interaction between water and sediments.

Unsteady-State Model. As shown earlier in this section, nitrogen and phosphorus are good examples of substances that reside in aquatic bed sediment. When released these enter the water column and move upward and eventually out. This same general process occurs with organics and metals.

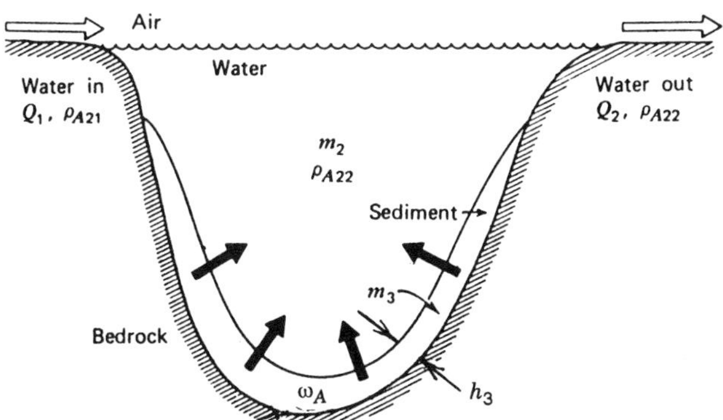

Figure 5.2-8. Lake-water and sediment system.

Indeed, the reverse process can occur whereby chemicals enter the lake and then move from the water column into the bed. This reverse process is undoubtedly how most bed sediment became contaminated initially.

The development and use of some simplistic lake and sediment models will help focus our attention on the chemical transport process. As shown, the minimum recovery time takes years, so the model time is long, and the annual cycle of events can be ignored. Assume that the lake water is of a uniform concentration ρ_{A22} and nutrient cycling does not occur. Assuming no nutrient recycling within the lake implies that when a chemical crosses the sediment–water interface, it eventually leaves the system in the overflow water. A component balance of chemical A on the lake water yields

$$Q_1\rho_{A21} - Q_2\rho_{A22} + n_{A0}A = \frac{d}{dt}(V\rho_{A22}) \tag{5.2-70}$$

where n_{A0} is the input rate from the sediment bed:

$$n_{A0} = {}^3K'_{A2}(\rho^*_{A2} - \rho_{A22}) \tag{5.2-71}$$

Equation 5.2-70 shows that the depletion of chemical A from the lake water is dependent on the input from the sediments as well as the input by hydraulic flow.

A component balance of chemical A on the mass of bottom sediments is expressed by

$$0 - An_{A0} = \frac{m_3 d\omega_A}{dt} \tag{5.2-72}$$

The equilibrium relationship between the sediments and the water can be expressed by

$$\rho^*_{A2} = \omega_A/K^*_{A32} \tag{5.2-73}$$

Assuming that ρ_{A22} is negligible, Eqs. 5.2-71, 5.2-72, and 5.2-73 are combined to give a differential equation, which can then be integrated to yield

$$t = \frac{m_3 K_{A32}^*}{{}^3K'_{A2}A} \ln \frac{\omega_A^0}{\omega_A} \tag{5.2-74}$$

If sediment cleansing of chemical A is assured at $\omega_A = 0.05\omega_A^0$, 95% removal, Eq. 5.2-74 becomes

$$t = \frac{3m_3 K_{A32}^*}{{}^3K'_{A2}A} \tag{5.2-75}$$

Equation 5.2-75 contains the important parameters that regulate sediment cleansing time and hence lake recovery time. See Example 5.2-3 for a calculation of the minimum recovery time of a lake sediment bed from contamination by a chemical.

Steady-State Model. Since recovery time could take centuries, in-lake water concentration of chemical A changes very slowly. It is now necessary to consider some results of a steady-state lake model. This model is also simplistic. Features include constant inflow, outflow, and sediment concentration. The only processes occurring are dissolution of a chemical A at the sediment–water interface and hydraulic dilution. The lake water is assumed to be completely mixed. A component balance on chemical A yields

$$Q_1 \rho_{A21} + {}^3K'_{A2}(\rho_{A2}^* - \rho_{A22})A = Q_2 \rho_{A22} \tag{5.2-76}$$

Substituting Eq. 5.2-73 for ρ_{A2}^* and solving for ρ_{A22} yields

$$\rho_{A22} = \frac{Q_1 \rho_{A21} + {}^3K'_{A2}A(\omega_A/K_{A32}^*)}{Q_2 + {}^3K'_{A2}A} \tag{5.2-77}$$

where $Q_1 \simeq Q_2$ is realistic. The in-lake or outflow concentration of chemical A is dependent on the inflow concentration ρ_{A21} and the concentration of A within the sediments. Elimination of A in the inflow may not significantly reduce ρ_{A22} because of the input from the sediment bed. As Q_1 and therefore Q_2 become small, the in-lake concentration approaches ω_A/K_{A32}^*, the equilibrium value.

Example 5.2-3. Minimum Cleansing Time of Chemical "Locked" in Lake Sediments. Table E5.2-3 contains data on the nutrient phosphorus (total P) within the sediments of the Upper Klamath Lake.[20] It appears that phosphorus tends to partition strongly and in a sense is locked into the sediment phase. The average partition coefficient at 10°C is 1640 and 540 at 23°C.

Table E5.2-3. Sediment–Interstitial Water Equilibrium[a]

Sample	Temperature (°C)	ω_A (μg P/g sed.)	ρ^*_{A2} (μg P/mL)	K^*_{A32} (L/kg)
1a	10	413	0.23	1800
1b	23	406	0.64	630
3a	10	793	1.66	477
3b	23	723	6.08	119
4a	10	504	0.19	2700
4b	23	499	0.58	860

Source: Ref. 20.
[a] $\rho_3 = 2.66$ g/cm^3.

Compute the minimum time for the natural cleansing of the lake sediment bed of phosphorus to occur (years). Assume a sediment phase mass transfer coefficient based on Eq. 5.1-40 of 0.63 m/yr and a sediment-bed depth of 2.0 cm and $\varepsilon = 0.5$.

SOLUTION The minimum time for natural cleansing can be estimated from Eq. 5.2-75. If the sediment bed is of uniform thickness, $m_3/A = h_3\rho_3$; Eq. 5.2-75 becomes

$$t = \frac{3h_3K^*_{A32}\rho_3(1-\varepsilon)}{{}^3K'_{A2}} \tag{E5.2-3}$$

The overall mass transfer coefficient is obtained from Eq. 5.1-43:

$$\frac{1}{{}^3K'_{A2}} = \frac{1}{{}^3k'_{A2}} + \frac{1}{{}^2k'_{A3}} \tag{5.1-43}$$

Using Example 5.2-1 as the source of the ${}^3k'_{A2}$ value, 33.4 m/yr, we obtain

$$\frac{1}{{}^3K'_{A2}} = 0.0299 + 1.6 = 1.63$$

The value ${}^3K'_{A2} = 0.61$ m/yr and the preceding calculation suggest that the sediment phase controls the rate of release of phosphorus. $K^*_{A32} = 1640$ since the bottom water is probably cold. Using Eq. E5.2-3 to calculate the minimum time yields

$$t = \frac{3|0.02\ \text{m}|1640|2.66|0.5}{0.61\ \text{m/yr}} = 215\ \text{yr}$$

It appears that the minimum cleansing time for phosphorus in this lake is on the order of centuries if nutrient recycling does not occur.

Attempts have been made at restoring entropic lakes toward an oligotropic state by reducing the inputs of nitrogen and phosphorus compounds. The case of Shagawa Lake on page 299 is an example. Most such attempts have proved futile because of the internal recycling process that occurs with these particular substances. This process is based on algae; it blooms, then dies, settling onto the bed. As shown in Example 5.2-3, even without recycling of phosphorus, the "cleansing" of lake sediment of refractory chemicals by natural flushing may involve centuries.

PROBLEMS

5.2A. Thermal and Phosphorus Profiles in Shagawa Lake

Water temperature and phosphorus concentrations for Shagawa Lake at sample station LBS for 1973 appear in Table 5.2A.

1. Make two graphs to visualize the seasonal temperature and phosphorus concentration developments. On one graph show temperature on the horizontal axis and depth on the vertical axis for each month. On another graph show phosphorus concentration on the horizontal axis and depth on the vertical axis for each month. Once the plots are made, confirm the description of the temperature profile and phosphorus concentration profile given on page 300.

2. Calculate turbulent diffusivity values representative of the reverse stratified period. Using Eq. 5.2-2, calculate $\mathscr{D}_{A2}^{(t)}$ values for January 3, February 13, and March 6. Choose $t = 0$ as December 10. Assume that $\rho_{A2}^{0} = 0.04$ mg/L and $\rho_{A2}^{*} = 0.25$ mg/L. Do not use data from 0 to 4.5 m. Report the answer in m^2/day.

3. Calculate turbulent diffusivity values representative of the normally stratified period. Using Eq. 5.2-2, calculate $\mathscr{D}_{A2}^{(t)}$ values for June 12, July 10, August 7, and September 11. Choose $t = 0$ as June 10. $\rho_{A2}^{0} = 0.03$ mg/L and $\rho_{A2}^{*} = 0.25$ mg/L. Exclude data from 0 to 4.5 m depth because algae may be active in this zone and reduce the phosphorus concentration. Report the answer in m^2/day. The depth to the lake bottom is 14 m.

5.2B. Phosphorus Upsurging from Sediments of Shagawa

The average quantity of phosphorus leaving the sediment–water interface and entering the photoactive zone of the lake for days 170 through 230 (mid-June to mid-August) was calculated to be 40 kg/day.[21]

Table 5.2A. Shagawa Lake Data[a]

	Jan. 3		Feb. 13		Mar. 6		Apr. 20	
y	T_2	ρ_{A2}	T_2	ρ_{A2}	T_2	ρ_{A2}	T_2	ρ_{A2}
0.1	1.0	0.053	—	0.048	1.1	0.052	5.1	0.058
1.5	1.1	0.050	1.5	0.048	1.5	0.053	5.0	0.059
3.0	2.5	0.052	2.5	0.052	2.4	0.061	4.9	0.059
4.5	3.0	0.043	—	0.047	2.8	0.049	—	0.059
6.0	3.2	0.043	3.5	0.048	3.2	0.043	4.9	0.063
7.5	4.0	0.047	3.8	0.051	3.9	0.059	4.9	0.063
9.0	4.5	0.049	4.2	0.089	4.0	0.130	4.9	0.064
10.5	4.9	0.085	4.5	0.145	4.0	0.154	4.9	0.059
12.0	5.0	0.162	4.5	0.192	4.1	0.143	4.9	0.061

	May 15		June 12		July 10		Aug. 7	
y	T_2	ρ_{A2}	T_2	ρ_{A2}	T_2	ρ_{A2}	T_2	ρ_{A2}
0.1	10.2	0.031	18.0	0.037	22.2	0.028	22.5	0.052
1.5	10.0	0.032	18.0	0.034	22.1	0.039	22.2	0.053
3.0	10.0	0.031	18.0	0.037	21.8	0.032	22.0	0.047
4.5	10.0	0.037	18.0	0.037	21.2	0.032	21.3	0.065
6.0	10.0	0.034	17.3	0.023	19.8	0.029	20.8	0.066
7.5	10.0	0.038	17.0	0.025	19.5	0.043	20.2	0.091
9.0	10.0	0.037	16.5	0.033	19.0	0.051	20.0	0.161
10.5	9.8	0.040	16.3	0.038	18.4	0.097	19.9	0.164
12.0	9.0	0.036	15.5	0.080	18.0	0.176	19.0	0.237

	Sept. 11		Oct. 10		Nov. 6		Dec. 18	
y	T_2	ρ_{A2}	T_2	ρ_{A2}	T_2	ρ_{A2}	T_2	ρ_{A2}
0.1	18.7	0.079	13.8	0.054	4.0	0.033	1.0	0.025
1.5	18.7	0.082	13.8	0.051	4.2	0.034	1.5	0.024
3.0	18.7	0.081	13.7	0.047	4.2	0.034	1.5	0.016
4.5	18.8	0.080	13.7	0.048	4.2	0.036	1.5	0.020
6.0	18.8	0.085	13.7	0.047	4.2	0.036	1.8	0.023
7.5	18.8	0.079	13.7	0.043	4.2	0.042	2.0	0.026
9.0	18.8	0.080	13.7	0.043	4.2	0.037	2.1	0.037
10.5	18.8	0.079	13.7	0.044	4.2	0.033	2.5	0.043
12.0	18.8	0.083	13.6	0.044	4.2	0.032	2.8	0.067

Source: Ref. 19.

[a] Depth (h) in meters; temperature (T_2) in °C; P concentration (ρ_{A2}) in mg/L.

1. Using Eq. 5.2-12, calculate the quantity of phosphorus leaving the sediment–water interface in kg/day.

2. Using Eq. 5.2-15, calculate the quantity of phosphorus entering the photoactive zone. Assume that Shagawa is a case I lake.

(Data: sediment–water interfacial area of 9.3 km^2, $\rho^*_{A2} = 0.36$ mg/L (laboratory analysis of interstitial water), $\rho^0_{A2} = 0.046$ mg/L (average lake surface water), $\mathscr{D}^{(t)}_{A2} = 0.085$ m^2/d, $h = 5.7$ m, $t = 230-170 = 60$ days.

5.2C. Likelihood of Fall Bloom

It has been observed in the field that both Shagawa Lake and Lake Fayetteville have indications of a fall bloom of algae. Use Eqs. 5.2-16 and 5.2-17 to confirm by calculation that a fall algae bloom is likely to occur on both lakes.

5.2D. Natural Convection Dissolution Mass Transfer Coefficients for Globs

Natural convection dissolution experiments like those described in Problem 5.1F were performed with globs consisting of pure furfural and chloroform. Using Eq. 5.2-31, 5.2-32, and 5.2-33, calculate the mass transfer coefficients and compare the values calculated with the experimental values.

1. Furfural: glob radius 6.5 cm, temperature 27.5°C, ${}^4k'_{A2}$ (experimental) 0.76 to 1.1 cm/h.

2. Chloroform: glob radius 7.05 cm, temperature 24.5°C, ${}^4k'_{A2}$ (experimental) 1.6 cm/h.

Use

$$\mathrm{Gr}_{A2} = \frac{d^3(\rho^*_{2A} - \rho_2)g}{\nu_2^2\bar{\rho}_2} \tag{5.2D}$$

where d is glob diameter and $\bar{\rho}_2 = (\rho^*_{A2} + \rho_2)/2$ for calculating the Grashoff number. Here ρ^*_{2A} is the density of water saturated with the chemical of interest.

5.2E. Sinker Spill in Lake

A bridge over a lake has prompted state authorities to study the consequences on water quality of chemical spills from tanker transport accidents. One study scenario involves the sinker perchloroethylene, where 10,000 kg occupies 100 ha of the bottom in globlike quantities.

1. Short-term analysis: Develop a model and estimate the maximum concentrations in the water column. Note that the lake is very large, deep, and strongly stratified in the summer.

2. Annual analysis: Develop a model to estimate the mass dissolved in the stratified period and in the unstratified period.

Table 5.2F. Chemical Concentrations in Buzzards Bay Sediment

Depth (cm)	DOC (mg/L)	TOC (%)	TON (%)	C/N Ratio	% H_2O
0–3	14.4	6.06	0.63	9.6	55.85
3–5	17.4	5.23	0.51	10.0	55.87
5–7	27.7	3.74	0.36	10.0	54.97
7–9	40.3	4.98	0.48	10.0	50.60
9–11	47.9	4.40	0.40	11.0	51.26
11–13	32.6	4.14	0.34	12.0	47.69

3. Long-term analysis: Develop a model to estimate the period of time the lake may be expected to be affected by this chemical.

4. Based in part on the results above, water quality is significantly compromised for its intended use. Propose a plan for dealing with such spills.

5.2F. Colloid Transport Within Bed Sediment

For Buzzard's Bay sediment Brownawell[67] reports the following concentration profiles, where DOC ≡ dissolved organic carbon, a measure of colloid concentration; TOC = total organic carbon; and TON = total organic nitrogen (Table 5.2F).

1. Choose an appropriate model to obtain an estimate of the first-order rate constant for colloid production within the bed. Brownawell estimates the particle biodiffusion coefficient to be 3.0E-7 cm^2/s for depths to 10 cm and 0.8E-7 cm^2/s for greater depths based on ^{210}Pb data. Assume that $v_3 = 0$.

2. Estimate the colloid flux to the water column.

3. Repeat parts 1 and 2 for a deposition velocity of 0.05 cm/yr. In all three cases explain and justify your model choice, beginning the analysis with the diagenetic equation. (Eq. 5.2-42).

5.2G. Transient Desorption from Semi-infinite Bed-Sediment Source

For years naphthalene from a chemical manufacturing facility has entered a small lake contaminating the bed to a level of 50 mg/kg. The source of input was virtually stopped in 1975. Estimate the mass of naphthalene reentering the water column to year 1992 and graph the concentration profile for:

1. Bioturbation from a semi-infinite source without a bed-surface film resistance.

2. Bioturbation from a semi-infinite source with a bed-surface film resistance.

Assume that the partition coefficient for naphthalene is 5.0 L/kg.

5.2H. Bed Contamination: Particle Deposition Versus Bioturbation

Contaminated particles settling onto the surface is an obvious mechanism for contaminating bed sediment. Compare it to the mechanism of particle reworking whereby benthic organisms transport clean particles from the depth to the surface, where they become exposed to water containing the contaminant in solution before being reinterred. Assume steady-state flux and that the lake water column is saturated with naphthalene. For the particles $K^*_{A32} = 5\ \mathrm{L/kg}$ and $v_3 = 0.05\ \mathrm{cm/yr}$. Assume that the particles are 100 μm in diameter.

5.2I. Water Column Concentration Profiles: Transient Versus Steady State

With increasing time the transient water column concentration profiles for a constant bed source of contamination change little if there is no loss by reaction. See, for example, the solution to Eq. 5.2-2 in Fig. 5.2-4 for $t = 75$ and 95 days. Considering the uncertainty in the data, one could wrongly assume that steady-state conditions exist and that Eq. 5.2-29 applies. Are the theoretical shapes similar? How can one avoid this trap?

5.2J. Sorption–Desorption: Film-Diffusion Model

For single spherical particles situated on or near the sediment–water interface bathed in gently flowing water, the outside film controls the transport rate.

1. Derive Eq. 5.2-66 assuming chemical adsorption.
2. For 2,3,7,8-TCDD, what value of surface coefficient is needed for a 30-μm-diameter particle if film diffusion controls the transport rate? Use $K^*_{A32} = 10{,}000\ \mathrm{L/kg}$.

5.3. CHEMICAL MOVEMENT AT BOTTOMS OF ESTUARIES AND OCEANS

Processes at the planetary boundaries control many of the properties of the atmosphere and the hydrosphere. The benthic boundary layer and the adjoining sediment layer are the least well known of these. The flux of various species across the sediment–water interface may be one of the more important processes controlling the chemical composition of seawater. The mechanism of the fluxes of various species is also important to the chemodynamics of anthropogenic substances that arrive for various reasons at this important environmental interface. In this section additional transport aspects of natural and synthetic chemicals in and above the ocean bottom sediments are presented. The processes of molecular diffusion, colloid diffusion, advection, deposition, bioturbation, and so on, presented in the preceding two sections apply here also.

The first section is concerned with the benthic boundary layer, those layers of water at the very bottom of the ocean. Next, chemical exchange processes within the sediment layer are presented. The final section is concerned with some important coupled phenomena.

Movement of Chemicals Through Benthic Boundary Layer

A chemical substance leaving the seafloor enters water where fluxes are controlled by molecular diffusion and passes into a region where fluxes are controlled by turbulent motions. Specifying the nature of the transition zone becomes one of specifying $\mathscr{D}_{A2}$ as a function of height above the bottom. It has been suggested that the nonturbulent boundary layer at the seafloor could substantially reduce chemical fluxes across this zone.

To model the effect of the benthic boundary layer on the diffusion of a dissolved chemical species, the analysis must extend into the water column above the sediments. Here the concentration of chemical A at steady state is described by

$$\frac{\partial}{\partial y}\left(\mathscr{D}_{A2}\frac{\partial c_{A2}}{\partial y}\right) = 0 \tag{5.3-1}$$

where $\mathscr{D}_{A2}$ is the coefficient of diffusivity specified as a function of height above the bottom. Descending into the boundary region, transport mechanisms are transformed from those dominated by turbulent processes to those dominated by molecular motion. The transformation is not a simple one. Turbulence is created by the drag of the seafloor on water derived by deep-sea currents. This turbulence mixes the water overlying the seabed, but the turbulent eddies are damped and become progressively less effective as we approach the interface. This system is further complicated by periodic variations in the velocity of deep-sea currents, so the regime is constantly changing its character.

Wimbush[22] distinguishes two regions in the vicinity of the seafloor: (1) just above the interface is the viscous sublayer, and (2) transitional to the sea above is the logarithmic layer. These regions are illustrated in Fig. 5.3-1. The viscous sublayer is distinguished from the zone above in that turbulent transfer of momentum has become less than molecular transfer of momentum. This does not require that turbulent fluxes be nonexistent, only that they be much less effective. In seawater (at 2°C) the thermal diffusivity, α_2 (1.4E-3 cm^2/s), and chemical molecular diffusivity, $\mathscr{D}_{A2}$ (5E-6 cm^2/s), are much less than the kinematic viscosity, ν_2 (1.7E-2 cm^2/s). Momentum transport in the viscous sublayer is dominated by viscosity, and hypothetically, according to the film theory, turbulent transport of heat and chemical tracers can still be accomplished by turbulent processes. Viscosity, then, becomes the major transmitter of momentum at a boundary layer thickness far greater than the hypothetical thickness of the layer in which molecular diffusion processes dominate turbulent diffusion. The diffusion sublayer, shown in Fig. 5.3-1, is therefore much thinner than the viscous sublayer (see Eq. 3.3C-1 and Fig. 3.3c).

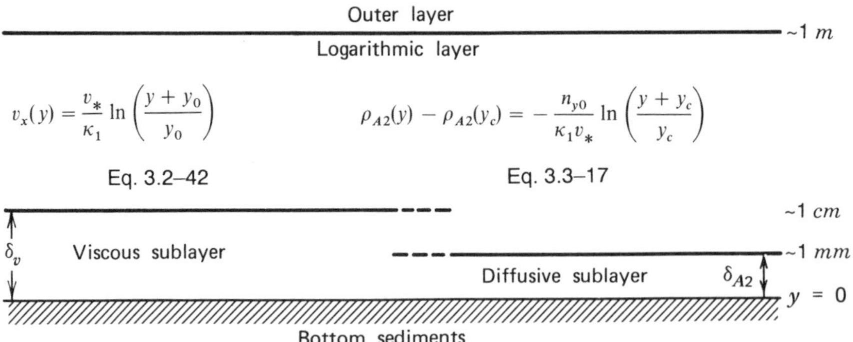

Figure 5.3-1. Dynamical and concentration structure of the benthic boundary layer.

Deissler[33] compiled a variety of data of momentum transfer in turbulent, isothermal flow in smooth tubes. From this work it is possible to establish that the outer limit at which viscous effects are prominent are at

$$\delta_v = \frac{26\nu_2}{v_*} \tag{5.3-2}$$

where δ_v is the viscous sublayer thickness. Wimbush[22] suggested a value $\delta_v = 12\nu_2/v_*$ for the viscous sublayer thickness of a smooth sea bottom. Using this result and Eq. 3.3C-1, the thickness of the diffusion sublayer can be estimated by

$$\delta_{A2} = \frac{26\nu_2}{v_* \cdot \mathrm{Sc}_{A2}^{1/3}} \tag{5.3-3}$$

For seawater at 2°C, Sc_{A2} for silica $\simeq 3000$, and the height of the diffusion sublayer is only one-fourteenth of the height of the viscous sublayer. The seafloor mass transfer coefficient, ${}^3k'_{A2}$, can be estimated by $\mathscr{D}_{A2}^{(l)}/\delta_{A2}$; Boudreau and Guinasso[82] present six similar formulas that are applicable for this coefficient.

In this analysis of the benthic boundary layer we assume that transport is by molecular diffusion in the diffusion sublayer and that eddy diffusion obeys a mixing rate law above this layer. Accordingly, $\mathscr{D}_{A2}$ in Eq. 5.3-1 becomes

$$\mathscr{D}_{A2} = \begin{cases} \mathscr{D}_{A2}^{(l)} & 0 < y < \delta_{A2} \\ \kappa_1 v_*(y + y_0) & y > \delta_{A2} \end{cases} \tag{5.3-4}$$

where κ_1 is the von Kármán constant (0.4) and y is the height above the bottom.

Measured values of v_* at a location in the deep Pacific Ocean were found to vary with the tidal cycle. Numerical values for v_* varied from about 0.005 to 0.4 cm/s. In a more energetic region, such as the Florida current, v_* varied from about 0.1 to 1.0 cm/s. Any diffusion gradient established when v_* was small was obliterated during the tidal cycle; the viscous sublayer was reduced to its least extent every 6 hr. Transient turbulent effects occasionally eroded even this minimum boundary layer. Accordingly, the effective boundary layer for chemical diffusion is a few millimeters at most (see Example 5.3-1). A simple calculation shows that concentration gradients disappear a short distance above the sublayer (see Problem 5.3C).

The preceding summary of the benthic boundary layer has been abstracted in part from investigations of Schink and Guinasso.[24,25] These investigations have addressed the movement of dissolved silica and calcium carbonate near the seafloor. It was found that the stagnant boundary layers do not play a significant role in regulating the flux of dissolved silica at the sea–sediment interface; however, turbulence in the boundary layer can have a significant influence on calcium carbonate movement patterns.

In estuarine flow the bottom water moves in the opposite direction of the surface flow; see Fig. 7.1-17b. Ippen[86] gives an equation for estimating the bottom shear stress for such arrested saline wedges. The equation is $\tau_0=(f/z)\rho_2 v^2$ based on the average tidal velocity v and depth h. The bottom friction factor is $f = 0.043/(vh/v_2)^{1/4}$. However, this equation underestimates τ_0 by a factor of three, according to Ippen.

Example 5.3-1. Viscous Sublayer and Diffusive Sublayer Thickness at Seafloor

(a) Based on the friction velocity range reported at a location in the deep Pacific Ocean, calculate the range of thickness of the viscous sublayer (mm) at the seafloor.

(b) Calculate the thickness of the diffusive sublayer for silica (mm) near the seafloor.

SOLUTION (a) The viscous sublayer thickness is obtained from Eq. 5.3-2. for $v_* = 0.4\,\text{cm/s}$.

$$\delta_v = \frac{26\nu_2}{v_*}$$

$$= 26\left|1.7\text{E-2}\,\frac{\text{cm}^2}{\text{s}}\right|\frac{\text{s}}{0.4\ \text{cm}}\left|\frac{10\ \text{mm}}{\text{cm}}\right. = 11.05\ \text{mm}$$

For $v_* = 0.05\,\text{cm/s}$, $\delta_v = 88.4\,\text{mm}$.

(b) The diffusive sublayer thickness is obtained by Eq. 5.3-3 or $\delta_{A2} = \delta_v/\text{Sc}_{A2}^{1/3}$ for $v_* = 0.4\,\text{cm/s}$ and $\text{Sc}_{A2} = 3000$.

$$\delta_{A2} = \frac{11.05\ \text{mm}}{3000^{1/3}} = 0.77\ \text{mm}$$

for $v_* = 0.05\,\text{cm/s}$ and $\delta_{A2} = 6.1\,\text{mm}$.

Movement of Chemicals Within Bottom Sediments

The distribution of chemicals in the watery portion of the benthic boundary layer below the water–sediment interface is a function of the chemistry of the underlying sediments, factors that disturb the sediment–water interface, and the physics of transport within the bottom water. As a result of a large ratio of solid surface to interstitial water volume (especially in fine-grained muds), concentrations of species in porewater may change appreciably and give rise to large concentration gradients between sediments and the overlying water. This in turn results in fluxes of dissolved constituents to and from the sediments. In this section some of the more important processes occurring in the upper few centimeters of sediment are discussed, and an attempt is made to show how transport between sediment and overlying water is brought about.

Chemicals in Sediment Porewater. Digenetic chemical reactions, those occurring during and after sediment burial, can be divided into two categories: biogenic and abiogenic. The criterion for classification is whether or not the reactions are dedicated by bacteria and other microorganisms. Berner[26,63] gives an excellent introduction to and a literature review of the chemical reactions and transport processes occurring within abyssal sediments.

Many of the geochemically important bacteria in sediments require the preexistence of organic compounds for their metabolism. Thus biogenic reactions are intimately tied to the deposition of organic matter. High rates of deposition of organic matter are favored by (1) high planktonic productivity in the overlying water and (2) quick settling and burial to avoid decomposition in the water column. Good examples of the situation are provided by many near-shore shallow-water muds. Some of the most important digenetic reactions resulting directly from the bacterial decomposition of organic matter are the removal of dissolved oxygen, the production of carbon dioxide, the reduction of nitrate, and the reduction of methane. An example of large differences between porewater and overlying seawater composition is shown Fig. 5.3-2. Billen[56,20] uses steady-state flux relations for a model of the upper sediment layers, with concentration profiles verified from field data, to estimate the release of dissolved nitrogen from the bed and the sediment oxygen demand (SOD).

Reactions that are not biochemically controlled, either directly or indirectly, are less numerous. Examples of these abiogenic reactions include the dissolution of opaline silica, the dissolution of calcium carbonate at the sediment–water interface, the recrystallization of carbonate minerals with the consequent uptake of magnesium and release of strontium, and various documented or suggested silica–seawater reactions. Figure 5.3-3 shows a typical concentration profile of SiO_2 in porewater of sediment.

Modeling Interstitial Silica (SiO_2) Concentrations. Sediments accumulating on the seafloor do not simply reflect what is delivered to the bottom of the

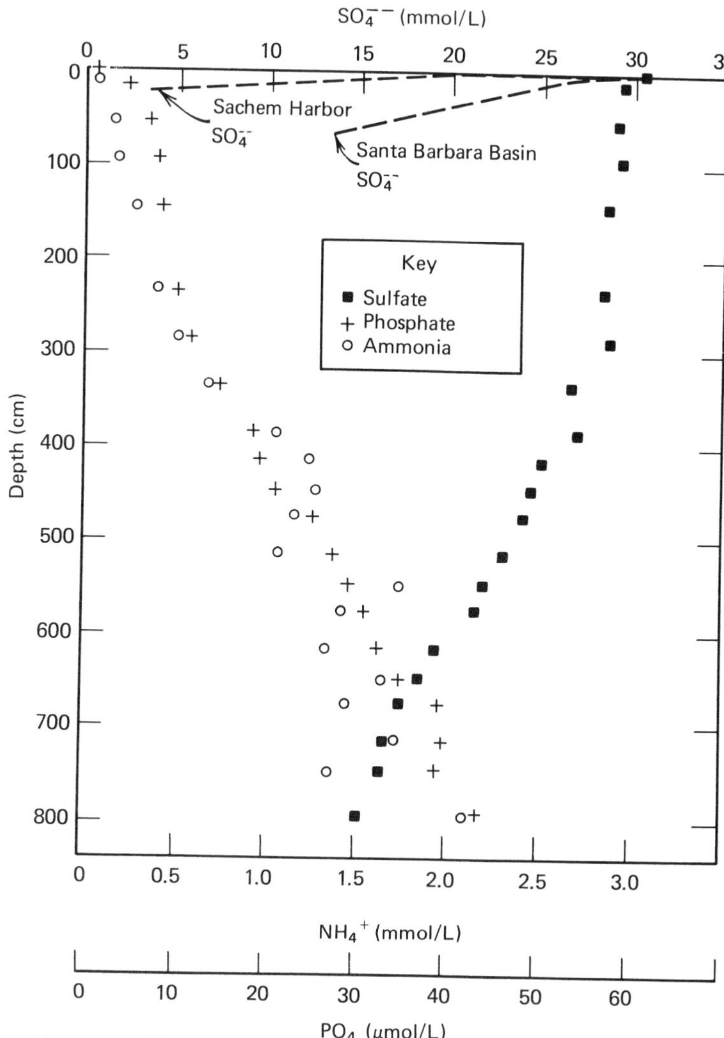

Figure 5.3-2. Sulfate, phosphate, and ammonia versus depth in porewaters of sediment from the West African continental borderland. (Reprinted by permission. From Ref. 26.)

ocean; rather, they represent a residuum—the net result of deposition, decomposition, dissolution, and reprecipitations. Reactions and transport in porewater play an important role in determining what stays deposited and what returns to the ocean cycles. Large gradients of dissolved silica are common in the uppermost layers of abyssal marine sediments. Silica has been studied extensively and provides a very good example to study the movement of chemicals within sediments. The silica model presented here is essentially that of Schink and Guinasso.[24]

Apparently, the concentrations of dissolved interstitial silica are controlled not by equilibrium processes between porewater and the associated solids, but

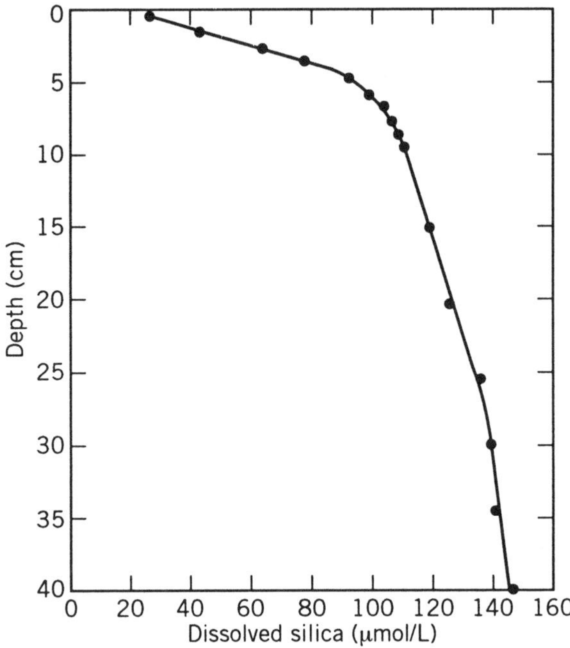

Figure 5.3-3. Dissolved silica versus depth in pore waters of sediment from the Bermuda Rise, North Atlantic. (Reprinted by permission. Source, Reference 26.)

rather, by a dynamic balance between dissolution of solids (primarily amorphous biogenic silica, but not necessarily exclusively) and the diffusion of the dissolution products out of the porewater into the overlying seawater. The dissolved silica comes from particles continuously stirred into the sediments by bioturbation.

The hypotheses that porewater is not in chemical equilibrium can be tested by calculations using component material balances with parameters chosen as accurately as possible. It is necessary (although not sufficient) that such calculations reproduce concentration profiles found in the seafloor. For these calculations and this model, we must include all the facts that have an important influence on pore-water composition. This include not only the aqueous phase processes but also those factors influencing the distribution and reactions of the solid particles that participate. The relevant processes can be incorporated into a fairly realistic model of the seafloor system by using the general diagenetic equation that is discussed in Section 5.2, of which Eq. 5.2-42 is the y-component.

Concentration gradients in reactive porous media do produce diffusive fluxes. The pores interconnect, and for abyssal surface sediments, porosity is 60 to 90%. The presence of nonreactive solid particles slows the diffusive flux by forcing the migrating species into more indirect routes and by reducing the cross-sectional area through which diffusion can proceed. In most marine sediments these two effects combine to slow aqueous diffusion by 30 to 40% (see Eq. 5.1-34).

Soluble particulate silica can be found at all depths in the sediment. Vertical mixing of solid particles at the seafloor is usually responsible for such burial, and this retards dissolution. Vertical mixing may be accomplished by physical movements of the waters at the seafloor. More important is the process of *bioturbation*; this process has a significant influence on the chemical interactions between sediment and seawater. See Fig. 3.2-9d for some bioturbation evidence on the seafloor.

Dissolution rates are hard to quantify. The actual processes whereby minerals interact with porewaters via dissolution, precipitation, adsorption, or ion exchange are abiogenic whether or not they are brought about ultimately by biogenic reactions. The kinetics of these reactions, unfortunately, is not at all well understood. Assumptions of first- and null-order mineral reactions make the equations mathematically simpler but may not be correct.

Berner[28] proposes the following general expression for mineral precipitation and dissolution in stagnant (or very-slow-flowing) porewater:

$$W_{A0} = \frac{a_v \mathscr{D}_{A2}}{r_0} (\rho^*_{A2} - \rho_{A2}) V\varepsilon \tag{5.3-5}$$

where ρ_{A2} is the concentration in mass per unit volume of pore-water, ρ^*_{A2} is the concentration in the layer immediately adjacent to the surface of the solid, $\mathscr{D}_{A2}$ is the diffusivity of A in the solution, a_v the surface area of solid per unit volume of porewater, and r_0 the average radius of solid particles. Hurd[29] has described dissolution from biogenic opal in terms of a mass transfer coefficient, ${}^3k_{A2}$ (mg/cm$^2 \cdot$ s), which involves consideration of surface area. This is probably the correct approach, but it leaves some uncertainty as to whether the surface area, as measured by nitrogen adsorption, is the appropriate surface area for dissolution.

Closure. The preceding was a semiquantitative description of the silica reactions and transport processes that occur in the seafloor sediments. Figure 5.3-3 is an example of a silica concentration profile. Although the description involved a "natural chemical" found at the seafloor environment, the same processes are active and a similar model will help describe the fate of an anthropogenic material introduced into this same environment.

Dissolved Chloride in Interstitial Waters of Chesapeake Bay. More often than not, conditions in the environment are transient. Because of the seasonal variations in the freshwater input into estuaries, the chemical content of these water bodies is constantly changing with time and provides ideal examples of transient transport. The chloride distribution in the bottom waters is constantly changing, and this produces a continually varying concentration gradient between bottom waters and interstitial waters. By relating the response of the chloride profile in the sediment to changes in the chlorinity of the overlying waters, an estimate of the net rate of transport in the sediment can be made.

The model presented here was developed by Holdren et al.[30] for analyzing the movement of chloride in the sediment of Chesapeake Bay. The bay is very productive biologically. This is reflected in the organic content of the sediments in the estuary,which is typically 2 to 3% on a dry weight basis. A large infaunal benthic community is supported by these organics. The resulting activity mixes the upper portion of the sediment and enhances the exchange of chemicals between the sediments and the overlying water. To investigate the magnitude of this mixing effect, along with other physical processes such as diffusion, the time-dependent changes that occur in porewater chloride concentration with depth beneath the sediment–water interface were studied.

Chloride is an ideal tracer to study these effects in an estuary such as the bay. It is essentially inert in terms of chemical reactivity in the estuarine environment. Figure 5.3-4d shows the results of chloride measurements within the sediment at a single station over a 1-year period. Easily measurable changes occurred in the chloride profile on a month-to-month basis. The surface sediments respond most quickly to the seasonal variations in the chloride concentration of the bottom waters. Seasonal variations result mainly from discharge of the Susquehanna River, which supplies between 90 and 97% of the fresh water to this portion of the bay.

If the primary mechanism for the transport of chloride is diffusional, the diffusion equation should adequately describe the shapes of the measured profiles. Neglecting the small lateral concentration gradients, the problem is reduced to a one-dimensional diffusion problem. Sediment deposition in the Chesapeake Bay at this location is on the order of 1 cm/yr, and this aspect should be considered in setting up the diffusion model for chloride. The sediment is derived from both shoreline erosion and suspended sediment discharge from the in-flowing rivers and streams.

The equation, incorporating a sedimentation term, is

$$\frac{\partial c_{A2}}{\partial t} = D_{A3}\frac{\partial^2 c_{A2}}{\partial y^2} - v_3\frac{\partial c_{A2}}{\partial y} \tag{5.3-6}$$

where c_{A2} is the interstitial chloride concentration, t the time, D_{A3} the aqueous diffusion coefficient that accounts for porosity and tortuosity, v_3 the sedimentation rate, and y the depth below the sediment–water interface. The boundary conditions are based on the physical observations made of the system.

The first boundary condition describes the chloride concentration in the overlying water as a function of time.

$$c_{A2i}(0, t) = \overline{c_{A2}} + c_{A2s}\cos\omega_1 t + c_{A2l}\cos\omega_2 t \tag{5.3-7}$$

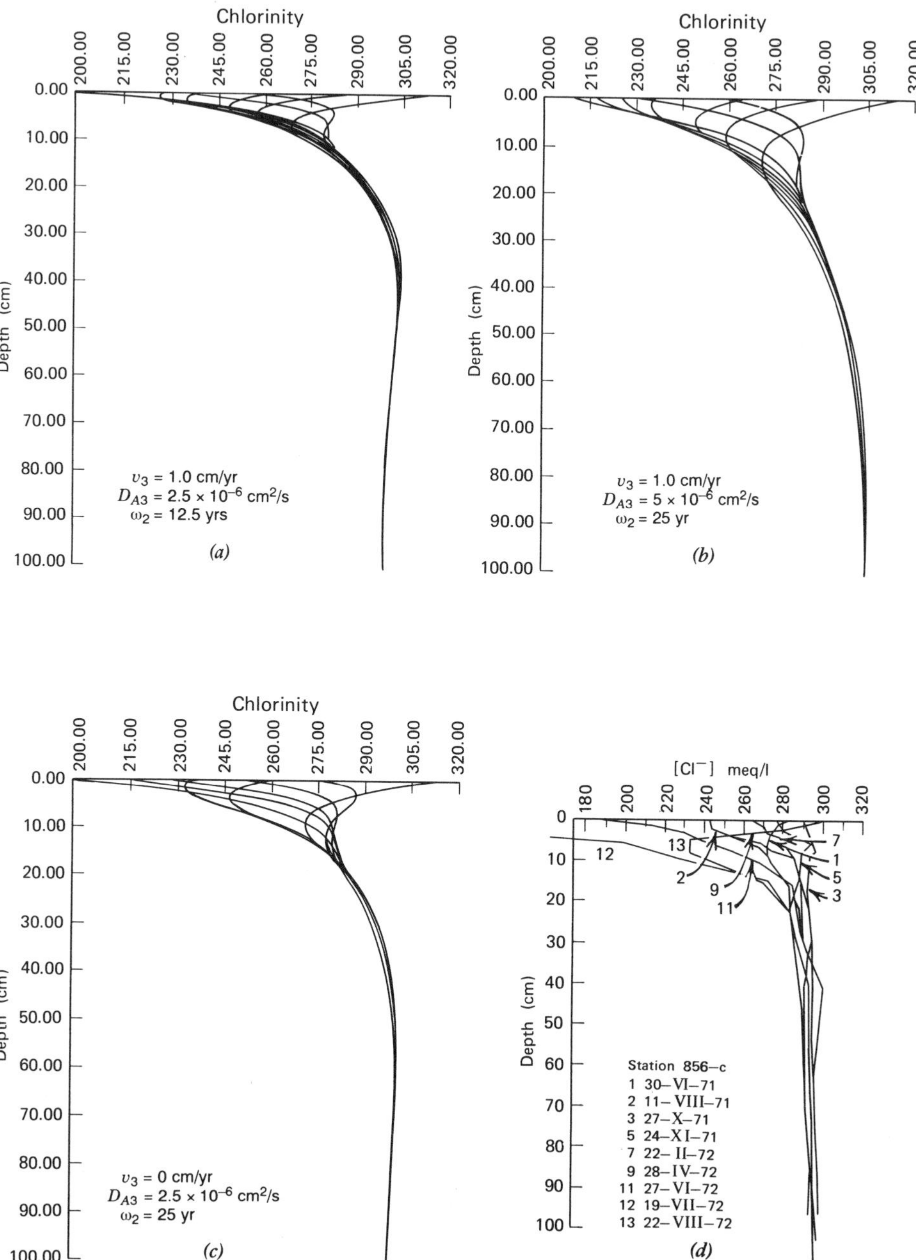

Figure 5.3-4. (*a* to *c*) Vertical profiles of chloride calculated by the diffusion model. These plots show the effect of varying D_{A3}, v_3. and ω_2 on the profiles. (*d*) For comparison to the model profiles, the field data are replotted. (Reprinted by permission, From Ref. 30).

where $\overline{c_{A2}}$ is the long-term mean chlorinity of the interface water, c_{A2s} and c_{A2l} are the short- and long-term chlorinity variations, respectively, and ω_1 and ω_2 are the frequencies of the short- and long-term variations, respectively. They account for the seasonal (month-to-month) chlorinity variations and the changes in the mean annual chlorinity, respectively. It is this last term that accounts for the skewing of the upper portion of the profile toward a more dilute concentration relative to the deeper pore waters.

The second boundary condition simply states that there is no net diffusional flux of chloride at great depth in the sediment. This is equivalent to saying that the estuary is "lined" by impermeable bedrock beneath the sediment.

For constant D_{A3} and v_3 the solution to this equation is found in Carslaw and Jaeger[31]:

$$c_{A2}(y,t) = \bar{c}_{A2} + c_{A2s}\cos\left[\omega_1 t - y a_1^{1/2}\sin\left(\frac{1}{2}\phi_1\right)\right]\exp\left[\frac{v_3 y}{2D_{A3}} - y a_1^{1/2}\cos\left(\frac{1}{2}\phi_1\right)\right]$$
$$+ c_{A2l}\cos\left[\omega_2 t - y a_2^{1/2}\sin\left(\frac{1}{2}\phi_2\right)\right]\exp\left[\frac{v_3 y}{2D_{A3}} - y a_2^{1/2}\cos\left(\frac{1}{2}\phi_2\right)\right] \tag{5.3-8}$$

where

$$a_1 = \left(\frac{v_3^4}{16D_{A3}^4} + \frac{\omega_1^2}{D_{A3}^2}\right)^{1/2} \qquad a_2 = \left(\frac{v_3^4}{16D_{A3}^4} + \frac{\omega_2^2}{D_{A3}^2}\right)$$

$$\phi_1 = \tan^{-1}\frac{-4D_{A3}\omega_1}{v_3^2} \qquad \phi_2 = \tan^{-1}\frac{-4D_{A3}\omega_2}{v_3^2}$$

and all other terms are as defined earlier.

By picking values for v_3, D_{A3}, and ω_2 ($\omega_1 = 2\pi$/yr), time-dependent chloride profiles can be calculated. Values of $\bar{c}_{A2}$, c_{A2s}, and c_{A2l} are available from the water analysis data, and the ranges of v_3 and ω_2 are available from independent data sources of other investigators. Therefore, an estimate of the diffusion coefficient typical of bay sediments can be made by matching calculated profiles to the field data.

The results of some representative calculations are shown in Fig. 5.3-4*a* to *c*. The close resemblance of the model results to the field data lends a degree of credence to the model. All three parameters, D_{A3}, v_3, and ω_2, were varied in the calculations to determine the net effect of each on the profiles. Changes in sedimentation rate had very little effect over the period of 1 year; however, there are provisions in the model to account for any long-term effects resulting from accumulation. Changes in the diffusion coefficient D_{A3} had the greatest effect of the three parameters. Comparison of the calculated profiles and the field data indicate that the best value for the constant D_{A3} is 5E-6 cm^2/s. This is in good agreement with values that have been reported in other sediment systems, such as the ocean floor.

It is apparent that the numerical model is capable of producing chloride concentrations profiles through time with the selection of reasonable constant parameters. However, the purpose here, as in much modeling work, is not to generate exact replicates of the observed chloride profiles in the bay sediments, but rather, to obtain a feeling for the processes and the magnitude of the combined effects of diffusion, bioturbation, and sedimentation of the distribution of any dissolved chemical of the interstitial waters. The simple model described here accomplishes this goal.

Steady-State Advection and Diffusion in Seabed. Maris and Bender[32] obtained calcium ion and magnesium ion concentration profiles in marine sediment porewaters. The data provided evidence of the upward advection rates of water through the sediment with the use of a steady-state transport model. One may model the shape of the Ca^{2+} and Mg^{2+} profiles assuming that the gradients reflect diffusion and advection only. For a species in porewater at steady state undergoing no reaction, the one-dimensional transport equation is

$$v_y \frac{d\rho_{A2}}{dy} = D_{A3} \frac{d^2\rho_{A2}}{dy^2} \tag{5.3-9}$$

where v_y is the vertical advection rate and D_{A3} is the effective diffusion coefficient. The solution of this differential equation is

$$\frac{\rho_{A2}(y) - \rho_{A2}(0)}{\rho_{A2}(h) - \rho_{A2}(0)} = \frac{\exp(y/y^*) - 1}{\exp(h/y^*) - 1} \tag{5.3-10}$$

where $\rho_{A2}(0)$ is the concentration at the sediment–water interface, $\rho_{A2}(h)$ is the "asymptotic" concentration at the "asymptotic" depth $y = h$, and $y^* \equiv D_{A3}/v_y$. Actually, $\rho_{A2}(h)$ is any other point on the profile.

The shape of the conservative species versus depth profile is dependent on the value of y^*. A small valve (advection-dominated transport) will result in a nonlinear profile. A large value of y^* (diffusion-dominated transport) will give a straight line connecting the end members. Therefore, given the depth variation of a conservative species and an experimental value of D_{A3}, the advection rate v_y can be calculated.

The Ca^{2+} and Mg^{2+} profiles in the Galapagos Hydrothermal Mound Field were nonlinear, yielding upward advection rates of 20 to 30 cm/yr. Equation 5.3-10 can be used to measure upward or downward v_y as low as 1 cm/yr by matching the model to the data in the presence of molecular diffusion. See Problem 5.3F for further information.

Important Coupled Phenomena

Models presented in Sections 5.1 and 5.2 can also be used for studying the fate of hydrophobic contaminants in marine beds sediment. These substances can be considered refractory in the short term because they degrade very slowly. In this part of the aquatic environment models are useful in providing order-of-magnitude estimates of flux, lifetime, and concentration. In this vein two such "vignette" models are presented for the release of contaminants to the water column.

Conveyor-Belt Bioturbation Model. In a series of three articles, Boudreau[83] explores tracer mixing in the upper sediment with mathematical models. Theoretical concentration profiles are generated assuming diffusive mixing, nonlocal mixing, and biological conveyor-belt phenomena. He notes that two conditions must be fulfilled before biological mixing of a tracer in sediments can be treated as a diffusive process. First, the frequency of mixing events must be greater than the rate of disappearance of the tracer. Second, the scale (i.e., distance) of material exchange must be smaller than the scale of the tracer profile and the thickness of the mixed layer. Infaunal macroorganisms are capable of exchanging sedimentary material over distances equal to or greater than the scale over which the concentration of a tracer changes substantially. This type of nondiffusive bioturbation is called *nonlocal mixing*. By considering mass balance in a small volume of sediment subject to biological reworking, a tracer conservation equation is derived in the form of an integrodifferential equation, which accounts for mixing on all scales, including local and nonlocal. The mixing effects of natural biological populations with their varied animal sizes and exchange distances are better represented by this integral formulation than is the widely employed diffusion model. Head-down deposit-feeders (i.e., the conveyor-belt phenomena) are special cases of the general theory presented by Boudreau. He relates the following tale.

An observer studying a small section of the seafloor notices that maldanid polychaetes, which dominate the local infauna, excrete black mud onto the mounds of sediment that surround the top ends of their burrows. This black mud contrasts sharply with the brown-gray mud in the upper oxidized layers a few centimeters thick. Curiosity aroused, our observer obtains a box core of this sediment. He notices that the worms in question are roughly 6 cm in length and that their heads are indeed buried in black sediment.

The observations made by our fictitious investigator can readily be duplicated by anyone studying near-shore marine sediments, and the tale illustrates two characteristics of this type of biological reworking of sediments. First, infaunal organisms such as the polychaetes ingest sediment at depth, advect this material through their guts, and defecate this material at the sediment–water interface. Figure 5.3-5b is an illustration of this. Presumably, physical compaction causes the sediment column to move downward, filling in any cavities created by feeding. Both process are strongly unidirectional.

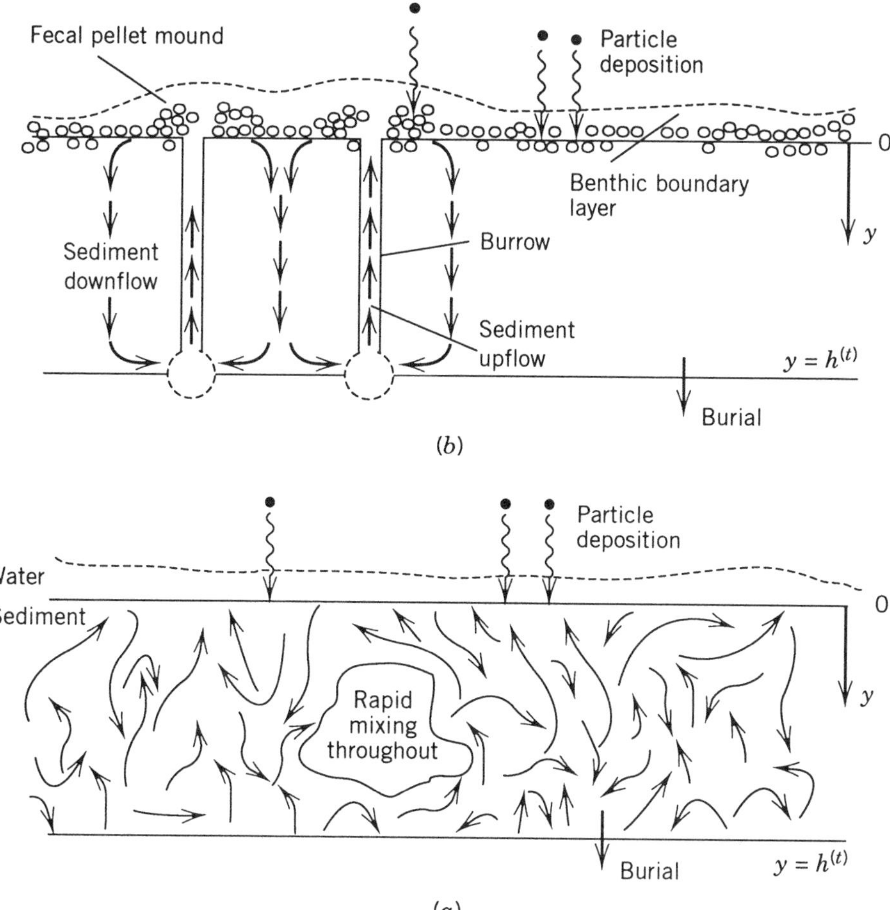

Figure 5.3-5. Particle movement for idealized bioturbation models: (*a*) box model; (*b*) conveyor-belt model.

Second, the material injected is of a different character than that deposited on the sediment–water interface. Such a process hardly fits the description of a diffusive phenomena and is more aptly described as an upward conveyor belt. In the following development this conveyor belt concept is applied to a soluble in-bed contaminant leaching into the water column.

The processes illustrated in Fig. 5.3-5*b* will be modeled as follows. Particle-bound quantities of chemical species A are transported up the burrow over the distance $h^{(t)}$. Particles of diameter d are deposited onto the sediment–water interface, where the fraction F_A is lost to the interface water. The now soluble species is then transported through the benthic boundary layer and to the overhead water column. The particles with the remaining sorbed fraction of A, $1 - F_A$, are mixed with cleaner particles depositing onto the surface and all subside downward into the bed. Eq. 5.2-54 defines F_A. The model is highly idealized but does provide the limiting case of nonlocalized mixing and in a sense is a circulation plug flow chemical desorber.

The particle residence time, τ_3, is the mass in the bioturbated zone divided by the sum of individual organisms reworking rates:

$$\tau_3 = \frac{h^{(t)}\rho_3(1-\varepsilon)}{\Sigma_i^n w'_{3i}N_i} \tag{5.3-11}$$

See p. 318 for definition of terms. In a similar fashion, if the depth of the fecal pellet layer for desorption is Δ, the pellet residence time is

$$\tau_p = \frac{\tau_3 \Delta}{h^{(t)}} \tag{5.3-11a}$$

As noted in the section on biodiffusion (see p. 325), Δ is a few pellet diameters in length. Typically, the pellets are much larger than the original sediment particles and approach 1 mm in diameter.[76] Since little is known about water pumping rates, only the solid portion is modeled. If the concentration at $y = h^{(t)}$ is ω_A, this level is also delivered to the interface, so that the flux equation is

$$n_A = {}^2K'_{A3}\rho_3(1-\varepsilon)(\omega_A - \omega_A^*) \tag{5.3-12}$$

where ω_A^* is the concentration on particles in equilibrium with the water column at ρ_{A2}. For the desorption step and the benthic boundary layer a resistance-in-series concept applies to obtain the coefficient

$$\frac{1}{{}^2K'_{A3}} = \frac{\rho_3(1-\varepsilon)}{F_A \Sigma_i^n w'_{3i}N_i} + \frac{\rho_3(1-\varepsilon)K^*_{A32}}{{}^3k'_{A2}} \tag{5.3-13}$$

where uniform surface coverage of fecal pellets is assumed.

The rate characterized by ω_A is constant through the first bed turnover time, τ_3. During this period of time burial of A occurs at the rate $v_3\rho_3(1-\varepsilon)\omega_A$, where v_3 is the particle burial rate. For constant bed density it is also the particle deposition rate at the surface. It is derived from fresh particulate matter of concentration $\omega_A = 0$ falling from the overhead water column. These depositing particles mix with the subsiding fecal pellets, yielding the sediment downflow concentration

$$\omega'_A = \omega_A(1-\mathrm{F}_A)\left[\frac{\Sigma_i^n w'_{3i}N_i}{\Sigma_i^n w'_{3i}N_i + v_3\rho_3(1-\varepsilon)}\right] \tag{5.3-14}$$

The $\omega_A(1-F_A)$ term is the concentration of A in the fecal pellet portion of the downflow stream. Figure 5.3-6b illustrates the stepwise decrease in concentration. Once the mixed particles of this concentration reach the feeding level, these particles appear at the surface. The flux drops at this time because ω'_A replaces ω_A in Eq. 5.3-12. Equation 5.3-13 applies unchanged and this rate continues for the same time period τ_3. The burial rate is $v_3\rho_3(1-\varepsilon)\omega'_A$. The concentration step-down process above continues at intervals of τ_3, and Fig. 5.3-6*b* shows the idealized profile through three time steps.

The model equations above are unsuited for estimating chemical lifetime in the sediment. As the bioturbated layer follows the upward-moving sediment–water interface, a remnant quantity of the contaminant remains at depth beyond $h^{(t)}$. Chemical transport here is by retarded molecular diffusion (see Problem 5.3D), which is a very slow process. The concentration profile of the remnant is roughly a bell-shaped curve, as shown at the top of Fig. 5.3-6*b*. The ultimate remnant faction of A associated with the original bioturbated

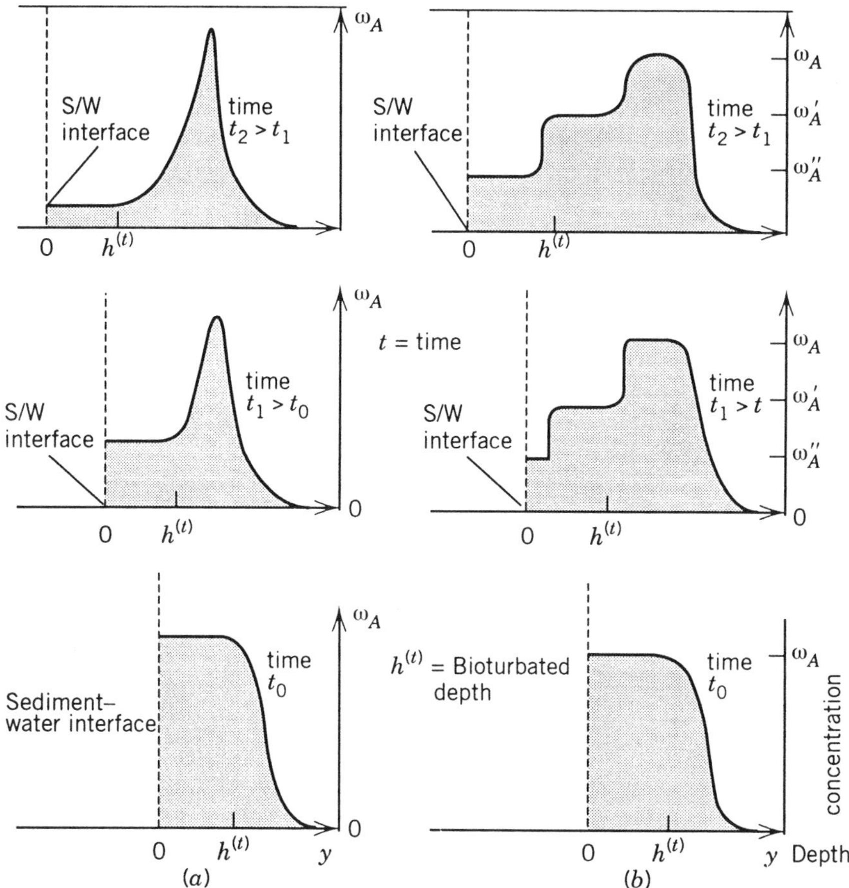

Figure 5.3-6. Contaminant profiles for idealized bioturbation models: (*a*) box models; (*b*) conveyor belt model.

depth is

$$F_{RA} = \frac{\rho_3(1-\varepsilon)v_3}{\Sigma_i^n w'_{3i} N_i} [1 + (1 - F_A)R_D + (1 - F_A)^2 R_D^2 + \cdots] \quad (5.3\text{-}15)$$

where R_D is defined as the particle mixing ratio, the term in brackets in Eq. 5.3-14. Diffusion occurs in both directions away from the plane of maximum concentration. The movement rate within the bed is very slow, as use of Eq. E5.1-2 in a calculation will show.

In essence the material in the paragraph above describes the process of natural capping of contaminated bed sediment. Particle deposition on the sediment–water interface dilutes the contaminant and gradually buries it. This process, accompanied by the change in the transport mechanism from bioturbation, which is very rapid, to molecular diffusion, which is very slow, does an effective job in isolating the contaminant remnant from the water column. Theoretically, the placement of an artificial cap achieves the same thing very rapidly.[74,81]

Bioturbation as Box Model. Berner[63] uses the term *box model* to describe bioturbation as a fast, random mixing process where all sediment properties are uniform from the sediment–water interface downward to the depth $y = h^{(t)}$. The biodiffusing coefficient model described in Section 5.2 is also a random mixing process, but the mixing rate of particles and porewater is much slower. The box-model approach represents one mixing extreme, the conveyor-belt model represents the other, and the biodiffusion coefficient model covers the middle ground. The biodiffusion coefficient model is also very useful. It covers many realistic bioturbation situations and its application commences with Eq. 5.2-42 and requires information on $\mathscr{D}_3^{(t)}$. Such data are given in Table 5.2-6. A simple use of this model appears in Example 5.2-2.

The bioturbation box model was first used in the interpretation of in-bed profiles of tracers delivered by deposition. According to this model, changes in the properties of sediment added at the sediment–water interface are immediately sensed at the bottom of the "box" but are damped by the mixing. For an insoluble solid tracer A undergoing no digenetic changes other than bioturbation, the mathematical representation of concentration in the box of volume $h^{(t)}A$ shown in Fig. 5.3-5*a* is

$$\frac{d\omega_A}{dt} = \frac{n_A(t) - v_3\rho_3(1-\varepsilon)\omega_A}{h^{(t)}\rho_3(1-\varepsilon)} \quad (5.3\text{-}16)$$

where $n_A(t)$ is the time-varying depositional flux of A onto the surface. The burial rate of A below the depth $y = h^{(t)}$ mimics this influx. Thus knowledge of $n_A(t)$ can be used via solution of Eq. 5.3-16 to predict the burial record, or more important, the burial record can be used to deduce $n_A(t)$.

The bioturbation box-model concept can be extended to study the fate of hydrophobic contaminants within the bed. Here the case of leaching of this refractory substance to the water column will be analyzed. Equilibrium is assumed between the concentration in the porewater, ρ^*_{A2}, and on the particles, ω_A, in the bed (i.e. box). Cleaner particles of concentration ω^*_A are deposited onto the surface at a constant rate v_3. This is accompanied by burial at constant velocity v_3 out the bottom face of the box (see Fig. 5.3-5*a*). The dissolution rate of A is controlled by the transport resistance in the benthic boundary layer and is

$$n_A = {}^3k'_{A2}(\rho^*_{A2} - \rho_{A2}) \tag{5.3-17}$$

Since mixing in the box is very rapid, particles of concentration ω_A are always available at the interface, so this model contains no sediment-side resistance.

Transient component mass balances for A in the porewater and on the particles, respectively, yield

$$\frac{d(h^{(t)}\varepsilon\rho^*_{A2})}{dt} = r_A h^{(t)} + v_3\varepsilon\rho_{A2} - {}^3k'_{A2}(\rho^*_{A2} - \rho_{A2}) - v_3\varepsilon\rho^*_{A2} \tag{5.3-18}$$

and

$$\frac{d(h^{(t)}\rho_B\omega_A)}{dt} = v_3\rho_B\omega^*_A - r_A h^{(t)} - v_3\rho_B\omega_A \tag{5.3-19}$$

where $-r_A$ is the rate of desorption of A from the particles in g $A/\text{cm}^3 \cdot \text{s}$ and $\rho_B \equiv \rho_3(1 - \varepsilon)$ is constant in the box. The equilibrium relationship is used to obtain ρ^*_{A2} as the only dependent variable, and the two equations are summed to yield to the total balance on A. Initially, the porewater concentration is $\omega_{A0} = K^*_{A32}\rho^*_{A20}$; *integrating* ρ^*_{A2} over time yields

$$\rho^*_{A2} = \rho_{A2} + (\rho^*_{A20} - \rho_{A2})e \tag{5.3-20a}$$

where

$$e \equiv \exp\left\{\frac{-[{}^3k'_{A2} + (\varepsilon + \rho_B K^*_{A32})v_3]t}{h^{(t)}(\varepsilon + \rho_B K^*_{A32})}\right\} \tag{5.3-20b}$$

This result indicates that concentrations in the box fall exponentially with time and the flux can be rewritten as

$$n_A = {}^3k'_{A2}(\rho^*_{A20} - \rho_{A2})e \tag{5.3-21}$$

where it behaves similarly.

Figure 5.3-6*a* illustrates the exponential variation of concentration in the bed, but in the box the concentration is uniform over the depth $0 \leqslant y < h^{(t)}$. The profile trace at the bottom of the box, $y = h^{(t)}$, falls exponentially with t and therefore with y because $|dy/dt| = v_3$ is the upward velocity of the box.

A remnant quantity of A exiting the bottom face of the box becomes "buried" for the same reasons as those given above for the conveyor-belt model. This results in the combined processes that is termed *natural capping*. The ultimate remnant fraction of A associated with the original bioturbated depth is

$$F_{RA} = \frac{(\varepsilon + \rho_B K^*_{A32})v_3}{{}^3k'_{A2} + (\varepsilon + \rho_B K^*_{A32})v_3} \tag{5.3-22}$$

This fraction includes A in the porewater as well as on particles.

Closure

Chemical transport within the upper layers of bed sediment is a very complex process that will continue to challenge the efforts of environmental chemist, benthic biologist, and engineers for decades. The laboratory and field data needed to verify models, which includes both fluxes and concentration profiles, appear very limited. The behavior of radiolabeled tracers and refractory organics are possibly the simplest, and aspects of their behavior can be predicted best. It appears that few generalities can be made since the presence or absence of benthic organisms at a particular site and its intensity ultimately control transport. The more information available at a site, the easier it is to choose a reasonably realistic model. The material in the chapter is tailored to yield specific tools, some more complex than others, to make crude estimates of chemical behavior. With an application in mind and possessing as much site-specific information as possible, the user should browse through this "boutique of models" choosing the one that best fits the situation. The advanced student should tailor-make models to fit specific sites and individual needs. Duursma and Smies[84] have presented some additional vignette models for bed-sediment chemical transport that includes a degradative reaction term.

PROBLEMS

5.3A. Flux Rate of Silica Through the Diffusive Sublayer

1. Assuming that the diffusive sublayer provides all of the resistance to the movement of silica from the sediment–seawater interface, calculate the flux rate of silica in μmol/cm$^2 \cdot$yr for $c_{A2i} = 150\,\mu$mol/L, $\rho_{A2} = 0.02$ mg/L, $\mathscr{D}_{A2} = 5\text{E-}6\,\text{cm}^2/\text{s}$, and $v_* = 0.40\,\text{cm/s}$.
2. Estimate the sea-bottom liquid-phase mass transfer coefficient (m/yr).

5.3B. Solution of Diffusion Equation in Benthic Boundary Layer

Solve the diffusion equation (Eq. 5.3-1) for the concentration profile of chemical A in the benthic layer. Use Eq. 5.3-4 for the variation of $\mathscr{D}_{A2}$ with y.

1. Obtain a relationship for $c_{A2}(y)$ in the diffusion sublayer. Assume that the flux rate N_{A0}, the concentration at the sediment–water interface c_{A2i}, and v_* are known.
2. Extend the relationship for the concentration profile $c_{A2}(y)$ into the logarithmic layer.

ANSWER:

1.

$$c_{A2}(y) = c_{A2i} - \frac{N_{A0}y}{\mathscr{D}_{A2}} \qquad \delta_{A2} \geqslant y \geqslant 0 \tag{5.3B-1}$$

2.

$$c_{A2}(y) = c_{A2}|_{\delta_{A2}} - \frac{N_{A0}}{\kappa_1 v_*} \ln \frac{y + y_0}{\delta_{A2} + y_0} \qquad y > \delta_{A2} \tag{5.3B-2}$$

5.3C. Concentration Profile of Silica in Benthic Boundary Layer

Calculate the concentration of dissolved silica at various heights above the sediment–seawater interface. Assume a silica flux rate of $15\,\mu\text{mol/cm}^2\cdot\text{yr}$ and an interface concentration of $c_{A2i} = 10.0\,\mu\text{mol/L}$, $v_* = 0.3\,\text{cm/s}$, $\mathscr{D}_{A2} = 5E\text{-}6\,\text{cm}^2/\text{s}$, and $y_0 = 0$. Use the results of Problem 5.3B in your calculations. Obtain silica (SiO_2) concentration (μmol/L) and make a graph of the concentration for each of the following distances from the sediment–seawater interface: 0.1, 0.2, 0.5, 1.0, 10.0, and 100 cm.

5.3D. Transport Beneath Bioturbated Surface Layer

Table 5.3D contains PCB concentration data for the bed sediment of New Bedford Harbor above and below the bioturbated layer.[74]

1. Propose a PCB discharge scenario and in-bed processor that could lead to such profiles. Estimate maximum concentrations.
2. Using the semi-infinite media model, estimate the effective diffusivity of each type of Aroclor beneath the bioturbated layer. Assume that $h^{(t)} = 15\,\text{cm}$ and a diffusion time of 42 years.

Site FX is 650 m from the waste discharge point, site DR is 1800 m away.

Table 5.3D. PCB Profile in NBH Bed Sediment for 1989

Site FX			Site DR		
Depth, y (cm)	Concentration, ω_A (ppm)		Depth, y (cm)	Concentration, ω_A (ppm)	
	A-1242	A-1254		A-1242	A-1254
0–1	1300	300	0–1	160	61
1–2	1100	230	1–2	210	80
2–3	1800	350	2–3	160	59
3–4	2900	390	3–4	130	50
4–6	3300	480	4–6	140	64
6–8	5600	740	6–8	210	95
8–10	6300	900	8–10	170	87
10–12	5700	870	10–12	140	58
12–14	2600	1400	12–14	110	49
14–16	3700	1400	14–16	90	54
16–20	1300	1200	16–18	49	34
20–24	480	1000	18–20	23	12
24–28	28	260	20–22	6.8	2.5
28–32	3.8	27	22–24	4.1	0.92
32–36	0.1	0.09	24–26	0.4	0.14
			26–28	1.2	—

Source: Ref. 74.

3. Is the box model for bioturbation consistent with the profile data in the upper layer? At these sites ${}^3k_{A2} \sim 0.59$ to 2.1 cm/h and $v_3 \sim 0.3$ to 1.0 cm/yr. $K^*_{A32} = 7920$ L/kg for A-1242 and 21,560 L/kg for A-1254.

5.3E. Fate of Aroclor 1254 in New Bedford Harbor Bed Sediment

The concentration profile data (ca. 1982) was homogenized over the entire Acushet River Estuary.[85] The average concentrations were 511 ppm between 0 and 4 cm depth, 2681 ppm between 4 and 8 cm, and 971 ppm beyond. Tidal current speed averages 7.5 cm/s and average water depth in 91.4 cm. The sediment area is 7.69E5 m^2, has 4% organic matter, dry bulk density 0.75 g/cm^3, and a particle density of 2.3 g/cm^3 . For A-1254 its solubility in seawater is estimated to be 0.012 mg/L, molecular diffusivity 0.45E-5 cm^2/s, and $K^*_{A32} = 21{,}560$ L/kg at 20°C.

1. Estimate the water-side transport coefficient. Answer in cm/h.
2. As isolated phenomena controlling transport through the boundary layer and the upper layers of the bed for each of the following, estimate the steady-state flux of A-1254 to water in kg/yr if the process is:
   (a) Controlled by the benthic boundary layer.
   (b) Molecular diffusion in the bed.

(c) Advection out of solute, $v_{2y} = 0.1$ cm/month.
(d) Brownian diffusion of colloids (see Problem 5.2F for data).
(e) Advection out of colloids.
(f) Bioturbation with $\mathscr{D}_3^{(t)} = 10 \pm 7\,\text{cm}^2/\text{yr}$.

3. As combined phenomena active across the interface, estimate the steady-state flux for:
   (a) The benthic boundary layer plus molecular diffusion and advection in the bed
   (b) The benthic boundary layer plus colloid diffusion and advection in the bed.
   (c) The benthic boundary layer plus biodiffusion of particles and porewater in the bed.

4. For the conveyor-belt bioturbation model determine the daily flux in kg/day for 1000 days. Assume that $h^{(t)} = 10\,\text{cm}$, an initial uniform sediment concentration of 1470 ppm, sediment reworking rate of 1 g/day per individual organism, 500 organisms per m^2, and particles of 30% clay and 70% silt sizes. Particle depositions yields a bed accretion rate of 0.3 cm/yr.

5. Repeat part 4 for the bioturbation box model.

6. Based on the analysis above, what mechanisms control the release rate of A-1254 to the water column? Prepare a memorandum report summarizing the findings of the analysis and developing your conclusions as to the mechanisms. Also give and justify a value for the rate in kg/yr. The report should not be less than two double-spaced typed pages of verbal material.

5.3F. Steady-State Advection and Diffusion

Using the following data in your calculation: Aroclor 1254, $K^*_{A32} = 11{,}310$ L/kg, $\varepsilon = 0.59$, $\omega_A = 2681$ mg/kg at 6 cm, and 113 mg/kg at surface.

1. Show that Eq. 5.3-10 is a solution to the steady-state advection-diffusion equation.

2. Show that the flux through the sediment–water interface is

$$n_A = v_y \left[\rho_{A2}(0) - \frac{\rho_{A2}(h) - \rho_{A2}(0)}{\exp(h/y^*) - 1} \right] \tag{5.3F}$$

3. Obtain porewater concentrations in μg/L at $y = 0$, 2, 4, and 6 cm for an upward advection velocity of 50 cm/yr. Repeat for a downward advection velocity of 50 cm/yr.

4. Compute the flux through the sediment–water interface for both conditions in part 3.

5.3G. Natural Cleansing Process for Kepone in James River

Between 1977 and 1979 the level of Kepone (decachlorooctahydro-1,3,4-metheno-2H-cyclobuta[cd]-pentalen-2-one), a pesticide, in the upper 2 cm of the bottom sediment extending over a 120-km reach fell from an average of 0.059 to 0.016 ppm. The partition coefficient is 6E4 L/kg, but a deposition velocity is not reported.[33]

1. If the solubility in fresh water is 2.2 mg/L at pH 7, can the loss be attributed to dissolution? Assume rapid mixing in the top 2 cm and ${}^3k'_{A2} = 15\,\text{cm/h}$.
2. What is the expected concentration in 1991 for a zero deposition velocity? What is it for $v_3 = 10\,\text{cm/ry}$?
3. What is the ultimate fraction of the mass in the top 2 cm in 1977 that becomes buried for $v_3 = 1\,\text{cm/yr}$? Repeat the calculation for $v_3 = 10\,\text{cm/yr}$.

5.3H. Inert Tracer Bioturbation via Box Model

An insoluble, refractory tracer on solid particles suddenly appears in the water, where it is deposited with other material, resulting in a bed accretion rate of 1.0 cm/yr. Its concentration is 1.0 μg/kg.

1. For a bed initially tracer-free, compute and graph the in-bed concentration profiles with depth for time 0.25, 0.5, 1.0, 2, 3, 5, 10, and 15 years. The mixing depth is 4.0 cm.
2. After 20 years the concentration in the bed is nearly 1.0 μg/kg, at which time incoming tracer ceases completely but particle deposition continues. Recompute and graph the in-bed tracer concentration profiles with depth for the times after the tracer input has stopped.

5.3I. Organic Carbon Reaction and Diffusion in Bed Sediment

The delivery of particulate organic matter onto the sediment–water interface results in its burial by bed accretion and bioturbation processes. As total organic carbon (TOC) in g C/g, it undergoes microbial degradation. One product of degradation that appears in the porewater is the dissolved organic carbon (DOC) in mg/L. See pages 271 and 314 for a discussion of this colloidal material. Table 5.2F contains an example of in-bed profiles.

1. The theoretical steady-state profile for TOC is

$$\omega_A = \omega_{A0} \exp\left\{y\left[\frac{v_3}{2\mathscr{D}_3^{(t)}} - \sqrt{\left(\frac{v_3}{2\mathscr{D}_3^{(t)}}\right)^2 + \frac{k_A'''}{\mathscr{D}_3^{(t)}}}\right]\right\} \tag{5.3I-1}$$

where $\omega_A(0)$ is the concentration at the sediment–water interface ($y = 0$), y the depth, v_3 the bed accretion rate, $\mathscr{D}_3^{(t)}$ the biodiffusion coefficient, and k_A'''

the degradation constant for an assumed first-order reaction equation. Beginning with the general digenetic equation, show the details of obtaining Eq. 5.3I-1 listing all the assumptions.

2. Using the data in Table 5.2F, verify the form of Eq. 5.3I-1, estimate ω_{A0}, and extract the rate constant. Brownawell[67] reports $\mathscr{D}_3^{(t)} = 3\text{E-7}\,\text{cm}^2/\text{s}$ based on ^{210}Pb profiles in the upper 10 cm and $v_3 = 0.05\,\text{cm/yr}$.

3. Show that the flux of TOC to the bed is

$$n_A = \rho_3(1 - \varepsilon)\omega_{A0}\sqrt{k_A'''\mathscr{D}_3^{(t)}} \qquad (5.3I\text{-}2)$$

and compute it in $\text{mg/m}^2 \cdot \text{h}$ for $\rho_3(1 - \varepsilon) = 0.75\ \text{g/cm}^3$.

4. Develop a profile equation similar to Eq. 5.3I-1 for DOC in the porewater.

5. Using the DOC data in Table 5.2F between $0.0 < y \leqslant 10\,\text{cm}$ with the equation from part 4, extract the rate constant for DOC production in the bed.

6. Obtain an expression for the flux of DOC to the water column and compute it in $\text{mg/m}^2 \cdot \text{h}$.

7. What fraction of the TOC delivered to the bed reemerges as DOC in the water column?

5.4. THERMAL ENERGY MOVEMENT ACROSS THE SEDIMENT–WATER INTERFACE

A net outward flow of heat across the sediment–water interface from geothermal sources deep within the earth averages about $0.042\ \text{J/m}^2 \cdot \text{s}$. Concerning the heat budget of lakes, Lerman[54] notes that on an annual basis this is the net amount of heat entering from the bottom and amounts to 0.1 to 0.5% of the input. Except in meromictic lakes this amount is insignificant and can be ignored in annual heat budget studies (see Section 4.3 on heat transfer to water bodies for estimates of the other heat sources to lakes). Seasonal fluxes are much greater than this average and can affect bottom-water thermal gradients to enhance or suppress chemical transport at the sediment–water interface.

Heat flows through the bottom of lakes in response to seasonal changes of water temperature. It flows from the water to the sediments in the summer and early fall and then back into the water during the winter. The amount involved in this seasonal exchange depends on the temperature range of the water in contact with the bottom and on the thermal properties of the bottom sediments or underlying rock. The seasonal heat exchange between lakes and their bottom media is exactly out of phase with the annual temperature cycle, withdrawing it during the summer and adding it during the winter. During the winter, for example, when the lake is covered with ice and the water temperature is below 4°C, the warm bottom sediments are a significant source of heat and vertical water mixing.[54]

Heat transport on the water side of the interface can be estimated by the use of heat transfer coefficients. These, through the concept of Newton's law of cooling, were introduced in Chapter 3 (see Eq. 3.3-3). Its use is illustrated in Example 5.4-1. The analogy theories, also introduced in Chapter 3, provide the means of estimating these coefficients from the available data on mass and momentum transport. Example 3.3-1 illustrates the process.

Example 5.4-1. Heat Loss Through Bottom. In Section 4.3 it was assumed that the heat losses through the bottom and sides of the lake, pond, or stream are negligible. The temperature difference between the water at the bottom and the mud surface is on the order of a tenth of a degree Celsius. Using the heat transfer coefficient of Example 3.3-1, estimate the bottom heat flux and decide whether it is negligible.

SOLUTION The heat exchange across the water–mud interface is by a conductive mechanism, and Eq. 3.3-3 can be used.

$$q_{23} = h_{23}\Delta T = \frac{120 \text{ J}}{\text{m}^2 \cdot \text{s} \cdot \text{K}} \, |\pm 0.1 \text{ K}| = \pm 12.0 \text{ J/m}^2 \cdot \text{s}$$

Even if ΔT was ± 1.0 K, the heat exchange would be small compared to the exchanges occurring at the air–water surface, which are on the order of ± 400 $\text{J/m}^2 \cdot \text{s}$.

One mode of heat transport on the sediment side of the interface is conduction. The general concept was introduced in Chapter 3 (see Eq. 3.3-1). Fourier's law applies in the case of porous media. These media consists of solid earthen materials and porewater. Thermal conductivities and heat capacities of these type materials and sediments are given in Table D.4. These data are also given for soils, which are very similar to sediments, in Table 6.4-1.

Water advection through the sediment–water interface is also an effective means of energy transport at the bottom of aquatic bodies. The combined steady-state processes of conduction and advection were used by Anderson et al.[80] to study the geothermal convection through oceanic crust and sediment. Temperatures were measured in the upper meters of the bed and the following relationship used to extract in-bed water advection velocities in the uniformly permeable media:

$$\frac{T_3(y) - T_3(0)}{T_3(h) - T_3(0)} = \frac{\exp(\text{Pe}_3\, y/h) - 1}{\exp(\text{Pe}_3) - 1} \qquad (5.4\text{-}1)$$

where $T_3(y)$ is temperature at depth y between those at $y = 0$ and $y = h$, $Pe_3 \equiv v_y h/\alpha_3$ is the Peclet number, where α_3 is the thermal diffusivity of the porous media. Independent measurements of thermal conductivity were needed; the values ranged from 0.65 to 0.72 $J/s \cdot m \cdot K$. The student should appreciate and use the transport analogies to develop additional models for thermal processes in bed sediment and across the benthic boundary layer (see Eqs. 5.3-9 and 5.3-10 and Problem 5.3F).

PROBLEMS

5.4A. Estimating Bottom-Water Heat Transfer Coefficients

Estimate the bottom-water heat transfer coefficient for a freshwater lake in winter if $T_{3i} = 4.1°C$ and $T_2 = 4.0°C$. Give the assumptions for the model used.

5.4B. Water Advection in Bed of the Seine

The bed of the Seine overlies a heavily used freshwater aquifer a few kilometers downstream from Paris. It is suspected that there may be a threat of water contamination from pesticides in the river. Based on the following three temperature measurements, estimate the direction and magnitude of the water advection rate: $(y, T_3) = (0\,cm, 16.2°C; 9\,cm, 15.9°C; 16\,cm, 15.3°C)$.

REFERENCES

1. L. J. Thibodeaux, "Spill of Soluble, High Density, Immiscible Chemicals in Water", CG-UOA-77-004, U.S. Department of Transportation, Washington D.C., 1978.
2. L. J. Thibodeaux, "Mechanisms and Idealized Dissolution Modes or High Density ($\rho > 1$), Immiscible Chemicals Spilled in Flowing Aqueous Environments," *Am. Inst. Chem. Eng. J.*, **23**(4), 544 (1977).
3. S. Hu and R. C. Kintner, "The Fall of Single Liquid Drops Through Water," *Am. Inst. Chem. Eng. J.*, **1**(1), 42 (1955).
4. I. Langmuir, "Oil Lenses on Water and the Nature of Monomolecular Expanded Films," *J. Chem. Phys.*, **1**, 756 (1933).
5. M. S. Yalin, *Mechanics of Sediment Transport*, Pergamon Press, New York, 1972, p. 204.
6. D. E. Simons, E. V. Richardson, and C. F. Nordin, Jr., *CER* 64*DES-EVR-CFN*-15, Colorado State University, Ft. Collins, Colo., 1964.
7. H. Kramers and P. J. Kreyger, "Mass-Transfer Between a Flat Surface and a Falling Liquid Film," *Chem. Eng. Sci.*, **6**, 42 (1956).
8. L. J. Thibodeaux, L.-K., Chang, and D. J. Lewis, "Dissolution Rates of Organic Contaminants Located at the Sediment Interface of Rivers, Streams and Tidal Zones", Chapter 16 in R. A. Baker, Ed., *Contaminants and Sediments*, Vol. 1, Ann Arbor Science, Ann Arbor, Mich., 1980.

9. V. Novotny, "Boundary Layer Effects on the Course of the Self-Purification of Small Streams," in S. H. Jenkins, Ed., *Advances in Water Pollution Research*, Pergamon Press, Elmsford, N.Y., 1969, pp. 39–50.
10. T. Fujii and H. Imura, "Natural Convection Heat Transfer from a Plate at Arbitrary Inclinations," *Int. J. Heat Mass-Transfer*, **15**, 755 (1972).
11. C. Chang, G. C. Vliet, and A. Saberian, "Natural Convection Mass-Transfer at Salt–Brine Interfaces," *Paper* 76-*HT*-33, ASME-AIChE Heat Transfer Conference, St. Louis, Mo., Aug. 1976.
12. P. S. Christy and L. J. Thibodeaux, "Spill of Soluble, High-Density Immiscible Chemicals on Water," *Environ. Prog.*, **1**(2), 126–129, (1982).
13. J. M. Smith, *Chemical Engineering Kinetics*, 2nd ed., McGraw-Hill, New York, 1970, p. 414.
14. E. J. Greskovich, J. M. Pommersheim, and R. C. Kenner, Jr., "Determination of Hindered Diffusivities for Nonadsorbing Pollutants in Mud," *Am. Inst. Chem. Eng. J.*, **21**(5), 1022 (1975).
15. G. W. Daigre, Dow Chemical Company, provided data on the chloroform spill and water sample analysis results, private communication, 1976.
16. J. C. Willmann, "PCB Transformer Spill, Seattle, Washington," *J. Hazard. Mater.*, **1**, 361–372 (1975–1977).
17. L. J. Carter, "News and Comment... Chemical Plants Leaves Unexpected Legacy for Two Virginia Rivers," *Science*, **198**, Dec. 9, 1977, pp. 1015–1020.
18. L. J. Thibodeaux and C. K. Cheng, "A Fickian Analysis of Lake Sediment Upsurge," *Water Resources Bull.*, **12**(1) (1976).
19. P. Larsen, Private communication, U.S. Environmental Protection Agency, Pacific NW Laboratory, Corvallis, Oreg., 1974.
20. G. Billen, "An Idealized Model of Nitrogen Recycling in Marine Sediment," *Am. J. Sci.*, **282**(Apr.), 512–541 (1982).
21. D. P. Larsen, H. T. Mercier, and K. W. Malueg, "Modeling Algal Growth Dynamics in Shawgawa Lake, Minnesota," in E. J. Middlebrooks, D. H. Falkenborg, and T. E. Maloney, Eds., *Modeling the Eutrophication Process*, Ann Arbor Science, Ann Arbor, Mich., 1974, pp. 15–31.
22. M. Wimbush, "The Physics of the Benthic Boundary Layer," in I. N. McCave, Ed., *The Benthic Boundary Layer*, Plenum Press, New York, 1976, pp. 3–10.
23. R. G. Deissler, *National Advisory Committee for Aeronautics Report* 1210, 1955.
24. D. R. Schink and N. L. Guinasso, Jr., "Effects of Bioturbation on Sediment–Seawater Interaction," *Mar. Geol.*, **23**, 133–154 (1977).
25. D. R. Schink and N. L. Guinasso, Jr., "Modeling the Influence of Bioturbation and Other Processes of $CaCo_3$ Dissolution at the Sea Floor," in N. R. Anderson and A. Malahoff, Eds., *The Fate of Fossil Fuel* CO_2 *in the Oceans*, Plenum Press, New York, 1976, pp. 375–399.
26. R. A. Berner, "The Benthic Boundary Layer from the Viewpoint of a Geochemist," in I. N. McCave, Ed., *The Benthic Boundary Layer*, Plenum Press, New York, 1976, pp. 33–55.
27. N. L. Guinasso and D. R. Schink, "Quantitative Estimates of Biological Mixing Rates in Abyssal Sediments," *J. Geophys. Res.*, **80**(21), 3032–3043 (1975).

28. R. A. Berner, "Kinetic Models for the Early Diagenesis of Nitrogen, Sulfur, Phosphorous and Silicon in Anoxic Marine Sediments," in E. D. Goldberg, Ed., *The Sea*, Wiley-Interscience, New York, 1974, pp. 427–450.

29. D. C. Hurd, "Factors Affecting Solution Rate of Biogenic Opal in Seawater," *Earth Planet. Sci. Lett.*, **15**, 411–417 (1972).

30. G. R. Holdren, Jr., O. P. Bricker, III and G. Matisoff, "A Model for the Control of Dissolved Manganese in the Interstitial Waters of Chesapeake Bay," in T. M. Church, Ed., *Marine Chemistry in the Coastal Environment*, American Chemical Society Symposium Series 18, ACS, Washington, D.C., 1975, pp. 365–381.

31. H. S. Carslaw and J. C. Jaeger, *Conduction of Heat in Solids*, Oxford University Press, London, 1959, p. 43.

32. C. R. P. Maris and M. L. Bender, "Upwelling of Hydrothermal Solutions Through Ridge Flank Sediments Shown by Porewater Profiles," *Science*, **216**, May 7 1982, p. 623–626.

33. R. J. Huggett and M. E. Bender, "Kepone in the James River," *Environ. Sci. Technol.*, **14**(8), 918–923 (1980).

34. W. B. Neely, G. E. Blau, and A. Turner, Jr., "Mathematical Model Predicts Concentration–Time Profile Resulting from Chemical Spill in River," *Environ. Sci. Technol.*, **10**, 72 (1976).

35. G. K. Rogers, "The St. Clair River Pollution Issue," *Water Pollut. Res J. Can.*, **21**(3), 283 (1986).

36. D. J. Mossman, J. L. Schnoor, and W. Stumm, "Predicting the Effects of Pesticide Release to the Rhine River," *J. Water Pollut. Control Fed.*, **60**(10), 1806–1812 (1988).

37. P. Ashworth, "Dispersion Model for Sinker Liquids Spilled into Waterways," *Proceedings of the Hazardous Materials Spills Conference*, Milwaukee, Wisc., Apr. 1982, pp. 404–413.

38. Y. L. Lau and J. Marsalek, "Movement of Perchloroethylene in Flowing Water," *Water Pollut. Res. Canada*, **21**(3), 303 (1986).

39. V. G. Levich, *Physicochemical Hydrodynamics*, Prentice Hall, Englewood Cliffs, N.J., 1962, p. 170.

40. V. P. Singh, D. Reible, and L. J. Thibodeaux, "Mathematical Modeling of Fine Sediment Transport," *Hydrol. J. International Association of Hydrogists*, **XI**(4), 1–24 (1988).

41. W. H. Schlesinger, *Biochemistry: An analysis of Global Change*, Academic Press, New York, 1991, p. 242.

42. E. M. Thurman, *Organic Geochemistry of Natural Waters*, Martinus Nijhoff/Dr. W. Junk, Publisher, Dordrecht, The Netherlands, 1985.

43. G. J. Thoma, A. C. Koulermos, K. T. Valsaraj, D. D. Reible, and L. J. Thibodeaux, Chapter 13 in R. A. Baker, Ed., *Organic Substances and Sediments in Water*, Lewis Publishers, Chelsea, Mich., 1991.

44. J. C. Means and R. Wijayaratne, "Role of Natural Colloids in the Transport of Hydrophobic Pollutants," *Science*, **215**; 968–970 1982.

45. E. G. Horn, L. J. Hetling, and T. J. Tofflemir, "The Problem of PCBs in the Hudson River," *Ann. NY. Acad. Sci.*, **320**, 591–609 (1979).

46. S. A. Savant, "Modeling of Pasive Contaminant Transport in River Sediments," PhD Thesis, Louisiana State University, Baton Rouge, Louisiana (1988).
47. A. J. Medine and S. C. McCutcheon, "Fate and Transport of Sediment-Associated Contaminants" in J. Saxena, Ed., *Hazardous Assessment of Chemicals*, Hemisphere, New York, 1989, 225–291.
48. A. Lerman, *Geochemical Processes*, Wiley, New York, 1979 pp. 90–92.
49. R. J. Millington and J. P. Quirk, *Trans. Faraday Soc.*, **57,** 1200 (1961).
50. W. G. Vaux, "Interchange of Stream and Intragravel Water in a Salmon Riffle," *Special Science Report: Fisheries* 505, *U.S. Fish and Wildlife Service*, Washington, D.C., 1962.
51. L. J. Thibodeaux and J. O. Boyle, "Bedform Generated Convective Transport Within Stable River Sediments," *Nature*, **325**; 341–343 (1987).
52. C. G. Delos, W. L. Richardson, J. V. DePinto, R. B. Ambrose, Jr., P. W. Rogers, K. Rygwelski, J. P. St. John, W. J. Shaughnessy, T. A. Faha, and W. N. Christie, "Toxic Substances," Chapter 3 in *Technical Guidance for Performing Waste Load Allocations*, Book II; *Streams and Rivers*. U.S. EPA, Washington, D.C., 1984.
53. R. E. Bird, W. E. Steward, and E. N. Lightfoot, *Transport Phenomena*, Wiley, New York 1960, p. 629.
54. A. Lerman, Ed., *Lakes: Chemistry, Biology and Physics*, Springer-Verlag, New York, 1978.
55. M. W. Lorenz, D. J. Smith, and L. V. Kimmel, "A Long-Term Phosphorous Model for Lakes: Applied to Lake Washington," in R. P. Canale, Ed. *Modeling Biochemical Processes in Aquatic Ecosystems*, Ann Arbor Science, Ann Arbor, Mich, 1976. pp. 75–91.
56. G. Billen, "Modeling the Processes of Organic Matter Degradation and Nutrient Recycling in Sedimentary Systems," Chapter 2 in D. M. Nedwell and C. M. Brown, Eds., *Sediment Microbiology*, Academic Press, London 1982.
57. L. J. Thibodeaux and B. B. Becker, "Chemical Transport Rates near the Sediment of Wastewater Impoundments," *Environ. Prog.*, **1**(4), 296–300 (1982).
58. S. O. Ryding and C. Forsberg, "Sediments as a Nutrient Source in Shallow Polluted Lakes," in H. L. Golterman, Eds., *Interaction Between Sediments and Fresh Water*, Dr. W. Junk, B.V. Publishers, The Hague, The Netherlands, 1977, pp. 227–234.
59. Y. Cohen, W. Cocchio, and D. Mackay, "Laboratory Study of Liquid-Phase Controlled Volatilization Rates in Presence of Wind Wavers," *Environ. Sci. Technol.*, **12**, (5), 553–558 (1979).
60. W. Lick, "Numerical Models of Lake Currents," Report EPA-6001-1-3-76-00020, U.S. EPA, Office of Research and Development, Duluth, Minn., Apr. 1976.
61. F. A. DiGiano and P. D. Snow, "Consideration of Phosphorus Release from Sediments in a Lake Model," in H. L. Goltermann, Eds., *Interactions Between Sediments and Freshwater*, Dr. W. Junk, The Hague, The Netherlands, 1977, pp. 318–323.
62. S. A. Formica, J. A. Baron, L. J. Thibodeaux, and K. T. Valsaraj, "PCB Transport into Lake Sediments: Conceptual Model and Laboratory Simulation," *Environ. Sci. Technol.*, **22**,(12), 1435 (1988).
63. R. A. Berner, *Early Diagnesis: A Theoretical Approach*, Princeton University Press, Princeton, N.J., 1980.

64. B. F. Jones and C. J. Bowser, "The Mineralogy and Related Chemistry of Lake Sediments," Chapter 7 in A. Lerman, Ed., *Lakes: Chemistry, Geology and Physics*, Springer-Verlag, New York, 1978.

65. E. Tipping, "Colloids in the Aquatic Environment," *Chem. Ind., Aug.* 1988, pp. 485–490.

66. S. A. Savant, D. D. Reible, and L. J. Thibodeaux, "Convective Transport Within Stable River Sediments," *Water Resources Res.*, **23**,(9), 1763–1768, (1987).

67. B. J. Brownawell, "The Role of Colloidal Organic Matter in the Marine Geochemistry, of PCBs," Ph.D. thesis, Massachusetts Institute of Technology, Cambridge, Mass 1986.

68. P. M. Gschwend, S. C. Wu, O. S. Madsen, J. L. Wilkin, R. B. Ambrose, Jr., and S. C. McCutcheon, "Modeling the Benthos–Water Column Exchange of Hydrophobic Chemicals," Final Report, U.S. EPA, Environmental Research Laboratory, Office of Research and Development, Athens, Ga., 1986.

69. H. Lee and R. C. Swartz, "Biological Processes Affecting the Distribution of Pollutants in Marine Sediments II. Biodeposition and Bioturbation," Chapter 29 in R. A. Baker, Ed., *Contaminants and Sediments;* Vol., 2 Analysis, Chemistry, Biology, Ann Arbor Science, Ann Arbor, Mich. 1980.

70. R. C. Aller, "The Influence of Macrobenthos on Chemical Diagenesis on Marine Sediments," Ph. D. dissertation, Yale University, New Haven, Conn., 1977, 600 p.

71. R. C. Aller, "Diagenetic Processes near the Sediment Water Interface and Long Island Sound I. Decomposition and Nutrient Element Geochemistry (S, N, P)," in B. Saltzman, Ed., *Estuarine Physics and Chemistry: Studies in Long Island Sound*, Advance in Geophysics, Vol, 22, Academic Press, New York, 1980.

72. D. R. Schink and N. L. Guinasso, Jr., "Redistribution of Dissolved and Adsorbed materials in Abyssal Marine Sediments Undergoing Biological Stirring," *Am. J. Sci.*, **278**, 687–702 (1978).

73. J. B. Fisher, W. J. Lick, P. L. McCall, and J. A. Robbins, "Vertical Mixing of Lake Sediments by Tubificid Oligochaetes," *J. Geophy. Res.*, **85**, 3997–4006 (1980).

74. L. J. Thibodeaux, D. D. Reible, W. S. Bosworth, and L. C. Sarapas, "The Theoretical Evaluation of the Effectiveness of Capping PCB Contaminated New Bedford Harbor Bed Sediment," Final Report, LSU Hazardous Waste Research Center, Baton Rouge, La., Nov. 14, 1990, 92 p.

75. A. P. Jackman and K. T. Ng, "The Kinetics on Ion Exchange on Natural Sediments," *Water Resources Res.*, **22**(12), 1664–1674 (1986).

76. S. W. Karickhoff and K. R. Morris, "Impact of Tubificid Oligochaetes on Pollutant Transport in Bottom Sediments," *Environ. Sci. Technol.*, **19**, 51–56 (1985).

77. J. Crank, *The Mathematics of Diffusion*, 2nd Ed., Oxford University Press, London, 1975, p. 36.

78. W. R. Davis and J. C. Means, "A Developing Model of Benthic–Water Contaminant Transport in Bioturbated Sediment," *Proceedings of the 21st EMBS* Polish Academy of Sciences, Institute of Oceanology, Gdansk, Poland, 1989, pp. 215–226.

79. S.-C. Wu and P. M. Gschwend, "Sorption Kinetics of Hydrophobic Organic Compounds to Natural Sediments," *Environ. Sci. Technol.*, **20**,(7), 717–724 (1986).

80. R. N. Anderson, M. A. Hobart, and M. G. Landseth, "Geothermal Convection Through Oceanic Crust and Sediments," *Science*, **204**, May 25, 1979, pp. 828–832.

81. X. Q. Wang, L. J. Thibodeaux, K. T. Valsaraj, and D. D. Reible, "Efficiency of Capping Contaminated Bed Sediments In Situ: Laboratory Scale Experiments of Diffusion–Adsorption in the Capping Layer," *Environ. Sci. Technol.*, **25**(9), 1578–1584 (1991).
82. B. P. Boudreau and N. L. Guinasso, Jr., Ch. 6 "The Influence of a Diffusive Sublayer on Accretion, Dissolution, and Diagenesis at the Sea Floor", in K. A. Fanning and F. T. Manheim, Eds., *The Dynamic Environment of the Ocean Floor*, Lexington Books, Lexington, Mass., 1982.
83. B. P. Boudreau, "Mathematics of Tracer Mixing in Sediments", *Am. J. Sci.*, I: **286**, 161–198 (1986); II: **286** 199–238 (1986); III: with D. M. Imboden, **287**, 693–719 (1987).
84. E. K. Duursma and M. Smies, Chapter 3 "Sediments and Transfer at and in the Bottom Interfacial Layer," in G. Kullenberg, Ed., *Pollutant Transfer and Transport in the Sea*, Vol. II, CRC Press, Boca Raton, Fl., 1982.
85. Applied Sciences Associates, "Selected Studies on PCB Transport in New Bedford Harbor", *ASA Report 86–18*, Narragansett, R.I. 1987.
86. A. T. Ippen, *Estuary and Coastline Hydrodynamics*, McGraw-Hill, N.Y., 1966, p. 570.

6

CHEMICAL EXCHANGE BETWEEN AIR AND SOIL

This chapter deals with transfer processes occurring in the lower atmosphere, across the interface, and in the uppermost layers of natural earthen solid materials. Aspects of micrometeorology, important to the subject of the movement and residence of chemicals in the air layers immediately above the soil interface, are presented. According to Sutton,[1] the subject of micrometeorology is worthy of serious study because such aspects as the dispersion of pollution from a factory chimney or the draining of cold air into a valley are significant and may profoundly affect human welfare and economy. He continues: "The physics of the lower strata of the atmosphere is both interesting and important, chiefly because of the large variation in conditions which are found in the layers of air nearest the ground. Such variations are significant, not only for meteorology, but for other sciences. The climate into which a plant first emerges is quite unlike that experienced by man and the larger animals a few feet higher up, for the layers of air within a fraction of an inch of the ground may experience both tropical heat and icy cold in the course of a single day."

The topic of thermal turbulence is presented and related to the physics of the lower air layers. The mixing-length theory is then extended to develop the Thornthwaite–Holzman equations, which are useful in calculating chemical flux rates from soil surfaces. The movement of pesticides and ammonia from agricultural land provides classic examples of the movement of chemicals through the air boundary layer. Atmospheric deposition processes, presented in Chapter 4 for water surfaces, are extended to chemicals and particles delivered to soil-like surfaces. Once again, radon is used as a tracer to aid in quantifying movement of chemicals in the upper soil layers. The chapter ends with a presentation of heat transfer processes at the air–soil interface.

6.1. THERMAL TURBULENCE ABOVE AIR–SOIL INTERFACE

As wind moves over natural surfaces, the friction with the surface generates turbulence. This is called *mechanical* turbulence and was the subject of Section 3.2. Turbulence is also generated when air is heated at a surface and moves

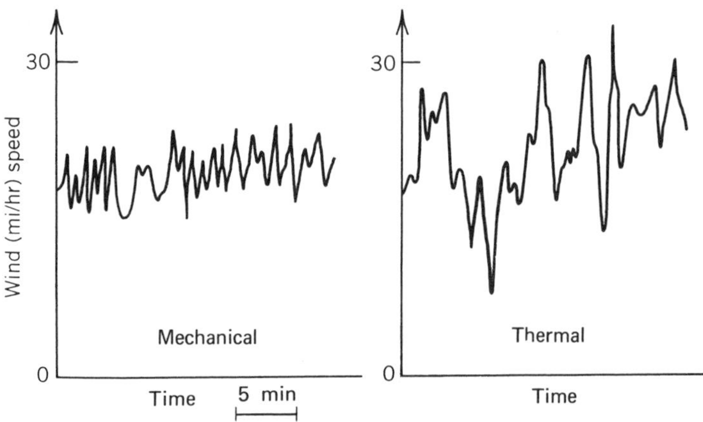

Figure 6.1-1. Typical traces of a fast-response wind sensor for conditions of pure mechanical turbulence and mechanical plus thermal turbulence. (Reprinted by permission, from Ref. 2.)

upward because of buoyancy. This is called *thermal* or *convective* turbulence. The size of the eddies produced by these two processes is different, and as a consequence the fluctuations from mechanical turbulence tend to be smaller and more rapid than the thermal fluctuations, as is shown in Fig. 6.1-1. A very good demonstration of these types of turbulence can be seen by watching the plume from a smokestack on a hot day. The plume seems to loop up and down. The plume is called a *looping plume* because in addition to the small-scale mechanical turbulence that tears the plume apart and spreads it with distance, the thermal updrafts and downdrafts cause parts of the plume to be transported upward or downward.

Chapter 4 contained information on the movement of air over water. Velocity profiles over water are similar to those over land. Mechanical turbulence is present over water also, but thermal turbulence effects are not as great over water because the temperature gradients are more moderate. For this reason the subject of thermal turbulence is presented now. The reader is referred to Section 3.2 for a review of mechanical turbulence.

Concept of Thermal Stability

Buoyancy plays a major role in the vertical transport and mixing of air, which, when it moves, carries with it the heat, water vapor, carbon dioxide, and other chemicals it contains. When air is stable, relatively little mixing occurs, whereas unstable air results in turbulent conditions with considerable mixing between adjoining vertical layers. Both the wind speed and temperature gradient determine the stability of air. Here we examine only the temperature gradient effect, as presented by Gifford.[3] For a detailed presentation of stability of the environment, see Eskinazi.[4]

Because of the vertical force exerted on any volume of air by gravity, the pressure of the atmosphere decreases with elevation. This vertical pressure variation implies a certain vertical structure governed by the ideal gas law. If temperature is known as a function of height, the pressure versus height relationship can be obtained. The temperature of a volume of dry air displaced upward by a process that does not add or remove sensible heat (i.e., an adiabatic process) decreases at the linear rate of 9.66 K/km; it is called the dry adiabatic lapse rate:

$$\Gamma_A \equiv \frac{-dT_1}{dy} = \frac{-g}{C_{p1}} \tag{6.1-1}$$

where C_{p1} is the heat capacity of dry air (see Problem 6.1B for more details).

Under certain circumstances the lower layers of the atmosphere may possess a dry adiabatic lapse rate; if so, a small isolated air parcel that is undergoing adiabatic vertical motion due to the eddies of mechanical turbulence will at all times adjust so that it will experience no buoyancy force, tending to restore it to its original level. It will always possess just the temperature of its environment. Figure 6.1-2 illustrates that the dry adiabatic lapse rate exists in the lower atmosphere for only a fraction of the total time. This figure shows that during a normal clear-day diurnal variation of temperature structure of the lower atmosphere, the adiabatic state can be expected just after dawn and at dusk and will last perhaps for several minutes. The reason is that the flow of heat to and from the underlying surface by radiation, conduction, and convection causes the lapse rate in the lower air layers to vary from day to night over wide limits. During the day vertical displaced volumes of air undergoing adiabatic expansion (thermals) must be acted on by positive buoyant forces, and as a result, turbulence is enhanced. During the night, the converse effects tend ordinarily to suppress turbulence sharply.

The buoyant force on an air parcel is easily calculated, being equal to the weight of the displaced air volume of density ρ_1' minus the weight of the original air parcel, ρ_1. Dividing by the mass of the original, the resulting acceleration of the displaced parcel is

$$a = g\,\frac{\rho_1 - \rho_1'}{\rho_1} \tag{6.1-2}$$

From the ideal gas law equation of state, assuming that the pressure of each parcel is equal, Eq. 6.1-2 can be written as follows:

$$a = g\,\frac{T_1' - T_1}{T_1} \tag{6.1-3}$$

where T_1 and T_1' are the temperatures of the original and displaced parcels, respectively, in kelvin. Since the air parcel is conceived of as acquiring

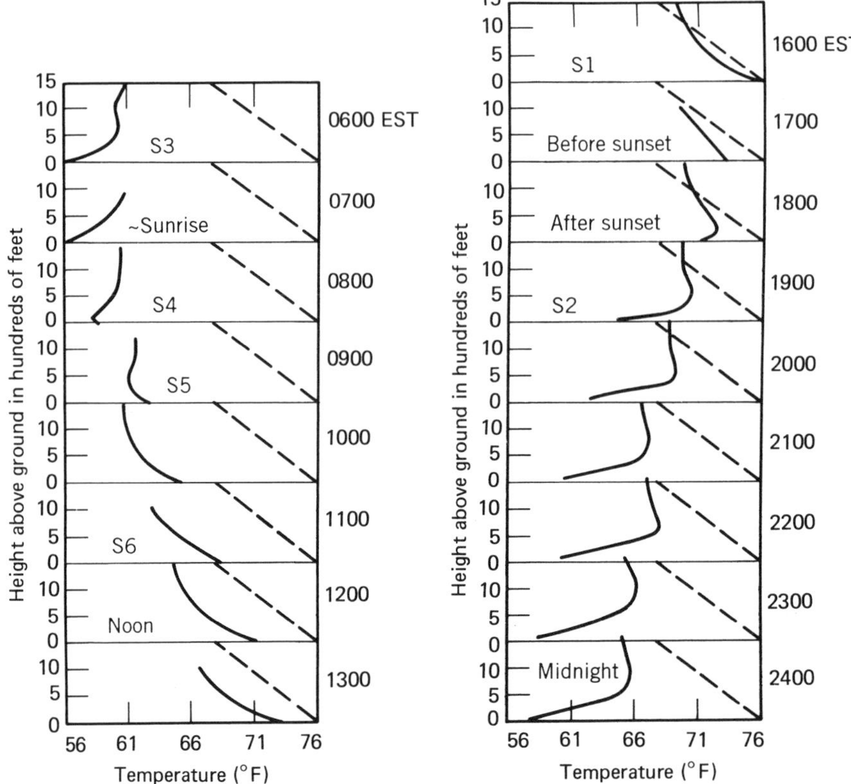

Figure 6.1-2. Average diurnal varation of the vertical temperature structure at the Oak Ridge National Laboratory during the period September to October 1950. The data were obtained from captive-balloon temperature soundings. The dashed line in each panel represents the adiabatic lapse rate. S1, S2, etc. correspond to the local time scale in Fig. 7.2-1. (Modified from Ref. 3.)

buoyancy by a change in the temperature between the dry adiabatic and that in the diabatic (i.e., nonadiabatic) environment, the last equation can also clearly be written as follows:

$$a = \frac{g(\gamma - \Gamma_A)\,\Delta y}{T_1} \tag{6.1-4}$$

where γ is the existing (in general, diabatic) lapse rate in the surrounding air and Δy is the height through which the process operates.

The adiabatic lapse rate can be used as a natural standard of vertical temperature stratification in the atmosphere. Since turbulent eddies are continuously rearranging parcels of air in the lower layers, vertical displacements of air parcels of diabatic lapse rate γ produce the following conditions of air stability: (1) neutral stability,

$$\gamma = \Gamma_A \qquad (T_1 = T_1') \tag{6.1-5}$$

and vertically displaced air parcels tend neither to fall nor to rise; (2) unstable

$$\gamma < \Gamma_A \qquad (T_1 < T_1') \tag{6.1-6}$$

and vertical displacements are unstable and are amplified by buoyancy; (3) stable,

$$\gamma > \Gamma_A \qquad (T_1 > T_1') \tag{6.1-7}$$

and vertical displacements are strongly damped.

In effect, strong heating of the air near the earth's surface causes vertical overturning of the air layers, with resultant increases in turbulence and mixing. Conversely, strong cooling of these layers suppresses mixing and turbulence. Figure 7.2-1 illustrates this for a typical 24-hr period. The time markers S1, S2, and so on, in Fig. 6.1-2 correspond to those in Fig. 7.2-1.

Richardson Number

As demonstrated earlier, instability results from buoyancy effects, which are a function of temperature differences between air at one level and the air above. However, buoyancy effects can be strongly counteracted by the wind stream, which in generating turbulence breaks up the rising eddies. These counteracting influences, the temperature gradient and the wind speed, are related in the dimensionless gradient Richardson number Ri:

$$\mathrm{Ri} \sim \begin{bmatrix} \text{(rate of consumption of turbulent energy by buoyancy forces)} \\ \div\text{(rate of production of turbulent energy by wind shear)} \end{bmatrix} \tag{6.1-8}$$

The ratio is approximated as

$$\mathrm{Ri} \equiv \frac{(g/T_1)(\partial T_1/\partial y)}{(\partial v_x/\partial y)^2} \tag{6.1-9}$$

where g is the acceleration due to gravity, $\partial T_1/\partial y$ and $\partial v_X/\partial y$ are the gradients of temperature and wind speed, and T_1 is the absolute temperature at a level usually defined by[5]

$$\bar{y} = (y_1 y_2)^{1/2} \tag{6.1-10}$$

where $y_2 > y_1$. For application to the first few meters near the ground, the Richardson number approximation above may be used. The exact Ri employs the potential temperature gradient instead of $\partial T_1/\partial y$ (see Problem 6.1B, part 4). Expressing the derivatives as finite differences, Eq. 6.1-9 becomes

$$\mathrm{Ri} \simeq \frac{g}{T_1} \frac{(T_{12} - T_{11})(y_2 - y_1)}{(v_{x2} - v_{x1})^2} \tag{6.1-11}$$

where T_{12} and T_{11} are temperatures at heights y_2 and y_1 and v_{x2} and V_{x1} are wind speeds at the same heights.

Typically, numerical values of the Richardson number varies from $+0.2 \gtrsim \mathrm{Ri} \geqslant -2.0$. It can be said that when $\mathrm{Ri} = 0$ the dynamic atmosphere is vertically neutral in the absence of buoyant forces; when $\mathrm{Ri} < 0$, turbulence increases because the temperature difference with the adiabatic is negative and thus unstable; and when $\mathrm{Ri} > 0$, turbulence is suppressed because $T_1' < T_1$. A comparison with the adiabatic lapse rate results given by Eqs. 6.1-5, 6.1-6, and 6.1-7 indicates that Ri relates the same stability information. The Richardson number has come to be used as a characteristic turbulence parameter rather than an absolute criterion of turbulence. That is, it is regarded as broadly indicating the nature and to some extent the intensity of the turbulence rather than specifying an exact criterion for turbulence to occur.

PROBLEMS

6.1A. Diurnal Variation of Diabatic Lapse Rate

The average diurnal variation of the vertical temperature structure for a period of 18 hr is shown in Fig. 6.1-2.

1. For each time period shown, compute the existing lapse rate γ in the lower atmosphere at the soil surface (°F/1000 ft). (*Hint:* Estimate dT/dy at $y = 0$ by constructing a tangent line to the temperature profile at the ground and determine the slope graphically.)

2. Using the stability criteria given in terms of the lapse rate, determine the stability for each time. Denote the stability as n = neutral, u = unstable, and s = stable.

Note that although a temperature gradient exists for the adiabatic lapse rate condition there is no vertical heat transfer in this atmosphere.

6.1B. Dry Adiabatic Lapse Rate

1. Beginning with Eq. 3.2-5, obtain a pressure versus height relationship for a dry atmosphere.

2. Beginning with the first law of thermodynamics as $dq = C_v\,dT + p\,dV$, derive Eq. 6.1-1.

3. Show that Eq. 6.1-1 does yield the correct numerical value and dimensions for the dry adiabatic lapse rate.

4. The absolute potential temperature, θ, is defined as the temperature an elevated air parcel, with actual T_1 and p, would have at standard sea level pressure of 1000 mbar, given by $\theta = T_1(1000/p)^{R/C_p}$. Use the first law in part 2 to obtain this form of the Poisson equation.

6.2. CHEMICAL FLUX RATES THROUGH LOWER LAYER OF ATMOSPHERE

The soil surface is both a source and a sink for anthropogenic and biogenic chemicals. In this section we present the mechanisms and rates of chemical movement within the layers of the atmosphere next to the soil interface, $y \lesssim 10$ m. In principle, transport here is similar to that within the atmospheric layers above water so that the information here is an extension of that presented in Chapter 4, with special emphasis on the soil–surface interactions.

Thornthwaite–Holzman Equation

The aerodynamic approach of estimating the flux of chemicals toward and away from a soil or vegetative surface, based on gradient measurements obtained in the air boundary layer, depends on an adequate expression for the transfer of momentum as a function of change of wind speed with height and a knowledge of the relationship between eddy transfer mechanisms for momentum and for the chemical of interest. The following is based on the development presented by Brooks and Pruitt.[6]

Under conditions of neutral atmospheric stability, wind speed can be described as a function of elevation over relatively smooth surfaces and short crops by Eq. 3.2-41 for $y > h > y_0$, where $v_* \equiv \sqrt{\tau_0/\rho_1}$ and τ_0 is the shear stress (the flux of horizontal momentum transferred vertically and absorbed by the ground and constant in the lower layers of the atmosphere), ρ_1 the density of air, and h the crop height. For rough surfaces (e.g., tall crops) the zero plane displacement d is introduced into Eq. 3.2-41, so that it becomes

$$v_x = \frac{v_*}{\kappa_1} \ln \frac{y - d}{y_0} \qquad \text{for } y > h > y_0 + d \tag{6.2-1}$$

Means of obtaining d, y_0, and v_* from measured wind-speed profiles under neutral or nearly neutral atmospheric stability conditions are given by Rosenberg.[5] He also gives relationships, developed from large accumulations of data in the micrometeorology literature, for estimating the magnitude of y_0 and d. In very short crops (e.g., lawns), y_0 describes the roughness, and little adjustment of the zero plane is necessary. For tall crops y_0 is related to crop height h by

$$\log y_0 = 0.997 \log h - 0.883 \tag{6.2-2}$$

In tall crops y_0 is no longer adequate to describe the roughness, and a value of d, the zero plane displacement, is needed. For a wide range of crops and heights, $0.02 < h < 25$, an equation giving the zero plane displacement has been obtained:

$$\log d = 0.979 \log h - 0.154 \tag{6.2-3}$$

Crop height h, roughness coefficient y_0, and displacement d in the preceding equations are in meters.

The aerodynamic approach to quantifying chemical transport to or from a soil surface requires the quantification of the advection and vertical diffusion components, as illustrated in Fig. 6.2-1. Masts are shown at points A, B, and C. At predetermined heights along the mast wind speed, temperature, relative humidity, and chemical concentrations in air are measured. A quantitative assessment of chemical losses is based on a specified soil surface area and a well-defined control volume in the air space above the soil surface. Assume that the soil surface is area A (m^2), is rectangular, of length l (m), width w (m), and the wind approach is perpendicular to one side. The control volume is then Ah (m^3), where h is the mast height in meters above the soil surface.

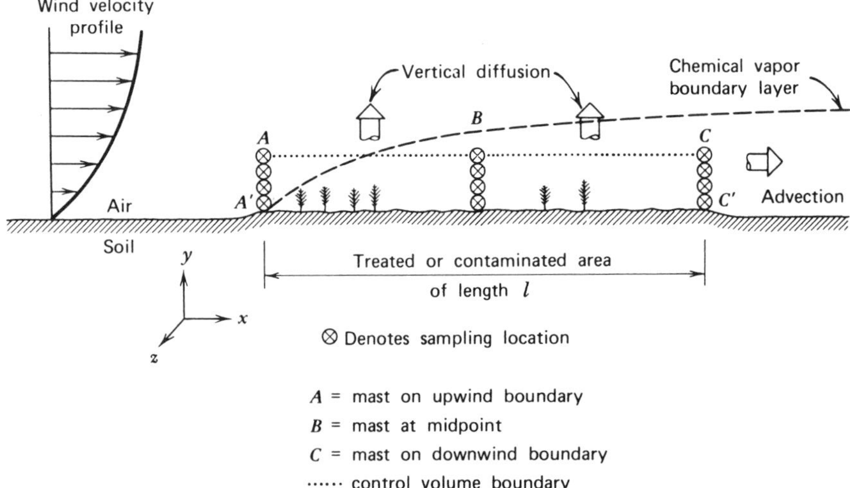

Figure 6.2-1. Aerodynamic method of quantifying chemical emission from a soil surface.

Vertical Diffusion. Under adiabatic or near-adiabatic conditions, the well-known logarithm law indicates a straight-line relationship between wind velocity and the logarithm of height when adjustments are made for the influence of the crop in displacing the effective surface upward. Considering the velocity at two heights above the soil surface, v_{x1} at y_1 and v_{x2} at y_2, Eq. 6.2-1 can be employed to yield

$$\frac{v_*}{\kappa_1} = \frac{v_{x2} - v_{x1}}{\ln[(y_2 - d)/(y_1 - d)]} \tag{6.2-4}$$

Now employing the definition of the friction velocity, $v_* \equiv \sqrt{\tau_0/\rho_1}$, Eq. 6.2-4 can be used to obtain the shear stress τ_0:

$$\tau_0 = \frac{\rho_1 \kappa_1^2 (v_{x2} - v_{x1})^2}{\{\ln[(y_2 - d)/(y_1 - d)]\}^2} \tag{6.2-5}$$

The shear stress is also expressed in the momentum transport equation (Eq. 3.2-33):

$$\tau_0 = -\rho_1 v_1^{(t)} \frac{dv_x}{dy} \tag{6.2-6}$$

where $v_1^{(t)}$ is the eddy viscosity or eddy transfer coefficient for momentum in L^2/t.

The equivalent transport expression for chemical A is as follows (Eq. 3.1-6 mass rate form):

$$n_{Ay} = +\mathscr{D}_{A1}^{(t)} \frac{d\rho_{A1}}{dy} \tag{6.2-7}$$

where n_{Ay} is the chemical flux in the positive y direction ($M/L^2 \cdot t$), $\mathscr{D}_{A1}^{(t)}$ the turbulent diffusion coefficient for the chemical in air (L^2/t), and ρ_{A1} the concentration of A in air (M/L^3). Dividing Eq. 6.2-7 by Eq. 6.2-6 and converting the gradients to finite differences, we obtain

$$\frac{n_{A0}}{\tau_0} = -\frac{\mathscr{D}_{A1}^{(t)}}{v_1^{(t)}} \frac{\rho_{A12} - \rho_{A11}}{v_{x2} - v_{x1}} \tag{6.2-8}$$

where ρ_{A12} and ρ_{A11} are the concentrations at y_2 and y_1, respectively. Substituting Eq. 6.2-5 for τ_0 and solving for n_{Ay} results in

$$n_{A0} = -\frac{\kappa_1^2 \mathscr{D}_{A1}^{(t)}}{v_1^{(t)}} \frac{(v_{x2} - v_{x1})(\rho_{A12} - \rho_{A11})}{\{\ln[(y_2 - d)/(y_1 - d)]\}^2} \tag{6.2-9}$$

If under adiabatic conditions it is assumed that $\mathscr{D}_{A1}^{(t)}/v_1^{(t)} = 1.0$, Eq. 6.2-9 becomes the mass flux equation developed by Thornthwaite–Holzman in 1939.

Modified Thornthwaite–Holzman Equation. As indicated earlier, the preceding approach is reliable only under near-adiabatic conditions and then only if $\mathscr{D}_{A1}^{(t)}/v_1^{(t)}$ does indeed equal 1.0. When surface heating or cooling produce a nonadiabatic temperature profile, the straight-line relationship v_x versus $\ln(y - d)$ no longer holds, especially under calmer conditions. In the development of Eq. 6.2-1, the neutral condition relation, Eq. 3.2-37, was used.

The neutral stability condition is an exception rather than the rule in the lower atmosphere. More often, the exchange of sensible heat, q_0, between the surface and the atmosphere leads to thermal stratification of the surface layer as shown in Section 6.1. The Monin–Obrikov similarity theory as outlined by Arya[7] is a suitable theoretical and semiempirical framework for a quantitative description of the mean and turbulence structure of the stratified layer. The simplifying assumptions of the theory are that the flow is horizontally homogeneous and quasi-stationary, the turbulent fluxes of momentum and heat are constant (independent of y), molecular transport is insignificant and the influence of surface roughness, boundary layer height, and wind is fully accounted for through v_*. Monin and Obukhov proposed that in the surface layer the mean flow and turbulence characteristics depend only on four independent variables: y, v_*, $q_0/\rho_1\hat{C}_{p1}$, and g/T_1. Equations 3.3-16 and 3.2-49 suggest the origin of the surface-sensible kinematic heat flux, q_0 being positive into the atmosphere, and the buoyancy variables, respectively. Since three fundamental dimensions are involved (L, t, T), Buckingham's theorem limits formulation of only one dimensionless group. It is traditionally expressed as

$$\zeta \equiv \frac{y}{L_m} \tag{6.2-10}$$

where

$$L_m = -\frac{\rho_1 \hat{C}_{p1} v_*^3 T_1}{\kappa_1 g q_0} \tag{6.2-11}$$

is known as the Obukhov length. The ratio y/L_m is an important parameter measuring the relative importance of buoyancy versus shear effects in the stratified surface, similar to the Richardson number (Ri) introduced earlier. Physically, L_m is the approximate height above the surface at which the production of turbulence by buoyancy effects becomes comparable to that by shear effects.

Thermal buoyancy will distort the velocity profiles from the simple, linear v_x versus $\ln(y)$ relationship. This distortion is accounted for by modifying Eq. 3.2-37, the velocity gradient in the neutral surface layer, with a correction factor to give

$$\frac{dv_x}{dy} = \frac{v_*}{\kappa_1 y} \phi_m \tag{6.2-12}$$

where ϕ_m is a function of y/L_m. With the proper value of ϕ_m, Eq. 6.2-12 can be used for all stability conditions. The Richardson number is closely related to y/L_m, and it has been observed that for Ri > 0, a good approximation is given by $\mathrm{Ri} = y/L_m$. There have been many efforts to define precisely the functional relationships between Ri and ϕ_m in stable and unstable conditions.

After reviewing field measurements and the associated uncertainties, Arya[7], suggests the following parameterization for the basic universal similarity function for momentum, ϕ_m versus ζ, for most practical situations:

$$\phi_m^2 = (1 - 15\zeta)^{(-1/2)} \qquad \text{for } \zeta < 0 \tag{6.2-13a}$$

and

$$\phi_m = 1 + 5\zeta \qquad \text{for } \zeta \geqslant 0 \tag{6.2-13b}$$

He also recommends the following relating ζ and Ri:

$$\zeta = \mathrm{Ri} \qquad \text{for Ri} < 0 \tag{6.2-14a}$$

and

$$\zeta = \frac{\mathrm{Ri}}{1 - 5\mathrm{Ri}} \qquad \text{for } 0 \leqslant \mathrm{Ri} \leqslant 0.2 \tag{6.2-14b}$$

Recall that $\zeta \equiv y/L_m$. These are based on experiments performed in a Kansas wheat field during 1968 for the specific purpose of determining such Monin–Obukhov similarity functions.

In a fashion similar to Eq. 6.2-12, hence the notion of similarity theory, the gradients of temperature and humidity are corrected as

$$\frac{dT_1}{dy} = \frac{q_0}{\kappa_1 \rho_1 \hat{C}_{p1} v_* y} \phi_h \tag{6.2-15}$$

where ϕ_h is the similarity function for heat transfer and

$$\frac{d\rho_{B1}}{dy} = \frac{n_{B0}}{\kappa_1 v_* y} \phi_w \tag{6.2-16}$$

where ϕ_w is the similarity function for water vapor ($B \equiv H_2O$) as a function of ζ. Arya[7] recommends $\phi_h = \phi_m^2$ for $\zeta < 0$ and $\phi_h = \phi_m$ for $\zeta \geqslant 0$. Development of a relationship for ϕ_w follows.

In analogy with Newton's law, Boussinesq in 1877 proposed Eq. 6.2-6 for the turbulent shear stress, where $\nu_1^{(t)}$ is called the eddy exchange coefficient for momentum. Similar expressions have been proposed for the turbulent fluxes of heat, Eq. 3.2H-3, water vapor, and other transferable constituents (e.g., pollutants), Eq. 6.2-7, which are analogous to Fourier's and Fick's laws, respectively. Here $\alpha_1^{(t)}$ and $\mathscr{D}_{B1}^{(t)}$ are called the *exchange coefficients*; separate K's are the common symbols used for these three coefficients in the micrometeorology literature and the flux relationships are those of the K-theory.

It is now possible to revisit the procedure leading to Eq. 6.2-9; when performed it yields

$$n_{BO} = -\frac{\kappa_1^2}{\phi_m^2}\frac{\mathscr{D}_{B1}^{(t)}}{\nu_1^{(t)}}\frac{(v_{x2} - v_{x1})(\rho_{B12} - \rho_{B11})}{[\ln(y_2 - d)/(y_1 - d)]^2} \tag{6.2-17}$$

This result is for water vapor evaporation from the topmost soil layers, a natural chemical tracer convenient for field testing. It is common to assume that $\alpha_1^{(t)} = \mathscr{D}_{B1}^{(t)}$. A few micrometeorological experiments have been performed and it appears that $\phi_w = \phi_h$, so that the similarity functions given for heat transfer are appropriate for water vapor. Further, it is easy to show that

$$\frac{\alpha_1^{(t)}}{\nu_1^{(t)}} = \frac{\phi_m}{\phi_h} \tag{6.2-18}$$

and Eq. 6.2-17 can be simplified further, yielding the modified Thornthwaite–Holzman equation. The final result provides a gradient or aerodynamic method, by which measurements of concentration in addition to those of windspeed and temperature at two levels, y_1 and y_2, in the surface layer near the interface may be used to determine chemical flux to or from the surface. The methodology has been further developed to reduce the number of measurements.[26]

Air stability also affects the profile equations. Equation 6.2-1 is applicable only under conditions of neutral atmospheric stability. The diabatic profile equations must now be related to stability and other parameters, as discussed earlier. The diabatic profile equations corresponding to the neutral stability profile equations are

$$v_x = \frac{v_*}{\kappa_1}\left(\ln\frac{y}{y_0} - \psi_m\right) \tag{6.2-19}$$

for wind velocity,

$$T_1 = T_1|_{y_0} - \frac{q_0}{\kappa_1\rho_1\hat{C}_{p1}v_*}\left(\ln\frac{y}{y_0} - \psi_h\right) \tag{6.2-20}$$

for temperature, and

$$\rho_{A1} = \rho_{A1}|_{y_0} - \frac{n_{A0}}{\kappa_1 v_*}\left(\ln\frac{y}{y_0} - \psi_h\right) \tag{6.2-21}$$

for the concentration of chemical A. The heat, q_0, and mass flux, n_{A0}, are positive away from the surface. Both $T_1|_{y_0}$ and $\rho_{A1}|_{y_0}$ are extrapolated values to $y = y_0$. The integral functions ψ_m and ψ_h are obtained from the original similarity functions; those corresponding to Eq. 6.2-13 are

$$\psi_m = -\frac{5y}{L_m} \qquad \text{for } \frac{y}{L_m} \geqslant 0 \tag{6.2-22a}$$

and

$$\psi_m = \ln\left[\left(\frac{1+x^2}{2}\right)\left(\frac{1+x}{2}\right)^2\right] + -2\tan^{-1}x + \frac{\pi}{2} \qquad \text{for } \frac{y}{L_m} < 0 \tag{6.2-22b}$$

where $x \equiv (1 - 15y/L_m)^{1/4}$. The corresponding ϕ_h integral functions are

$$\psi_h = \psi_m \qquad \text{for } \frac{y}{L_m} \geqslant 0 \tag{6.2-23a}$$

and

$$\psi_h = 2\ln\frac{1+x^2}{2} \qquad \text{for } \frac{y}{L_m} < 0 \tag{6.2-23b}$$

Recall that $y/L_m \equiv \zeta$. Under stable conditions the profiles are log linear, becoming linear for large values of y/L_m. Under unstable conditions the profiles in the surface layer are expected to become more and more curvilinear as stability increases. See Eqs. 3.3-15 to 3.3-17 for comparison with the neutral atmosphere conditions. See deVries and Afgan[8] on the aerodynamics of vegetated surfaces, plant and forest canopies.

Because of the effect of stability on the wind profile, measurements taken to permit accurate estimation of y_0, the roughness parameter, and d, the zero place displacement, must be done during periods of near-neutral stability. There are other precautions that must be taken when using the aerodynamic method of estimating vertical flux of chemicals. Adequate upwind distance (fetch) must be guaranteed so that the boundary layer developed is that equilibrated with the surface of interest. This means that there must be a long expanse of uniform cover upwind from the site at which the wind profile is measured. Parmele[9] relates that the height/fetch ratio requirement should be 1:40 for minor discontinuities and at least 1:100 or 1:200 if the discontinuity is severe. Campbell[2] reports that the wind can usually be assumed to be 90% or more equilibrated with the new surface to heights of 0.01 times fetch. Thus at a distance 1000 m downwind from the edge of a uniform field of grain, we might expect our wind profile equations to be valid to heights of 10 m.

Example 6.2-1. Characteristics of Pentachlorophenol (PCP) Transport in Lower Atmospheric Boundary Layer. The surface soil of a 4-ha site is contaminated with PCP. For a wind speed of 2 m/s at 2-m elevation and $y_0 = 5$ cm, estimate the friction velocity and the PCP flux/concentration difference ratio at 25°C for (a) stable (Ri = 0.1), (b) neutral (Ri = 0), and (c) unstable (Ri = −0.5) conditions.

SOLUTION The integral similarity functions, ψ_h and ψ_m, are most useful for flux determinations with Eqs. 6.2-19 and 6.2-21. The former equation yields v_*, the friction velocity. It, combined with the latter equation, yields the required ratio,

$$\frac{n_{A0}}{\rho_{A1} - \rho_{A1}|_{y_0}} = \frac{\kappa_1 v_x}{\ln(y/y_0) - \psi_h}$$

where A is for water vapor. With Ri, Eqs. 6.2-14, 6.2-22, and 6.2-23 yield the following numerical values:

Stability	Ri	ζ	ψ_m	ψ_h	v_* (m/s)	$n_{A0}/\Delta\rho_{A1}$ (cm/h)
(a)	0.1	0.11	−0.55	−0.55	0.188	2560
(b)	0	0	0	0	0.217	3390
(c)	−0.5	−0.5	0.77	1.3	0.274	6600

The entries in the last column are for water vapor (≡ A). In the case of PCP (≡ B) these values should be corrected by $(\mathscr{D}_{B1}/\mathscr{D}_{A1})^n$, according to Eq. 3.1-25, with $n = \frac{2}{3}$. This will reduce the values by approximately 0.5. This example illustrates that as boundary layer instability increases, so does the transport coefficient for PCP.

Advection Component. Quantifying the vertical diffusion component, illustrated in Fig. 6.2-1, is made possible by use of the Thornthwaite–Holzman modified equation (Eq. 6.2-17). By its use the vertical flux of chemical A in the vicinity of each mast is obtained, yielding $\overline{n_{A0}}$, representative of the A–B–C plane in Fig. 6.2-1. The mass rate of chemical A leaving through the top of the control volume is then

$$w_{A,\text{diff}} = \overline{n_{A0}}lw \tag{6.2-24}$$

To complete the material balance on the control volume, the advective component must be obtained.

The advective component is the mass rate of movement due to the bulk airflow. It is obtained by

$$n_{Ax} = \frac{1}{h} \int_{y=0}^{y=h} v_x \rho_{A1} \, dy \tag{6.2-25}$$

where n_{Ax} is the horizontal component (x-direction) flux of chemical A at point x, v_x is the wind profile, ρ_{A1} is the chemical concentration profile, and h the mast height. Measurements of v_x and ρ_{A1} at specified heights can be used in Eq. 6.2-25 to calculate the advective component of the flux at the $A'-A$ plane and the $C'-C$ plane shown in Fig. 6.2-1. This will yield the respective fluxes $n_{A,x=A}$ and $n_{A,x=c}$. The mass rate due to advection is then

$$w_{A,\text{adv}} = (n_{A,x=c} - n_{A,x=A})hw \tag{6.2-26}$$

where w is the width of the soil surface.

Quantifying the mass rate of chemical A leaving the air control volume through the $A-B-C$ plane, and the difference between that entering the $A'-A$ plane and leaving the $C'-C$ plane determines the amount desorbing from the soil surface of area wl. The total quantity is obtained from the sum of the advective and diffusive components given by Eqs. 6.2-26 and 6.2-24, respectively. If the samples are closely spaced vertically and the gradient adjacent to the soil surface is used in computation, the advective component may be negligible.

Selected Field Observations of the Movement of Chemicals Between Air and Soil Surfaces

Volatilization of Pesticides from Field with Vegetation.[10] The postapplication losses of pesticides by volatilization is a pathway for general environmental contamination and is a process limiting their effectiveness. The work described here was designed to measure the volatilization of dieldrin and heptachlor, two chemically persistent insecticides, over a period of 3 weeks of warm summer weather after their application to field vegetation. A uniform grass pasture, freshly mowed to 10-cm height, was chosen.

The site was a 3.34-ha rectangular field. Between 0930 and 1030 EDT on July 12, dieldrin and heptachlor were applied together as a single uniform spray containing 5.6 kg/ha of both chemicals. The application was made with a regular farm spray rig equipped with a 21-ft spray boom mounted at about 70 cm in height. The insecticides were applied to a rectangular 2.00-ha area (82 by 244 m) within the total experimental area, leaving untreated strips 27 m wide along the northern and southern boundaries. These areas were left to ensure a smooth wind fetch over the boundary of the treated field without interference from fences or changes in vegetation height.

The vertical flux intensities were calculated from the concentration gradients by the aerodynamic method, using wind-speed profile data obtained from anemometer masts. On each sampling date, insecticide concentrations were measured at five heights (10, 20, 30, 50, and 100 cm) above the grass surface at two locations in the treated area, one in the center and the second on the downwind edge. Differential measurements of the air temperature were taken at 20- and 50-cm heights so that atmosphere stability corrections could be made (see Problem 6.2A for details).

Vertical flux intensities for days 3 and 6 are plotted in Fig. 6.2-2*a* and *b*, respectively. Total quantities of pesticides lost by volatilization on each sampling day are presented in Table 6.2-1. These results were obtained by integration of the hourly flux values with the assumption that volatilization was small and could be neglected before 0600 and after 2200 to 2300 EDT.

Marked diurnal variations were observed in the volatilization of both insecticides during the period of greatest loss early in the experiment, the rates closely following the diurnal variation in solar radiation. Flux intensities were controlled by the rate of evaporation from plant surfaces. Dispersion by turbulent diffusion was never limiting. As overall volatilization decreased because of depletion of the residues remaining on plant surfaces, diurnal variations were less marked. Up to 1300 EDT on the first day, about 40% of the dieldrin and 58% of the heptachlor applied can be accounted for as being in or having evaporated from the target area, the remainder being lost directly to the atmosphere as vapor or spray drops that never reached the target area. Estimates of the total postapplication losses were made up to the twenty-third day. The dieldrin loss was 1900 ± 250 g/ha, of which about 35% was lost on the first day and 90% in the first 7 days. The heptachlor estimates for the

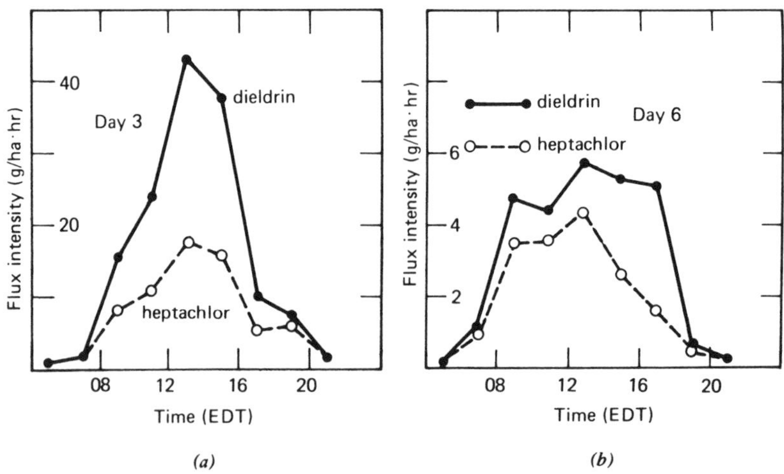

Figure 6.2-2. Vertical flux intensities of dieldrin and heptachlor during 2-hr sampling periods on (*a*) day 3 and (*b*) day 6. (Reprinted by permission from Ref. 10.)

Table 6.2-1. Observed Daily Volatilization Losses of Dieldrin and Heptachlor (g/ha · day)

Day	Daily Losses by Volatilization (g/ha · day)	
	Dieldrin	Heptachlor
1	654	2554
2	(325)[a]	(335)[a]
3	282	132
6	53.7	33.5
9	40.0	24.1
14	9.2	6.9
23	6.2	7.4

Source: Reprinted by permission from Ref. 10.

[a]Estimates assuming loss between 0400 and 1200 EDT is 30% of total.

23-day period were 3200 $\pm$ 250 g/ha, with 75% of this on the first day and 95% in the first week.

Ammonia Flux into the Atmosphere from Grazed Pasture. Volatile inorganics will also move from the ground surface into the lower atmospheric boundary layer. The statement of Problem 6.2B describes the details on measurements of ammonia in the atmospheric boundary layer above an alfalfa field.

Closure. The material above was concerned with a chemodynamic model by which numerous measurements in air for profiles in the boundary layer were used to estimate vaporization flux from a surface. In the next section a technique for estimating this flux will be developed based on the air-side mass transport coefficient and chemical concentrations in the uppermost soil layer.

Evaporation of Volatile Chemicals Spilled or Otherwise Placed on Land

The following discussion is confined to a process that is fairly common in the lower atmosphere, the removal of chemical vapors from free-liquid pools or soil surfaces containing quantities of volatiles in pure or sorbed form. The rate of evaporation from the surface is an important process when considering the fate of a chemical in this soil locale. The fate here is also controlled by irreversible absorption onto and wind erosion of soil solids, degradation to other species in place, and downward leaching of porewater. Evaporative losses are nearly always significant in cases of oil, fuel, and solvent spills, decontamination by microorganisms (i.e., bioremediation), pesticide application, and during excavation of hazardous substances, for example. The evaporation rate is key in assessing the duration and intensity of the volatile chemical hazard in air directly above or downwind from the site.

A solution of the general problem requires a determination of the rate of evaporation of a given chemical spill as a function of temperature, wind speed, atmospheric conditions, solar radiation, ground conditions, the dimensions of the chemical spill, the volatility, and diffusion characteristics of the chemical. This estimate involves a description of the total mass transfer process from the soil to the atmosphere. This section is concerned mainly with the vapor-phase resistance, for which a theoretical model and an empirical model are presented. In some cases a description of the soil phase resistance may or may not be necessary; depending on conditions and this and other aspects of the general problem are discussed only qualitatively, with a detailed presentation in a separate section.

Sutton[1] presents a theoretical study of evaporation by focusing on the diffusional aspects of the removal of water vapor from a free liquid or permanently saturated soil surface. The problem is concerned with limited areas such as pools of liquid and patches of saturated soil. The two-dimensional evaporation problem is illustrated in Fig. 6.2-3.

A systematic approach to the chemical evaporation problem begins with the general equation for the fate of chemical A, Eq. 3.0-3. The expanded version for the air layers above a soil surface in mass concentration dimensions is

$$\frac{\partial \rho_{A1}}{\partial t} + \rho_{A1}\frac{\partial v_x}{\partial x} + \rho_{A1}\frac{\partial v_y}{\partial y} + \rho_{A1}\frac{\partial v_z}{\partial z} + v_x\frac{\partial \rho_{A1}}{\partial x} + v_y\frac{\partial \rho_{A1}}{\partial y} + v_z\frac{\partial \rho_{A1}}{\partial z}$$
$$= \frac{\partial}{\partial x}\left(\mathscr{D}_{A1}^{(t)}\frac{\partial \rho_{A1}}{\partial x}\right) + \frac{\partial}{\partial y}\left(\mathscr{D}_{A1}^{(t)}\frac{\partial \rho_{A1}}{\partial y}\right) + \frac{\partial}{\partial z}\left(\mathscr{D}_{A1}^{(t)}\frac{\partial \rho_{A1}}{\partial z}\right) + r_{A1} \qquad (6.2\text{-}27)$$

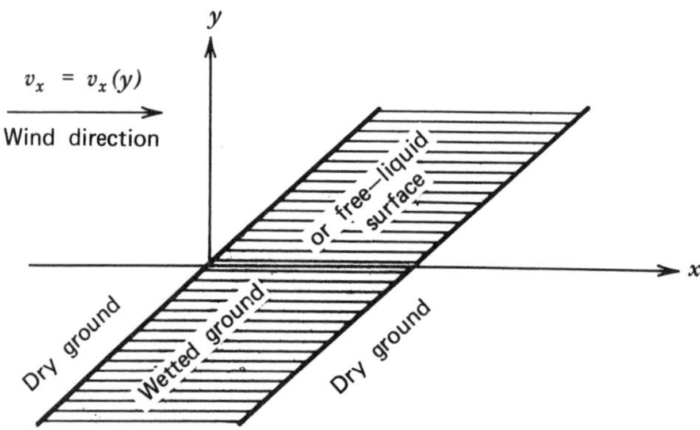

Figure 6.2-3. Two-dimensional evaporation into a steady wind. (From Ref. 1.)

The coefficient of eddy diffusivity $\mathscr{D}_{A1}^{(t)}$ is used rather than the combined eddy and molecular diffusivity because the applications to be presented are concerned with the turbulent regions of the air boundary layer. If the diffusion process is specified as a constant-source and constant-density problem, Eq. 6.2-27 may be simplified by use of the continuity equation, Eq. 3.2-2, to give

$$v_x \frac{\partial \rho_{A1}}{\partial x} + v_y \frac{\partial \rho_{A1}}{\partial y} + v_z \frac{\partial \rho_{A1}}{\partial z} = \frac{\partial}{\partial x}\left(\mathscr{D}_{A1}^{(t)} \frac{\partial \rho_{A1}}{\partial x}\right) + \frac{\partial}{\partial y}\left(\mathscr{D}_{A1}^{(t)} \frac{\partial \rho_{A1}}{\partial y}\right) + \frac{\partial}{\partial z}\left(\mathscr{D}_{A1}^{(t)} \frac{\partial \rho_{A1}}{\partial z}\right) + r_{A1} \qquad (6.2\text{-}27a)$$

By using a single-component wind and neglecting the downwind diffusion term and the crosswind diffusion term and assuming no reaction, Eq. 6.2-27*a* can be further simplified to

$$v_x \frac{\partial \rho_{A1}}{\partial x} = \frac{\partial}{\partial y}\left(\mathscr{D}_{A1}^{(t)} \frac{\partial \rho_{A1}}{\partial y}\right) \qquad (6.2\text{-}27b)$$

The final equation contains a horizontal advective component and a vertical diffusive component.

The boundary conditions must be specified for a solution to the two-dimensional problem. It is evident that $\rho_{A1} = \rho_{A1b}$, the background concentration of A in air, if $x < 0$, for all y, and that $\rho_{A1} \rightarrow \rho_{A1b}$ as $y \rightarrow \infty$ for $0 \ll x \ll L$. The value of ρ_{A1} at all points on the evaporating surface is a constant concentration identified with the saturation value ρ_{A1i}, which depends only on the temperature of the surface and the nature of the evaporating liquid.

At this point a mathematical difficulty enters. In conditions of various stabilities, the velocity profile is most accurately represented by a logarithmic function, Eq. 6.2-19, and $\mathscr{D}_{A1}^{(t)}$ is proportional to y and ϕ_h^{-1}, but the analytical solution of Eq. 6.2-27*b* for these two conditions has not yet been found. If a power law relationship for velocity is used, this aids in the solution and its form is

$$v_x = v_x(y_1)\left(\frac{y}{y_1}\right)^m \qquad (6.2\text{-}28)$$

where $v_x(y_1)$ is the value of v_x at a fixed reference height y_1, which by general agreement is 10 m. This formula is used with values of the parameter m given in Table 6.2-2 for six air stability classes (see Table 7.2-1 for definitions of A through F) and two general surface roughness conditions. The power law is less accurate than the theoretical formula and should not be used for $y > 200$ m. In Eq. 6.2-28, m is a constant:

$$m = \frac{n}{2 - n} \qquad (6.2\text{-}29)$$

Table 6.2-2. Estimates of the Power *m*

	Stability Class					
	A	B	C	D	E	F
Urban	0.15	0.15	0.20	0.25	0.40	0.60
Rural	0.07	0.07	0.10	0.15	0.35	0.55

Source: Ref. 23.

This relation enables n to be found from observations on wind structure near the ground. If the seventh-root profile is used, neutral conditions are assumed and the value of n is $\frac{1}{4}$. The eddy diffusivity is also known to be a function of velocity and height. Sutton develops an expression that enables the eddy viscosity to be evaluated for flow near an aerodynamically smooth surface, based on the power-law velocity profile.

The problem of evaporation from a saturated or free-liquid plane surface, extended indefinitely in the cross-wind and finite length downwind, has been solved by Sutton[1] for the wind profile (Eq. 6.2-28) and the eddy diffusivity expressed by the eddy viscosity. The boundary conditions are

$$\lim_{y \to 0} \rho_{A1}(x, y) = \rho_{A1i} \qquad \text{for } 0 \leqslant x \leqslant L \tag{6.2-30a}$$

$$\lim_{x \to 0} \rho_{A1}(x, y) = \rho_{A1b} \qquad \text{for } y > 0 \tag{6.2-30b}$$

$$\lim_{y \to \infty} \rho_{A1}(x, y) = \rho_{A1b} \qquad \text{for } 0 \leqslant x \leqslant L \tag{6.2-30c}$$

The solution of Eq. 6.2-27b subject to the preceding boundary conditions is given by Sutton.[1] It gives the vapor concentration over the wetted area as a function of distance from the leading edge x and height above the surface y as $\rho_{A1}(x, y)$. The total rate of evaporation is

$$w_{A0} = \int_0^\infty v_x(\rho_{A1}(L, y) - \rho_{A1b})\, dy \tag{6.2-31}$$

since the integral obviously represents the total mass of vapor carried across the plane $x = L$ by the wind. Employing Eq. 6.2-28 and $\rho_{Ai}(x, y)$, the total rate of evaporation per unit crosswind length is shown to be[1]

$$w_{A0} = av_x(y_1)^{(2-n)/(2+n)} L^{2/(2+n)}(\rho_{A1i} - \rho_{A1b}) \tag{6.2-32}$$

where a is also a complex function of n, y_1, v_1, and κ_1. This expression shows that the rate of evaporation from a smooth surface can be calculated knowing these parameters in addition to the wind velocity at y_1, the interface, and background concentration of A in air. Employing Eq. 6.2-32 for obtaining an expression of the gas-phase mass transfer coefficient yields

$$^3k'_{A1} = cv_x(y_1)^{(2-n)/(2-n)}L^{-n/(2+n)} \tag{6.2-33}$$

where c is a constant. The exponent n is a function of ground roughness and temperature profile in the atmosphere, and typical values are from 0.25 to 1.00. For average atmospheric conditions a value of 0.25 for n was assumed and the equation becomes

$$^3k'_{A1} = c[v_x(10\text{ m})]^{0.78}L^{-0.11} \tag{6.2-34}$$

The analysis above indicates that the transport in the neutral boundary layer is primarily a function of wind speed and length of the evaporating surface parallel to the wind direction, L, more commonly termed the *fetch*. The important result is the function of L; as it increases the coefficient decreases, accounting for the buildup of A in the boundary layer and therefore effectively compensating for ρ_{A1b} being constant in the flux expression, Eq. 6.2-32.

A brief introduction to boundary layers appears in Chapter 3. Equation 3.1-24 is a semitheoretical coefficient for parallel flow past a sharp-edged flat plate. Here the buildup of both concentration and momentum boundary layers begins at the tip of the plate; nevertheless, the velocity and plate length functions are similar to those in Eq. 6.2-34. A review of such correlations for air–soil exchange coefficients comparing laboratory and field data has been made.[24]

Evaporation studies provide a means of bridging the theoretical voids and obtaining pragmatic results for estimating such coefficients. The evaporation data for numerous volatile organics and water in wind tunnels and similar apparatus are correlated well by

$$^3k'_{A1} = \frac{0.04\mathscr{D}_{A1}\text{Re}^{0.78}\text{Sc}_{A1}^{1/3}}{L} \tag{6.2-35}$$

in the range $1500 < \text{Re} < 300{,}000$, where $\text{Re} \equiv v_\infty L/\nu_1$ is the plate Reynolds number. The use of evaporating pans outdoors yielded a very similar empirical correlation,[1,12]

$$^3k'_{A1} = \frac{0.0292v_x^{0.78}}{\text{Sc}_{A1}^{2/3}L^{0.11}} \tag{6.2-36}$$

where the coefficient is in m/h, the wind speed, v_x, is in m/h measured at a height of 10 m, and L is in m.

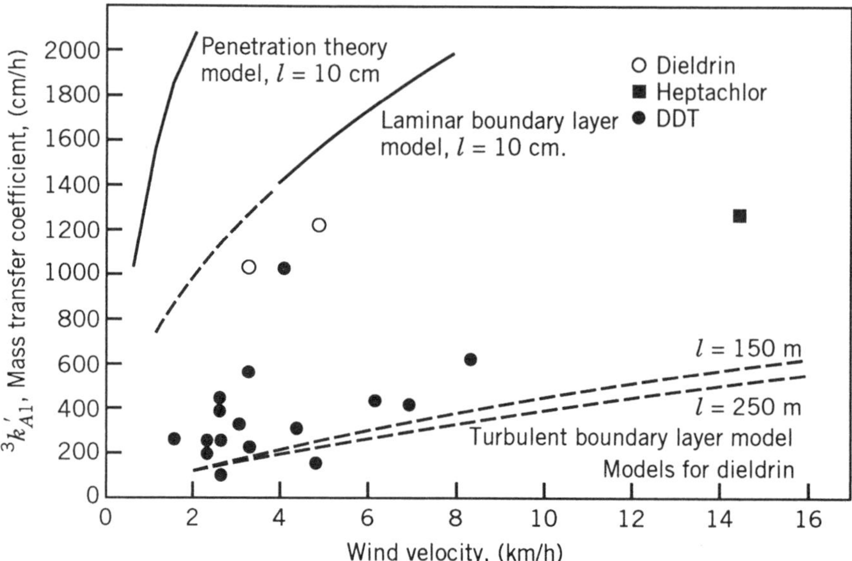

Figure 6.2-4. Pesticide air/soil coefficiants for evaportation as a function of wind speed and "pool" fetch. (From Ref. 24.)

Results of laboratory and field experiments on pesticide evaporation from surface soils provide a realistic test of the algorithms.[24] Figure 6.2-4 contains a few coefficient $^3k'_{A1}$ measurements and results of model predictions as a function of wind speed. If applicable, laminar flow models such as Eq. 3.1-23 are appropriate only for ridiculously short pool fetch values and therefore display exceedingly high coefficients. Equation 6.2-35 was used for the turbulent model. In general, the measured coefficients are higher than the model predicted values. The DDT values were with cotton plants 50 cm in height.[25] It has been noted in laboratory experiments that in the case of rough surfaces the coefficient is higher than for smooth surfaces. For roughness protrusions that do not extend into the turbulent fluid, the increase in $^3k'_{A1}$ is very slight; however, for large protrusions, increases of three to four times might be expected.

The effect of air stability on the transport coefficient is unaccounted for since the correlations above are for smooth surfaces and neutral conditions. Assuming that a very thin diffusive sublayer exists adjacent to the surface, thermal instabilities, which normally generate turbulence here, are of little significance since y is very small and therefore $\zeta \simeq 0$ in this layer. Using Eqs. 6.2-19 and 6.2-21 it is easy to show that the effective exchange coefficient, $^3k_{A1}(y)$, in the turbulent zone for $y > y_0$ is

$$^3k'_{A1}(y) = \frac{\kappa_1^2 v_x}{[\ln(y/y_0) - \psi_h][\ln(y/y_0) - \psi_m]} \tag{6.2-37}$$

Numerical values of this coefficient, using the data in Example 6.2-1, are 2560, 3390, and 6600 cm/h, respectively. These large coefficients indicate that the turbulent zone transport resistance is small in comparison to that of the diffusive sublayer and further, that decreasing thermal stability is of secondary importance for wind speeds of 1.8 km/h and higher. In the case of no wind the air stability in the boundary layer is very important, however. Free convection correlations such as those used for the air–water exchange (Eq. 4.1-32) can be used to estimate ${}^3k'_{A1}$ under these conditions. Free convection values of 400 to 800 cm/h were estimated for dieldrin on a summer day in Mississippi.[24]

Although much more research needs to be done on this air-side mass transfer coefficient in order to quantify the effects of various conditions of surface type, roughness, canopy factors, stability, and so on, the algorithms above are capable of providing reasonable estimates for smooth soil or similar solid surfaces. Provided that the chemical concentration in air is known with accuracy at the soil surface, ρ_{A1i}, reasonable estimates of the evaporation rate can be made knowing only the wind speed and pool fetch. Using the form of Eq. 6.2-32, the rate is

$$w_{A0} = {}^3k'_{A1} Lw(\rho_{A1i} - \rho_{A1b}) \tag{6.2-38}$$

where w is the width of the contaminated surface. Relating ρ_{A1i}, in an equilibrium way to the concentration of A at the soil surface, ω_{Ai}, including the effects of temperature and moisture, is extremely important for reasonably accurate predictions (see the section entitled: "Equilibrium Partitioning of Organic Chemicals Between Air/Soil and Air/Particles" in Chapter 3).

Dry Deposition of Gases and Particles to Ground

This subject of dry deposition was presented in some detail in Chapter 4 (see pp. 219–231). Basically, the process to the ground is the same as to water. In addition to the turbulent diffusion and surface layer resistances, it is appropriate to include a transfer resistance onto earthen surfaces. This additional resistance occurs upon the contact of the molecule or particle with the solid surface and depends on the physicochemical interaction between the material and the surface. If the surface is not a perfect sink, all the impinging matter does not adhere to the surface. For the water surface it was neglected, assuming that all the material that arrives sticks. In the uptake of many gases by vegetation the rates are determined by biological factors such as stomal resistances and surface wetness. Seinfeld[27] compares the relative magnitude of the three resistances on the overall deposition velocity and concludes that the transfer resistance is a key parameter. McRae and Russell[28] note that during the daytime the deposition rate is much more likely to be influenced by chemical interactions with the surface, where uptake by vegetation is expected to be much greater when leaf stoma are likely to be open.

Table 6.2-3. Experimental Deposition Velocities (cm/s) and Receptor Surface

Species	Soil, Cement	Grass, Soil	Soybean field	Maize	Alfalfa
O_3	—	0.10–2.1	0.29–0.84	0.20–0.80	1.67
NO	0.10–0.20	—	—	—	0.10
NO_2	0.30–0.80	—	0.05–5.6	—	1.90
PAN[a]	—	0.14–0.30	—	—	0.63

Source: Ref. 28.

[a] Peroxyacyl nitrate.

Experimentally determined deposition velocities for some gaseous species are summarized in Table 6.2-3, where surface type is characterized. Seinfeld[27] presents data on gaseous species, ranking the deposition velocity in order of reactivity. There is a great need for additional field and laboratory measurements that explore and characterize the influence of surface chemistry. Considering the paucity of reported results, it is difficult to separate whether turbulent transport or the chemical and physical nature of the underlying surface is controlling the deposition velocity. It appears that the effect of surface type on particle deposition is just as problematic. Increasing surface roughness and surface area of vegetation cause field measured values of particulate deposition velocities to be higher than those measured under similar conditions in wind tunnels. Field measurements of the deposition velocity of particles show significant variations, and although the bulk of the evidence from field data indicates values on the order of 0.1 cm/s, the existence of a number of studies yielding values approaching or exceeding 1.0 cm/s makes it difficult at this time to prescribe a specific value. See Refs. 28 and 29 for data and model comparisons of dry deposition values. These sources should also be consulted on the subject of wet deposition, a topic introduced in Chapter 3 (see p. 161).

PROBLEMS

6.2A. Pesticide Evaporation from Field Surface

Figure 6.2A-1 shows an air sampling network above a soil surface where postapplication volatilization of trifluralin and heptachlor is occurring. Figure 6.2A-2 shows the field air sampling results at mast B, which is 50 m downwind from the edge of the field. The wind velocity profile above the soil is specified by the logarithm height profile with $v_* = 26$ cm/s, $y_0 = 0.1$ cm, and $d = 0$. Assume that a neutral atmospheric stability exists. As part of a project of estimating the volatilization rate of heptachlor 1 hr and 24 hr after application,

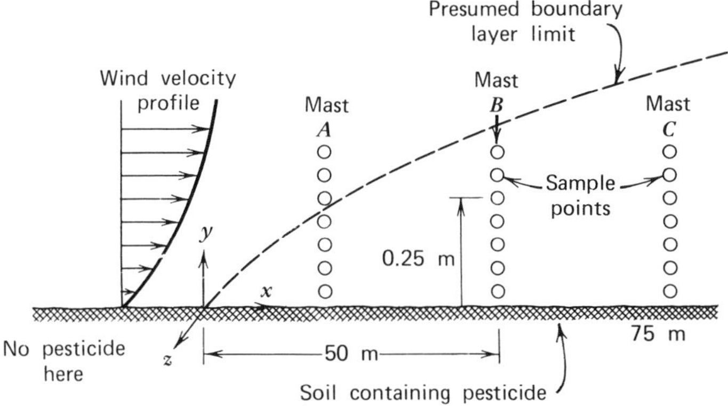

Figure 6.2A-1. Pesticide sampling network.

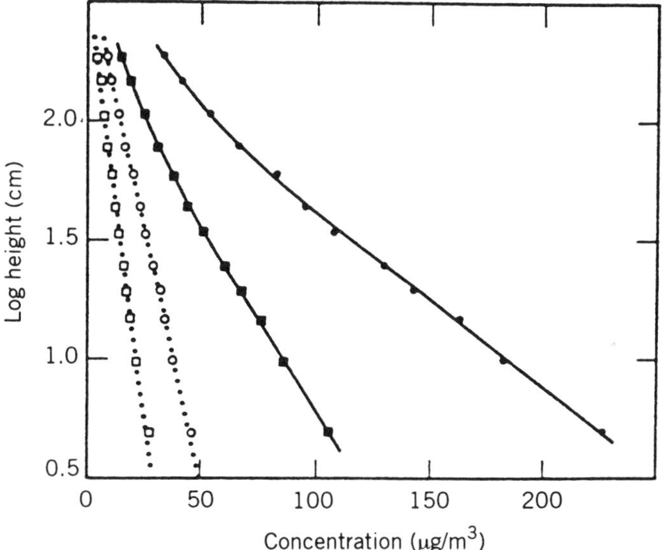

Figure 6.2A-2. Trifluralin (■) and heptachlor (●) concentration profiles in air following soil surface application; 1 hr (—) and 24 hr (···) after application. (Reprinted by permission from Ref. 13. Copyright by the American Chemical Society.)

the following flux rate calculations must be made in the vicinity of mast *B*:

1. Compute the advection flux rate ($\mu g/m^2 \cdot s$) of heptachlor from $y = 0$ to $y = 0.25$ m in the vicinity of mast B at 1 hr and 24 hr after application.

2. Compute the vertical diffusion flux rate ($\mu g/m^2 \cdot s$) of heptachlor at a point $y = 0.25$ m in the vicinity of mast *B* at 1 hr and 24 hr after application.

3. Assume that the treated field area is 50 m wide. Estimate the evaporation loss of heptachlor (μg/s) from the 50 by 50 m area upwind of mast *B* for each time period.

6.2B. Ammonia Flux over Grazed Alfalfa Field

Near Canberra, Australia, during March 1974, an area of approximately 4 ha was evenly grazed by 200 sheep. The surface layers of soils in the area were slightly acidic. The proportion of ammonia varied from day to day, and in cool, humid weather virtually pure ammonia was found.

Two representative profiles measured at about the same hour on consecutive days are shown in Fig. 6.2B. The very large differences in concentration

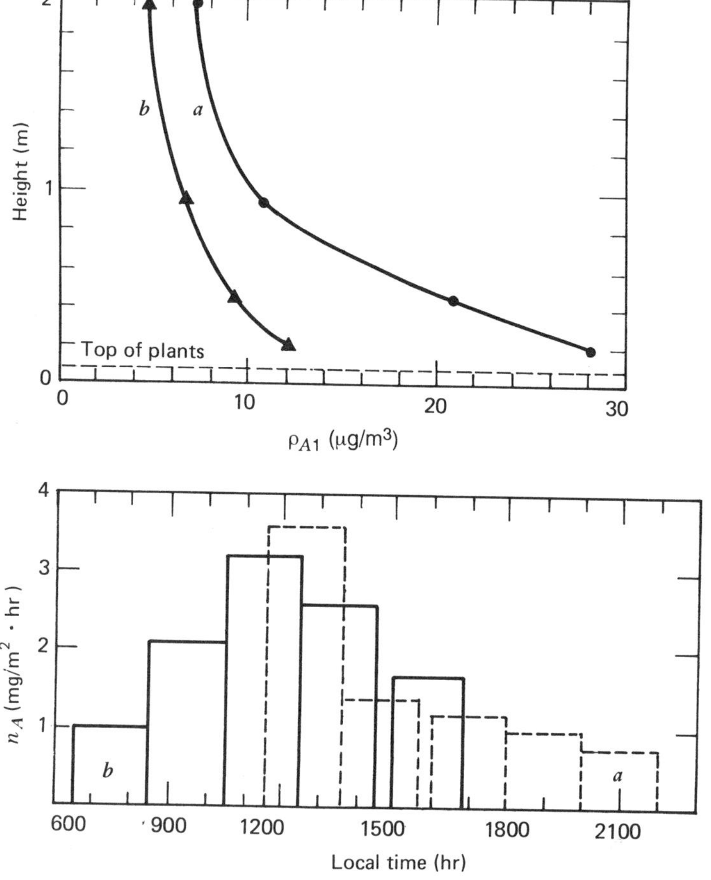

Figure 6.2B. (*Top*) Profiles of ammonia concentration ρ_{A1} over an alfalfa pasture: Profile *a*, 1600–1800 hr on March 14, 1974; profile *b*, 1500–1700 hr on March 15, 1974, (*Bottom*) Flux densities of nitrogen (as ammonia and related compounds) from pasture: Profile *a*, March 14, 1974; profile *b*, March 15, 1974. (Reprinted by permission, Copyright 1977 by the American Association for the Advancement of Science.) From Ref. 11.

and gradient are due to differences in the turbulence of the air. Profile a, for instance, was measured in light winds about 0.9 m/s at 1 m above the ground; profile b was measured in strong winds about 3 m/s at the same height. The general character of the flux determinations is also shown in Fig 6.2B.

1. Show how the data on top of Fig. 6.2B can be linearized and estimate the concentration near the soil surface.
2. Obtain values for n_{A0}/v_* for each profile.
3. Using the flux data in the lower part of Fig. 6.2B, estimate the roughness height, y_0, for each profile.

6.2C. Similarity Theory Developments

The material on pages 378 to 382 covers some basic aspects of similarity theory. Equations 6.2-19 to 6.2-21 are velocity, temperature, and concentration profiles in the atmospheric boundary layer.

1. Show details in arriving at these, beginning with Eqs. 6.2-12, 6.2-15, and 6.2-16.
2. Relate ζ to Ri.
3. Show the origins of and develop Eq. 6.2-18.

6.2D. Power Law Velocity Profile

Using the air velocity data of Problem 3.3A, show that the power law formula (Eq. 6.2-28) is a reasonable expression for the wind profile near the soil surface.

1. Make a graph of log v_x versus log y (v_x in cm/s and y in cm).
2. Determine m from the graph. Compute n from Eq. 6.2-29. Note that n is normally between 0.25 and 1.0.
3. Using $y = 1.0$ m as the reference point, obtain the final expression for the power law profile.

6.2E. Measurements of α-Pinene Fluxes from Loblolly Pine Forest[15]

Estimates of the flux of α-pinene emanating from a stand of loblolly pine trees have been obtained by measuring the net radiation and the vertical gradients of α-pinene, temperature, and water vapor above the forest using an energy balance–Bowen ratio approach.

The area of study was a uniform 19-year-old loblolly pine plantation in rural Alamance County in the central piedmont of North Carolina. Equipment for micrometeorological measurements included a 25-m scaffold tower that extends 10 m above the forest canopy. The measurements were made around

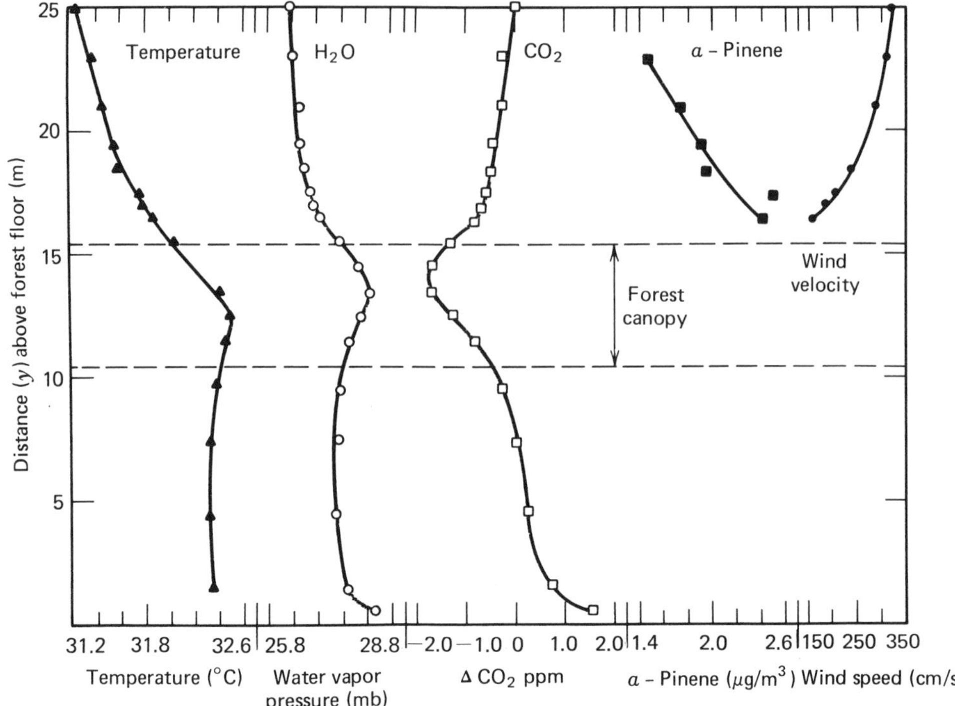

Figure 6.2E. Loblolly pine forest vertical gradients, July 18, 1977, 1035–1100 hr EST. (Reprinted by permission from (*Proceedings of the 4th Joint Conference of Sensing of Environmental Pollutants*. Ref. 15.)

solar noon when the atmosphere was in quasi-steady state. Air samples were pumped into 20-L FEP Teflon bags. The α-pinene analysis was made by chromatograph. Figure 6.2E contains a typical set of profiles.

Using the data, compute the emission rate of α-pinene ($\mu g/m^2 \cdot min$) for the period from 1035 to 1100 hr on July 18, 1977. The emission rate as determined by the energy balance–Bowen ratio approach was 46.9 $\mu g/m^2 \cdot min$. Use the aerodynamic method for computing the flux.

6.2F. DDT Volatilization from Cotton[25]

DDT [1,1,1-trichloro-2,2-bis(*p*-chlorophenyl)ethane] data have been collected over a 3-day period from the center of a 24-ha cotton field containing plants 50 cm in height. A trace of rain occurred between 1700 and 1800 hr on August 28, 1976 and on August 29 between 0500 and 0900 hr a total of 3.6 mm rain fell on the site. Table 6.2F contains some field measurements; it includes the soil surface ($y = 0$) temperature, T_3 (°C), wind speed at $y = 70$ cm, $\bar{v}_x$ (cm/s), soil water, ω_B (g/g), and the flux of DDT from the ground, n_A (g/ha · h), as a function of time.

Table 6.2F. DDT Flux and Field Data

Time (day/hour)	T_3 (°C)	v_x (cm/s)	ω_B (g/g)	n_A (g/ha · hr)
27/1100	44	200	0.045	0.50
1300	53	170	0.035	0.30
1500	51	170	0.030	0.55
1700	42	130	0.030	0.50
1900	33	140	0.035	0.30
2100	29	90	0.040	0.20
2300	27	60	0.045	0.15
28/0100	25	40	0.050	0.15
0500	23	70	0.060	0.05
0700	24	70	0.060	0.15
0900	29.5	90	0.050	0.15
1100	41.5	120	0.040	0.95
1300	53	130	0.030	1.80
28/1500	55	160	0.040	2.50
1700	55	160	0.040	0.90
1900	28	120	0.060	0.25
2100	27	160	0.060	0.35
2300	26	60	0.060	0.15
29/0100	25.5	70	0.060	0.25
0300	25	80	0.060	0.20
0500	24.5	70	0.060	0.15
0700	24	90	0.26	0.25
0900	27.5	100	0.24	0.80
1100	29	190	0.205	0.45
1300	27	230	0.19	0.50

1. Assuming negligible DDT concentration in the background air, develop a single theoretical equation for the maximum flux of DDT as a function of T_3 and v_x. Use $\log p_A^0 = 13.781 - 6.0408E3/T_3$ with p_A^0 in mmHg and T_3 in K, for the pure component vapor pressure. Answer in g/ha · h.
2. Graph the theoretical flux and the measured flux as a function of time and day.
3. Comment on the predicted versus measured results as appropriate.

6.2G. Dry Deposition Rate of Iodine-131

Iodine-131, emitted during the Soviet nuclear accident of 1986, was measured at 500 pCi/L (STP air) at Richland, Washington.

1. Estimate the dry deposition rate in curies/day in the state of Washington with $A = 2.6E11\ \text{m}^2$.

2. To assess the potential health hazard of this radioactivity, estimate the natural background radon exhalation from the soil. Refer to Table 6.3-3.
3. Based on the above, is the I-131 fallout likely to affect the health of residents of Washington compared to background radiation? Why?

Clearly record assumptions in your calculations.

6.3. CHEMICAL FLUX RATES THROUGH UPPER LAYER OF EARTHEN MATERIAL

This section is related to Section 5.3, which focused on the movement of chemicals within bottom sediments, in that both are concerned with interstitial fluid passages within a solid matrix as a route for chemical movement. Unlike chemicals that lie on the soil surface and desorb directly into the lower layers of the atmosphere, chemicals that originate in the soil column below the interface must find open passages within the soil in order to move upward and eventually escape to the atmosphere. After a brief consideration of the nature of the upper soil layer, the processes by which chemicals are introduced therein will be covered. This will be followed by sections on steady-state and transient transport phenomena aimed at quantifying the fate of chemicals in these layers.

Brief Introduction to Soil Structure

Natural earth materials, whether consolidated or unconsolidated, contain varying amounts of internal space not occupied by mineral material. This space is due to the presence of individual pores and structural features such as joints and bedding planes. Generally, this internal space is interconnected, permitting movement of water through the material and the associated transport of chemical species dissolved or suspended in the water. The bulk of the soil and rock material near the Earth's surface contains interconnected pore spaces that allow circulation of air or other gases and the transport of water and associated materials.

The individual particles composing granular material vary from irregular spheroids to flat plates. The shape and configuration of the intervening pores are dependent on the arrangement and relative size of the particles. Porosity is a measure of the total pore space contained in a given volume of material and is dependent more on the arrangement and size distribution of the particles than on their absolute size. A wide range of particle sizes tends to reduce the porosity by filling spaces between large particles with smaller particles. In addition to this interstitial porosity, secondary structures such as joints, fractures, or soil aggregates contribute to the total porosity of a material.

The curves in Fig. 6.3-1 illustrate the particle size distribution in soils representative of three textural classes. Note the gradual change in percentage composition in relation to particle size. This figure emphasizes that there is no sharp line of demarcation in the distribution of sand, silt and clay fractions,

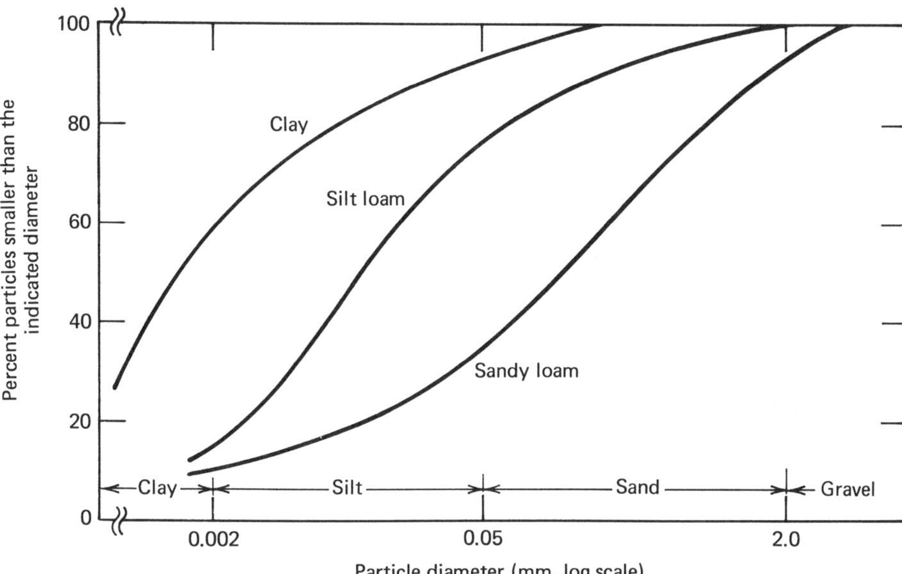

Figure 6.3-1. Particle size distribution in three soils varying widely in their textures. Note that there is a gradual transition in the particle size distribution in each of these soils. (Reprinted by permission from Ref. 16. Copyright ©1974 by Macmillian Publishing Co., Inc.)

which also suggests a gradual change of properties with change in particle size. There is considerable difference in the total pore space of various soils, depending on conditions. Sandy surface soils show a range of from 35 to 50%, whereas medium- to fine-textured soils vary from 40 to 60% or even more in cases of high organic matter and marked granulation. Pore space also varies with depth; some compact subsoils drop to as low as 25 to 30%. Cultivation and cropping can appreciably reduce the total pore space. Table 6.3-1 shows

Table 6.3-1. Effect of Continuous Cropping for at Least 40 to 50 Years on Total Pore Space and Macro- and Micropore Spaces in a Houston Black Clay from Texas

Sampling Depth (in.)	Soil Treatment	Organic Matter (%)	Pore Space Total (%)	Pore Space Macro (%)	Pore Space Micro (%)
0–6	Virgin	5.6	58.3	32.7	25.6
	Cultivated	2.9	50.2	16.0	34.2
6–12	Virgin	4.2	56.1	27.0	29.1
	Cultivated	2.8	50.7	14.7	36.0

Source: Ref. 16.

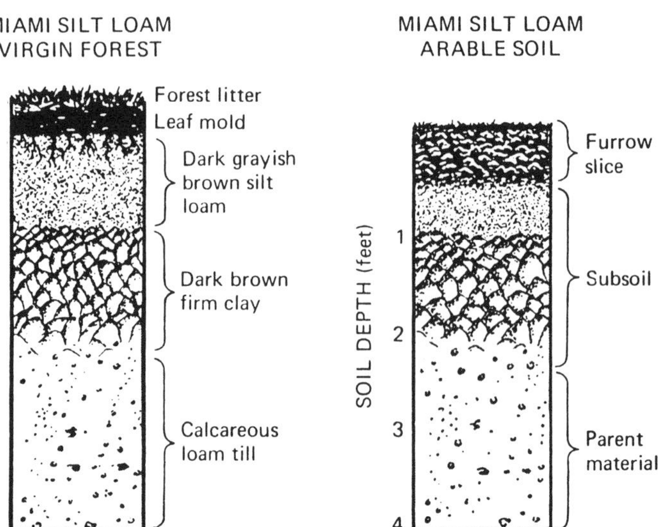

Figure 6.3-2. Generalized profile of the Miami silt loam, one of the alfisols or gray-brown podzolic soils of the eastern United States. A comparison of the profile of the virgin soil with its arable equivalent shows the changes that may occur as the land is plowed and cultivated. The surface layers are mixed by tillage. If erosion occurs, they may disappear, at least in part, and some of the *B* horizon will be included in the furrow slice. (Reprinted by permission from Ref. 16. Copyright ©1974 by Macmillian Publishing Co., Inc.)

the effect of cropping on pore space, and Fig. 6.3-2 shows typical profiles of virgin and arable soils. This reduction is usually associated with a decrease in organic matter content and a consequent lowering of granulation.

Two types of individual pores in general occur in soils: macro and micro. Although there is no sharp line of demarcation, the macropores characteristically allow the ready movement of air and percolating water. In contrast, in the micropores air movement is greatly impeded and water movement is restricted primarily to capillary movement. Thus, in a sandy soil, despite the low total porosity, the movement of air and water is surprisingly rapid because of the dominance of the macropore spaces. Typical distributions of total pore space and macro- and micropore spaces are given in Table 6.3-1.

The presence of water in earth materials greatly reduces the space filled with air. The pore spaces in granular material are irregularly shaped, with cusps or necks between adjacent pores, as illustrated in Fig. 6.3-3. Water placed in contact with granular materials tends to displace the air or any other gases present in the pores. The term *saturation* is used to describe a condition where all of the pore spaces are filled with water. When some portion of the pore spaces is only partially water filled, the material is termed *unsaturated* or *partially saturated.* In its natural state, the soil is normally unsaturated, and it contains both water and air. In truth, it is exceedingly difficult to accomplish saturation of any granular material, in part because of the presence of pores that are not interconnected.

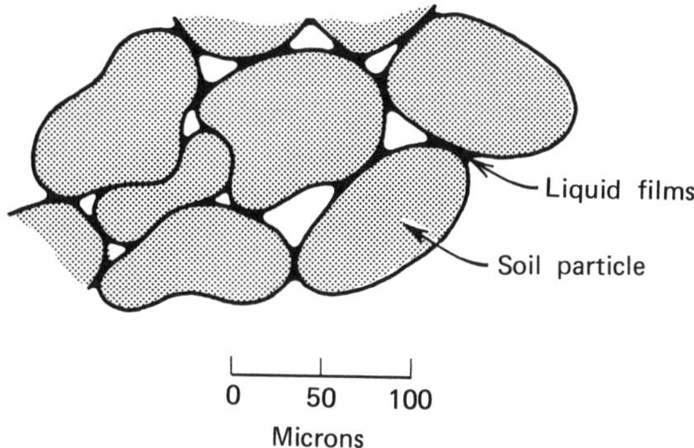

Figure 6.3-3. Soil water and internal pore structure. (Reprinted by permission from Ref. 16. Copyright © 1974 by Macmillan Publishing Co., Inc.)

Two forces tend to hold water in contact with granular material: adhesion between the water and the solid surface and cohesion between the water molecules. This situation is identical to water rising in a thin capillary tube. If water is allowed to drain from an initially saturated material, it will be removed first from the center of the individual pores, leaving behind water in the cusps around the edges of the pores and a relatively thin film of water on the surface of the grains. In Fig. 6.3-3 the expanded view of the partially filled pores reveals various radii of curvature around the perimeter of the pores, dependent on the grain shape and the dimensions of the various cusps. As for capillary tubes, the smaller the radius of curvature, the more tenaciously the water is held. As greater amounts of water are removed from the material, the water surfaces retreat farther into the cusps and the films become thinner. Eventually, these films become discontinuous, leaving only isolated pockets of water in the material. In some minerals (such as clays) water can be incorporated into their crystal lattice or chemical reactions may occur that incorporate water into the mineral structure. This 'bound' water does not participate in most intergranular flow and is not of significance to our discussion of pore space.

Some Processes of Soil Contamination

The placement of chemicals, both solid and liquid forms' on and within soil can occur by several processes. These include the primitive processes of dumping onto the surface, thereby creating *dump* or *tip* sites. Covering the waste, which has been dumped or placed within an excavation or depression in the ground results in burial. Landfills of various types, from the very primitive to engineered storage vaults, fall within the category of buried waste. Subsurface injection of liquids or sludges through nozzle-fitted plows also places waste and chemical pesticides within the soil a few centimeters below the surface.

In addition to direct placement, inadvertent spillage occurs as solids and liquids are manufactured, transported, and used. For unsaturated soils, bulk liquids, immiscible and miscible alike, infiltrate downward fairly rapidly. In the case of highly viscous organic oils moving downward, a very sharp front develops. For miscible organics, mobile porewater is displaced and pushed ahead. In the less well defined drainage tail that follows, residual saturation levels of the organic remain as the slug moves past. The process is transient, with the organic displaying intrinsic permeability values in damp soils very close to those obtained with water at saturation. The organic is initially less wetting than water; thus when it infiltrates the soil it slides over the film of water that surrounds the particles.[30] Normally, the infiltration and achievement of residual saturation levels occurs within hours to a day in the top 1 to 2 m.

Chemodynamic Models for Upper Soil Layers

Once in place, the material undergoes other slower processes that control its long-term fate in the soil. These include diffusion, vaporization, reaction, phase partitioning, and advection. In the following sections vignette models are developed to describe selected aspects for engineering prediction purposes with both organics and metals. The emphasis is on chemical substances in the upper soil layers, which includes the air–soil interface. Typically, this region is the topmost centimeters downward 1 to 2 m at most, including both saturated and unsaturated (i.e., H_2O) soil conditions. Once in the upper layers, chemical movement further downward occurs; however, in the case of volatiles, the emphasis is primarily on the movement upward to and through the air–soil interface. Vignette models, both steady-state and transient, will then be presented to illustrate some principles of chemodynamics in this zone. Section 7.3 is concerned with transport of chemicals within the underground well away from the influence of the air–soil interface and addresses the case of chemical movement downward and interacting with groundwater.

Steady-State Processes. The fluid in the macroporous channels within the soil are the primary pathways for chemical transport. In the case of the gas-filled channels, the steady-state flux of species A is

$$n_{Ay} = \bar{v}_y \rho_{A1} - D_{A3} \frac{d\rho_{A1}}{dy} \tag{6.3-1}$$

where $\bar{v}_y$ is the volumetric gas flow per unit soil cross-sectional area, and ρ_{A1} is concentration in the pores. The effective diffusivity of species A in the porous media is D_{A3}.

The integrated results of Eq. 6.3-1 assume several forms depending upon the volatility of A and the apparent gas velocity $\bar{v}_y$ through the media.[31] If A is less than 5 vol% of the gas mixture and apparent velocity is zero, it assumes the simple form

$$n_{Ay} = \frac{D_{A3}(\rho^*_{A1} - \rho_{A1i})}{h} \tag{6.3-2}$$

where ρ_{A1}^* is the concentration within the soil at depth h and ρ_{A1i} is the concentration at the interface. The flux is positive for A moving from soil to air. Farmer et al.[32] first used this expression for hexachlorobenzene emission control studies in land disposal of the waste product. If A is of higher concentration, $y_A > 0.05$, the diffusion process itself creates a significant apparent velocity in the gas-filled pores. The appropriate flux equation is

$$n_{Ay} = \frac{D_{A3} p_T M_A}{RTh} \ln \frac{p_T - p_{Ai}}{p_T - p_A^*} \tag{6.3-3}$$

where p_A^* and p_{Ai} are the partial pressure of A at depth h and at the surface, respectively; p_T is total pressure. Other gases, including atmospheric air, carbon dioxide, methane, and additional volatile species, contribute to $\bar{v}_y$: positively when directed from soil to air. Barometric pressure changes and biogas generation within the earth affect the direction and magnitude of $\bar{v}_y$.[33,34] Both of these factors can significantly enhance the emission of hazardous volatile chemicals buried beneath the soil; however, in the case of barometric pressure pumping a large subterranean air volume below the cover is needed for this contribution to be significant. This is usually the case with landfills.

The parent equation that combines both the effect of partial pressure of A on velocity in addition to the velocity of other gaseous constituents $\bar{v}_y$ is

$$n_A = \frac{\bar{v}_y \rho_{A1}^*}{1 - \dfrac{\rho_{A1}^*}{\rho_1} - \dfrac{\rho_{A1}^* - \rho_{A1i}}{\rho_{A1}^* \exp[(n_A/M_A + \bar{v}_y \rho_1/M_1) h M_1/\rho_1 D_{A3}] - \rho_{A1i}}} \tag{6.3-4}$$

where ρ_1 and M_1 is the density and average molecular weight of the gas, excluding A. It should be noted that the flux n_A appears on either sides of the expression. A few iterations, using either Eq. 6.3-2 or 6.3-3 to initiate the process, are required for convergence in the trial-and-error solution.

The effective diffusivity, D_{A3}, is known to be a complex function of soil type, soil compaction, water content and direction of movement, temperature, porosity, chemical species, and so on. Table 6.3-2 contains some measured values for various soils. These values are less than the molecular diffusivities in air, suggesting that the presence of the media retards the mobility of the species. The soil-gas diffusion coefficient is usually equated to the species diffusion coefficient in air, $\mathscr{D}_{A1}$, multiplied by a factor to account for the reduced flow area and another for the increased path length of diffusing A molecules in soil. A model that has proven useful for describing volatile species effective diffusivities is that of Millington and Quirk[36]:

$$D_{A3} = \frac{\varepsilon_1^{10/3}}{\varepsilon^2} \mathscr{D}_{A1} \tag{6.3-5a}$$

where ε_1 is the volumetric air content of the soil and ε is the total soil porosity. The formula does not have calibration constants. In a landfill cover simulation

Table 6.3-2. Effective Diffusion Coefficients in Surface Soils

Chemical	Temperature (°C)	Soil Type, Condition	Total Porosity	D_{A3} (cm^2/s)
Carbon disulfide	15.4	Sand, dry	0.374	0.193
Ethanol	21.3	Quartz sand, dry	0.415	0.0415
Oxygen	25	Soil, 32.7% moisture	0.38	0.105
	25	Soil, 45.3% moisture	0.27	0.056
Ethylene dibromide	20	Garden soil	0.389	0.0151
	20	Ashhurst soil	0.199	0.00506
	20	Ashhurst soil	0.303	0.0112
Sulphur hexafluoride	22	Compacted sand, dry	0.45	0.031
	22	Sieved sand, dry	0.48	0.032
	22	Sieved sand, 10% moisture	0.41	0.026
	22	Sieved sand, 15% moisture	0.32	0.019
	22	Sieved sand, 20% moisture	0.11	0.0037

Source: Refs. 14 and 35.

experiment with five organic volatiles and mercury vapor, the measured D_{A3} values were larger than those predicted by Eq. 6.3-5*a* by a factor from 1.6 to 4.9, with a weighted average of 2.7.[37] Reible and Shair[14] reported measured effective diffusivities on moist porous media, 8.5 to 20% moisture, lower than predicted by Eq. 6.3-5*a*. They speculate that dead-end pores may be formed due to the presence of pockets of moisture that result in reduction of effective pore volume. They find that $D_{A3}/\mathscr{D}_{A1} = \varepsilon^{1.43}$ is the best-fit relationship for both moist and dry porous media.

Similarly, the soil-water effective diffusivity of dissolved species A is set equal to its molecular diffusivity in water multiplied by the appropriate from of the Millington–Quirk correction factor, to yield

$$D_{A3} = \frac{\varepsilon_2^{10/3}}{\varepsilon^2} \mathscr{D}_{A2} \tag{6.3-5b}$$

where ε_2 is the volumetric water fraction of the soil ($\varepsilon_2 = \varepsilon - \varepsilon_1$).

One application for the vapor transport models above is volatile chemical emissions to air from buried waste. The ratio h/D_{A3} in Eqs. 6.3-2 to 6.3-4 is the cover resistance, P_{A3}. Figure 6.3-4 illustrates a typical landfill with waste types in individual cells, a cover consisting of a clay layer, a plastic geomembrane, and an upper surface soil layer. The total chemical transport

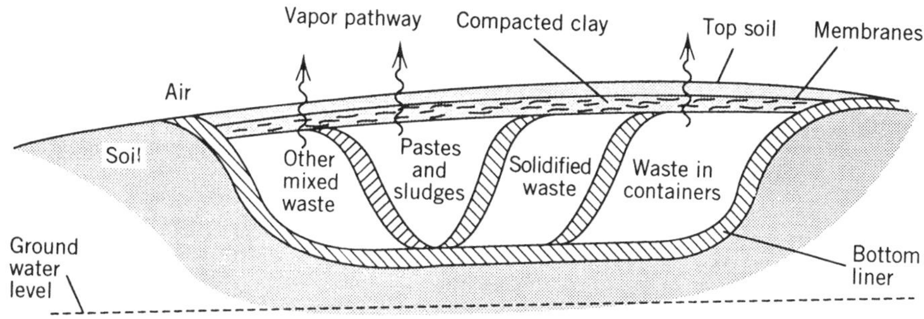

Figure 6.3-4. Landfill with composite cap, and waste types isolated into separate cells.

resistance of the three-layer cover is estimated by the resistance-in-series expression[31]

$$\mathrm{P}_{A3} = \sum_{i=1}^{3} \frac{h_i}{D_{Ai}} \tag{6.3-6}$$

where h_i and D_{Ai} are the respective layer thickness and corresponding effective diffusivity of the material for species A.

For most buried waste scenarios, estimating p_A^* and hence ρ_{A1}^* beneath the cover or within the gas-occupied void spaces of the cells is very problematic. Equilibrium partial pressure of the species between the waste form and the gas phase is probably a reasonable assumption. There are numerous equilibrium laws that cover nearly all mixture cases by which one can estimate partial pressures; see Groves et al.[38] and Chapter 2. Use of any of these estimating techniques presumes that one knows the composition of the material by chemical species, the number of solid and liquid phases, and the mass (or mole) fractions of each species, along with the water content, pore gas content, temperature, and total pressure. Rarely is all this known concerning the waste delivered to the landfill. Matters are made worse because often the original waste is mixed with other waste or fill material prior to placement in the cell so that the last resort is to use the pure component vapor pressure based on the cell temperature.

Example 6.3-1. Benzene in Landfill. Benzene is co-disposed with other material and buried under 1 m of cover soil with 51% porosity. Compute the flux to the atmosphere with zero background concentration assuming that cell conditions are 25°C and 1 atm pressure if benzene is (a) 10 mol% of an oily sludge and cover is dry, (b) 1800 mg/L in cell leached water and cover is 75% water-filled porosity, and (c) as (b) mixed with municipal waste and biogas flux through cover is 3E-6 $\mathrm{cm^3/cm^2 \cdot s}$.

SOLUTION (a) Raoult's law applies for equilibrium; $p_A = 0.125/10 = 0.0125$ atm; therefore, use Eq. 6.3-2. $\rho^*_{A1} = 0.0399$ g/L from the ideal gas law. $D_{A3} = 0.088(0.51)^{4/3}$ from Eq. 6.3-5*a* for 0.0359 cm²/s.

$$n_A = \frac{0.0359\ \text{cm}^2}{\text{s}}\left|\frac{1}{100\ \text{cm}}\right|\frac{39.9\ \text{mg}}{0.001\ \text{m}^3}\left|\frac{\text{m}}{100\ \text{cm}}\right|\frac{3600\ \text{s}}{\text{h}} = 516\ \text{mg/m}^2\cdot\text{h}$$

(b) Henry's law applies; benzene is at the solubility limit and $\rho^*_{A1} = 0.399$ g/L. $y^*_A = 0.125$ and Eq. 6.3-3 is appropriate. $\varepsilon_1 = 0.128$ and $D_{A3} = 3.53\text{E-4}$ cm²/s. $p_T M_A/RT = 3.19$ g/L.

$$n_A = \frac{0.0000353}{100}\left|\frac{3190}{0.001}\right| 36(0.134) = 54.3\ \text{mg/m}^2\cdot\text{h}$$

(c) With $\bar{v}_y = 3\text{E-6}$ cm/s, Eq. 6.3-4 applies. $n_A = 91.8\ \text{mg/m}^2\cdot\text{h}$.

The steady-state models presented above also apply to water-filled soil channels and can therefore be used for organics, inorganics, and metal species in solution as well as colloidal matter and the species associated (i.e., sorbed) with it. In the case of water-filled pores, the effective diffusivities are much smaller because of the low values of the molecular diffusivities in water. Radon, a natural tracer in the upper earthen layers, is used next to illustrate the chemodynamics of steady-state diffusion and reaction. The discussion is excerpted from Junge[20]. The information suggests some things to be expected when studying the transport and fate of synthetic chemicals in the soil near the air interface.

The Earth's crust contains the radioactive nuclides uranium-238, uranium-235, and thorium-232, which by decay produce isotopes of the noble gas radon. After formation in the ground, radon diffuses into the atmosphere. The mechanism of radon exhalation is strongly influenced by the varying local conditions of the soil and the atmosphere. Therefore, it is difficult to establish quantitative relations in individual cases, but the basic processes involved are fairly well understood. Israël[19] gave a simple but useful model for these processes, which demonstrates all the essential features and parameters involved and which was developed further by Junge.[20]

If the soil is sufficiently porous, diffusion proceeds as if the soil were replaced with small tubes filled with air. The solution of the diffusion transport equations for emanations toward the Earth's surface can be expressed with the equation

$$c_{A1} = c^{**}_{A1} - (c^{**}_{A1} - c_{A1i})\exp\left(-\sqrt{\frac{k'''_A}{D_{A3}}}\,y\right) \tag{6.3-7}$$

where y is depth from the surface, k'''_A the radon decay rate, c^{**}_{A1} the asymptotic concentration of radon emanation in undisturbed soil air in deep layers, and

c_{A1i} the radon concentration at the air–soil interface. The exhalation rate is

$$N_{A0} = \sqrt{k_A'''D_{A3}}(c_{A1}^{**} - c_{A1i}) \tag{6.3-8}$$

The radon decay constant is 2.1E-6 s^{-1}. Israël assumed D_{A3} to be 0.05 cm^2/s for radon. Values of c_{A1}^{**} can be calculated from uranium and thorium content of the soil if one considers that only a fraction of the equilibrium production of radon escapes into the soil air prior to decay within the soil particles. Assuming this fraction to be 0.10, it is possible to calculate c_{A1}^{**} and N_{A0}. The calculated values and observed values appear in Table 6.3-3.

The large variation of radium concentrations in soil and in soil conditions is reflected in the wide range of radon concentrations in soil. In certain geologically disturbed areas the concentration of radon in soil can be extremely high, as, for example, in Nauheim. The variation around the average is much less if only normal conditions are considered. The exhalation rate N_{A0} and the radon concentration near the ground vary over a range expected from the variation of soil and meteorological conditions. The average values are in satisfactory agreement with the calculated values.

It can be expected that the exhalation rate will depend on soil conditions. Decreases of 70% are observed during rainfall. Table 6.3-3 shows that radon concentrations over land are higher than those over the ocean by about two orders of magnitude. In Section 5.3 the Tappan Zee region of the Hudson River sediment displayed a radon diffusion coefficient of $D_{A2} = 2.8\text{E-5}\ cm^2/s$. The

Table 6.3-3. Radon Exhalation from Soil

	c_{A1}^{**} (Ci/cm^3)	N_{A0} ($Ci/cm^2 \cdot s$)
Calculated values	1.3E-13	4.0E-17
Observed values	0.5E-13 to 10,000E-13	0.1E-17 to 25E-17
	Average ~3E-13	Average ~4E-17

Radons Concentrations in Air near the Ground

Location	c_{A1} (Ci/cm^3)
Normal continental areas	70E-18 to 330E-18
Disturbed continental areas	400E-18 to 600E-18
Innsbruck; Nauheim	
Ocean	0.5E-18 to 3E-18
South America	20E-18 to 70E-18
Antarctic	0.2E-18 to 2E-18

Source: Ref. 20.

ratio $\sqrt{D_{A3}/D_{A2}}$ is 42, which is two orders of magnitude. Generally, variations in soil moisture result in large fluctuations of N_{A0} but do not much affect the average values. In the eastern Alps there is an indication of a yearly variation with a maximum in late spring, and it is suggested that this is caused by the accumulation of radon during the winter when the soil is frozen and by its escape when the soil thaws.

Fumigation of soil by subsurface injection, applying ammonia or methyl bromide, a fertilizer and a pesticide, respectively, is possible. A similar process places oily waste below the surface to reduce volatile emissions to air during the land treatment of waste. The soil layers provide the dominant transport resistance. Although a large fraction of the volatile material eventually enters the atmosphere, some fraction is transformed by soil microbial processes, and a fraction moves downward away from the active surface layers eventually to groundwater, possibly. Means by which to estimate the time scales for transport of the volatile fraction to air are presented in the next section.

Transient Processes. In this section four vignette models will be developed that describe some transient chemodynamic processes for chemicals in the upper soil layers. The first considers the highly idealized case of a volatile species existing in the air-filled void spaces only. The second addresses the case of volatile trace contaminants in the air-filled voids in equilibrium with quantities in both the porewater and sorbed onto soil solids. The third is tailored to volatile liquids or solids in pure form. The fourth model is not concerned with volatility but addresses the phase redistribution and transport of metals in near-surface soils.

Trace Volatiles in Soil Gas Without Sorption. Assume that, upon entering the soil, chemical A vaporizes instantly and fills the pore spaces with a saturated vapor to a depth h as shown in Fig. 6.3-5. This situation may occur with the spill of a low-boiling liquid onto a soil column during a hot summer day. For this diffusion problem the multicomponent continuity equation for the void spaces only reduces to

$$\varepsilon_1 \frac{\partial \rho_{A1}}{\partial t} = D_{A3} \frac{\partial^2 \rho_{A1}}{\partial y^2} \qquad 0 \leqslant y \leqslant h \tag{6.3-9}$$

where D_{A3} is a constant diffusion coefficient, defined by Eq. 6.3-1, that characterizes the movement of chemical A as a vapor within the porous solid, ρ_{A1} is the mass concentration of A within the pore spaces, and ε_1 is the fraction of vapor-filled pore spaces per volume of soil. Initially, the pore spaces are saturated with the contaminant: $\rho_{A1} = \rho_{A1}^*$ at $t = 0$ for all values of y. Boundary conditions: (1) at the soil–air interface, $y = h$; the pore air concentration is maintained at $\rho_{A1} = \rho_{A1i}$; and (2) at the bottom of the contaminated

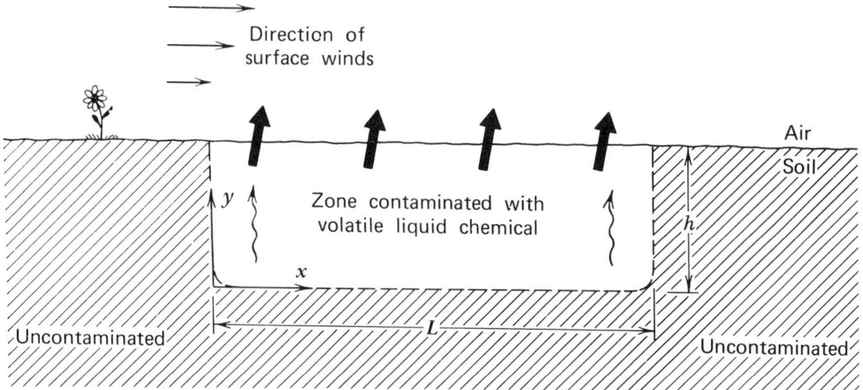

Figure 6.3-5. Soil contaminated with a volatile liquid.

zone, $y = 0$; the flux is assumed to be zero to give $\partial \rho_{A1}/\partial y = 0$. The original zone of contamination is depth h and areal extent A. Although unrealistic, the boundary conditions do not allow downward or lateral migration; the only escape route of the volatile species is through the air–soil interface.

This problem has been solved by Carslaw and Jaeger[17] for the analogous problem of heat conduction in a solid. For the case of linear flow of heat in a solid bounded by two parallel planes, the following solution is offered. The concentration profile of chemical A in the pore space as a function of distance from the bottom, y, and time after contaminant is in place, t, is

$$\rho_{A1} = \rho_{A1i} + (\rho_{A1}^{*} - \rho_{A1i}) \frac{4}{\pi} \sum_{n=0}^{\infty} \frac{(-1)^n}{2n+1} \exp\left[-\frac{D_{A3}(2n+1)^2\pi^2 t}{\varepsilon_1 h^2} \right] \cdot \cos \frac{(2n+1)\pi y}{2h} \tag{6.3-10}$$

This series solution converges slowly for values of $D_{A3}t/\varepsilon_1 h^2 < 0.01$, and Carslaw and Jaeger present alternative series involving error functions or their integrals. Some numerical results for this problem are given in Fig. 6.3-6. The average concentration in the pore space of the slab (or contaminated zone) at time t is

$$\bar{\rho}_{A1} = \rho_{A1i} + (\rho_{A1}^{*} - \rho_{A1i}) \frac{8}{\pi^2} \sum_{n=0}^{\infty} \frac{1}{(2n+1)^2} \exp\left[-\frac{D_{A3}(2n+1)^2\pi^2 t}{\varepsilon_1 4h^2} \right] \tag{6.3-11}$$

The quantity of contaminant A remaining in the zone per unit area at time t is $\varepsilon_1 h \bar{\rho}_{A1}$. The fraction of A remaining is

$$F_{A1} \equiv \frac{\bar{\rho}_{A1} - \rho_{A1i}}{\rho_{A1}^{*} - \rho_{A1i}} \tag{6.3-12}$$

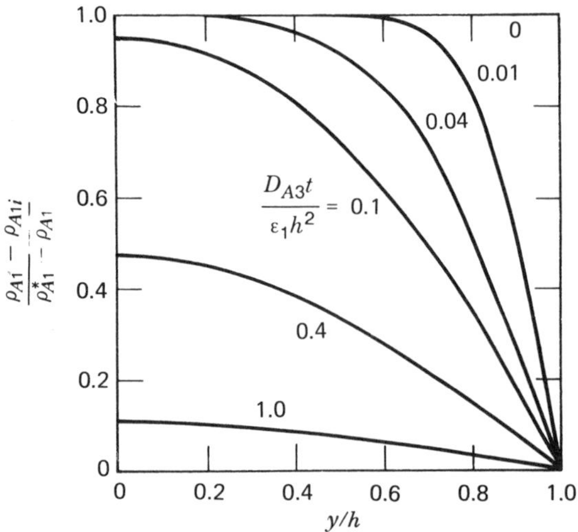

Figure 6.3-6. Concentrations in the contaminated zone. (Reprinted by permission from Ref. 17.)

and Fig. 6.3-7 is a useful graphical interpretation of the solution from which lifetime and half-life can be estimated. The dashed lines on the figure correspond to the half-life, $\tau_{1/2}$, and 90% evaporation time, $\tau_{0.9}$.

The flux of chemical A at the surface is

$$n_{A0} = \frac{2D_{A3}}{h} \sum_{n=0}^{\infty} \exp\left[-\frac{D_{A3}(2n+1)^2\pi^2 t}{\varepsilon_1 4h^2}\right](\rho^*_{A1} - \rho_{A1i}) \tag{6.3-13}$$

This solution can be used to determine D_{A3} (and α_3 for the case of heat transfer) of earthen materials.

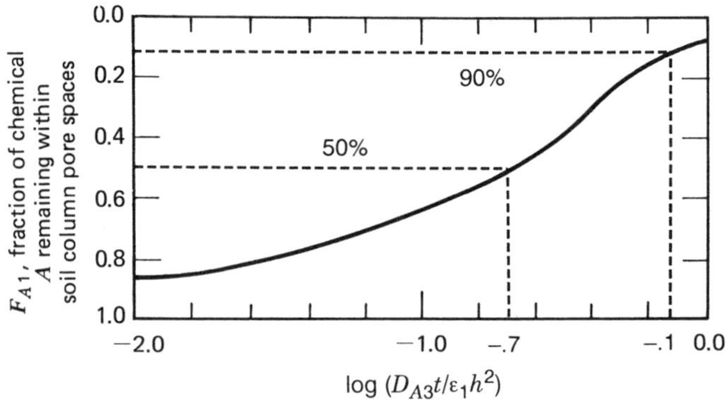

Figure 6.3-7. Fraction of contaminant remaining versus dimensionless time.

Trace Volatiles in Soil Gas with Sorption. It is common for volatile, trace contaminants to be present in all three phases of the soil. A reconsideration of the processes that resulted in Eq. 6.3-9 is required. Assuming that constituent A is present in the gas-filled spaces, in the soil water, and associated with the solid fraction, its mass per unit volume soil is

$$\rho_{A3} = \rho_{A1}\varepsilon_1 + \rho_{A2}\varepsilon_2 + (1 - \varepsilon_1 - \varepsilon_2)\rho_3\omega_A \tag{6.3-14}$$

If chemical equilibrium between phases is assumed to exist at all times throughout the column, then

$$\rho_{A3} = \rho_{A1}\left(\varepsilon_1 + \frac{\varepsilon_2}{H_{A\rho}} + \frac{\rho_B K^*_{A32}}{H_{A\rho}}\right) \tag{6.3-15}$$

where $\rho_B \equiv (1 - \varepsilon_1 - \varepsilon_2)\rho_3$. Neglecting advection, reaction, and diffusion in all but the gas phase yields for the continuity equation

$$\left(\varepsilon_1 + \frac{\varepsilon_2}{H_{A\rho}} + \frac{\rho_B K^*_{A32}}{H_{A\rho}}\right)\frac{\partial \rho_{A1}}{\partial t} = D_{A3}\frac{\partial^2 \rho_{A1}}{\partial y^2} \tag{6.3-16}$$

Clearly, Eq. 6.3-9 is special case of this one. The results presented for it can be used if modified as follows. In Eqs. 6.3-10, 6.3-11 and 6.3-13 the D_{A3}/ε_1 ratio appearing behind the summation sign is replaced with $D_{A3}/(\varepsilon_1 + \varepsilon_2/H_{A\rho} + \rho_B K^*_{A3}/H_{A\rho})$. The same replacement applies to Figures 6.3-6 and 6.3-7.

In the case of evaporation half-lives, $\tau_{A1/2}$, from Fig. 6.3-6, $\log(D_{A3}\tau_{A1/2}/\varepsilon_1 h^2) \simeq -0.7$. For the three-phase soil system it becomes

$$\tau_{A1/2} \simeq \frac{0.2h^2(\varepsilon_1 + \varepsilon_2/H_{A\rho} + \rho_B K^*_{A32}/H_{A\rho})}{D_{A3}\varepsilon_1} \tag{6.3-17}$$

The three terms in parentheses give the air-phase fraction and the air-phase equivalent fractions for the species is porewater and on the soil solids. Table 6.3-4 gives these equivalents for six pesticides. Obviously, the evaporation time and flux rate from the soil (see Eq. 6.3-13) will be strongly dependent on the phase containing the dominant fraction. If the soil is dry, $\varepsilon_2 = 0$ and the soil–air partition coefficient, K^*_{A31}, replaces the $K^*_{A32}/H_{A\rho}$ ratio. For a trace contaminant on dry soil the BET isotherm (see p. 64) yields

$$K^*_{A31} = \frac{\omega^*_A B_1}{\rho^*_{A1}} \tag{6.3-18}$$

where B_1 is the BET constant (see Eq. 2.1-30) and ρ^*_{A1} is the saturated vapor density of species A.

Table 6.3-4. Vaporization Parameters for Pesticides in Wet Soil at 20°C[a]

Chemical	K^*_{A32} (L/kg)	H_{Ap} (L_w/L_a)	ε_1	$\varepsilon_2 H_{Ap}$	$\rho_B K^*_{A32}/H_{Ap}$
Mirex	2.4E5	2E-2	0.4	3.3	1E7
Toxaphene	2.1E3	9.0	0.4	1E-2	2.9E2
Methoxychlor	8E2	3E-5	0.4	3.3E3	3.3E7
Lindane	1.3E1	5E-5	0.4	2E3	3.3E5
Malathion	2.3	4.9E-6	0.4	2.0E4	5.9E5
Di–Br–Cl–propane	4E-1	1E-2	0.4	10	50

Source: Ref. 39.

[a] $\varepsilon_2 = 0.1$, $\rho_B = 1.25$ kg/L, $\omega_c = 0.01$.

The preceding analysis does give some insight into the transient pore diffusion process that occurs below the soil surface and eventually dispels the chemical to the overlying air. It is likely that most chemicals associated with a solid phase for an extended period of time are in equilibrium with the vapor in the pore spaces during the desorption process. See page 325 for information of the subject of kinetics of chemical transport within soil or sediment particles. However, it is unlikely that downward or lateral chemical migration is absent. Because of downward movement, the preceding model, based on the initial h, will give an estimate of the minimum half-life or lifetime. An analogous gradientless vapor pore diffusion model is presented in Problem 6.3E.

Another major limitation of the model above is that it contains no provisions for the existence of the chemical as a pure substance, either solid or liquid, in the soil pore spaces. This evaporation and diffusion model appears next.

Pure Substance Evaporation and Diffusion in Soil Column: Vignette Model. Assume that the spilled liquid, a pure substance, soaks into the dry soil and contaminates the soil column to a depth h, as shown in Fig. 6.3-8. The liquid coats the pore walls and particle junction sites in the same manner as water (see Fig. 6.3-3). Once it is in place, the chemical evaporates from the interstitial soil surfaces, and the vapor diffuses through the pores upward toward the air–soil interface. In a very short time a hypothetical "dried-out" zone develops near the surface, and liquid vaporization occurs from the plane formed between this zone and the remaining contaminated zone. The term *dried-out* refers only to the pure material, called the *free phase* quantity. It is assumed further that the soil column is isothermal, that no vertical liquid movement occurs by capillary action, that there is no absorption on soil particles, and that there is sufficient thermal energy entering the soil column to sustain the evaporation process.

As vaporization occurs, the dried-out zone increases in depth and the contaminated zone decreases. If the limiting mechanism is vapor diffusion

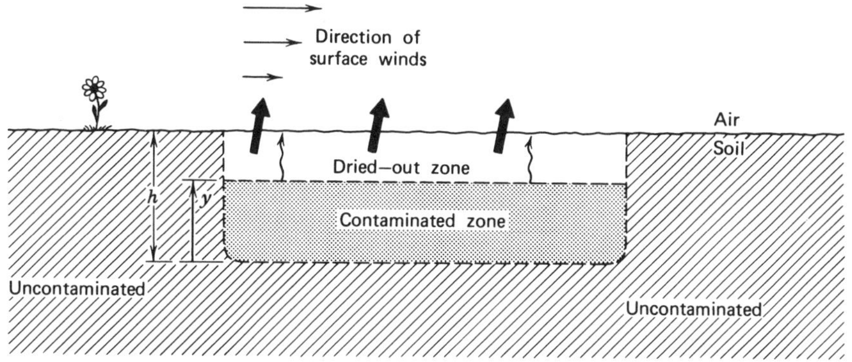

Figure 6.3-8. Evaporation and diffusion within pore spaces.

within the pore spaces of the dried-out zone, the diffusion path length increases as the chemical dissipates. The rate of vaporization is

$$n_{A0} = \frac{D_{A3}}{h - y}(\rho^*_{A1} - \rho_{A1i}) \tag{6.3-19}$$

where ρ^*_{A1} is the concentration of chemical A within the pore spaces at the evaporating plane y and ρ_{A1i} is the concentration at the air–soil interface. The flux expression can be modified to include an air-side coefficient, ${}^3k'_{A1}$, thereby eliminating the concentration at the interface, ρ_{A1i}. A component balance for chemical A in the contaminated zone yields the time for the dried-out zone to replace the contaminated zone, τ_A.

$$\tau_A = \frac{[(h^2/2D_{A3}) + (h/{}^3k'_{A1})]\rho_{A3}}{\rho^*_{A1} - \rho_{A1}} \tag{6.3-20}$$

where ρ_{A1} is the concentration of A in the atmosphere. For the evaporation–pore diffusion period the interface flux rate is

$$n_{A0} = \frac{\rho^*_{A1} - \rho_{A1}}{\{(1/{}^3k_{A1})^2 + [2(\rho^*_{A1} - \rho_{A1})t/D_{A3}\rho_{A3}]\}^{1/2}} \tag{6.3-21}$$

where A is the surface area of the contaminated region and $\rho_{A3} \equiv m_A/hA$ is the bulk soil concentration of chemical A as a pure substance. It therefore excludes that fraction in the gas phase, in solution in soil water, and associated with the soil solids as expressed by Eq. 6.3-15. This completes the evaporation and diffusion model.

The results of this model are more pleasing to the senses. Equation 6.3-20 suggests that the time to dispel the pure contaminant increases with the mass applied and decreases with chemical vapor pressure. Both these characteristics reflect the quantity and character of the chemical.

Example 6.3-2. Evaporative Fate of Lindane in Soil. One thousand kilograms of lindane, $M_A = 290.8$, was applied to a 1-ha plot of soil and down to 20 cm depth. At 20°C the pure vapor density is 518 ng/L and diffusivity in air is 0.0541 cm^2/s. For the soil in Table 6.3-4, estimate (a) the fraction of pure lindane, (b) its evaporation time, and (c) the evaporative half-life for the trace quantity. Assume that ${}^3k'_{A1} = 1000$ cm/h and a lindane-free atmosphere.

SOLUTION (a) As in Eq. 6.3-21, the bulk soil concentration is ρ_{A3} (pure) and Eq. 6.3-15 is the trace quantity, ρ_{A3}(tr.). The sum is the total, ρ_{A3} (tot.) = 1000 kg/ha $\cdot$ 0.2 m = 500 mg/L.

$$\rho_{A3}(\text{tr.}) = 518\text{E-}6\,\frac{\text{mg}}{\text{L}}\,(0.4 + 2\text{E}3 + 3\text{E}5)\,\frac{\text{L}}{\text{L}} = 172\ \text{mg/L}$$

$$\text{mass fraction pure} = 1 - \frac{172}{500} = 0.656$$

(b) From Eq. 6.3-20 with

$$D_{A3} = \frac{0.0541(0.4)^{10/3}}{(0.5)^2} = 0.0102\ \text{cm}^2/\text{s}$$

$$\tau_A = (1.96\text{E}4\,\text{s} + 72\,\text{s})\left(500 - 172\,\frac{\text{mg}}{\text{L}}\right)\frac{1\text{E}6\ \text{ng}}{\text{mg}}\left|\frac{\text{L}}{518\ \text{ng}}\right.$$

$$= 1.25\text{E}10\,\text{s (395 years)}$$

(c) From Eq. 6.3-17,

$$\tau_{A1/2} = \frac{0.2(20\ \text{cm})^2(3.32\text{E}5)}{(0.4)(0.0102\ \text{cm}^2/\text{s})}$$

$$6.51\text{E}9\ \text{s (206 years)}$$

Note that the air-side resistance is absent in the trace constituent model. As seen in part (b), its contribution is 72/1.96E4 fraction or 0.4% and decreases in significance as h increases. The evaporative fate of low-vapor-pressure chemicals such as lindane, $\rho_A^* = 3.2\text{E-}5$ mmHg at 20°C, at modest depths will occur over a period of decades; therefore, other biotic and abiotic processes such as chemical and biochemical degradations plus leaching downward are also likely to effect fate in this soil zone.

Closure and Complications on Volatiles Transport in Soils. The free-phase evaporation-plus-diffusion model presented above is a simple mathematical construct that attempts to mimic a complicated physicochemical process in the soil column. In reality the evaporation plane is a diffuse boundary that moves with time, and therefore the concentration profile within the soil column changes continuously.[41] With no soil this simple model has been used to evaluate gas diffusion coefficients from Arnold cell experimental data.[40] It has been applied in the laboratory with some success for soils containing pure liquids.[41,42]

The concept and the results, Eqs. 6.3-20 and 6.3-21, are applicable to trace contaminants in the upper soil layers also. In this case ρ_{A3} is given by Eq. 6.3-15 and the model has been applied in the laboratory with success.[43,44] It also gives reasonable flux predictions when compared to field data.[44]

More rigorous analytical approaches for trace contaminants have been developed and verified in the laboratory.[45,46] These are represented by numerical and analytical solutions of Eq. 6.3-16. Simple analytical solutions have been presented in the preceding section, where Eqs. 6.3-10 and 6.3-13 are useful results along with that shown in Figs. 6.3-6 and 6.3-7. From this more rigorous analysis, Eq. 6.3-17 is appropriate for estimating the evaporation half-life. However, when used in the trace contaminant form, the evaporation-plus-diffusion model gives comparable results. For example, it can be shown that Eq. 6.3-20 can be transformed to the form of Eq. 6.3-17 with a numerical constant of 0.25 instead of 0.2!

It is tempting to conceptualize the entire vaporization process in a soil column that receives large quantities of a liquid or solid in a two-part scenario. In the first part the free-phase fraction is modeled with Eqs. 6.3-20 and 6.3-21 where ρ_{A1}^* is constant and given by the gas law, with m_A/hA being the concentration loading of A onto the soil column as free-phase material. Once the free-phase material has evaporated the soil gas, porewater, and solids still contain A but only as trace quantities. In the second part, the trace fraction is modeled. Here the concentration loading is given by Eq. 6.3-15 and the exact evaporation model is Eqs. 6.3-16, 6.3-17, or the approximate model can be used, Eqs. 6.3-20 and 6.3-21, with ρ_{A3} as the trace concentration. The logic used in invoking this two-part evaporation scenario is that until all free-phase A has disappeared the other three phases are at their equilibrium capacities. The author is not aware of any experimental data that support or refute the two-part evaporation scenario.

It was pointed out as we began to develop the pore diffusion transport model that it is a highly idealized quantitative description of a complex process. A glimpse of the complexity of the movement of a volatile chemical in the upper layers of the soil may be appreciated in a brief study of water movement in this region. The review presented here is extracted in part from Jackson et al.[18]

Water is not free in the thermodynamic sense because of capillarity, absorption, and electrical double layers. Capillarity is dominant in wet coarse-textured media, and adsorption assumes its greatest importance in dry media. Double-layer effects may be significant in fine-textured media exhibiting colloidal properties.

In addition to the diffusion term in Eq. 6.3-9 there should appear an advective term. This term is particularly important at moisture saturations of 40% and greater. The advective term is usually quantified by a form of Darcy's law (see p. 518). The diffusivity employed in Eq. 6.3-9 is not a constant but a strong function of the moisture saturation; as soil moisture content increases the importance of the vapor-phase pathway decreases while the liquid-phase pathway increases. These effects are highly nonlinear.

The upper layers of soil are not isothermal as assumed in the pore diffusion model. During the course of a typical summer day it is possible for the soil temperature to vary from 17 to 28°C at 1 cm and 18 to 22°C at 20 cm. It is the rule rather than the exception that both heat and moisture are being transferred simultaneously in the upper soil layers. Water movement in soil is influenced by temperature and temperature gradients (*Soret effect*), and heat flow is simultaneously influenced by the movement of soil water (*Dufour effect*). The coupling of heat and moisture transfer needs to be modeled and accounted for by simultaneous equations that include the energy balance and soil moisture balances.

It is likely that volatile chemicals placed in dry soils will behave similar to water. Detailed analytical models of the transport can be constructed to aid in determining the fate of such chemicals. However, because of the complexities and interrelationships of the transport mechanisms in soils it cannot be inferred with any degree of certainty what effect water content has on chemical movement. Until a complete understanding of all the important factors in volatile chemical transport and fate are known and readily quantifable, the simple models probably give results that are just as defensible as the more rigorous ones.

Processes for Metals and Inorganics

Metal ions, such as Pb^{2+}, Cu^{2+}, Cd^{2+} or Zn^{2+}, can be deposited on the soil surface via sewage sludge, fertilizers, and wastewater applications or atmospheric deposition. In addition, a variety of radionuclides, such as $^{137}Cs^{+}$, $^{90}Sr^{2+}$, or $^{60}Co^{2+}$, can be deposited on soils from radioactive fallout of discharges from the nuclear industry. In a similar manner, inorganic cations and anions also find their way onto surface soils. These include the cations NH_4^{+}, K^{+}, Ca^{2+}, Mg^{2+}, Cr^{3+}, and Al^{2+} as well as the inorganic anions NO_3^{2-}, F^{-}, and $H_2PO_4^{-}$. Within the soil these species create a most complex chemical system, with unnumbered reactions occurring at any moment between mineral surfaces, organic matter, and the aqueous phase. The material presented here provides a very brief glimpse of this complexity and is extracted from more complete works on the subject.[47]

Soil reactions are generally classified according to the nature of the main chemical processes involved: adsorption, ion exchange, dissolution, and so on. However, to assess the kinetics one should consider the nature of the rate of the transport processes associated with the chemical reactions: flow and diffusion in the soil solution, transport across the solid–liquid interface, diffusion in liquid-filled pores and micropores, and surface diffusion penetration into the solid. An expression for the kinetics of soil reactions can be devised by assigning rate equations to transport and chemical processes and combining these equations. The expressions finally obtained have to be validated by comparison to experimental results.

Transport Processes in Soil Reaction. These are shown schematically in Figure 6.3-9 for both the nonactivated and the activated transport processes. The process is termed *activated* if it takes place at the solid surface. As keyed in the figure, the *nonactivated* processes are (1) transport in the soil solution of the macropores, (2) transport across a liquid film at the solid–liquid interface, and (3) transport in a liquid-filled micropore. Except for charged species, processes 1 to 3 have been described previously with regard to solutes in sediment (see p. 280).

If the diffusing solute is an ion and the pore walls are charged surfaces, the flux may be determined by the gradient of the electrical potential as well as by the concentration gradient. A term for the gradient of the electrical potential is added to Fick's first law to obtain the Nernst–Plank equation:

$$j_A = -D_{A3}\left[\frac{\partial \rho_{A2}}{\partial y} - \frac{\rho_{A2} z_e F}{RT}\left(\frac{\partial \phi_e}{\partial y}\right)\right] \tag{6.3-22}$$

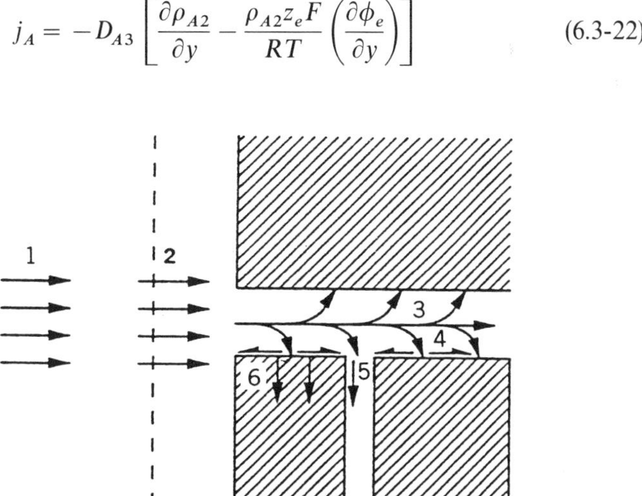

Figure 6.3-9. Transport processes in solid-liquid soil reactions. See pages 417–418 for an explanation of process numbers. (Reprinted by permission from Ref. 47.)

where $\partial\phi_e/\partial y$ is the gradient of the electrical potential, z_e the electrochemical valence of the diffusing ion, and F the Faraday constant. The chemical potential can generally be related to the concentration of the ions in the system and the equation is then reduced to a function relating the concentration to distance and time.

Transport process 4 in Fig. 6.3-9 is the diffusion of the sorbate on the surface of the solid. Surface diffusion results from the fact that an adsorbed species attached at a surface site is capable of moving to a neighboring empty site without desorbing. A jump to a neighboring site implies weakening and breaking of a surface bond while a new bond is being formed and the jump frequency is determined by the activation energy. The process is possible if the activation energy of the jump is small enough so that the frequency is significant and if the activation energy for desorption is high enough so that desorption does not occur. Jump direction is random from regions of high surface concentration to regions of low surface concentration. Fickian-type equations are used to describe the process.

Diffusion of sorbate occluded in micropores is process 5 in Fig. 6.3-9. If the solute species penetrates a narrow pore with width only slightly larger than its diameter, it cannot be assumed to be in the dissolved state. It is always assumed to be in a sorbed state and moves along the pore by a process similar to surface diffusion. The last process, number 6 in Fig. 6.3-9, is diffusion in the bulk of the solid. Here ions migrate along lattice vacancies or along crystal fractures.

Chemical Interactions in Soil Reactions. This occurrence at the solid phase may comprise formation or rupture of a bond between sorbate and surface, further reaction between adsorbed species and rearrangements of the solid structure, and formation and disappearance of solid species. Simple kinetic models for such interactions often fail because reacting solid surfaces are rarely homogeneous and because effects of transport phenomena and chemical reactions are difficult to separate experimentally. Reactions can also occur concurrently and consecutively in these systems. Figure 6.3-10 illustrates some of the types of reactions that occur in soil systems and the time ranges required to obtain equilibrium by these reactions.

The ion association, multivalent-ion hydrolysis, and mineral-crystallization reactions are all homogeneous because they occur within a single phase. The first two of these occur in the liquid phase and the last occurs in the solid phase. The other reaction types are heterogeneous because they involve transfer of chemical species across interfaces between phases. Ion association reactions refer to ion pairing, complexation (inner and outer sphere), and chelation-type reactions in solution. Gas–water reactions refer to the exchange of gases across the air–liquid interface. Ion-exchange reactions refer to electrostatic ion-replacement reactions on charged solid surfaces. Sorption reactions refer to simple physical adsorption, surface-complexation (inner and outer sphere), and surface-precipation reactions. Mineral-solution reactions

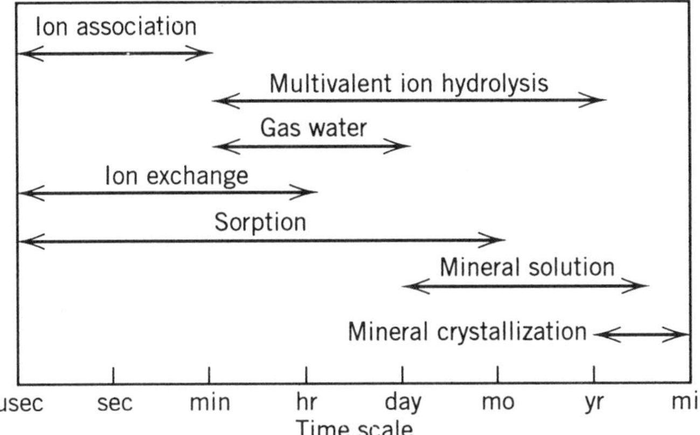

Figure 6.3-10. Time ranges required to attain equilibrium by different types of reactions in soils. (Reprinted by permission from Ref. 47.)

refer to precipitation/dissolution reactions involving discrete mineral phases and coprecipitation reactions by which trace constituents can become incorporated into the structure of discrete mineral phases.

As shown in Fig. 6.3-10, the reactions encompass a wide range of time scales. The complexity of reactions over time defies a simple analysis of the kinetics involved. Numerous methods have been developed to isolate and study the various types of reactions in soils. Many experiments are usually required to establish a mechanism, which is largely a trial-and-error process of testing and retesting to rule out alternative explanations of the rate functions observed. A lifetime of investigation may be required to establish a particular complicated mechanism, and in some cases a solution may never by reached.[47]

Local Equilibrium Assumption (LEA). The assumption of local equilibrium during solute transport suggest that the chemical, physical, and biological interactions of the solute with the porous media are instantaneous, or that the solute resident time in the media is sufficiently large to achieve negligible concentration differences between pore classes. It has been suggested that the validity of the LEA depends on the degree of interaction between macroscopic transport processes such as water flow and hydrodynamic dispersion, and the microscopic processes of diffusion and sorbed-solute distribution in conjunction with soil aggregate size. When the rate of change of solute mass during microscopic sorption processes is fast relative to the bulk flow, the interaction is generally considered instantaneous, therefore conforming to the LEA. Deviations from local equilibrium occur as the interactions of the solute with the porous media becomes increasingly time dependent. Divergence also occurs as soil aggregates increase in size and the pore-class heterogeneity increases. Kinetic limitations during solute transport have been shown to become significant as the Damkohler number decreases below 10 (Jardine in Ref. 47). See Problem 6.3F for definition of the Damköhler number.

Frequent use has been made of the retarded diffusion transport model for metal and inorganics in both the laboratory and the field soil plot. It is very unlikely in most cases that this LEA-based model reflects the actual soil-solute processes. Nevertheless, often this two-parameter algorithm does a good job curve fitting the data and yielding realistic parameters so that it can be used to interpret the evolution of chemical species profiles in soils. However, using it to predict future occurrences in all but identical conditions is unwise. For example, it was used successfully to model the movement of Cu^{2+} and Mn^{2+} in soil above buried (45-cm depth) human bones which served alternatively as a sink and source of the metals over a 1400-year period.[48] The LEA between solid surfaces and solution yielded metal partition coefficients in rough agreement with published values. In this field-application case the imprecision of the data made the use of more complex kinetics impractical. With due regard to its limitations, the LEA-based model should be considered for several reasons: numerous analytical solutions reflecting realistic boundary conditions are readily available, it has simplicity in parameter interpretation, and its inherent first-order kinetics apparently mimics many complex soil reactions.

PROBLEMS

6.3A. Simplification of Multicomponent Equation of Continuity for Soil Contaminant Problem

Beginning with the multicomponent equation of continuity:

1. Simplify it to yield Eq. 6.3-9. State reasons for each modification or simplification made and/or term excluded.
2. Reformulate the problem to account for transport of the chemical by groundwater movement in the vertical direction.

6.3B. Contaminant Lifetime for Acetaldehyde Spilled on Soil

A quantity of acetaldehyde (CH_3CHO) was spilled onto dry soil. The soil column was contaminated to a depth of 35 cm. This particular soil is estimated to be nonadsorptive to acetaldehyde and to have a porosity of 30%.

1. Estimate the contaminant half-life (hr) for the acetaldehyde spill if the soil temperature is uniform at 25°C.
2. Estimate the time (hr) required for 90% of the acetaldehyde to dissipate into the atmosphere at 25°C.

6.3C. Concentration Profile and Flux for Di–Br–Cl–Propane in Soil

Concentration profiles for trace volatiles in soil can be estimated using Eq. 6.3-10 provided that the interface concentration is known. Grasses and other vegetation on the surface can hinder transport, reducing ${}^3k'_{A1}$ significantly from the values obtained for bare soil with Eq. 6.2-36. Although the background concentration, ρ_{A1}, may be insignificant compared to, ρ^*_{A1}; that at the interface, ρ_{A1i}, may be significant in comparison.

1. A related model for a semi-infinite soil body that includes a fluid-side resistance is given by Eqs. 5.2-48 to 5.2-50. Reformat these for the transport of volatiles in surface soils.
2. Determine the concentration profiles, ω_A versus y of Di–Br–Cl–propane based on the data in Table 6.3-4 for an initial uniform concentration $\omega_A^0 = 100$ mg/kg. Perform calculations for the top meter including the interface for $t = 1$ day, 1 week, 1 month and 1 year. Assume that ${}^3k'_{A1} = 50$ cm/h and $\rho_{A1\infty} = 0$.
3. Compute the flux to air in μg/m$^2\cdot$h at each time.

6.3D. Benzene Spill and Evaporation to Air

H.-G. Schecker speculates that elevated levels of benzene in the air around a railyard near Dortmund, Germany in the 1980s might have been due to ongoing evaporation from the ground of a large quantity spilled from a tankcar in the 1940s.

1. Invoke a soil structure, construct a spill scenario, and select an evaporation model that may be used to support Schecker's notion.
2. Estimate the benzene flux to air in the 1980s.
3. Using the D. D. Reible large-ground-surface area-source equation,

$$\rho_{A1} = \frac{27n_A}{\bar{v}_x} \tag{6.3D}$$

estimate the enhanced concentration in air due to the spill.

In this northern latitude assume 7.6°C average temperature, 40 mm vapor pressure, spill penetration depth of 2m, and benzene residual saturation of 0.50.

6.3E. Gradientless Vapor Pore-Diffusion Model

The vapor pore-diffusion model presented in Section 6.3 involved a partial differential equation (Eq. 6.3-9). The solution yielded vapor concentration gradients within the soil column (Eq. 6.3-10 and Fig. 6.3-6), average concentration (Eq. 6.3-11), fraction chemical remaining (Fig. 6.3-7), and flux rate at the air–soil interface (Eq. 6.3-13).

A gradientless vapor pore-diffusion model may be developed as follows. Assuming that the length of the diffusion path is $h/2$, the flux rate from the pores is

$$n_{A0} = \frac{2D_{A3}}{\mathrm{h}}(\rho_{A1} - \rho_{A1i}) \tag{6.3E-1}$$

where ρ_{A1} is a uniform concentration of chemical vapor in pore space, D_{A3} the diffusion coefficient, h the depth of the contamination zone, and ρ_{A1i} the concentration at the soil surface. The gradientless concentration within the contaminated zone pore spaces is

$$\rho_{A1} = \rho_{A1i} + (\rho^*_{A1} - \rho_{A1i})\exp\left(\frac{-2D_{A3}t}{\varepsilon_1 h^2}\right) \tag{6.3E-2}$$

The half-life is

$$t_{1/2} = \frac{0.693h^2\varepsilon_1}{2D_{A3}} \tag{6.3E-3}$$

and the flux rate at the surface is

$$n_{A0} = \frac{2D_{A3}}{h}\exp\left(\frac{-2D_{A3}t}{\varepsilon_1 h^2}\right)(\rho^*_{A1} - \rho_{A1i}) \tag{6.3E-4}$$

The gradientless model equations for concentration and flux rate can be compared to those for the gradient model (Eqs. 6.3-10 and 6.3-13).

1. Verify Eqs. 6.3E-2 to 6.3E-4 by first performing a component balance on chemical A to obtain a simple differential equation of ρ_{A1} in t.

2. Using Fig. 6.3-7 as a guide, obtain an expression similar to Eq. 6.3E-3 for the half-life of chemical A as predicted by the gradient model. Which model will predict the longer vapor half-life?

3. Rework Problem 6.3B using the gradientless model.

6.3F. Damköhler Number Limitations on Local Equilibrium Assumption

In situations where convective transport can be neglected, the Damköhler number can be used to scale the limits of the LEA. In terms of characteristic quantities in the soil microenvironment it is defined as

$$\mathrm{Da} \equiv \frac{r_{Aj}l^2}{\rho_{Aj}D_{Aj}} \tag{6.3F-1}$$

where r_{Aj} is a characteristic rate at which the mass of species A is produced per unit volume by homogeneous chemical reactions within phase j and j = 1,

2, or 3. The other terms are characteristic length, concentration, and diffusion coefficient, respectively. Using the information in Fig. 6.3-10:

1. Determine the upper time-scale limit at which the LEA is applicable if a first-order reaction is occurring within the aqueous phase. Assume that $l = 0.1$ and 1.0 mm.

2. Repeat part 1 for reaction and diffusion in the solid phase where the particle or aggregate porosity is 15%.

3. Determine the lower limit of reaction rates at which the LEA is applicable for the two cases above. Assume that the concentration A is 1 mg A/kg. Answer in mg/cm$^3 \cdot$h.

6.3G. Controlling Hexachlorobenzene Evaporation from Soil Surface

Large quantities of pure hexachlorobenzene are lying on a soil surface. A soil cover is being considered to control the vaporization rate and therefore the concentration in air.

1. Compute the flux without a cover at 30°C. The vapor pressure of HCB is 1.1E-5 mm at 20°C and 1.0 mm at 114°C. Use the data in Fig. 6.2-4 for a 10-km/h wind.

2. If 99.9% flux reduction is desired, what thickness of soil cover, dry at 0.5 porosity, is needed?

3. For a 30-cm cover, compute the flux under the following soil-water saturation values: 0% (dry), 30%, 70%, and 95%.

4. Is moisture an effective control measure for volatile losses? Present and discuss the case for volatile and soluble releases at $\geqslant$100% saturation. $M_A = 285$ and solubility (assumed) $\simeq$6 μg/L at 30°C.

6.3H. Doing Is Believing

Using first principles, develop the following model relationships and list all assumptions in the process.

1. Derive Eqs. 6.3-2 to 6.3-4 and the associated concentration profiles, ρ_{A1}, in the layer $h \geqslant y \geqslant 0$.

2. Develop Eq. 6.3-6.

3. Derive Eq. 6.3-14.

4. Derive the transient models represented by Eqs. 6.3-20 and 6.3-21.

5. Show that the term ρ_{A3} in Eqs. 6.3-20 and 6.3-21 is inclusive of Eq. 6.3-14.

6. Is Eq. 6.3-20 a less restrictive evaporative-life result than Fig. 6.3-7 or Eq. 6.3-17? For evaporative half-life, convert it to the form of Eq. 6.3-17.
7. Put Eq. 6.3-21 in a form similar to Eqs. 6.3-13 and 6.3E-4.
8. Verify the units of the terms in the Nernst–Plank equation, Eq. 6.3-22.

6.4. HEAT TRANSFER AT AIR–SOIL INTERFACE

Soil surface and column temperatures down to a depth of 50 to 100 cm can have a pronounced effect on chemical movement from and within this region of the air–soil interface. In the case of a pure substance in soil, for example, its temperature regulates vapor pressure, p_A^0, significantly, as shown by Eq. 2.1-7. This, in turn, determines the concentration, ρ_{A1}^*, in Eq. 6.3-21, and therefore influences the flux of this volatile substance in and from the soil surface.

Fundamental concepts of heat transfer were presented in Section 3.3. The reader should have a firm understanding of these concepts prior to attempting to apply the results of this section to a chemodynamics problem. The first topic to be considered is the basic mechanisms of heat exchange at the air–soil interface. The subject of simultaneous heat and moisture transport through the soil surface is a complex process. It often controls the surface temperature, the energy flux, and temperatures within the soil layers. Although the processes are key to quantifying aspects of chemical transport and fate in this region, the detailed procedures and algorithms are beyond the scope of this book. Despite this limitation some degree of qualitative understanding of heat penetration into the ground is needed and it will be covered briefly.

The energy balance equation at the air–soil interface may be written as

$$q_{13} = q_{\text{lw}} + q_e + q_c + q_s \tag{6.4-1}$$

where q_{lw} is the net longwave radiation flux, q_e the evaporative (water) heat flux, q_c the sensible heat flux (conduction) between the surface and air, q_s the shortwave solar radiation, and q_{13} is the sensible heat flux into the soil column. Here q_{13} is directly proportional to the rate at which the column is warming. The sign convention adopted is that energy additions to the soil column are positive. All energy or heat flux relations are in $\text{J/m}^2 \cdot \text{s}$ ($1\ \text{Btu/ft}^2 \cdot \text{hr} = 3.15\ \text{J/m}^2 \cdot \text{s}$, $1\ \text{J/m}^2 \cdot \text{s} = 4.187\ \text{cal/m}^2 \cdot \text{s}$). With the sign convention adopted, if the right-hand side of Eq. 6.4-1 is positive, the soil column is heating up; if it is negative, the soil column is cooling off.

Equation 6.4-1 for the air–soil interface is similar in form and content to Eq. 4.3-2, which is for the air–water interface. In general, the mechanisms for transfer at the respective interfaces are similar; however, there are some important differences in the details of the algorithms. The soil surface is more diverse and complex than most water surfaces, so that the individual components on the right side of Eq. 6.4-1 are slightly or drastically more complex than those in Eq. 4.3-2. In particular, this includes the shortwave radiation; the soil is generally more reflective.[21] The sensible and evaporative exchanges are considerably more complex for the soil surface, and the reader should consult Kreith and Sellers[22] or similar works for these details.

Penetration of Heat into the Ground

The thermal diffusivity of soil is small. It is considerably less than that for air at rest, as shown in Table 6.4-1. Thermal diffusivity of a soil is a parabolic function of increasing moisture content. A small amount of water reduces the insulating effect of the pore space filled with air, but further increases in water content markedly increase the heat capacity. Soil organic matter lowers thermal diffusivity because of its influence in increasing porosity. Compaction increases the thermal diffusivity by decreasing the volume of the insulating pore space.

The computation of daily and seasonal patterns of soil temperature below the surface is beyond the scope of this book; however, a brief overview of the nature and magnitudes of variation is in order. Soil temperature changes can be observed as waves during the course of a day. The amplitude of the temperature wave at the soil surface is great but diminishes with depth below the surface. The pattern of soil temperature profiles changes rapidly during a normal day, as illustrated in Fig. 6.4-1 for measurements made at Argonne National Laboratory. The pattern of decreasing amplitude of the soil column temperature wave between summer and winter is illustrated in Fig. 6.4-2. In general, soil temperature decreases with depth during the daytime in summer. Temperature gradients direct heat into the soil. At night, however, the temperature is highest between 20 and 40 cm; from that level heat is directed both upward and downward. In winter the daily amplitude of the surface temperature is very small. The 81-cm level is warmest, and heat is transferred upward from this level throughout the day and night.

Table 6.4-1. Density and Thermal Properties of Some Soil and Reference Materials[a]

Material	Density, ρ_3 (g/cm^3)	Specific Heat, $\hat{C}_{p3}$ ($cal/g \cdot K$)	Thermal Conductivity, k_3 ($cal/cm \cdot K \cdot s$)	Thermal Diffusivity, α_3 (cm^2/s)
Clay	1.8	0.8	2.88E-3	2E-3
Light soil with roots	0.3	0.3	2.70E-4	3E-3
Wet sandy soil	1.6	0.06	6.40E-3	0.01
Dead air	1.3E-3	0.24	4.99E-5	0.16

Source: Reprinted by permission from Ref. 5.

[a]J = 4.1868 cal.

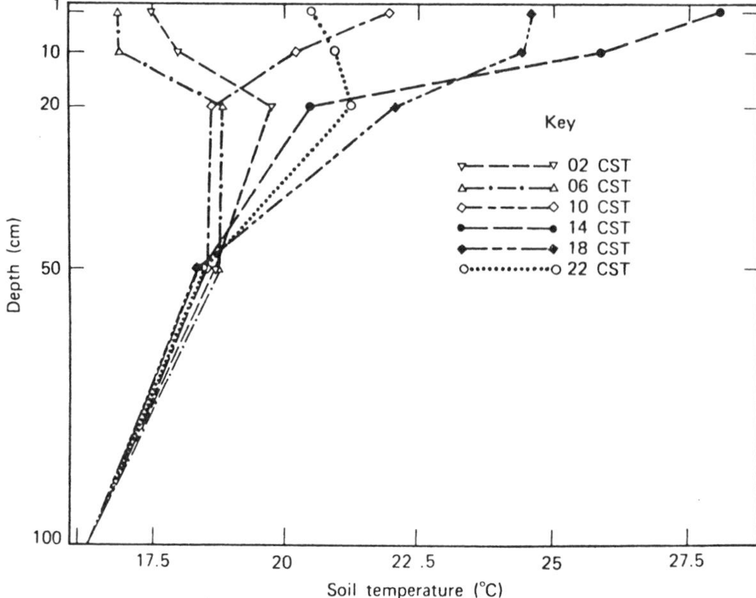

Figure 6.4-1. Vertical temperature profiles in soil during the course of a typical summer day at Argonne, Ill, July 27, 1955. (Reprinted by permission from Ref. 5.)

Once the range of soil surface temperature, ΔT_{3i}, has been estimated, it is also possible to estimate the range of subsurface temperature variation. Rosenberg[5] suggested that the range of temperature at any depth in the soil can be obtained by

$$\Delta T_3(y) = \Delta T_{3i} \exp\left[-y\left(\frac{\omega}{\alpha_3}\right)^{1/2}\right] \tag{6.4-2}$$

where $\Delta T_3(y)$ is the temperature range at depth y, α_3 is the thermal diffusivity of the soil material, and ω the period of oscillation ($\pi/12\ \text{hr}^{-1}$). Equation 6.4-2 assumes that the soil properties, including porosity, water content, and organic matter, were uniform with depth so that a constant value of the thermal diffusivity may be used.

Heat Transport Limitations to Vaporization of Buried Waste[49]

Volatile chemicals in liquid or solid form within a landfill cell, such as illustrated in Fig. 6.3-4, or otherwise buried within the earth must vaporize prior to being transported through the soil. A source of energy is needed for this to occur. Although some minor reactions (biochemical, hydrolysis, etc.) may occur, regulatory limitations on the placement of highly reactive materials in landfills reduces the likelihood of this process being a significant energy source. The worldwide terrestrial geothermal heat flux from within the earth is

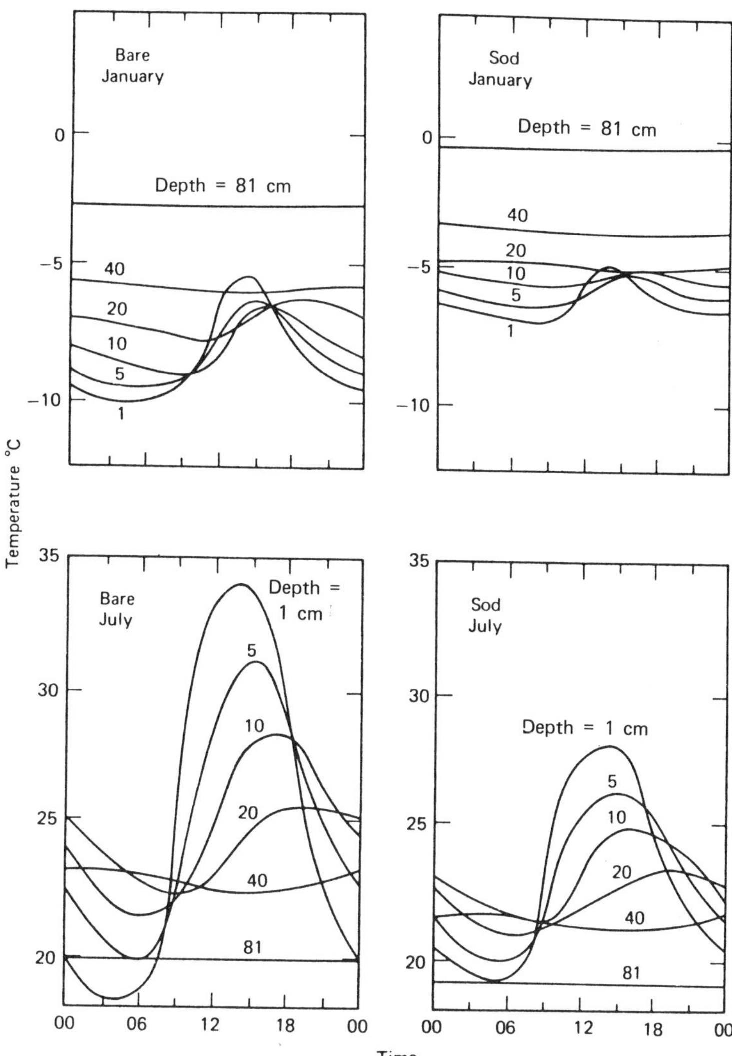

Figure 6.4-2. Average hourly soil temperature under bare and sod-covered soil at St. Paul, Minnesota, in January (*top*) and July (*bottom*) 1961. Soil depth is shown in cm. (Reprinted by permission from Ref. 5.)

6.3E-6 W/cm^2. It is only one-thousandth of the average influx of solar heat that is received during the day and reradiated during the night. Solar energy is the dominant source; the rate of sensible heat transfer through the soil cover material must equal the rate of energy needed to vaporize (or desorb) the chemical

$$-k_3 \frac{dT}{dy} = \lambda_A n_A \tag{6.4-3}$$

where k_3 is the average thermal conductivity of the soil-like material in the cap, dT/dy the temperature gradient across the cap, λ_A the latent heat of vaporization at the cell temperature, and n_A the steady-state volatilization rate. If the chemical is sorbed onto the soil, an additional heat-of-desorption term must be added to λ_A (see Eq. 2.1-31). The above equation places an upper bound on the volatiles emissions rate possible from buried wastes.

Closure

Soil surface and subsurface temperature is an important parameter with respect to chemodynamics in the soil column near the air interface. Although a general treatment of heat and moisture transfer in soil is beyond the scope of this book, the preceding material does provide a qualitative description of the major process occurring to obtain order-of-magnitude estimates of temperature ranges.

PROBLEMS

6.4A. Energy-Limited Vaporization Rate

The benzene emission rate from a certain landfill computed by Eq. 6.3-4 is 17.2E-9 $g/cm^2 \cdot s$.

1. If the summer temperature gradient is $-0.05°C/cm$, can this emission rate occur? The cap is made of coarse gravel and clay soil with 5 to 10% moisture and 25 to 50% porosity.
2. Can the average terrestrial, geothermal heat flux maintain this emission rate?

6.4B. Penetration of Heat into the Ground

The range of subsurface temperature variation can be estimated by Eq. 6.4-2 provided that the surface temperature variation and the soil thermal diffusivity are known. In general, the subsurface temperature oscillates around some seasonal, slowly changing temperature $T_3(\infty)$ that is somewhat constant beyond a certain depth. The deviation of the maximum subsurface temperature, $T_{3,MX}(y)$ is related to $T_3(\infty)$ and $T_{3,MX}(0)$ by

$$T_{3,MX}(y) = \exp\left[-y\left(\frac{\omega}{\alpha_3}\right)^{1/2}\right][T_{3,MX}(0) - T_3(\infty)] + T_3(\infty) \qquad (6.4B\text{-}1)$$

The deviation of the minimum subsurface temperature, $T_{3,MN}(y)$, is obtained from

$$T_{3,MN}(y) = T_{3,MX}(y) - \exp\left[-y\left(\frac{\omega}{\alpha_3}\right)^{1/2}\right][T_{3,MX}(0) - T_{3,MN}(0)] \qquad (6.4B\text{-}2)$$

From the preceding equations it is possible to calculate the subsurface temperature variations with depth.

1. For a light soil with roots known to have a surface temperature of $T_{3,MX}(0) = 30°\text{C}$ and $T_{3,MN}(0) = 17°\text{C}$, compute the subsurface temperature variations for $T_3(\infty) = 19°\text{C}$. Use $y = 0$, 1, 3, 5, 9, 15, 25, and 30 cm in your calculations. Construct a graphical representation similar to Fig. 6.4-1.
2. Repeat part 1 for $T_3(\infty) = 16°\text{C}$.
3. Based on the data shown in Fig. 6.4-1, estimate the thermal diffusivity (cm^2/s) of the Argonne, Illinois soil.

6.4C. Subsurface Heat Flow

1. Show that Eq. 3.3-1 may be integrated to obtain

$$q_y = \frac{k_3}{h}(T_{si} - T_s) \tag{6.4C-1}$$

where T_{si} is the soil surface temperature and T_s is the subsurface temperature at depth h. Assume that heat flow is constant and thermal conductivity k is independent of depth.

2. Calculate the heat flux rate for Argonne, Illinois soil temperature profile shown in Fig. 6.4-1. Use the data between $y = 1$ cm and $y = 20$ cm (i.e., $h = 19$ cm) and calculate the flux rate ($\text{J/cm}^2 \cdot \text{s}$) and the direction of heat transfer for each time shown. Assume that $\alpha = 4.4\text{E-}2\ \text{cm}^2/\text{s}$, $\rho_3 = 1.0\ \text{g/cm}^3$, and $C_{p3} = 0.5\ \text{cal/g} \cdot °\text{K}$.

REFERENCES

1. O. G. Sutton, *Micrometeorology*, McGraw-Hill, New York, 1953.
2. G. S. Campbell, *An Introduction to Environmental Biophysics*, Springer-Verlag, New York, 1977, p. 34.
3. F. A. Gifford, Jr., "An Outline of Theories of Diffusion in the Lower Layers of the Atmosphere," in D. H. Slade, Ed., *Meteorology and Atomic Energy*, U.S. Atomic Energy Commission, Oak Ridge, Tenn., 1968, pp. 66–116.
4. S. Eskinazi, *Fluid Mechanics and Thermodynamics of Our Environment*, Academic Press, New York, 1975, pp. 83–129.
5. N. J. Rosenberg, *Microclimate: The Biological Environment*, Wiley-Interscience, New York, 1974, p. 105.
6. F. A. Brooks and W. O. Pruitt, "Investigations of Energy, Momentum and Mass Transfer near the Ground," DA Task IVO-14501-B53A-08, U.S. Army Electronics Command, Atmospheric Science Laboratory, Research Division, Fort Huachuca, Ariz., 1965, Chapter 4.

7. S. Pal Arya, *Introduction to Micrometeorology*, Academic Press, New York, 1988. Chapter 11.
8. D. A. deVries and N. H. Afgan, Eds., *Heat and Mass Transfer in the Biosphere*, Part 1, *Transfer Processes in the Plant Environment*, Scripta, Washington, D.C., 1975.
9. L. H. Parmele, E. R. Lemon, and A. W. Taylor, "Micrometeorological Measurement of Pesticide Vapor Flux from Bare Soil and Corn Under Field Conditions," *Water Air Soil Pollut.*, **1**, 433–451 (1972).
10. A. W. Taylor, D. E. Glotfelty, B. C. Turner, R. E. Silver, H. P. Freeman, and A. Weiss, "Volatilization of Dieldrin and Heptachlor Residues from Field Vegetation," *Agric. Food Chem.*, **25**(3), 542 (1977).
11. O. T. Denmead, Jr. R. Simpson, and J. R. Freney, "Ammonia Flux into the Atmosphere from a Grazed Pasture," *Science*, **185**, 609 (1974).
12. D. Mackay and R. S. Matsugu, "Evaporation Rates of Liquid Hydrocarbon Spills on Land and Water," *Can. J. Chem. Eng.*, **51**, 434–439 (1973).
13. B. C. Turner and D. E. Glotfelty, "Field Air Sampling of Pesticide Vapors with Polyurethane Foam," *Anal. Chem.*, **49**(1), 7–10 (1977).
14. D. D. Reible and F. H. Shair, "A Technique for the Measurement of Gaseous Diffusion in Porous Media," *J. Soil. Sci.* **33**, 165–174 (1982).
15. R. R. Arnts, R. L. Seila, and R. L. Kuntz, "Measurements of α-Pinene Fluxes from a Loblolly Pine Forest," *Proceedings of the 4th Joint Conference on Sensing of Environmental Pollutants*, American Chemical Society, Washington, D.C., 1978, p.831.
16. N. C. Brady, *The Nature and Properties of Soils*, 8th ed., Macmillan, New York, 1974.
17. H. S. Carslaw and J. C. Jaeger, *Conduction of Heat in Solids*, Oxford University Press, London, 1959, p. 96.
18. R. D. Jackson, B. A. Kimball, R. J. Reginato, S. B. Idso, and F. S. Nakayama, "Heat and Water Transfer in a Natural Soil Environment," in D. A. deVries and N. H. Afgan, *Heat and Mass Transfer in the Biosphere*, Part 1 *Transfer Processes in the Plant Environment*, Scripta, Washington, D.C., 1975, pp. 67–76.
19. H. Israël, "Die natürliche Radioaktivitat in Boden, Wasser and Air," *Beitr. Phys. Atmos.*, **30**, 177–188 (1958).
20. C. E. Junge, *Air Chemistry and Radioactivity*, Academic Press, New York, 1963, pp. 209–221.
21. D. R. DeWalle, "An Agro-Power-Waste Complex for Land Disposal of Waste Heat," *Res. Publ. 68*, University of Pennsylvania, Philadelphia, 1974.
22. F. Kreith and W. D. Sellers, "General Principles of Natural Evaporation," in D. A. deVries and N. H. Afgan, *Heat and Mass Transfer in the Biosphere*, Part 1 *Transfer Processes in the Plant Environment*, Scripta, Washington, D.C., 1975, pp. 207–214.
23. S. R. Hanna, G. A. Briggs, and R. P. Hosker, Jr., *Handbook on Atmospheric Diffusion*, DE82002045(DOE/TIC-11223), U.S. Department of Commerce, National Technical Information Service, Springfield, Va., 1982.
24. L. J. Thibodeaux and H. D. Scott, "Air/Soil Exchange Coefficients," Chapter 4 in W. B. Neely and G. E. Blau, Eds., *Environmental Exposure from Chemicals*, Vol. 1, CRC Press, Boca Raton, Fl., 1985.

25. L. A. Harper, L. L. McDowell, G. H. Willis, S. Smith, Jr., and L. M. Southwich, "Microclimate Effects on Toxaphene and DDT Volatilization from Cotton Plants," *Agron. J.*, **75**; 295–302 (1983).

26. S. T. Hwang and L. J. Thibodeaux, "Measuring Volatile Chemical Emission Rates from Large Waste Disposal Facilities," *Environ. Prog.*, **2**(2), 81–86 (1983).

27. J. H. Seinfeld, *Atmospheric Chemistry and Physics of Air Pollution*, Wiley, New York, 1986, Chapter 16.

28. G. J. McRae and A. C. Russell, "Dry Deposition of Nitrogen-Containing Species," Chapter 9 in B. B. Hicks, Ed., *Deposition Both Wet and Dry*, Butterworth, Woburn, Mass., 1984.

29. H.-W. Georgii and J. Pankrath, *Deposition of Atmospheric Pollutants*, D. Reidel, Boston, 1982.

30. D. D. Reible, T. H. Illangasekare, D. V. Doshi, and M. E. Malhiet, "Infiltration of Immiscible Contaminants in the Unsaturated Zone," *Groundwater*, **28**(5), 685–692 (1990).

31. L. J. Thibodeaux, K. T. Valsaraj, C. Springer, and G. Hildebrand, "Mathematical Models for Predicting Chemical Vapor Emissions from Landfills," *J. Hazard. Mater.*, **19**, 101–118 (1988).

32. W. J. Farmer, M. Yang, J. Letey, and W. F. Spencer, "Land Disposal of Hexachlorobenzene Waste: Controlling Vapor Movement in Soil," EPA-600/12-80-11, Final Report, U.S. EPA, Office of Research and Development, Washingtion D.C., Aug. 1980.

33. L. J. Thibodeaux, C. Springer and L. M. Riley, "Models of Mechanisms for the Vapor Phase Emission of Hazardous Chemicals from Landfills," *J. Hazard. Mater.*, **1**, 63–74 (1982).

34. L. J. Thibodeaux, "Estimating the Air Emissions of Chemicals from Hazardous Waste Landfills," *J. Hazard. Mater.*, **4**, 235–244 (1981).

35. C. A. I. Goring, and J. W. Hamaker, *Organic Chemicals in Soil Environment*, Vol. 1, Marcel Dekker, New York, 1972, pp. 352–355.

36. R. J. Millington and J. M. Quirk, "Permeability of Porous Solids," *Trans. Faraday Soc.*, **57**, 1200–1207 (1961).

37. G. J. Thoma, G. Hildebrand, K. T. Valsaraj, L. J. Thibodeaux, and C. Springer, "Transport of Chemical Vapors Through Soil: A Landfill Cover Simulator," *J. Hazard. Mater.*, **30**, 333–342 (1992).

38. F. R. Groves, D. D. Reible, and L. J. Thibodeaux, "Estimation of Physical and Chemical Properties of Waste Organic Mixtures Associated with Land Pollution," Section 3 in J. D. Dean Ed., *Exposure Assessment Involving Mixtures of Environmental Pollutants*, U.S. EPA, Office of Health and Environmental Assessment, Office of Research and Development, Washington, D.C., Dec. 1984.

39. D. C. Bomberg, J. L. Gwinn, W. R. Mabey, D. Tuse, and T.-W. Chow, "Environmental Fate and Transport at the Terrestrial–Atmospheric Interface," *Proceedings, Division of Pesticide Chemistry, American Chemical Society Meeting*, Kansas City, Mo., Sept. 1982.

40. J. R. Welty, C. E. Wicks, and R. E. Wilson, *Fundamentals of Momentum, Heat and Mass Transfer*, 3rd ed., Wiley, New York, 1984, pp. 530–533.

41. R. W. Buff, "Transport Processes in Organic Vapor Emissions from Sand," M. S. thesis, University of Arkansas, Fayetteville, Ark., Aug. 1984.
42. J. Carvanos, "Validation of Mathematical Models Predicting the Emission Rates of Selected Organic Solvents from Saturated Soils," Ph.D. thesis, Columbia University, New York, 1984.
43. L. J. Thibodeaux and S. T. Hwang, "Landfarming of Petroleum Waste-Modeling the Air Emission Problem," *Environ. Prog.*, **1**(1), 42–46 (1982).
44. R. R. DuPont and J. A. Reineman, "Evaluation of Volatilization of Hazardous Constituents at Hazardous Waste Land Treatment Sites," *EPA/600/2-86/071*, U.S. Department of Commerce, National Technical Information Service, Springfield, V., 1986.
45. W. A. Jury, D. Russo, G. Streile, and H. El Abd, "Evaluation of Volatilization by Organic Chemicals Residing Below the Soil Surface," *Water Resources Res.*, **26**, 13–20 (1990).
46. P. A. Ryan and Y. Cohen, "Diffusion of Sorbed Solutes in Gas and Liquid Phases of Low-Moisture Soils," *Soil Sci. Soc. Am. J.*, **54**, 341–346 (1990).
47. D. L. Sparks and D. L Suarez, Eds., *Rates of Soil Chemical Processes*, Soil Science Society of America, Madison, Wisc., 1991, Chapter 1.
48. J. Thomas, L. J. Thibodeaux, A. F. Ramenofsky, S. P. Field, B. J. Miller, and A. M. Whitmer, "Archaeological Chemistry," *Environ. Sci. Technol.*, **22**(5), 480–487 (1988).
49. C. Springer, P. D. Lunney, K. T. Valsaraj, L. J. Thibodeaux, and S. C. James, "Emission of Hazardous Chemicals from Surface and Near Surface Impoundments to Air, Part B, Landfills," U.S. EPA, Hazardous Waste Engineering Research Laboratory, Office of Research and Development, Cincinnati, Ohio, May 1986.

7

INTRAPHASE CHEMICAL TRANSPORT AND FATE

The waste products of our civilization must be disposed of. Receptacles for this debris are the Earth's land masses, water bodies, and atmosphere. Some fate and transport topics for waste that is released to the fluid medium water and air and to the solid plus fluid subterranean media are the subject of this chapter.

Wastes that are released to the atmosphere consist of particles and gases, and wastes released to water and the ground consist of solids, liquids, and substances in aqueous solution. Residence times for some of these materials may be very short—hours or even minutes—while for others it may be measured in years and even centuries. Regardless of the residence time, the movement of molecules and particles in the respective media is, in large measure, governed by the motions of the medium. Some motions dictate the paths to be followed by the contaminants, and other motions determine the extent to which the contaminants are diluted.

It is to be expected that in the future the designs of disposal systems for gaseous, liquid, and solid wastes will meet increasingly stricter pollution control regulations aimed at limiting specific chemical contaminants. The decisions involved will have to be based on actual data (e.g., field and/or laboratory) and extensions such as reasonably accurate, quantitative predictions (i.e., models) of the behavior of chemicals introduced into the atmosphere, ocean, lakes, and rivers.

Intraphase transport and fate of chemicals is a topic concerned with the movement and reaction of chemicals from point to point within earthen media, whereas interphase transport is concerned with movement across the natural interfaces. The earthen media involved are the surface waters of aquatic bodies, the air in the lower atmospheric boundary layer, and the land mass of shallow geologic formations. A most common environmental, chemodynamic, intraphase transport process is the dispersion of air pollutants from an elevated point source. This much studied topic is presented briefly along with similar chemical and physical processes in the aquatic and land media. Although it is impossible to cover all such intraphase processes in this chapter, the few presented will illustrate most of the mechanisms. The student should then recognize the similarities in the processes within air, land, and water relevant to the subject of chemodynamics. A major practical result of this chapter will be methods for estimating concentrations of contaminants from continuous and transient sources. As was done in earlier chapters, example and exercise problems are presented in a context that extends the text material and provides realistic exercises for application of the theory.

7.1. CHEMICAL TRANSPORT AND FATE IN SURFACE WATERS

Annual Thermal Structure of Deep Reservoirs and Lakes

Lakes and reservoirs display seasonal temperature cycles. The energy from the sun causes lakes to become warmer in the summer. Diminished solar energy in the winter results in energy lost to the atmosphere and colder water during that period. The spring and fall seasons are transition periods in the direction of lake heat transfer. In the spring, lakes are in a state of receiving heat, and they lose it in the fall. This annual cycle plays an important role in their chemodynamics. Details of temperature profiles and thermal stratification of lakes follow. Some chemodynamic aspects related to the annual thermal structure appear in Section 5.2.

Water bodies such as natural lakes and artificial reservoirs in the temperate zones of the world are in an isothermal condition in the early spring. The water temperature at all depths is constant and somewhere between 4 and 10°C. At approximately the vernal equinox (March 21) the net heat transfer becomes positive, and the upper layers of water become warmer than the colder bottom (winter) water. Mixing between hot and cold does not occur easily because the density of water above 4°C decreases with temperature. As spring proceeds into summer, a definite thermal structure develops (see Fig. 7.1-1). The upper layers of water contain warm water at a somewhat uniform temperature; this region is called the *epilimnion*. Located below the epilimnion is a smaller layer of water that contains some large temperature gradients. In this region the water temperature decreases rapidly with increasing depth, and it is called the *metalimnion*. The particular point within the metalimnion where the temperature gradient is maximum is defined as the thermocline. The region below the metalimnion is the *hypolimnion*; this region contains cold water. This water remains cold throughout the year, increasing above 4°C only by a few degrees. In very deep lakes the bottommost water may not deviate from 4°C. Data of the annual thermal history for a typical reservoir are presented in Problem 7.1A.

Around the time of the autumnal equinox (September 23), the net heat transfer becomes negative and the lake begins to cool. The upper layers cool rapidly, especially on clear nights when the sky is a nearly perfect blackbody absorber. The upper layers are cold and more dense, so that the water "plunges" from the surface and causes a high degree of mixing in the epilimnion. As fall proceeds into winter, the cooling process continues. In late November the lake is at nearly isothermal conditions. While at isothermal

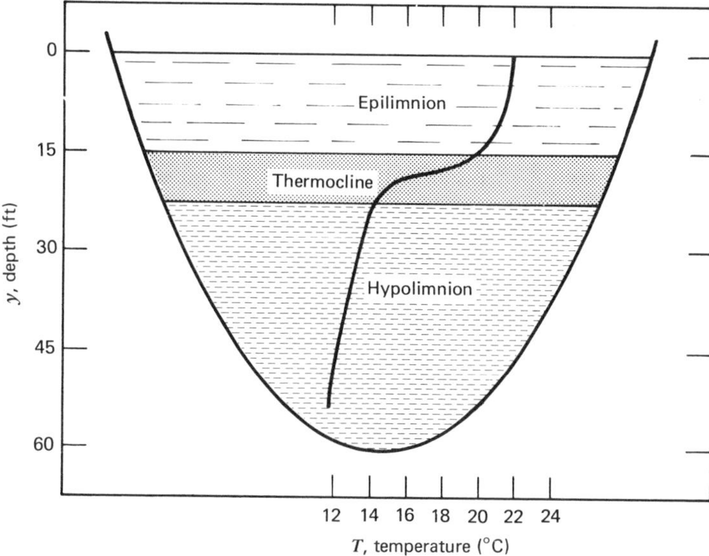

Figure 7.1-1. Typical temperature stratification in a reservoir.

conditions the water is of uniform density throughout the column. Surface winds can usually exert enough force to cause complete or nearly complete mixing of the water. This happening is commonly called lake turnover. Many turnovers can occur throughout the winter until the spring, when the heating cycle begins again and a stratified water column redevelops. A graphical illustration of the buildup and collapse of the thermal structure of Hungry Horse Reservoir in 1965 is shown in Fig. 7.1-2 for May through October. The annual thermal history data for Keowee Reservoir is presented in Problem 7.1A. The thermal histories of these two reservoirs provide excellent examples of the seasonal cycles of lakes.

Understanding the heat exchange processes in lakes is important. Human beings intrude on the natural process when they employ lakes and similar bodies of water as heat sinks. Artificial lakes and reservoirs frequently become the source of cooling water for power plants and also become the receiver of the hot water. The quantity of additional heat that a lake can accept (assimilative capacity) without altering the ecosystem to a critical degree can be partially determined by a study of the energy balance.

In the following section a simplistic lake model is developed. The lake is visualized as a very large mass with no flow of water in or out. The heat exchange with the surroundings is through the air–water interface. No heat is gained or lost from the sides or bottom. The lake is infinite in depth and has a nonchanging bottom temperature. Seemingly farfetched, these assumptions are very good for large, deep lakes and are a rough approximation for small lakes. The model is not good for shallow ponds and run-of-the-river lakes.

Semi-infinite Solid Model.[2] A lake is visualized as a solid body occupying the space from $y = 0$ to $y = \infty$ at an initial temperature T°. At time $t = 0$ the surface at $y = 0$ is suddenly heated (or cooled) by an external source at rate $q_{12}(t)$. The temperature at $y = \infty$ remains at T° during the heating (and cooling) process. If the thermal diffusivity α_2 is replaced by the coefficient of eddy thermal diffusivity $\alpha_2^{(t)}$ (assumed constant), the general energy equation can be put in the following familiar form:

$$\frac{\partial T(y, t)}{\partial t} = \alpha_2^{(t)} \frac{\partial^2 T(y, t)}{\partial y^2} \tag{7.1-1}$$

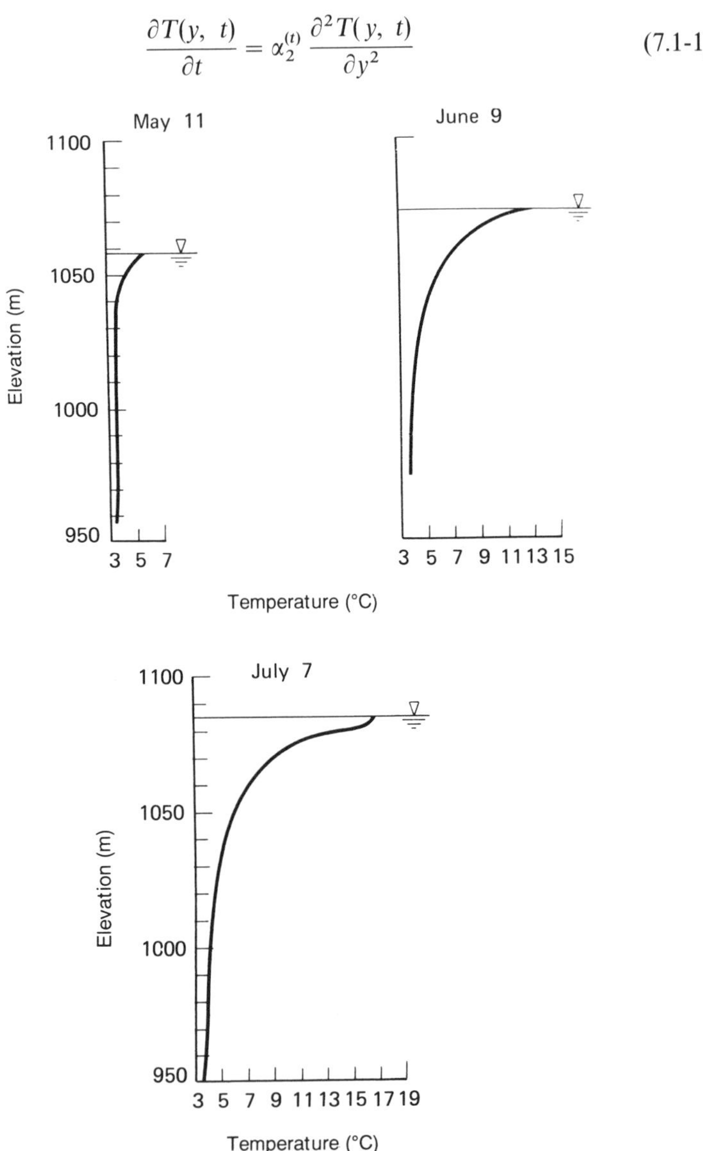

Figure 7.1-2. Observed temperature profiles, Hungry Horse Reservoir, 1965. (From Ref. 1.)

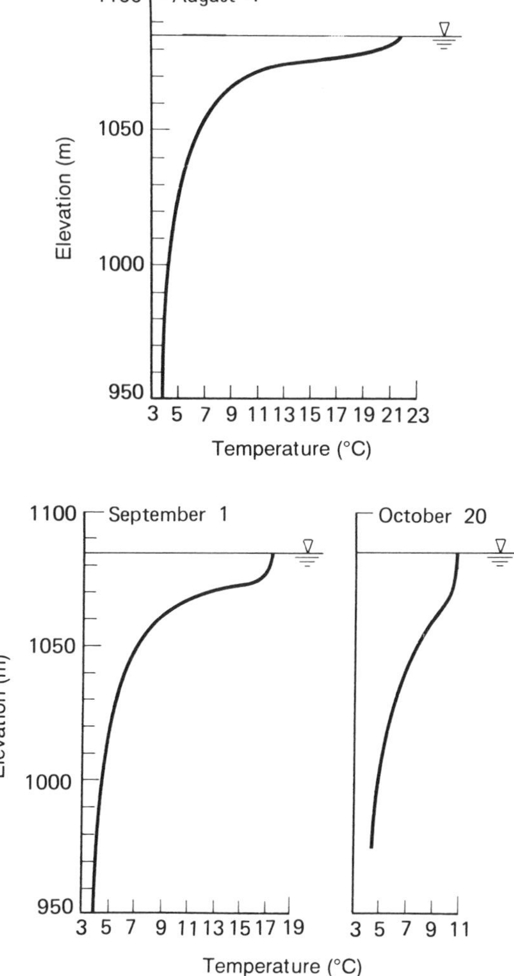

Figure 7.1-2. (*Continued*).

where $\alpha_2^{(t)} \equiv k_2^{(t)})/\rho_2 \hat{C}_{p2}$, $k_2^{(t)}$ is the turbulent coefficient of thermal conductivity and $T(y, t)$ is the water temperature as a function of depth y and time t. The initial and boundary conditions are

$$\text{IC: at } t \leqslant 0,\ T = T^\circ \text{ for all } y \tag{7.1-2a}$$

$$\text{BC 1: at } y = 0,\ \rho_2 \hat{C}_{p2} \frac{\alpha_2^{(t)} \partial T}{\partial y} = q_{12}^{(t)} \text{ for all } t > 0 \tag{7.1-2b}$$

$$\text{BC 2: at } y = \infty,\ T = T^\circ \text{ for all } t > 0 \tag{7.1-2c}$$

Additional assumptions in this simplistic lake model described by Eq. 7.1-1 are no enthalpic energy, biochemical reaction, and radiant energy absorption beneath the surface. The only source or sink term for energy addition is at the upper surface. The movement of energy within the lake occurs in the vertical direction only (i.e., one-dimensional model) and is characterized by one coefficient, $\alpha_2^{(t)}$, which is assumed constant for purposes of obtaining an analytical solution. Carslaw and Jaeger[3] give the following expression for the solution of Eqs. 7.1-1 to 7.1-2c:

$$T(y,\, t) = T^\circ + \frac{1}{(\alpha_2^{(t)}\pi)^{1/2}\rho_2\hat{C}_{p2}} \int_0^t q_{12}(t-\tau) \exp\left(\frac{-y^2}{4\alpha_2^{(t)}t}\right) \frac{d\tau}{\tau^{1/2}} \tag{7.1-3}$$

where τ is a dummy variable of integration.

Dodd[17] noted that the denominator in the exponential term of Eq. 7.1-3 in the second edition of the Carslaw and Jaeger book τ appears as a t. This means that the subsequent equations are approximations of the mathematically correct solution. In the correct form an additional term involving a double integral containing $q_{12}(\tau)$ is needed (see Problem 7.1E) and it is not possible to get closed-form results such as Eqs. 7.1-6 and 7.1-7. However, in practice it appears that Eqs. 7.1-4 and 7.1-5 can give realistic numerical results! Therefore, without the correction the closed-form result provides a mathematically simple one-parameter $\alpha_2^{(t)}$, model for temperature profiles in deep water bodies.

In removing the exponential term from the integral in Eq. 7.1-3 the lake surface temperature can be defined and expressed as follows:

$$T(0,\, t) \equiv T^\circ + \frac{1}{(\alpha_2^{(t)}\pi)^{1/2}\rho_2\hat{C}_{p2}} \int_0^t q_{12}(t-\tau) \frac{d\tau}{\tau^{1/2}} \tag{7.1-4}$$

where $T(0, t)$ is the surface temperature. Equation 7.1-3 then expresses the temperature at any depth as a function of the surface temperature and becomes

$$T(y,\, t) = T^0 + [T(0,\, t) - T^\circ] \exp\left(\frac{-y^2}{4\alpha_2^{(t)}t}\right) \tag{7.1-5}$$

This final expression yields an equation for the time locus of the thermocline position:

$$y_{tc} = \sqrt{2\alpha_2^{(t)}t} \tag{7.1-6}$$

after the operation $\partial^2 T/\partial y^2 = 0$ is performed (see Problem 7.1B for an exact definition of the thermocline). The term y_{tc} is the depth of the thermocline from the surface of the lake. Eqs. 7.1-4 to 7.1-6 are useful working forms of the semi-infinite solid model.

Equation 7.1-5 is a useful interpretative expression for evaluation of $\alpha_2^{(t)}$ values from field temperature profile data when rearranged to

$$\alpha_2^{(t)} = \frac{y^2}{4t \ln\{[T(0, t) - T^\circ]/[T(y, t) - T^\circ]\}} \tag{7.1-7}$$

Model time t used in all the preceding equations is the lapse time since the lake was isothermal. The time of this isothermal condition varies with geographic location but typically occurs during March and April.

The coefficients of eddy thermal diffusivity as computed by Eq. 7.1-7 from field data are not constant. The coefficients vary with time and depth and are closely related to the turbulence in the lake. Turbulence in a lake is induced by bulk flow of the water, water movement due to the wind and waves (i.e., Langmuir circulations), and water density variations. Coefficients of eddy thermal diffusivity display some general characteristics, as reported by Orlob et al.[1] The coefficients typically display large values in the epilimnion, small values in the metalimnion, and large values in the hypolimnion. Plunging cold water caused by a net heat loss at the surface results in large epilimnion coefficients in the fall (see Problem 7.1A). An average $\alpha_2^{(t)}$ should be used with Eqs. 7.1-4 and 7.1-5.

Figures 7.1-3 and 7.1-4 show the results of a simulation with a single value of $\alpha_2^{(t)}$. Figure 7.1-3 is a simulation of the surface temperature of Beaver Reservoir in northwestern Arkansas by Eq. 7.1-4. The flux at the surface is a combination of solar radiation, longwave radiation, natural convection, and evaporation, that is,

$$q_{12}(t) = q_s(t) + q_{1w}(t) + q_c(t) + q_e(t) \tag{4.3-2}$$

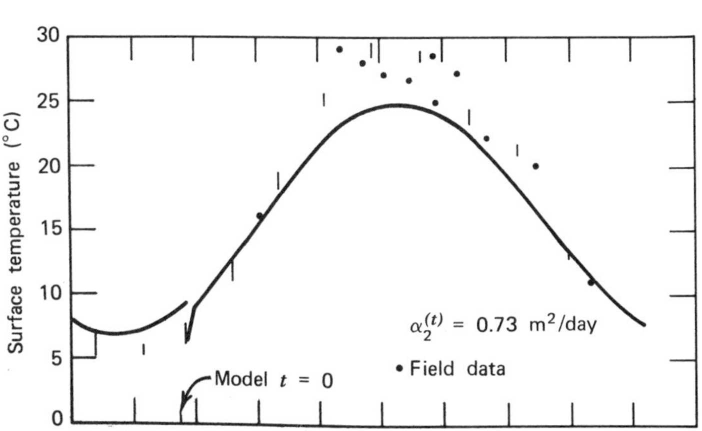

Figure 7.1-3. Semi-infinite slab model, simulation temperature, Beaver Reservoir. (Reprinted by permission of American Water Resources Association, Minneapolis, Minn., from Ref. 2.)

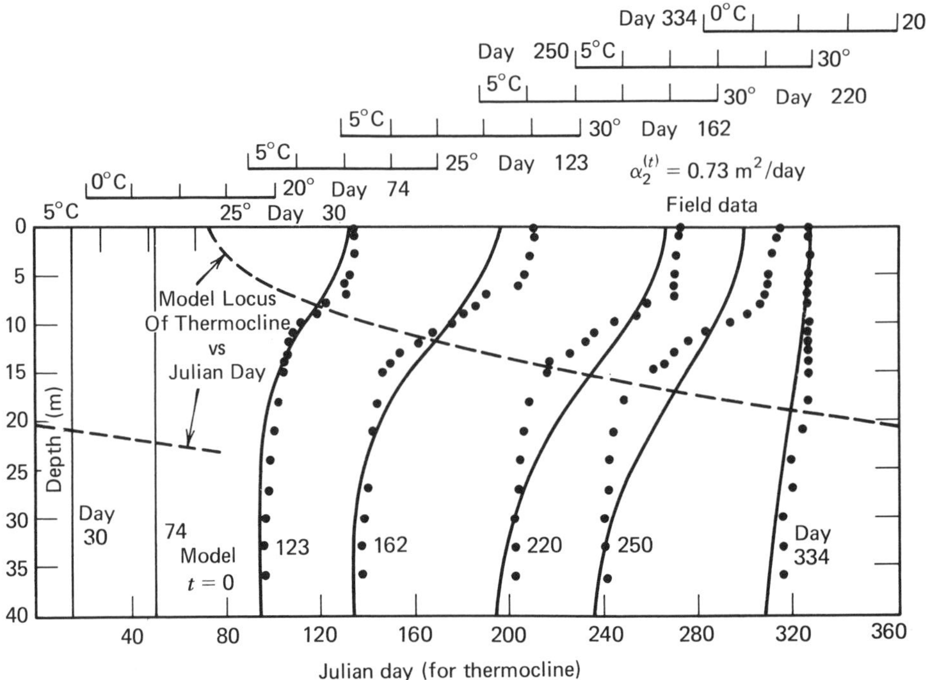

Figure 7.1-4. Semi-infinite slab model, simulation temperature, Beaver Reservoir. (Reprinted by permission of American Water Resources Association, Minneapolis, Minn., from Ref. 2)

All the heat flux terms in Eq. 4.3-2 are time dependent (i.e., Julian day). Figure 7.1-4 shows the in-lake temperature profiles for both the field data and the simulation.

As a realistic lake model, the semi-infinite solid approach is very simplistic and incapable of producing a perfect simulation. As with all models, it has both advantages and disadvantages. Concise, explicit representation of the surface and water column temperature is a major advantage of the model. Another advantage of the model is the qualitative simulation of the gross features of the thermal changes with depth and time within and at the surface of the water body. A major implication of this result is that the general shape of the temperature profile in deep freshwater bodies is structured primarily by the heat input at the air–water interface and the time lapse since an isothermal state existed. The action of surface winds, precipitation, condensation, bulk warm water flow, and plunging cold surface water in the fall on the shape of the profile in the epilimnion and the effects of bulk water movement on reshaping the profile in the hypolimnion are but minor modifications compared to the interfacial heat exchange, induced thermal regime.

The science of modeling lake thermal behavior is quite advanced. Sophisticated predictive models exist that have the capability of a high degree of precision in simulation lake water temperatures. Parker et al.[4] have reviewed these types of models and present essential features, advantages, and disadvantages of each.

Thermocline in Oceans

Unlike lakes, the oceans have a permanent thermocline. This feature is so named because its character is virtually unchanged seasonally. In the arctic and antarctic regions the water is cold from top to bottom. As this dense water flows south and north, respectively, it sinks beneath warmer water that moves outward from the equator. This gives rise to the temperature discontinuity known as the *permanent thermocline.* The top of the permanent thermocline is quite shallow at the equator, reaches maximum depth at midlatitudes, and becomes shallow again at about 50° latitude (see Fig. 7.1-5). The thermocline disappears between 55 and 60° N or S.

As with lakes, the world's oceans have a seasonal thermocline. This feature is a summer phenomenon found at shallower water depths than the permanent thermocline in all the world's oceans except those perennially ice infested. As air temperatures rise above ocean temperatures in the spring season and the sea surface receives more heat than it loses by radiation and convection, the surface water begins to warm so that a negative temperature gradient develops in the first few feet. The surface waters are mixed by transfer of energy from the wind. Although this mixing serves to lower the surface temperature, the net effect is a downward transport of heat and formation of an isothermal layer whose temperature is warmer than the underlying water. A strong temperature

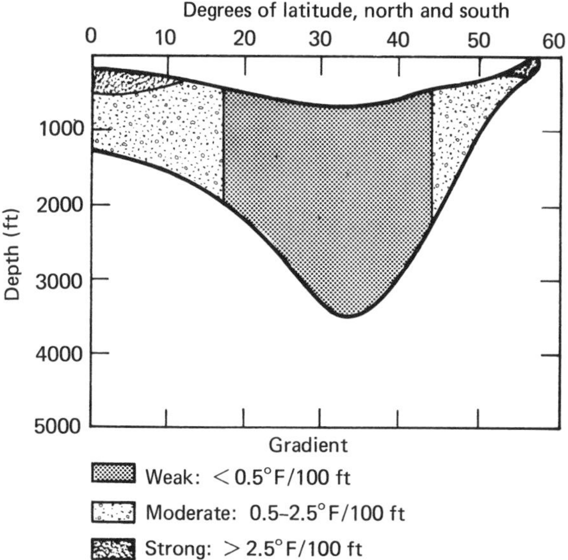

Figure 7.1-5. Permanent thermocline, based on averages for depth, thickness, and gradient within thermocline (From Encyclopedia of Environmental Science. Copyright © 1974 by McGraw-Hill, Inc. Used with permission of McGraw- Hill Book Company.)

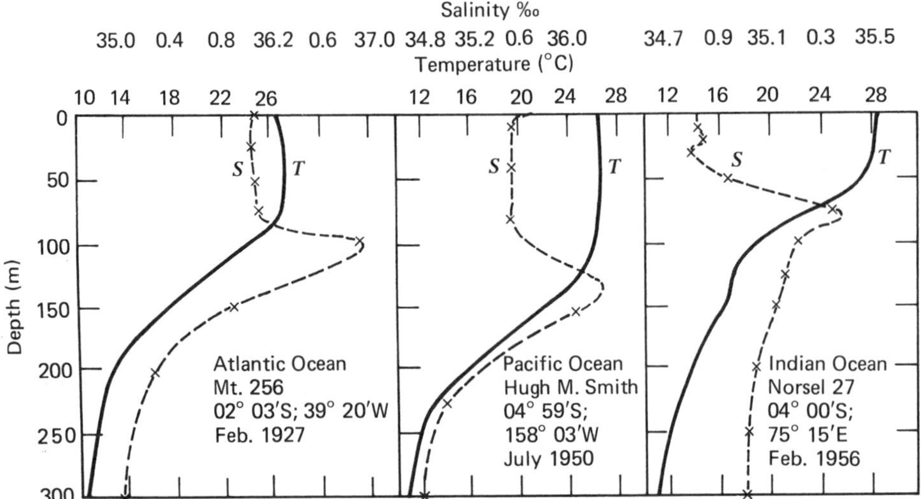

Figure 7.1-6. Examples of temperature and salinity distribution with depth in the upper 300-m layer of the tropical part of the oceans. (From Gerhard Neumann, and Willard J. Pierson, Jr., Principles of Physical Oceanography, © 1966, p. 447. Reprinted by permission of Prentice Hall, Englewood Cliffs, N.J.)

gradient, or seasonal thermocline, is thus formed between the isothermal surface layer and the water beneath. From July through September a surface layer of mixed water underlain by a strong negative temperature gradient is found in most of the ocean. As air temperatures fall in autumn, the water loses heat to the atmosphere by convective and radiative processes, and the surface layer is cooled to the temperature below. The seasonal thermocline breaks up to form again the following spring. Figure 7.1-6 shows three examples of the oceanic seasonal thermocline. The salinity profile is also shown in the figure. The thermohaline structure of the upper strata in the oceans is governed by the thermocline.

Thermoclines are structures semipermeable to the vertical movement of chemicals in lakes and oceans. This physical structure plays a dominant role in the movement of chemicals up and down the water column. Aspects of chemical movement through thermoclines are presented in Section 5.2 and in a subsequent section of this chapter.

Vertical Turbulent Diffusivity in Lakes and Oceans

One result of the semi-infinite solid lake model was a simple analytical expression for obtaining the vertical component of the turbulent thermal diffusivity $\alpha_2^{(t)}$ from field temperature profile data. The analytical expression to use is Eq. 7.1-7. The turbulent thermal diffusivities as computed by this equation or others are not constant but a function of depth and time. The variation of $\alpha_2^{(t)}$ with depth is illustrated in Fig 7.1-7 for two lakes and a

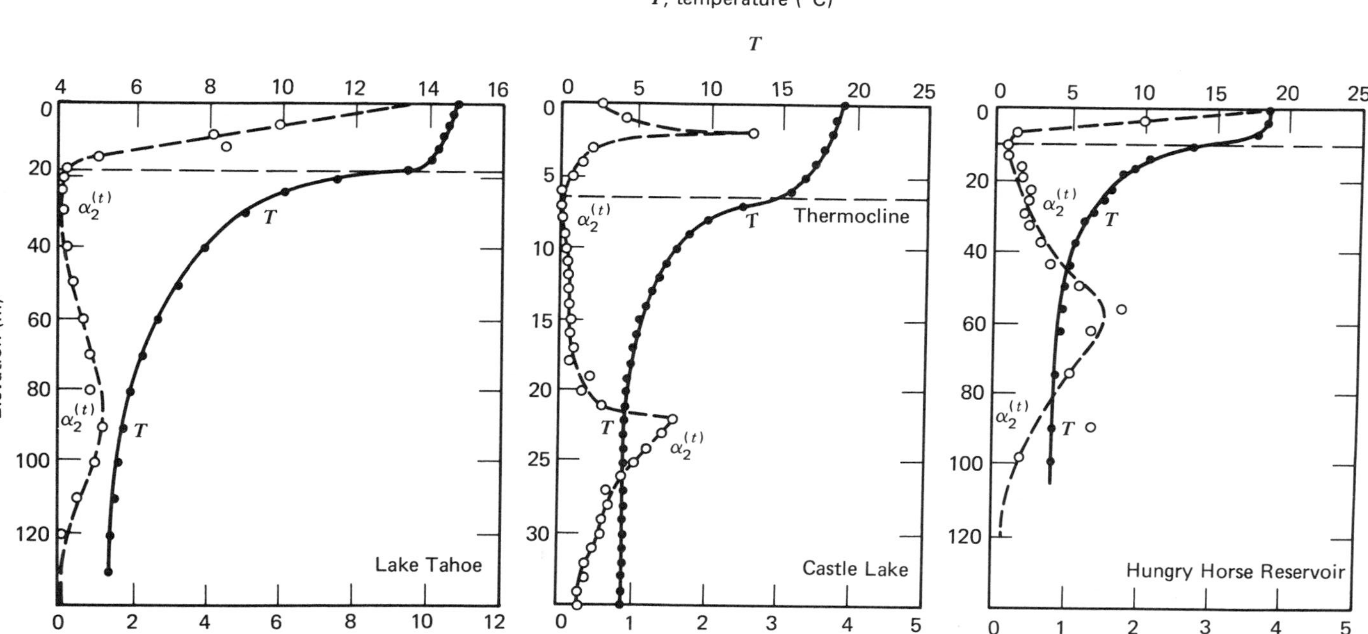

Figure 7.1-7. Effective diffusion and temperature profiles for selected impoundments. (From Ref. 1.)

reservoir. The trend of high diffusion coefficients in the epilimnion, low coefficients at the thermocline, and high coefficients in the hypolimnion is typical of most deep lakes and the oceans.

In general, $\alpha_2^{(t)}$ has its maximum value in the surface layer; in the open ocean $\alpha_2^{(t)}$ at the surface varies between 10 and 100 cm^2/s; in coastal areas, 10 to 50 cm^2/s; in lakes, approximately 10 cm^2/s. Below the surface mixed layer (or epilimnion) $\alpha_2^{(t)}$ drops to a minimum in the thermocline, on the order of 1 cm^2/s in the open ocean; in lakes $\alpha_2^{(t)}$ may drop as low as 0.05 cm^2/s. Below the thermocline $\alpha_2^{(t)}$ may increase again.

The presence of density stratification tends to suppress the vertical exchange of thermal energy. It is obvious from a casual study of the data in Fig. 7.1-7 that in the regions of steep density gradients, $\partial\rho_2/\partial y$, that $\alpha_2^{(t)}$ is lowest. Therefore, one expects the vertical diffusivity to be a decreasing function of density stratification. The presence of shear tends to be destabilizing and increases vertical exchange. It is to be expected that for similar flows the vertical diffusivity should be related to the Richardson number, Ri, defined by Eq. 6.1-9. Water density is a function of temperature and salinity. An empirical expression for the density of fresh water as a function of temperature is given in Appendix D. Numerous proposed relations between $\alpha_2^{(t)}$ and Ri are of the form

$$\alpha_2(t) = \alpha(1 \pm b \cdot \mathrm{Ri})^{\pm 1/n} \tag{7.1-8}$$

where b and n are constants, determined by experiment and in part by theory. The term α is the turbulent diffusivity at Ri $= 0$, the stable condition. Reference 7 summarizes the results of various investigations of the relation between $\alpha_2^{(t)}$ and Ri. These relations are useful when water velocity profiles are available. Unfortunately, this information is usually not available for quiescent water columns. An alternative development is desirable.

Since water density gradients exist throughout a water body such as a lake or the ocean, it is not altogether absurd to consider the property of density ρ_2 as a diffusing species. Beginning with Eq. 3.0-3 for $A \equiv 2$ and assuming no reaction, $r_2 = 0$, and only diffusion due to turbulence, so that molecular diffusion may be ignored, we obtain

$$\frac{\partial\rho_2}{\partial t} + v_x\frac{\partial\rho_2}{\partial x} + v_y\frac{\partial\rho_2}{\partial y} + v_z\frac{\partial\rho_2}{\partial z} = \frac{\partial}{\partial x}\left(\alpha_{2x}^{(t)}\frac{\partial\rho_2}{\partial x}\right) + \frac{\partial}{\partial y}\left(\alpha_{2y}^{(t)}\frac{\partial\rho_2}{\partial y}\right) + \frac{\partial}{\partial z}\left(\alpha_{2z}^{(t)}\frac{\partial\rho_2}{\partial z}\right) \tag{7.1-9}$$

where v_x, v_y, and v_z are mean currents in the x, y, and z directions, respectively. Because of the strong relationship between density and temperature, eddy thermal diffusivities are employed, and $\alpha_{2x}^{(t)}$, $\alpha_{2y}^{(t)}$, and $\alpha_{2z}^{(t)}$ are the variables for the x, y, and z directions, respectively. Since the horizontal variations of ρ_2 are usually much smaller than the vertical variations, we assume that $\partial\rho_2/$

$\partial x = \partial\rho_2/\partial z = 0$; also, we assume that $v_x = v_y = v_z = 0$. The Eq. 7.1-9 becomes

$$\frac{\partial\rho_2}{\partial t} = \frac{\partial}{\partial y}\left(\alpha_{2y}^{(t)}\frac{\partial\rho_2}{\partial y}\right) \tag{7.1-10}$$

The term $\partial\rho_2/\partial t$ is usually very small except for the near-surface waters, which may undergo some diurnal changes. Considering an instantaneous thermal profile we can assume steady state; then

$$\alpha_{2y}^{(t)} = \frac{a}{d\rho_2/dy} \tag{7.1-11}$$

where a is a constant.

This final expression indicates that the turbulent diffusivity is related to the density gradient of water, and it is also possible to attempt a correlation of $\alpha_2^{(t)}$ with

$$\Omega_2 \equiv \left|\frac{1}{\rho_2}\frac{d\rho_2}{dy}\right| \tag{7.1-12}$$

the density normalized gradient. Intuitive considerations also suggest that the eddy diffusivity should to some extent depend on the steepness of the density gradient. When the density gradient is strong, the degree of turbulence, and hence the magnitude of the eddy diffusivity in the pycnocline, may be expected to be low or the gradients would be destroyed by turbulent eddies. The steepness of the density gradient within the water layer determines the stability of stratification: the greater the density gradient, the more stable is the stratification. Koh and Fan[7] collected data on $\alpha_{2y}^{(t)}$, where Ω_2 is measured simultaneously to construct the graphical relationship shown in Fig. 7.1-8. It can be seen that almost all the data fall within a factor of 10 of the empirical relation

$$\alpha_{2y}^{(t)} = \frac{10^{-4}}{\Omega_2} \tag{7.1-13}$$

for 4E-7 $m^{-1} \leqslant \Omega_2 \leqslant$ E-2 m^{-1}, where $\alpha_{2y}^{(t)}$ is in cm^2/s and Ω_2 is in m^{-1}. Unless independent field data are available, Eq. 7.1-13 can be used to estimate $\alpha_2^{(t)}$ from a single observation of the temperature profile.

The relationship between density gradient and the degree of stability of stratification can also be expressed by the quantity known as the Brunt–Valisada stability frequency v:

$$v^2 \equiv \left|g\frac{1}{\rho_2}\frac{d\rho_2}{dy}\right| \tag{7.1-14}$$

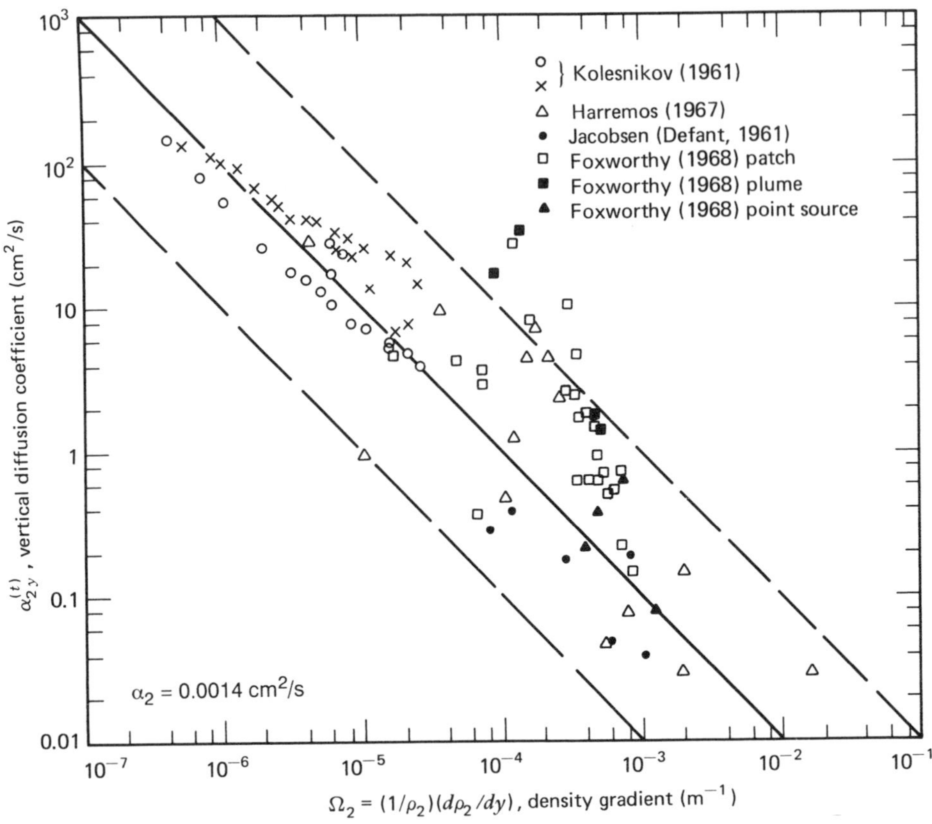

Figure 7.1-8. Correlation of $\alpha_{2y}^{(t)}$ with density gradient. (From Ref. 7.)

where g is the acceleration due to gravity and the dimension of v is t^{-1}. It is claimed that Eq. 7.1-14 is valid for a water column up to several hundred meters deep, where the effect of adiabatic compressibility on the density gradient within the water column can be ignored. Eckart[8] has discussed the derivation and physical significance of the concept of stability frequency.

Example 7.1-1. Estimating Vertical Turbulent Diffusivities. Equation 7.1-13 and Fig. 7.1-8 provide a means of estimating $\alpha_{2y}^{(t)}$ values when only a single temperature profile is available for a particular body of water. Using these, estimate $\alpha_{2y}^{(t)}$ at the thermocline for the three water bodies of Fig. 7.1-7 and compare the results to the $\alpha_{2y}^{(t)}$ values observed.

SOLUTION It is possible to express Ω_2 as a function of the coefficient of thermal expansion of water, β_2, and the thermal gradient by use of the chain rule to yield

$$\Omega_2 = \left| -\beta_2 \frac{\partial T}{\partial y} \right| \tag{E7.1-1a}$$

Table E7.1-1

	Lake		
	Tahoe	Castle	Hungry
T (°C)	12.5	14.5	14.5
$-\Delta T/\Delta Y$ (°C/m)	0.75	3.125	1.136
ρ_2 (g/cm^3)	0.9995	0.9992	0.9992
$-\partial\rho_2/\partial T$ (g/cm$^3\cdot$°C)	1.1879E-4	1.14784E-4	1.14784E-4
$-\beta_2$ (°C^{-1})	1.1885E-4	1.1488E-4	1.1488E-4
Ω_2 (m^{-1})	8.9137E-5	3.5899E-4	1.3053E-4
$\alpha_{2y}^{(t)}$ (cm^2/s)	1.12	0.216	0.766
$\alpha_{2y}^{(t)}$ (cm^2/s)	0.1–10.0	0.03–3.0	0.06–6.0
$\alpha_{2y}^{(t)}$ (cm^2/s)	0.2	0.05	0.1
Estimate ÷ actual	5.6	4.3	7.7

where $\beta_2 \equiv [(-1)/\rho_2](\partial\rho_2/\partial T)$. The density of pure water, ρ_2, as a function of temperature is given in Appendix D. From this empirical relationship $\partial\rho_2/\partial T$ can be obtained as a function of temperature:

$$\frac{\partial\rho_2}{\partial T} = 0.1546919\text{E-5} + 2(0.2141986\text{E-5})T - 3(0.6508630\text{E-6})T^2$$
$$+4(0.1975524\text{E-7})T^3 - 5(0.1894802\text{E-9})T^4 \qquad \text{(E7.1-1}b\text{)}$$

Thermal gradients at the thermocline may be estimated graphically from Fig. 7.1-7. Thermocline temperature and thermal gradients appear in lines 1 and 2 of Table E7.1-1. Other pertinent calculated values appear in succeeding lines. Line 7 contains $\alpha_{2y}^{(t)}$ values calculated from Eq. 7.1-13, and line 8 contains the range of values obtained from Fig. 7.1-8. Line 9 contains the $\alpha_{2y}^{(t)}$ values observed in the field. In all cases the observed values of $\alpha_{2y}^{(t)}$ are within the lower end of the range of values suggested by Fig. 7.1-8. Line 10 indicates that Eq. 7.1-13 overestimates the actual value of $\alpha_{2y}^{(t)}$ by a factor of approximately 6 for these lakes.

It is unlikely that the preceding empirical information concerning $\alpha_{2y}^{(t)}$ applies throughout the surface mixed layer. It has been observed that in the surface mixed layer of the ocean the density gradient is often zero. The preceding empirical relation is certainly invalid since it implies an infinite $\alpha_{2y}^{(t)}$. In this case the vertical transport is governed primarily by the vertical turbulence created by waves and wind. Koh and Fan[7] summarize equations proposed by Golubeva and Isayeva for relating the vertical diffusion coefficient in the mixed layer and the surface-wave characteristics. The equation proposed is

$$\alpha_{20}^{(t)} = \frac{0.02h^2}{\tau} \qquad (7.1\text{-}15)$$

Table 7.1-1. Beaufort Scale: British and U.S.

No.	Descriptive Term	Wind Speed (mi/h)	No.	Descriptive Term	Wind Speed (mi/h)
0	Light	<1	5	Fresh	19–24
1	Light	1–3	6	Strong	25–31
2	Light	4–7	7	Strong	32–38
3	Gentle	8–12	8	Gale	39–46
4	Moderate	13–18	9	Gale	47–54

Source: Encyclopedia Britannica (1966), Vol. 3, p. 339.

where $\alpha_{20}^{(t)}$ is the vertical diffusivity at the surface, h the wave height, and τ the wave period.

There exist simple means of estimating wave height and period.[18] These are related to wind speed, water depth, and fetch. In general, fetch is defined as a region of length and width in which the speed and direction are reasonably constant. In mathematical models it is a length. Whereas the effect of fetch on limiting ocean wave growth may usually be neglected, in inland waters (bays, rivers, lakes, and reservoirs) fetches are limited by land forms surrounding the body of water. Hsu[18] gives simple methods of estimating h and t in deep water.

In 1806, Admiral Sir Frances Beaufort quantified the *sea state* by a series of numbers from 0 to 12 that relates to the strength of the wind. The original scale had no reference to the speed of the wind; however the Beaufort force numbers have been updated continually. Table 7.1-1 is an abbreviated version of the U.S. scale in use today. Figure 7.1-9, provides a rough means of estimating $\alpha_{20}^{(t)}$ based on sea state.

Closure. The gross, stratified features produced by water density gradients and illustrated in Fig. 7.1-1 suggest that for these conditions two or more compartments separated by imaginary boundaries is a reasonable physical description. This structure for simplified, transient chemical modeling is considered in the next section. In addition, in situ temperature profile measurements on a particular water body are much easier to obtain than chemical concentration profile measurements, and simple algorithms exist to estimate $\alpha_{2y}^{(t)}$. Furthermore, under most conditions of turbulence in these water bodies $\mathscr{D}_{A2y}^{(t)} \simeq \alpha_{2y}^{(t)}$ can be assumed as a reasonable first approximation (see Table 5.2-1. In essence much information can be extracted from the energy balance plus its consequences for a water body and then used as an aid in chemical modeling.

Intraphase Chemical Transport Processes in Presence of Stratification

In the preceding section the general thermal structure of lakes and the ocean was examined. Based in large part on the thermal structure, these water bodies stratify vertically into a series of at least three layers. Ignoring the fine details

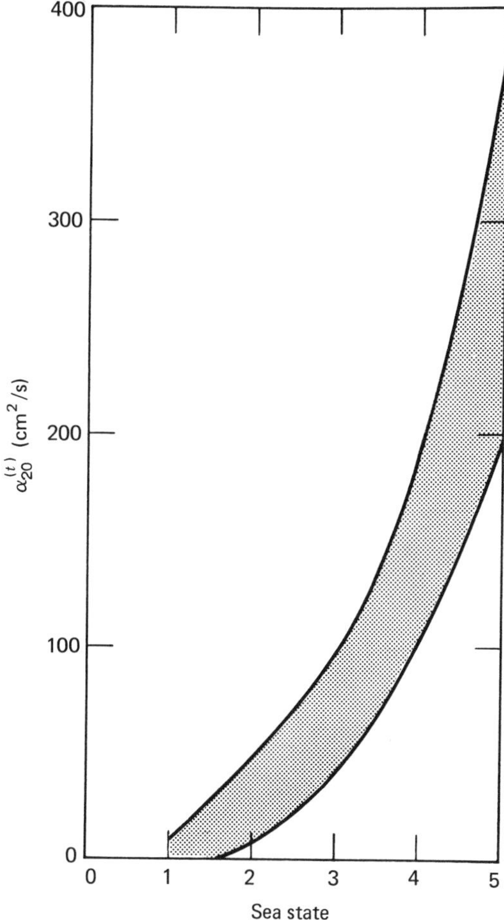

Figure 7.1-9. Dependence of $\alpha_{20}^{(t)}$ on sea state. (From Ref. 7.)

of the thermal and density gradients, illustrated in Fig. 7.1-1, a gross description of the stratified system consists of an upper well-mixed layer, a layer with a more or less pronounced density gradient (*pycnocline*) below it, and a well-mixed layer below the pycnocline. This particular structure has important consequences for the chemodynamic behavior of lakes, oceans, and estuaries. This section is devoted to a study of the rates of chemical movement, chemical lifetimes, and chemical profiles in stratified water bodies by the use of simple models.

Simple Intraphase Transport Models

Molecular Diffusion Within Homogeneous Medium. The movement of a dilute chemical species between two points within a homogeneous medium by molecular diffusion processes is well established and easily quantified. Figure

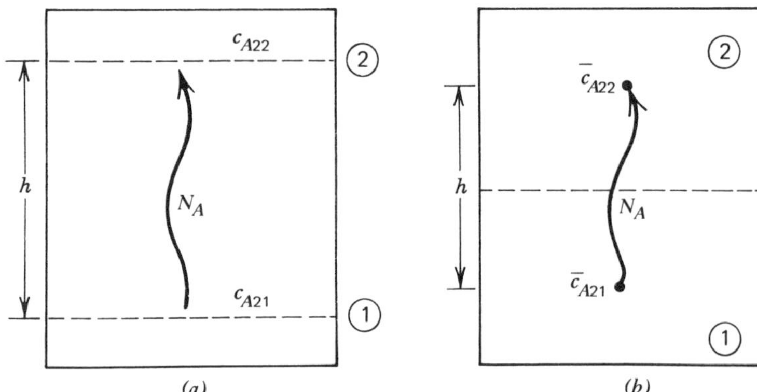

Figure 7.1-10. Basis for turbulent intraphase diffusion rate expression: (*a*) molecular diffusion in a homogeneous medium, (*b*) turbulent diffusion between mixed compartments.

7.1-10*a* shows a section of any homogeneous medium in which a concentration gradient exists and component A is being transported by molecular processes. Species A moves from the lower plane, 1, of high concentration, denoted by c_{A21}, to the upper plane, 2, of lower concentration, denoted by c_{A22}. The rate of movement is quantified by Fick's first law, Eq. 3.1-18:

$$N_A = \frac{\mathscr{D}_{A2}}{h}(c_{A21} - c_{A22}) \tag{7.1-16}$$

where h is the length of the diffusion path or the distance between planes 1 and 2.

Turbulent Diffusion Between Mixed Chambers. Figure 7.1-10*b* illustrates the case for turbulent diffusion of chemical species A between two mixed chambers separated by an arbitrary, permeable plane. The lower chamber contains A at a high average concentration $\bar{c}_{A21}$, and the upper chamber contains A at a lower average concentration $\bar{c}_{A22}$. Although the chambers are mixed within by the same turbulent process, a concentration gradient exists and the average path length for movement is from the midpoint of the lower chamber to the midpoint of the upper chamber. It is desirable in many instances to have a rate expression for the movement of chemical A, and a logical expression to use is an extension of the molecular diffusion equation:

$$N_A = \frac{\mathscr{D}_{A2}^{(t)}}{h}(\bar{c}_{A21} - \bar{c}_{A22}) \tag{7.1-17}$$

where $\mathscr{D}_{A2}^{(t)}$ is a well-chosen, average, turbulent diffusion coefficient that is a characteristic of each chamber and h is the distance between the midpoints of the chambers. The movement of oxygen from the epilimnion to the hypolimnion of a lake is an example of the use of this turbulent diffusion rate equation.

The epilimnion is often referred to as a *mixed layer*. This is not an altogether incorrect description. It is known that the vertical eddy thermal diffusivity in the epilimnion is large compared to the eddy thermal diffusivity in the thermocline region. It is also known that surface winds exert shear stresses upon the air–water interface that causes bulk water flow and turbulence. The constant change in direction and magnitude of the wind also aids in mixing the surface layer. The thermocline is a thin zone of relative calm water compared to the surface mixed layer. Below the thermocline, the turbulent diffusivity again increases dramatically, as illustrated in Fig. 7.1-7. In light of this physical evidence, Eq. 7.1-17 is reasonable and will be used to study chemical movement rates between mixed chambers.

Turbulent Diffusion from Phase Boundaries into Mixed Chambers. It is desirable to express the movement of a chemical from the region of a phase interface into the bulk fluid within a chamber. Examples are the movement of oxygen from the air–water interface into the epilimnion and the movement of phytoplankton-producing nutrient chemicals from the mud–water interface into the overlying hypolimnic waters. Figure 7.1-11 illustrates two such transport processes.

For the specific case of oxygen transfer the rate expression takes the form

$$N_A = \frac{\mathscr{D}^{(t)}_{A22}}{h_2/2}\,(c^*_{A22} - \bar{c}^*_{A22}) \tag{7.1-18}$$

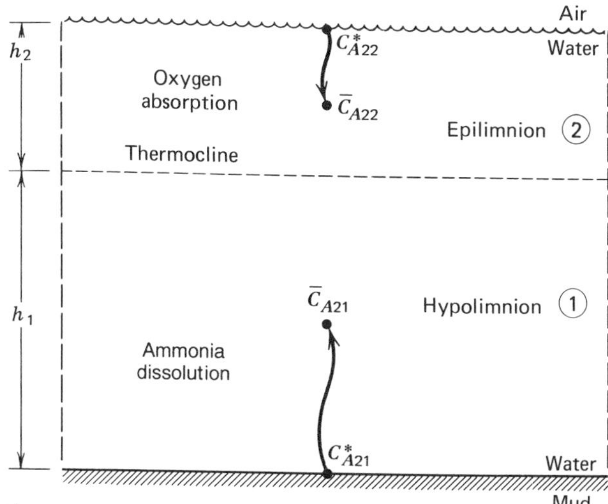

Figure 7.1-11. Turbulent diffusion from phase boundary.

where c^*_{A22} is the oxygen concentration in water at the air–water interface and $\bar{c}_{A22}$ is the average concentration in the upper chamber. In the case of O_2 the group $\mathscr{D}^{(t)}_{A22}$ consisting of the effective turbulent diffusivity of the chamber and the diffusion path length is, by definition, the *lake surface reaeration coefficient* ${}^1k'_{A2}$.

For the case of ammonia release from the lake mud-water interface shown in Fig. 7.1-11, the rate expression takes the form

$$N_A = \frac{\mathscr{D}^{(t)}_{A21}}{h_1/2}(c^*_{A21} - \bar{c}_{A21}) \tag{7.1-19}$$

where c^*_{A21} is the ammonia concentration at the mud interface and $\bar{c}_{A21}$ is the average concentration of ammonia in the overlying well-mixed chamber. The group $\mathscr{D}^{(t)}_{A21}$ is by definition the benthic boundary layer coefficient ${}^3k'_{A2}$. (See Section 5.2 for further information concerning this coefficient.

Time to Chemical Steady States in Lakes and Ocean. In natural systems of large dimensions, such as bodies of water, sediments, and the atmosphere, many chemical processes are controlled by the transport of reacting species through the system. The distribution of chemical species in natural systems is only too often not homogeneous; concentration gradients and more or less abrupt changes in abundance from one part of an environment to another are commonplace. In general, the nonhomogeneous distributions of chemical species are a combination of (1) the geometry of the environment: its shape and location of the sources and sinks of the chemical species; (2) physics: mechanisms of transport of matter through the system; and (3) chemistry: the nature and rates of the chemical reactions in which the species are involved.

Knowledge of these three facets of a natural system is indispensable when we need to understand its present chemical state and also to predict quantitatively the changes in the chemical state and their duration that would occur when the present characteristics of the system undergoes a change.

The large variation in the values of the turbulent diffusion coefficients reported in the literature for chemical species in different environments and the laboriousness of their determination in natural environments makes it difficult in many cases to obtain accurate estimates of the time required for a certain chemical process to go to completion. However, when the turbulent diffusivities are not well known, it is still possible in some systems to choose "reasonable" lower and upper limits of the diffusion coefficients and thereby to bracket the model in short and long time estimates.

In view of the primary significance of turbulent diffusional processes in the transport of dissolved matter in water column, these mechanisms and their bearing on a number of chemical processes are presented in this section. The effects of the magnitude of the eddy diffusivity and the presence of a relatively calm metalimnion on the transport of dissolved species in a stratified body of water are discussed in some simplified lake models.

Model I: Two Adjoining, Mixed Layers. The first model is highly idealized and illustrates the slowness of the intraphase diffusion process by considering chemical movement between two adjoining, mixed chambers separated by a plane that has no resistance. Adjoining mixed layers in the atmosphere or the hydrosphere may also be approximated by the crude model to obtain a first-order approximation of the real environmental exchange processes.

A general two-box model is developed so that the final result will have several applications. Figure 7.1-12 illustrates the model status at the start. A certain chemical has been placed into the lower chamber 1, and the rapid mixing processes have resulted in an average concentration, c_{A21}^{0}. The chemical is also present in the upper chamber, denoted 2, but at a residual background level, c_{A22}^{0}. The chambers have depths of h_1 and h_2, respectively, the quantity of chemical A placed in chamber 1 is

$$m_A = h_1 A(c_{A21}^{0} - c_{A22}^{0})$$

where A is the cross-sectional area of the layers.

The movement of chemical A between chambers is a transient process. Initially, the rate is rapid because of the large concentration difference, but as time progresses the rate decreases, and c_{A21} decreases while c_{A22} increases. Eventually, the concentrations are equal, and net chemical movement ceases. A simultaneous solution of component A material balances on each chamber can be manipulated to yield the concentration–time history of chamber 1:

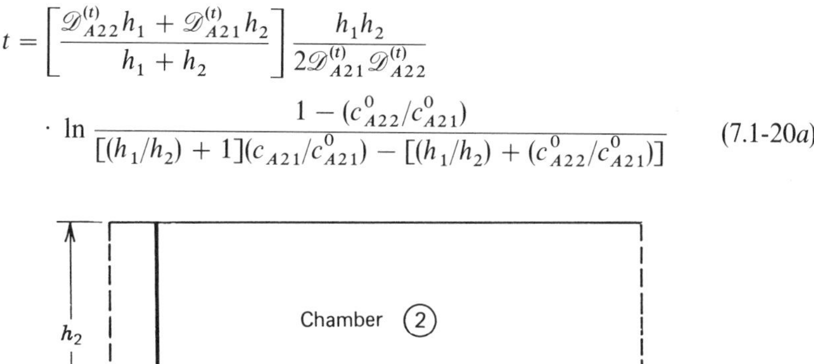

$$t = \left[\frac{\mathscr{D}_{A22}^{(t)} h_1 + \mathscr{D}_{A21}^{(t)} h_2}{h_1 + h_2}\right] \frac{h_1 h_2}{2\mathscr{D}_{A21}^{(t)} \mathscr{D}_{A22}^{(t)}} \cdot \ln \frac{1 - (c_{A22}^{0}/c_{A21}^{0})}{[(h_1/h_2) + 1](c_{A21}/c_{A21}^{0}) - [(h_1/h_2) + (c_{A22}^{0}/c_{A21}^{0})]} \qquad (7.1\text{-}20a)$$

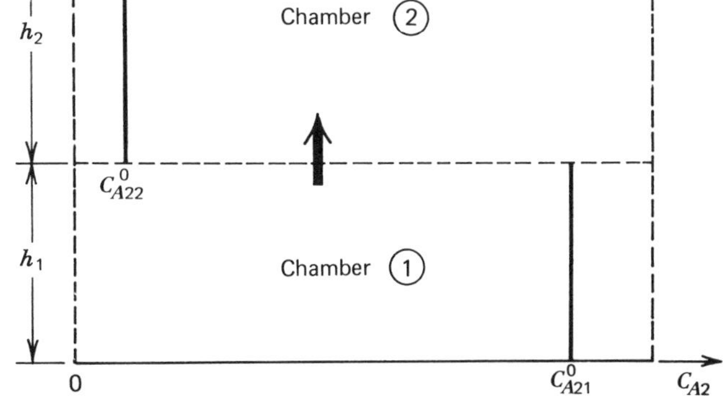

Figure 7.1-12. Chemical movement between two adjoining, well-mixed layers.

where $\mathscr{D}^{(t)}_{A21}$ and $\mathscr{D}^{(t)}_{A22}$ are turbulent diffusivities, characteristic of chambers 1 and 2. The bars above the concentrations denoting average are omitted for clarity. Equation 7.1-20*a* gives the time required for achieving a concentration of c_{A21} in chamber 1. The only restriction on the equation is $c^0_{A22} < c^0_{A21}$. The equation can be used for the subcases $h_1 = h_2$, $\mathscr{D}^{(t)}_{A21} = \mathscr{D}^{(t)}_{A22}$, $c^0_{A22} = 0$, or any combination of these subcases.

The usual form of a concentration–time history equation is to have c_{A21} as the dependent variable and t as the independent variable. The equation can be obtained by solving Eq. 7.1-20*a* for c_{A21}. (see Example 7.1-2.) The concentration in chamber 2 can be obtained from the concentration in chamber 1 and the relationship

$$c_{A22} = c^0_{A22} + \frac{h_1}{h_2}(c^0_{A21} - c_{A21}) \tag{7.1-20b}$$

Sample calculation results using the preceding model appear in Fig. 7.1-13. The model conditions and parameters are shown in the figure. These are typical values; Zison et al.[19] give several algorithms for estimating the vertical dispersive transport coefficient, $\mathscr{D}^{(t)}_{A2y}$, that reflect actual lake conditions. Using a broad range of turbulent diffusivities, characteristic of water environments, it appears that approximately 4 to 36 years is required before the equalization of the concentration of A occurs in well-mixed adjoining layers.

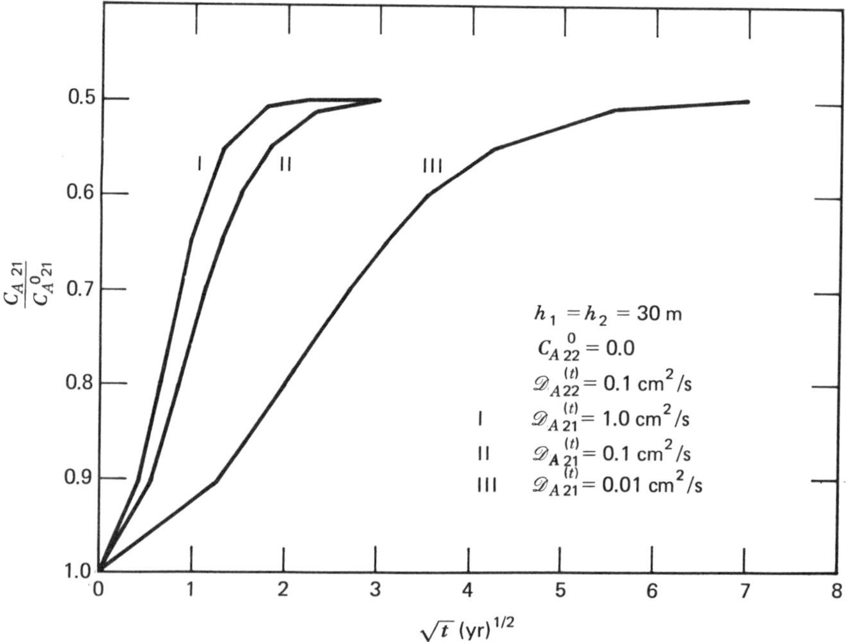

Figure 7.1-13. Sample calculation for model I.

Example 7.1-2. Movement of Pesticide Between Two Layers of Reservoir. Assume that the epilimnion of Beaver Reservoir becomes uniformly contaminated at 5 ppb with a pesticide in a very short period of time because of agricultural runoff following intense rain during the first week in May. It is expected that the pesticide will biodegrade to harmless products within 50 days. The regional water supply intake is below the thermocline, and there is concern about the movement of the pesticide into the hypolimnion. Estimate the maximum pesticide concentration likely to be observed in the hypolimnion. Use Fig. 7.1-4 as a source of data for Beaver Reservoir.

SOLUTION A study of Fig. 7.1-4 indicates that during May and for a period of about 125 days the reservoir is stratified and three layers are present. Assuming that the resistance to chemical movement through the thermocline region is replaced by resistances in the epilimnion and hypolimnion equal to the lake average, $\alpha_2^{(t)}$, Eq. 7.1-20*a* can be used. Solving the equation for c_{A21} yields

$$c_{A21} = \frac{c_{A21}^0}{(h_1/h_2) + 1} \left\{ \frac{h_1}{h_2} + \frac{c_{A22}^0}{c_{A21}^0} + \left(1 - \frac{c_{A22}^0}{c_{A21}^0}\right) \cdot \exp\left[\frac{-2\mathscr{D}_{A21}^{(t)}\mathscr{D}_{A22}^{(t)}t}{h_1 h_2}\left(\frac{h_1 + h_2}{\mathscr{D}_{A21}^{(t)}h_2 + \mathscr{D}_{A22}^{(t)}h_1}\right)\right]\right\} \tag{E7.1-2a}$$

Let chamber 1 ≡ epilimnion and chamber 2 ≡ hypolimnion. From Fig. 7.1-4, $h_1 = 10$ m, $h_2 = 30$ m, $\mathscr{D}_{A22}^{(t)} = \mathscr{D}_{A21}^{(t)} = \alpha_2^{(t)} = 0.73$ m^2/day, and $t = 50$ days. Substituting into Eq. E7.1-2*a* with $c_{A22}^0 = 0$ yields

$$c_{A21} = \frac{5}{(10/30) + 1}\left\{\frac{10}{30} + 0 + (1 - 0)\exp\left[\frac{-2(0.73)50}{(10)(30)}\right]\right\} = 4.19 \text{ ppb}$$

This is the concentration of the pesticide in the upper mixed layer. The concentration in the lower mixed layer is obtained from Eq. 7.1-20*b*:

$$c_{A22} = 0 + \frac{10}{30}(5.00 - 4.19) = 0.270 \text{ ppb}$$

The maximum concentration likely in the hypolimnion after 50 days is 0.27 ppb.

The preceding model and models presented later in this section contain significant assumptions that allow the development of closed-form equations such as Eqs. 7.1-20*a* and 7.1-20*b*. The use of these models to estimate concentrations and lifetimes of chemicals in real-world situations should be tempered with knowledge of the limitations and assumptions, which are:

1. Layers are closed; that is, there is no inflow or outflow.
2. Surface area of the air–water interface is the same as the mud–water interface.
3. Layers have constant depths.
4. There is no chemical movement across the air–water or sediment–water interfaces.
5. Introduced chemicals become uniformly distributed in a short period of time.
6. No chemical or biochemical reaction consumes or produces the chemical species.

Model II: Two Well-Mixed Layers Separated by a Calm (i.e., Relatively Unmixed) Middle Layer. An idealized picture of a stratified body of water is a well-mixed layer at the surface, a layer with a more or less pronounced density gradient (pycnocline) below it and a well-mixed layer below the pycnocline. Lerman[9] developed a simple model of this situation by assuming that the pycnocline, of depth Δh and eddy diffusivity $\mathscr{D}_{A2}^{(t)}$, provided all the resistance to the movement of a dissolved chemical species between two well-mixed chambers. Figure 7.1-14 illustrates the model concentration status at the start of the process.

The simplest way to estimate how long it takes for the concentrations in the lower and upper layers to become equal is to assume that the flux across the pycnocline is at all times proportional to the concentration difference between

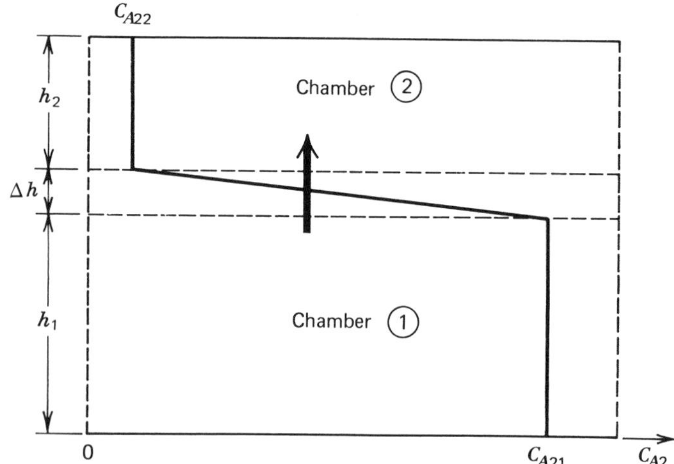

Figure 7.1-14. Chemical movement between two adjoining well-mixed layers separated by a calm layer.

the upper and lower layers:

$$N_A = \frac{\mathscr{D}_{A2}^{(t)}}{\Delta h}(c_{A21} - c_{A22}) \tag{7.1-21}$$

where $\mathscr{D}_{A2}^{(t)}$ is the eddy diffusion coefficient characteristic of the middle layer and Δh is the thickness of the layer. A component balance of species A in the lower chamber yields

$$\frac{dc_{A21}}{dt} = -\frac{\mathscr{D}_{A2}^{(t)}}{\Delta h h_1}(c_{A21} - c_{A22}) \tag{7.1-22}$$

The total amount of the chemical species per unit surface area of the lake is:

$$\dot{m}_A = c_{A21}h_1 + c_{A22}h_2 + (c_{A21} + c_{A22})\frac{\Delta h}{2} \tag{7.1-23}$$

The total amount may also be expressed in terms of the initial concentrations in the two mixed layers and the mean concentration in the pycnocline:

$$\dot{m}_A = c_{A21}^0 h_1 + c_{A22}^0 h_2 + (c_{A21}^0 + c_{A22}^0)\frac{\Delta h}{2} \tag{7.1-24}$$

Combining the last three equations and integrating gives

$$c_{A21} = \frac{\dot{m}_A}{h_1 + h_2 + \Delta h}\left\{1 - \exp\left[-\frac{\mathscr{D}_{A2}^{(t)}t}{\Delta h}\left(\frac{1+\alpha}{h_1}\right)\right] + c_{A21}^0 \exp\left[-\frac{\mathscr{D}_{A2}^{(t)}t}{\Delta h}\left(\frac{1+\alpha}{h_1}\right)\right]\right\} \tag{7.1-25}$$

and

$$c_{A22} = \frac{\dot{m}_A - c_{A21}(h_1 + \Delta h/2)}{h_2 + \Delta h/2} \tag{7.1-26}$$

where

$$\alpha \equiv \frac{h_1 + \Delta h/2}{h_2 + \Delta h/2} \tag{7.1-27}$$

The model equations are written so that they can be used for any two well-mixed chambers separated by any unmixed chamber. When the initial concentration in the upper chamber is zero, Eq. 7.1-26 may be put in the form

$$c_{A22} = c_{A21}^0 \frac{h_1 + \Delta h/2}{h_1 + h_2 + \Delta h}\left\{1 - \exp\frac{\mathscr{D}_{A2}^{(t)}t}{\Delta h}\left[\left(\frac{1+\alpha}{h_1}\right)\right]\right\} \tag{7.1-28}$$

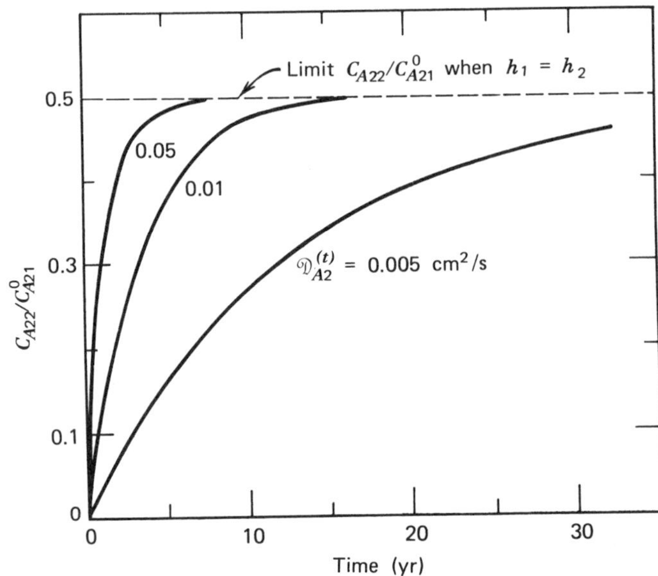

Figure 7.1-15. Sample calculation for model II. (Reprinted with permission American Chemical Society publication: *Non-equilibrium Systems in Natural Water Chemistry*, Ref. 9.)

It is instructive to perform a sample calculation employing Eq. 7.1-28. Assume that a dissolved species has been introduced into the lower mixed layers and that the initial concentration in the upper layer is nil. When the three-layer system remains closed and the dimensions of the water layers do not change, a conservative chemical species in one of the mixed layers will redistribute itself between the two layers because of the diffusional flux down the concentration gradient from one mixed layer into the other. For a case of transport from the lower into the upper, change in the concentration in the upper as function of time is shown in Fig. 7.1-15. The curves have been calculated for a 60-m deep water column (lower layer $h_1 = 25$ m, pycnocline $\Delta h = 10$ m, and upper layer $h_2 = 25$ m) for three different eddy diffusion coefficients in the pycnocline (0.005, 0.01, and 0.05 cm^2/s). These values of the eddy diffusion coefficient are in the range reported for pycnoclines in stratified lakes (Table 3.1-4). The curves show that the concentrations in the two layers would equalize in 10 to 40 years. The time required to attain certain concentration levels in such a model lake depends on the eddy diffusivity in the pycnocline and on the vertical dimensions of the individual layers.

Model III: Two Adjoining Unmixed Layers.[9] In this model the two fluid layers are not well mixed; each can be characterized by a different value of the turbulent diffusivity. The initial status of this model is as shown in Fig. 7.1-12. Concentration gradients can develop within each layer. Initially, a conservative species is homogeneously distributed within the lower layer, its concentration being c^0_{A21}. Migration across the boundary between the two layers and subsequent dispersal within the upper layers would eventually equalize the concentrations. At the limit when all gradients disappear, the concentration of the species would be homogeneous throughout the two layers and equal to $c^0_{A21}h_1/(h_1 + h_2)$.

In the two-layer system, when the diffusion coefficients in the two layers are equal, the concentration of a dissolved substance originally confined to one layer is given by the following relationship:[10]

$$c_{A2} = c^0_{A22} + \frac{1}{2} c^0_{A21} \sum_{n=-\infty}^{\infty} \left[\operatorname{erf} \frac{h_1 + 2n(h_1 + h_2) - y}{2\sqrt{\mathscr{D}^{(t)}_{A2} t}} + \operatorname{erf} \frac{h_1 - 2n(h_1 + h_2) + y}{2\sqrt{\mathscr{D}^{(t)}_{A2} t}} \right] \tag{7.1-29}$$

where c^0_{A21} is the initial concentration in one layer ($0 < y < h_1$), h_1 and h_2 are the boundaries of the two layers, and y is the vertical dimension ($0 \leqslant y \leqslant h_2$).

For the case when the two diffusion coefficients in the two layers are not equal, derivation of a closed-form relationship c_{A2} is difficult. As an alternative, Lerman[9] presents a simpler method that gives the mean concentration as a function of time in the upper layer into which the substance diffuses from the lower layer. He reports that the mean concentrations computed by this method are within a few percent of the values obtainable by use of a complete expression for c_{A2}, such as Eq. 7.1-29, and he uses the model to obtain concentration–time curves for the upper layer in a two-layer model with the values of $h_1 = 30$ m, $h_2 = 30$ m, $\mathscr{D}^{(t)}_{A21} = 0.1\ \text{cm}^2/\text{s}$ and $\mathscr{D}^{(t)}_{A22} = 0.01$ and $1.0\ \text{cm}^2/\text{s}$. As in the previous example calculations, the values of the turbulent diffusivities are taken to represent the range reported for stratified lakes. The calculated curves appear in Fig. 7.1-16. The conclusion that may be drawn from these curves is essentially the same as for the two models presented earlier: The time required to attain equal concentrations is relatively short, 1 to 16 years.

Closure. The preceding discussion of the three simple models suggests that in closed water bodies with stationary stratification of the water column, a change in the chemical composition of one of the layers produces a transient condition that can persist for 5 to 50 years. The transient phenomenon is relatively short in comparison to geologic time scales, but it is long when considering the annual thermoclinic cycle in lakes. To maintain a steady concentration gradient in either water column, the water bodies must be open, such that the input of solute is balanced by its removal. An example of such a case is presented in the next section.

Quasi-Steady-State Models of Chemical Profiles in Lakes.[11] Because of certain similarities between lakes and oceans, the models described in this section are also appropriate for studying chemical distributions in the coastal

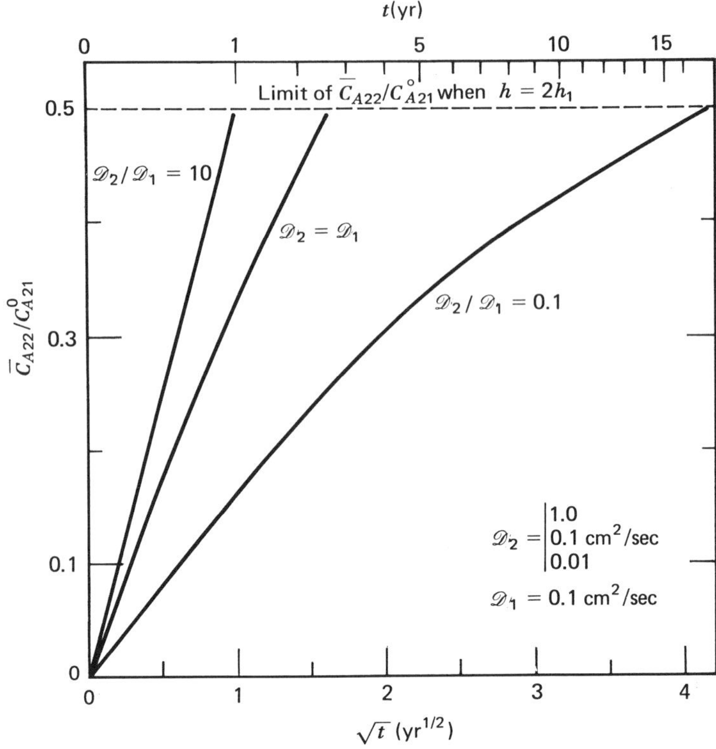

Figure 7.1-16. Sample calculation for model III. (Reprinted with permission. American Chemical Society publication: *Non-equilibrium Systems in Natural Water Chemistry*, Ref. 9.)

and offshore regions of oceans. In general, chemical distribution problems in lakes and oceans are time variable in two or three space dimensions. The deep, relatively slow-moving hydraulic regime over broad horizontal areas is an important factor in this regard. Because of the greater depths involved, thermal differences and density structures are encountered seasonally. The vertical distribution of many chemicals associated with these regimes can be a significant water quality problem. The variable nature of the wind-driven currents in both lakes and oceans is an additional factor that may preclude the use of simplified mathematical approaches. This is not to suggest that the steady-state approach is never useful, simply that its application is more restrictive and the analysis more approximate. Despite these limitations the one-dimensional steady-state analysis applied under appropriate conditions does offer reasonable insight and understanding of the nature of chemical distributions and provides a limiting condition or the basis for justifying the construction of a more sophisticated analysis.

A most pronounced vertical variation in chemical concentrations occurs in the central zone of most lakes. The variation of concentration is one of the most characteristic chemodynamic features of the water in lakes and reservoirs.

Both time-variable and steady-state analyses are valid for the vertical distribution. Although there may be some question about the application of the steady-state analysis to the temperature distribution, it may be appropriate for the dissolved oxygen analysis. During the period of maximum temperature or shortly thereafter, observations indicate in many cases a reasonable constancy of the vertical temperature gradient. Although this condition may not truly reflect a thermal steady state, it may persist for sufficiently long periods to justify the use of the steady-state analysis for other constituents.

Quasi-Steady-State Model Definition. The use of a steady-state model for a fixed period of time, before significant changes to the system occur nullifying the prime assumption, is termed a quasi-steady-state model. In the case of oxygen, nutrient and temperature profiles in lakes, for example, steady-state conditions may be appropriate for a month or two during the stratified period. In this and similar contexts the models that follow are quasi-steady-state models.

Conservative Substances. A logical starting place for considering steady-state vertical profiles of chemicals in lakes is Eq. 3.2-11 in mass concentration form:

$$\frac{\partial \rho_{A2}}{\partial t} + \nabla \cdot (\rho_{A2}\mathbf{V}) = \nabla \cdot (\mathscr{D}_{A2}^{(t)} \nabla \rho_{A2}) + r_{A2} \tag{7.1-30}$$

If steady state and no advection is assumed and only the vertical profile of a conservative substance is considered, it simplifies to

$$0 = \frac{d}{dy}\left(\mathscr{D}_{A2}^{(t)} \frac{d\rho_{A2}}{dy}\right) \tag{7.1-31}$$

$\mathscr{D}_{A2}^{(t)}$ is assumed to be a function of y. Two integrations yield

$$\rho_{A2} = c_1 \int_0^y \frac{dy}{\mathscr{D}_{A2}^{(t)}} + c_2$$

where c_1 and c_2 are constants of integration. To evaluate the constant c_2, the following concentration boundary condition is applied at the surface:

$$\rho_{A2} = \rho_{A20} \qquad \text{at} \quad y = 0 \tag{7.1-32}$$

This gives $c_2 = \rho_{A20}$.

The constant c_1 is a flux rate of emission of chemical A from either the air–water or sediment–water interface. Since there are no sources or sinks for the constituent within the water column and steady state prevails, it follows that the water is of a uniform concentration ρ_{A20} and $c_1 = 0$. This simple model of the steady-state distribution of a conservative substance does not yield equations of a practical value. There are three oversimplifications that usually account for this lack of practical utility, any one of which may control.

The first factor is related to the time-variable nature of the phenomena. The time required to achieve a steady-state condition for a conservative species may extend over a long period in many lakes (see the preceding section). The spatial distributions observed in this case are various stages of a slow-moving transient, which may never reach the steady state. This is identical to the question relating to the temperature distribution and the upsurge of nutrients from the mud–water interface (see Sections 7.1 and 5.2, respectively, for a detailed analysis).

The second factor that affects the distribution of conservative substances in lakes emanating from the bed is related to the boundary conditions at this location. The problem is slightly more complex, involving the interaction between two separate systems—the vertical water column and an associated segment of the bed—each with its own differential equation. With respect to the water column, the boundary condition at the mud–water interface is normally transient.

The third is the fact that a small amount of advection, either horizontal or vertical, can produce observable concentration gradients; consequently, the simplistic dispersion model may be inadequate to describe the distribution of a conservative substance, and the more encompassing advective-dispersion analysis is required. Although it may be desirable to employ the more general model for a conservative substance, it does not necessarily follow that it is required for a nonconservative substance, for which the dispersive equation subsequently described in this section may be quite adequate.

Nonconservative Substances. The vertical distribution of a nonconservative substance in lakes and reservoirs under steady-state conditions is similar in origin to Eq. 7.1-31.

$$0 = \frac{d}{dy}\left(\mathscr{D}_{A2}^{(t)} \frac{d\rho_{A2}}{dy}\right) + r_{A2} \tag{7.1-33}$$

The reaction term r_{A2} accounts for the net production of A within the vertical column.

The following analysis applies specifically to the vertical distribution of dissolved oxygen in a lake water column. Although the analysis is specific, the general technique is applicable to all nonconservative chemical species: radon, methane, ammonia, orthophosphate, mercury, organochlorides, and so on (see Problems 7.1I and 7.1J).

When applying Eq. 7.1-33 to the vertical distribution of dissolved oxygen in a lake, the internal sources and sinks must be defined and the boundary conditions specified. If one considers the more general case: both a source of oxygen, caused by either chemical or biochemical oxidation, and a sink must be assumed, both of which are a function of depth:

$$r_{A2} = r_{A2}|_p - r_{A2}|_c \tag{7.1-34}$$

where the p and c subscripts denote oxygen production and consumption.

HOMOGENEOUS OR UNSTRATIFIED LAKE. The boundary conditions may be specified at the surface ($y = 0$) and the bed ($y = h$). The air–water interface, which allows for the transfer of oxygen through the surface layer, is the first boundary condition:

$$\left[-\mathscr{D}_{A2}^{(t)} \frac{d\rho_{A2}}{dy}\right]_{y=0} = {}^1k'_{A2}(\rho^*_{A2} - \rho_{A2})|_{y=0} \tag{7.1-35}$$

where ρ^*_{A2} is dissolved oxygen saturation value at a given temperature. Zison et al.[19] review a limited number of expressions for estimating ${}^1k'_{A2}$ for oxygen in stratified lakes or estuaries. The benthal oxygen demand at the sediment–water interface provides the second boundary condition:

$$\left[-\mathscr{D}_{A2}^{(t)} \frac{d\rho_{A2}}{dy}\right]_{y=h} = n_{Ah} \tag{7.1-36}$$

Here n_{Ah} is the sediment oxygen demand. Rearranging Eq. 7.1-33 and integrating twice yields the following:

$$\mathscr{D}_{A2}^{(t)} \frac{d\rho_{A2}}{dy} = -\int_0^y r_{A2}\,dy + C_1 \tag{7.1-37}$$

and

$$\rho_{A2} = -\int_0^y \frac{\int_0^y r_{A2}\,dy}{\mathscr{D}_{A2}^{(t)}}\,dy + C_1 \int_{y=0}^y \frac{dy}{\mathscr{D}_{A2}^{(t)}} + C_2 \tag{7.1-38}$$

The unknown constants, C_1 and C_2, may be evaluated by applying two boundary conditions. At the surface $\rho_{A2} = \rho_{A20}$ and Eq. 7.1-35 applies to yield

$$C_2 = \rho_{A20} \quad \text{and} \quad C_1 = -{}^1k'_{A2}(\rho^*_{A2} - \rho_{A20}). \tag{7.1-39}$$

Substituting these into Eq. 7.1-38 yields

$$\rho_{A2} = \rho_{A20} - \int_0^y \frac{\int_0^y r_{A2}\,dy}{\mathscr{D}_{A2}^{(t)}}\,dy - {}^1k'_{A2}(\rho^*_{A2} - \rho_{A20}) \int_0^y \frac{dy}{\mathscr{D}_{A2}^{(t)}} \tag{7.1-40}$$

The remaining unknown, ρ_{A20}, may be evaluated by application of the second boundary condition, Eq. 7.1-36, at the bed, $y = h$, which yields

$$\rho_{A20} = \rho^*_{A2} - \frac{1}{{}^1k'_{A2}}\left(n_{Ah} - \int_0^h r_{A2}\,dy\right) \qquad (7.1\text{-}41)$$

The final equation is obtained by substitution this into Eq. 7.1-40 and simplifying to give

$$\rho_{A2} = \rho^*_{A2} - \left(n_{Ah} - \int_0^h r_{A2}\,dy\right)\left(\frac{1}{{}^1k'_{A2}} + \int_0^y \frac{dy}{\mathscr{D}^{(t)}_{A2}}\right) - \int_0^y \frac{\int_0^y r_{A2}\,dy}{\mathscr{D}^{(t)}_{A2}}\,dy \qquad (7.1\text{-}42)$$

This equation gives the vertical distribution of dissolved oxygen or any chemical affected by comparable surface and bed conditions, such as chlorinated organics (i.e, PCBs) in the air for which the bed is a sink. Students should hone their skills, all the while demonstrating to themselves the generality of the process, by redoing the analysis above and obtaining a result comparable to Eq. 7.1-42 for the case of the bed being the chemical source with volatilization to air.

A simple application of the nonconservative, homogeneous lake model is considered next. The actual conditions reflected are admittedly not widely applicable since variable diffusivity and reaction are usually encountered in most lake environments. However, these simplified conditions provide some insight into the nature of the exchange and distribution of such constituents as dissolved oxygen. Furthermore, they may be used as an approximate solution to the more realistic situations, particularly in the central zone of broad lakes. Given the conditions that the source and sink are the surface and bed transfers and there is no reaction within the water column, Eq. 7.1-42 reduces to

$$\rho_{A2} = \rho^*_{A2} - n_{Ah}\left(\frac{1}{{}^1k'_{A2}} + \frac{y}{\mathscr{D}^{(t)}_{A2}}\right) \qquad (7.1\text{-}43)$$

for constant $\mathscr{D}^{(t)}_{A2}$. A close study of this final equation yields two significant terms: one, $\rho_{A20} = \rho^*_{A2} - n_{Ah}/{}^1k'_{A2}$, which is the dissolved oxygen at the surface $(y = 0)$, and the other, the dimensionless number ${}^1k'_{A2}h/\mathscr{D}^{(t)}_{A2}$, in which $y = h$. Eq. 7.1-43 can be transformed to

$$\rho_{A2} = \rho^*_{A2} - (\rho^*_{A2} - \rho_{A20})\left(1 + \frac{{}^1k'_{A2}y}{\mathscr{D}^{(t)}_{A2}}\right) \qquad (7.1\text{-}44)$$

It can be seen from this result that if ${}^1k'_{A2}h/\mathscr{D}^{(t)}_{A2}$ is small and much less than unity, as is the case for large eddy diffusivity, the concentration is uniform and equal to ρ_{A20}. A common range for ${}^1k'_{A2}$ is 1.3 to 6.4 cm/h, and for $\mathscr{D}^{(t)}_{A2}$ is 390

Table 7.1-2. Oxygen Absorption-Dispersion Parameter for a Homogeneous Lake, ${}^1k'_{A2}h/\mathscr{D}^{(t)}_{A2}$

${}^1k'_{A2}$ (cm/h):	1.3		6.4	
$\mathscr{D}^{(t)}_{A2}$ (cm^2/h):	390	1900	390	1900
h (m)				
3	1.0	0.2	5.0	1.0
15	5.0	1.0	25	5.0

Source: Modified from Ref. 11.

to 1900 cm^2/h in the equilimnion. For depths of 3 to 15 m, the dimensionless parameter ${}^1k'_{A2}h/\mathscr{D}^{(t)}_{A2}$ has the numerical values shown in Table 7.1-2. It is apparent that in most practical cases for the conditions considered, significant gradients of dissolved oxygen develop except in the extreme case of ${}^1k'_{A2} = 1.3$, $\mathscr{D}^{(t)}_{A2} = 1900$, and $h = 3$. It should be emphasized that no sources, such as photosynthesis-generated O_2, were considered in this analysis.

Example 7.1-3. Vertical Oxygen Profiles in Unstratified Lakes. This problem illustrates, among other things, the importance of the oxygen consumption rate at the mud–water interface (i.e., the SOD) on the oxygen concentration profile. Consider two unstratified lakes containing little or no phytoplankton that produce oxygen and little or no dissolved organic and/or inorganic material to consume dissolved oxygen within the water column (i.e., $r_{A2} = 0$). One lake is a large, artificial reservoir with bottom muds rich in organic matter (i.e., decaying vegetation) left over from the prefilling days. The other lake is a smaller, artificial "run-of-the-river" reservoir in which the detention time is low and the flow is relatively rapid. For a water temperature of 30°C, compute and graph the vertical oxygen concentration profile as a function of depth for each lake type. Table E7.1-3 contains pertinent chemical and physical data for both lakes.

Table E7.1-3

Description	n_{Ah} [a] (g O_2/m$^2\cdot$day)	${}^1k'_{A2}$ (m/day)	$\mathscr{D}^{(t)}_{A2}$ (m^2/day)	h (m)
Large reservoir, mud bottom	0.2	0.3	0.7	20
Run-of-the-river reservoir, sandy bottom	0.05	1.5	4.0	10

[a]From Ref. 12.

SOLUTION Since $r_{A2} = 0$, Eq. 7.1-43 applies. At 30°C, $\rho^*_{A2} = 7.63 \text{ mg } O_2/\text{L}$.
For the large reservoir:

$$\rho_{A2} = 7.63 - \frac{0.2 \text{ g } O_2}{\text{m}^2 \cdot \text{day}} \left[\frac{\text{d}}{0.3 \text{ m}} + \frac{y(\text{m})\text{d}}{0.7 \text{ m}^2} \right]$$

$$= 6.96 - 0.286y, \qquad y \leqslant 20 \text{ m} \tag{E7.1-3a}$$

For the run-of-the-river reservoir:

$$\rho_{A2} = 7.63 - 0.05 \left(\frac{1}{1.5} + \frac{y}{4} \right)$$

$$= 7.60 - 0.0125y, \qquad y \leqslant 10 \text{ m} \tag{E7.1-3b}$$

Figure E7.1-3 is a graph of each concentration profile. The surface oxygen concentration in both cases is less than ρ^*_{A2} and decreases with depth. *A* large reaeration coefficient and large turbulent diffusivity coupled with a low oxygen consumption rate at the mud–water interface and a shallow depth results in a nearly vertical profile displaying a high, uniform oxygen concentration. The profile displayed for the large reservoir reflects a significant oxygen concentration gradient.

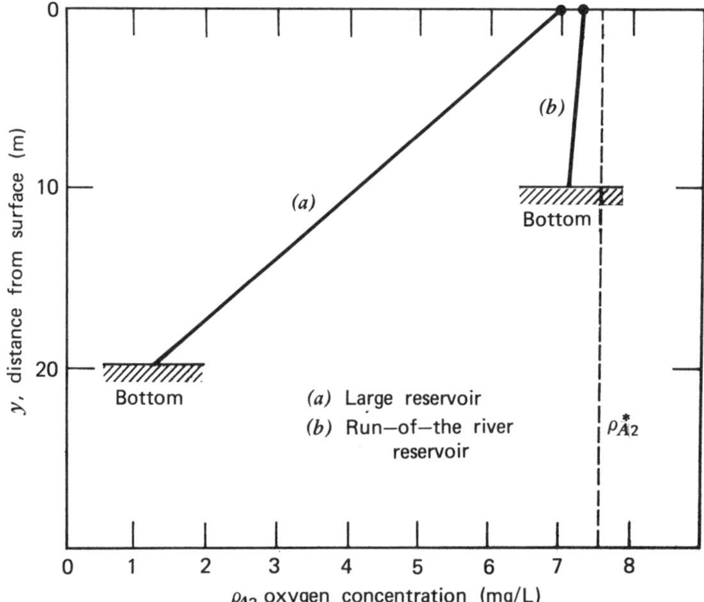

Figure E7.1-3. Oxygen concentration profiles in two reservoirs.

TWO-LAYER OR STRATIFIED LAKE. A more realistic condition is that of a lake segmented vertically into two zones, representing the epilimnion and hypolimnion, each with its characteristic eddy diffusivity. This case is a representation of a stratified lake normally encountered during the summer months. The thermocline is located at $y = h_1$ and the sediment bed at $y = h_2$. The depth of the thermocline is thus h_1 and that of the hypolimnion $h_2 - h_1$.

The solution proceeds along lines similar to those for the unstratified lake. The solution for the upper layer is nearly identical to that for the previous case. The concentration profile in layer 1 can be determined for values of $0 \leqslant y \leqslant h_1$, by

$$\rho_{A21} = \rho_{A2}^{*} - \left(n_{Ah2} - \int_0^{h_1} r_{A21}\,dy - \int_{h_1}^{h_2} r_{A22}\,dy\right)\left(\frac{1}{{}^1k'_{A2}} + \int_0^y \frac{dy}{\mathscr{D}_{A21}^{(t)}}\right) - \int_0^y \frac{\int_0^y r_{A21}\,dy}{\mathscr{D}_{A21}^{(t)}}\,dy \tag{7.1-45}$$

Developing the concentration profile for the lower layer begins with

$$0 = \frac{d}{dy}\left(\mathscr{D}_{A22}^{(t)} \frac{d\rho_{A22}}{dy}\right) + r_{A22} \tag{7.1-46}$$

where $\mathscr{D}_{A22}^{(t)}$ and r_{A22} are the eddy diffusivity and reaction rate terms characteristic of the lower layer. By specifying the flux and the concentration at the thermocline $y = h_1$, the two needed boundary conditions are obtained:

$$\mathscr{D}_{A22}^{(t)} \frac{d\rho_{A22}}{dy} = \mathscr{D}_{A21}^{(t)} \frac{d\rho_{A21}}{dy} \quad \text{at} \quad y = h_1 \tag{7.1-47}$$

and

$$\rho_{A22} = \rho_{A2h_1} \quad \text{at} \quad y = h_1 \tag{7.1-48}$$

Equation 7.1-46 is now integrated twice to yield

$$\mathscr{D}_{A22}^{(t)} \frac{d\rho_{A22}}{dy} = -\int_{h_1}^{y} r_{A22}\,dy + C_3 \tag{7.1-49}$$

and

$$\rho_{A22} = -\int_{h_1}^{y} \frac{\int_{h_1}^{y} r_{A22}\,dy}{\mathscr{D}_{A22}^{(t)}}\,dy + C_3 \int_{h_1}^{y} \frac{dy}{\mathscr{D}_{A22}^{(t)}} + C_4 \tag{7.1-50}$$

Applying the boundary condition represented by Eqs. 7.1-47 to 7.1-49 yields

$$\mathscr{D}_{A21}^{(t)} \left.\frac{d\rho_{A21}}{dy}\right|_{h_1} = C_3 \tag{7.1-51}$$

Now the left-hand side can be obtained from Eq. 7.1-37 knowing $C_1 = -{}^1k'_{A2}(\rho^*_{A2} - \rho_{A20})$ to yield

$$-\int_0^{h_1} r_{A21}\,dy - {}^1k'_{A2}(\rho^*_{A2} - \rho_{A20}) = C_3 \tag{7.1-52}$$

Applying the boundary condition represented by Eq. 7.1-48 yields $\rho_{A2h_1} = C_4$. Substituting the appropriate expressions for C_3 and C_4 along with Eq. 7.1-41 for ρ_{A20} yields

$$\rho_{A22} = \rho_{A2h_1} - \int_{h_1}^{y} \frac{\int_{h_1}^{y} r_{A22}\,dy}{\mathscr{D}^{(t)}_{A22}}\,dy - n_{Ah_2}\int_{h_1}^{y} \frac{dy}{\mathscr{D}^{(t)}_{A22}} \tag{7.1-53}$$

where $h_2 \geqslant y \geqslant h_1$.

This final relationship can be used to calculate the oxygen concentration in the hypolimnion of a lake. It should be noted that ρ_{A2h_1} is computed from Eq. 7.1-45 at $y = h_1$. Proper utilization of Eqs. 7.1-45 and 7.1-53 requires that $\mathscr{D}^{(t)}_{A21}$, $\mathscr{D}^{(t)}_{A22}$, r_{A21}, and r_{A22} be known functions of y. Problem 7.1J is a realistic concentration profile calculation that includes oxygen production and utilization rates.

Closure. This section introduces basic concepts one should employ when estimating concentration profiles of nonconservative chemicals in water. The specific illustration involved oxygen; however, an analogous procedure can be used to formulate quasi-steady-state models for any chemical influenced by sources or sinks within the water column or at the interfaces. Descriptions of situations involving other chemicals appear in the problems at the end of Section 7.1.

Steady-State Longitudinal Profiles of Chemicals in River and Estuaries. An instructive but highly restrictive model for chemical behavior in a flowing stream was presented in Section 1.2. It was the classical Streeter and Phelps O_2 and organic waste model with longitudinal varying concentrations downstream from a single waste discharge point. In what follows, a more complete analysis will be presented that has application to partially mixed estuaries.

Estuaries. These are areas of interaction between fresh and salt water. The definition most commonly used is: "a semienclosed coastal body of water which has free convection exchange with the open sea and within which sea water is measurably diluted with fresh water derived from land drainage."[20] Water circulation and mixing processes are driven by the density differences between the two waters. The density of seawater depends on both salinity and temperature; in estuaries the salinity range is large, whereas the temperature range is generally small and has a relatively small influence on the density. At the sea surface the salinity average is close to 35‰ (parts per thousand), and at 20°C the density is 1.025 g/cm^3 (see Table D.3 for seawater properties).

Estuaries are formed in the narrow boundary zone between the sea and the land. Most that have been studied in detail fall within the coastal plain category. These display large differences in circulation patterns, density stratification, and mixing processes. A classification of estuaries based on salinity distribution and flow characteristics has resulted and there are four main types: highly stratified or salt wedge, fjords, partially mixed, and homogeneous. Examination of only the partially mixed estuary will lead to a general understanding of how circulation and mixing of the waters are maintained (see Dyer[20] for the others).

Partially Mixed Estuary. First, let's consider an idealized plug flow non-constant cross-sectional estuary emptying into a tideless sea. Under these conditions the river water, being less dense, flows outward over the saline layer (see Fig. 7.1-17*a*.). The velocity in the surface layer decreases toward the mouth

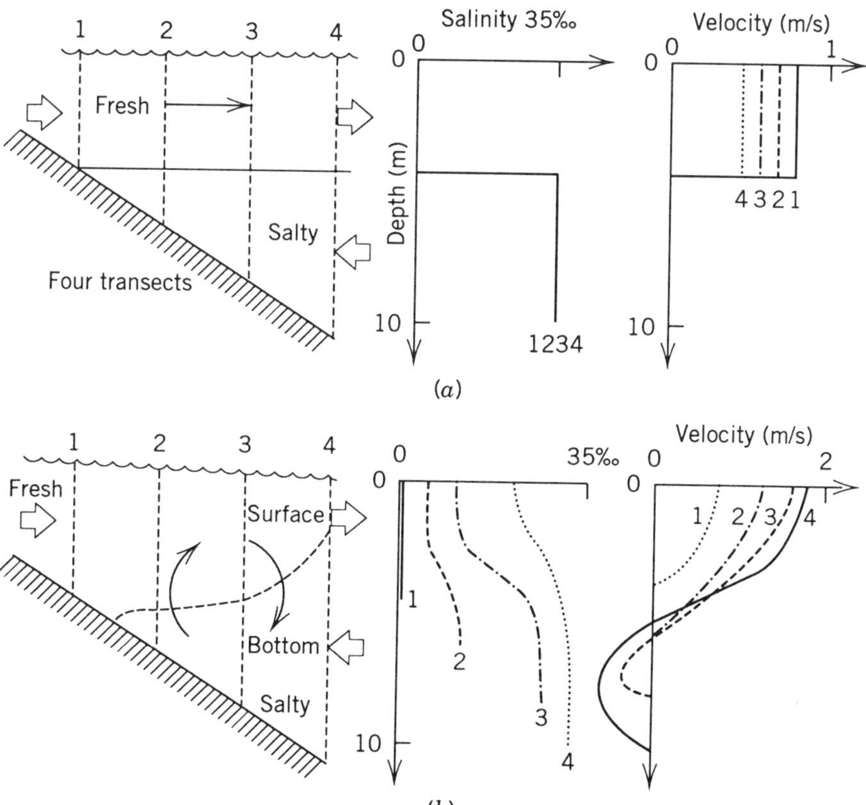

Figure 7.1-17. Estuarine salinity and velocity profiles: (*a*) idealized estuary; (*b*) partially mixed estuary. (Adapted from Ref. 20.)

as the estuary widens. Without viscosity (again idealized) and entrainment or diffusive transport of salt upward, the interface between the fresh and saltwater would be horizontal and would extend a distance up-estuary as far as mean sea level. There is no motion in the saline wedge. Because of the Coriolis force the seaward-flowing river water would be concentrated on the right-hand side (looking downstream) in the northern hemisphere. In effect, the interface between fresh and saline waters is tilted to the right. Along the centerline of the ideal estuary the velocity and salinity profiles are shown in Fig. 7.1-17*a*. The mouth of the Mississippi River at Southwest Pass, where river flow is large compared to tidal flow, has some of the foregoing characteristics because of the small tidal range ($\sim 70\,\text{cm}$).

Let us now introduce friction in the form of viscosity. There will be shear at the fresh–salt interface and the salt wedge will be pushed downstream until its upper surface has a slope sufficient to resist this force. The tip of the wedge becomes blunted and the water surface will slope more steeply toward the sea. Coriolis forces will now affect the lateral water slopes, with the interface sloping downward toward the right and the sea surface downward toward the left in the northern hemisphere. If we now introduce tides into the estuary, the entire contents of the estuary will oscillate back and forth. Only a small tidal range is required to make this occur when the river flow is low compared to the tidal prism. Typical average centerline estuaries salinity and velocity profiles are shown in Fig. 7.1-17*b*.

The energy involved in tidal movements is large and mostly dissipated by working against frictional drag on the bottom, producing turbulence (see Chapter 3). In addition to turbulence, mixing also occurs by entrainment. This is a one-way process in which a less turbulent water mass becomes drawn into a more turbulent layer. When the velocity shear across the fresh–salty interface is sufficiently intense, waves form and break and the quieter salty water is mixed into the surface fresh water. The rate of entrainment increases with increasing velocity differences between the layers. Note that turbulent diffusion is a two-way process in which equal volumes of water are exchanged between the two layers. These processes mix both salty water upward and fresh water downward. This raises the salinity of the surface layer, but in order to discharge the fresh water volume equal to the river flow, the seaward surface flow is enhanced. This causes an increase in the volume of the compensating landward bottom flow. Note the negative velocities in the bottom layer of Fig. 7.1-17*b*. Consequently, a distinct two-layer flow system is developed. The surface salinity increases steadily down the estuary and freshwater occurs only very near the head of the estuary, as illustrated in Fig. 7.1-17*b*. Vertically, there is normally a zone of high salinity gradient at about middepth and the surface and bottom layers are almost homogeneous. If the two layers are distinct, based on measured salinities so that averaging yields surface and bottom salinities, the seaward and landward flows across a vertical transect can be estimated knowing only the fresh river volumetric flow (see Problem 7.1M).

The presence of tidal movements and of turbulence introduce problems that can be resolved only by averaging. There is a need to separate the influences

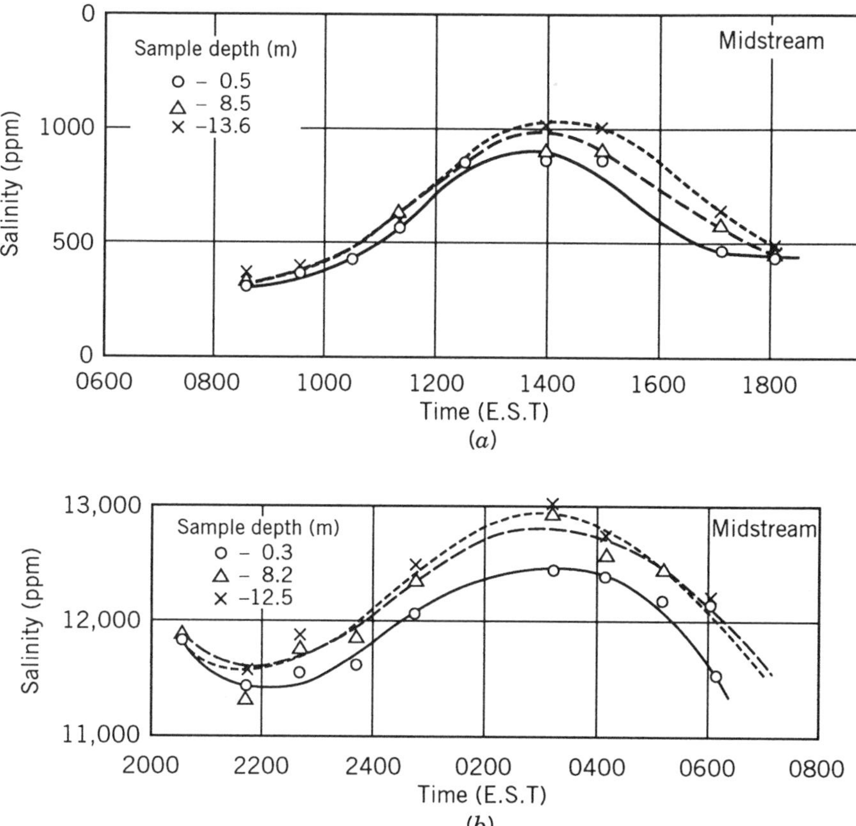

Figure 7.1-18. Midstream salinity measurements in the Hudson River: (*a*) Poughkeepsie station; (*b*) Dobbs Ferry station. (From Ref. 21.)

of river flow, tidal oscillation, and turbulent fluctuations in order to understand the interactions providing conditions of dynamic equilibrium in an estuary. Variations of salinity across two typical cross sections within the salt-intruded reach of the Hudson River are shown in Fig. 7.1-18. The Dobbs Ferry station is down river from the Poughkeepsie station and displays higher salinities. In this and many estuaries the tidal variation is velocity, and concentration can be represented moderately realistically by simple trigonometrical functions. The incoming tidal current causes the rise in salinity and salinity fall as the current goes out. While tidal elevation and current maximums are almost out of phase; tidal elevation and salinity maximums are almost in phase.

Instantaneous measurements of x component of velocity, the longitudinal component, v_x, can be represented by three terms:

$$v_x = \bar{v}_x + V_x + v'_x \tag{7.1-54}$$

where $\bar{v}_x$ is the mean over the tidal cycle, V_x the tidal variation, and v_x the short-period turbulent contribution. In many cases

$$V_x = V_x^0 \sin \omega t \tag{7.1-55}$$

where V_x^0 is a constant, ω the tidal frequency, and t the time. It is difficult for normal measuring instruments to record the instantaneous velocity and $\overline{v_x'}$ is zero. The vertical, v_y, and lateral, v_z, velocity components behave similarly. The salinity is generally about 90° out of phase with velocity and represented by *cos* ωt; see Eq. 5.3-7 for an example.

Lawler developed a very useful river–estuary model for dissolved oxygen and organic matter.[21,22] It is an excellent piece of work available in the gray literature; it is presented here so that it remains gray. It is a one-dimensional steady-state model for area-averaged and tidal-smooth concentration. Tidal and area averaging is done to make the model tractable. The major objective of developing this model is to represent a lengthy reach of estuary and be able to predict variations in chemical concentration along the longitudinal axis of the reach. Variations across a section are normally due to local conditions, and variations in time are equally complex, so that models using averages are more useful, generally. The area-averaged tidal-smooth concentration means the concentration of the entity in question, obtained by averaging point values of that entity over the river's cross section area, and throughout the tidal cycle.

To establish the tidal-smoothed area-averaged model for concentration, start with the time-smoothed equation of continuity for a single chemical species in which concentration is a function of three space dimensions and real time. This is Eq. 3.2-29*a* in mass concentration form. Many analyses of estuaries begin with this equation, neglecting the tidal fluctuation cross-products and the salt balance for values averaged over a tidal cycle (Ref. 21, p. 67). This can be done because with simple trigonometrical cyclic behavior the tidal-smoothed concentration and x-component of velocity become $\bar{\rho}_{A2}$ and $\bar{v}_x$, respectively. Assuming that the estuary is of constant cross-sectional area and there is no net tidal average flow in the lateral and vertical directions, $\bar{v}_z = \bar{v}_y = 0$. Due to the highly turbulent character of this natural system, molecular diffusion is neglected. Under the assumptions above, Eq. 3.2-39*a* becomes

$$\frac{\partial \bar{\rho}_{A2}}{\partial t} + \bar{v}_x \frac{\partial \bar{\rho}_{A2}}{\partial x} = -\left(\frac{\partial}{\partial x} j_{Ax}^{(t)} + \frac{\partial}{\partial y} j_{Ay}^{(t)} + \frac{\partial}{\partial z} j_{Az}^{(t)}\right) - k_{A2}''' \bar{\rho}_A \tag{7.1-56}$$

The three turbulent fluxes (i.e., $j_A^{(t)}$'s) are due to the fluctuating velocity–concentration products shown in Eqs. 3.2-29*b*, *c*, *d* in molar form. Note that $j_A = J_A M_A$ and $\rho_A = c_A M_A$ where M_A is the molecular weight of constituent A. It is common practice to use first-order-reaction decay kinetics for most waterborne constituents.

The turbulent fluxes are generally written in a form analogous to Fick's law for molecular diffusion (see Eq. 3.1-6) and are known as eddy-diffusion terms. The equations used in closure for the turbulent fluxes are these:

$$\overline{v_x'\rho_{A2}'} = -\mathscr{D}_{A2x}^{(t)}\frac{\partial\bar{\rho}_{A2}}{\partial x}, \quad \overline{v_y'\rho_{A2}'} = -\mathscr{D}_{A2y}^{(t)}\frac{\partial\bar{\rho}_{A2}}{\partial y}, \quad \text{and} \quad \overline{v_z'\rho_{A2}'} = -\mathscr{D}_{A2z}^{(t)}\frac{\partial\bar{\rho}_{A2}}{\partial z} \tag{7.1-56a, b, c}$$

where the $\mathscr{D}_{A2}^{(t)}$'s are effective eddy-diffusion coefficients. With the respective gradients the products represent the mixing conditions averaged over a tidal cycle and will have different physical meaning from those which result using shorter averaging times. Although it is possible to measure the contributions of the advection terms, the eddy-diffusion terms are all unknown. Consequently, considerable modification of Eq. 7.1-56 is necessary before it can be used in realistic situations.

Lawler offers the following equation for channels of constant cross section:

$$\frac{\partial\bar{\rho}_{A2}}{\partial t} + \bar{v}_x\frac{\partial\bar{\rho}_{A2}}{\partial x} = \frac{\partial}{\partial x}\left(\mathscr{D}_{A2x}^{(t)}\frac{\partial\bar{\rho}_{A2}}{\partial x}\right) - k'''\bar{\rho}_{A2} \tag{7.1-57}$$

Here it is particularly bothersome, in light of the two-layer estuarine model presented above, to discard the term for vertical, eddy-diffusion transport. In Eq. 7.1-57 both vertical, turbulent diffusion and saltwater entrainment must be accounted for with $\mathscr{D}_{A2x}^{(t)}$. At this juncture it becomes little more than an adjustable parameter. Often it is treated as a constant and calibrated to a particular estuary or portion of an estuary. Nevertheless, with this interpretation it has useful application in water quality and toxic pollutant modeling.

Steady-State Profile for a Reactive, Degrading Pollutant. Here it will be assumed that the waste stream enters as a single plane source and is discharged continuously into a tidal river of infinite length and constant cross-sectional area A. For convenience the waste can be visualized to enter some point between transects 2 and 3 in Fig. 7.1-17*b*. Lawler[21] developed the following equation specifically for waste materials that exert a biochemical oxygen demand (i.e., BOD) on the receiving stream (see p. 18); however, the resulting model has many more applications. Equation 7.1-57 becomes

$$D_{2x}\frac{d^2\bar{\rho}_{A2}}{dx^2} - \bar{v}_x\frac{d\bar{\rho}_{A2}}{dx} - k_{A2}'''\bar{\rho}_{A2} = 0 \tag{7.1-58}$$

where D_{2x} is the longitudinal dispersion coefficient. Analysis of the salinity profile with the appropriate solution of Eq. (7.1-58) yields a useful method for estimating D_{2x} in the saline-intruded reach of an estuary (see Problem 7.1R).

Since longitudinal dispersion transport is the same order of magnitude as advective transport, the incoming waste moves up-river. The overbars denoting average values of variables will be eliminated at this point to simplify the nomenclature. Two domains of solution to Eq. 7.1-58 are required; these are $+x$ (downstream) and $-x$ (upstream) from the waste entry point. Four boundary conditions are therefore necessary:

$$\rho^{\text{II}}_{A2}(\infty) = 0 \tag{7.1-58a}$$

constituent disappears before leaving the estuary,

$$\rho^{\text{I}}_{A2}(-\infty) = 0 \tag{7.1-58b}$$

constituent disappears before reaching the upper end of the estuary,

$$\rho^{\text{I}}_{A2}(0) = \rho^{\text{II}}_{A2}(0) \tag{7.1-58c}$$

constituent concentration is continuous across the plane of discharge, and

$$w_A = D_{2x} A \left[\frac{d\rho^{\text{I}}_{A2}(0)}{dx} - \frac{d\rho^{\text{II}}_{A2}(0)}{dx} \right] \tag{7.1-58d}$$

for the material balance across the plane of discharge. To reduce the number of parameters the equation and boundary conditions can be written in dimensionless form. Lawler[21] chose $P \equiv \rho_{A2}/(w_A/A\sqrt{k'''_A D_{2x}})$ for concentration and $\chi \equiv x\sqrt{k'''_A/D_{2x}}$ for distance; this gave rise to the parameter $N \equiv v_x/\sqrt{k'''_A D_{2x}}$. The solution for concentration in the up-river section (I) and the down-river section (II) is

$$\frac{P^{\text{I}}}{P^{\text{II}}} = \frac{1}{\sqrt{N^2+4}} \exp\left[\frac{(N \pm \sqrt{N^2+4})\chi}{2} \right] \tag{7.1-59}$$

where (+) is for the up-river section and (−) is for the down-river section, x being the distance from the discharge point with negative values and positive values, respectively. A graphical solution of Eq. 7.1-59 for four values of N is given in Fig. 7.1-19.

Example 7.1-4. BOD Profile in Estuary. A certain estuary has $k'''_A = 0.5$/per day for O_2 demanding (i.e., biochemical oxygen demand) waste, $D_{2x} = 3\text{E}6$ cm^2/s, $v_x = 8.33$ cm/s, and $w_A/A = 0.831$ g O_2/cm$^2 \cdot$day (0.17 lb/ft$^2 \cdot$day). Estimate the maximum concentration and at $\chi = \pm 1$.

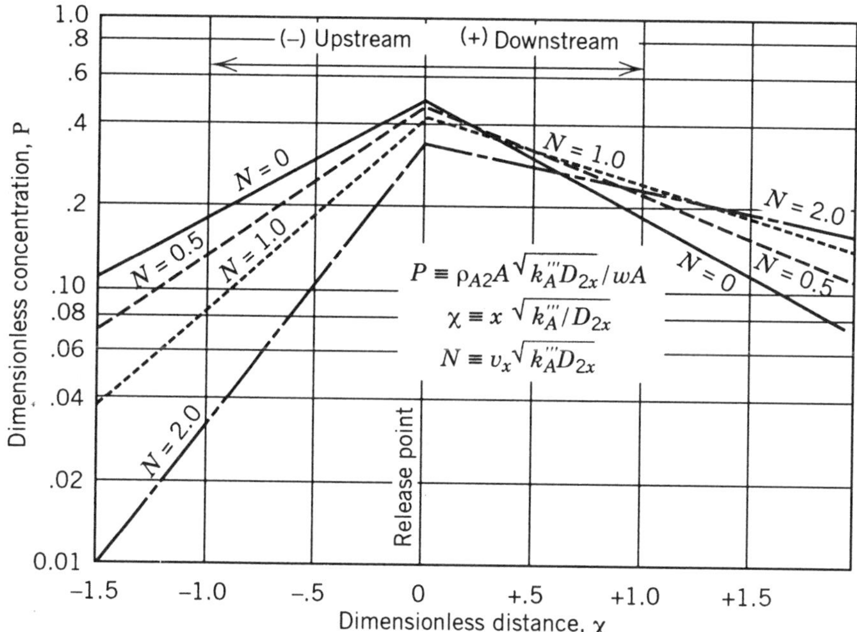

Figure 7.1-19. Pollutant concentration profiles in a partially mixed estuary. (From Ref. 21.)

SOLUTION From Fig. 7.1-19 the group

$$\sqrt{k_A''' D_{2x}} = \sqrt{\left(\frac{0.5}{24 \cdot 3600 \text{ s}}\right)(3\text{E}6 \text{ cm}^2/\text{s})} = 4.17 \text{ cm/s}$$

$$N = \frac{v_x}{\sqrt{k_A''' D_{2x}}} = \frac{8.33}{4.17} = 2.0$$

With these at $\chi = -1$, $P^{\text{I}} = 0.032$, and at $\chi = +1$, $P^{\text{II}} = 0.23$. $w_A/A = 831$ mg/cm^2 $(24 \cdot 3600\,\text{s}) = 0.00962$ mg/cm$^2 \cdot$s.

$$\rho_{A2} = \frac{P(w_A/A)}{\sqrt{k_A''' D_{2x}}} = P\left(\frac{0.00962 \text{ mg/cm}^2 \cdot \text{s}}{4.17 \text{ cm/s}}\right)$$

$$= P(2.3 \text{ mg/L})$$

$$x = \chi \sqrt{\frac{D_{2x}}{k_A'''}} = \chi(7200 \text{ m})$$

So upstream at $x = -7200$ m, $\rho_{A2} = 0.074$ mg/L, and downstream $x = +7200$ m, $\rho_{A2} = 0.53$ mg/L. At $x = 0$ and $N = 2$, $P^{\text{I}} = P^{\text{II}} = 0.35$, giving $\rho_{A2}(\text{max.}) = (0.35)(2.3 \text{ mg/L}) = 0.81$ mg/L.

Certain limiting forms of Eq. 7.1-59 are useful. For example, when advection is small in comparison to dispersive transport, Eq. 7.1-59 can be transformed to

$$\begin{matrix}\rho^{\mathrm{I}}_{A2}\\ \rho^{\mathrm{II}}_{A2}\end{matrix} = \frac{w_A}{2A\sqrt{k'''_A D_{2x}}} \exp\left(\pm x\sqrt{\frac{k'''_A}{D_{2x}}}\right) \qquad (7.1\text{-}60)$$

The profiles in this case are symmetrical about the release point. In the general case, as in Example 7.1-4, concentration decreases rapidly upstream but not as rapidly downstream. In the upstream region advection opposes dispersion, while downstream the two mechanisms are additive, thereby keeping the concentration high.

For the case where advective transport dominates, the limiting result is

$$\rho^{\mathrm{I}}_{A2} = 0 \qquad x < 0$$

$$\rho^{\mathrm{II}}_{A2} = \frac{w_A}{Av_x}\exp\left(-\frac{xk'''_A}{v_x}\right) \qquad x \geqslant 0 \qquad (7.1\text{-}61)$$

This result is for a plug flow river and was the assumption for the original Streeter–Phelps model presented in Chapter 1 (see p. 28 and Problem 1.2A).

Other uses of this model include the following. Since the processes result in a linear ordinary differential equation, the computed concentrations based on release points separated by a distance $\pm L$ can be added to give the total combined effect. The quotient ${}^1K'_{A2}/h$ can replace k'''_A or be added to it for evaporation if the constituent is volatile. The student is encouraged to find other applications that are consistent with the assumptions, boundary conditions, and mathematics of the model. It should be noted that the model is not applicable to the salt in the estuary because $k'''_A = 0$, (but see Problem 7.1R).

Steady-State Profiles for Dissolved Oxygen. (The following is abstracted from Lawler.[22]) The model for oxygen-demanding substances (i.e., BOD) has a corresponding model for dissolved oxygen (DO). Just as was done in developing the previous model, the DO analysis of a partially mixed estuary begins with the time-smoothed equation of continuity for a single species, Eq. 3.2-39*a*. A similar development with the same assumptions plus a term for oxygen transport from air to water yields the following steady-state counterpart to Eq. 7.1-58:

$$D_{2x}\frac{d^2\Delta_0}{dx^2} - v_x\frac{d\Delta_0}{dx} - \frac{{}^1k'_{02}}{h}\Delta_0 + k'''_A\rho_{A2} = 0 \qquad (7.1\text{-}62a)$$

where $0 \equiv 0_2$ and Δ_0 is the DO deficit given by Eq. 1.2-6. The boundary conditions are:

$$1.\ \Delta_0^{\mathrm{I}}(-\infty) \neq \infty \tag{7.1-62b}$$

$$2.\ \Delta_0^{\mathrm{II}}(+\infty) \neq \infty \tag{7.1-62c}$$

$$3.\ \Delta_0^{\mathrm{I}}(0) = \Delta_0^{\mathrm{II}}(0) \tag{7.1-62d}$$

$$4.\ \int_{-\infty}^{+\infty} k_A''' \rho_{A2} A\, dx = \int_{-\infty}^{+\infty} \frac{{}^1k_{02}'}{h} \Delta_0 A\, dx \tag{7.1-62e}$$

The first two conditions state that the DO deficit is finite at both extremes, $x = \pm\infty$, of the estuary. The third states that the DO deficit is continuous across the transect of waste discharge. The fourth states that at steady state, the oxygen consumption rate is equal to the rate oxygen is transferred from air to water. Details of the mathematical solution of Eq. 7.1-62 are given by Lawler.[22] The profile equations for the upstream section (I) and downstream section (II) are

$$\begin{matrix}\Delta_0^{\mathrm{I}} \\ \Delta_0^{\mathrm{II}}\end{matrix} = \frac{w_A k_A'''}{Q(k_A''' - {}^1k_{02}'/h)} \left\{ \frac{\exp[(v_x/2D_{2x})(1 \pm C_0)x]}{C_0} - \frac{\exp[(v_x/2D_{2x})(1 \pm C_A)x]}{C_A} \right\} \tag{7.1-63}$$

where

$$C_0 \equiv \sqrt{1 + \frac{4\mathscr{D}_{2x}\,{}^1k_{02}'}{hv_x^2}} \quad \text{and} \quad C_A \equiv \sqrt{1 + \frac{4\mathscr{D}_{2x} k_A'''}{v_x^2}}$$

When longitudinal dispersion is negligible ($D_{2x} = 0$) the equation reduces to the classical Streeter–Phelps model, Eq. 1.2-9. Details of the procedure to obtain the equation for and the location of the maximum deficit are given by Lawler. These, however, can be determined numerically with Eq. 7.1-63 (see Problem 7.1S).

Turbulent Diffusion in Rivers and Estuaries. The general subject of turbulent diffusion is developed in Section 3.2. Specifically, Eq. 3.2-29 is the theoretical origin of many of our concepts of turbulence and suggests a statistical approach of quantifying and interpreting it through fluctuating velocity and concentration measurements. Much remains unknown about the process, so that many concessions and compromises are made along the way in applying the theory to real-world, turbulence-influenced chemodynamic processes. The type of justifications made in the application of Eq. 3.2-29 to the partially mixed estuary, beginning with Eq. 7.1-56 and ending with Eq. 7.1-63, is not atypical. However, there is a need to relate these pragmatic,

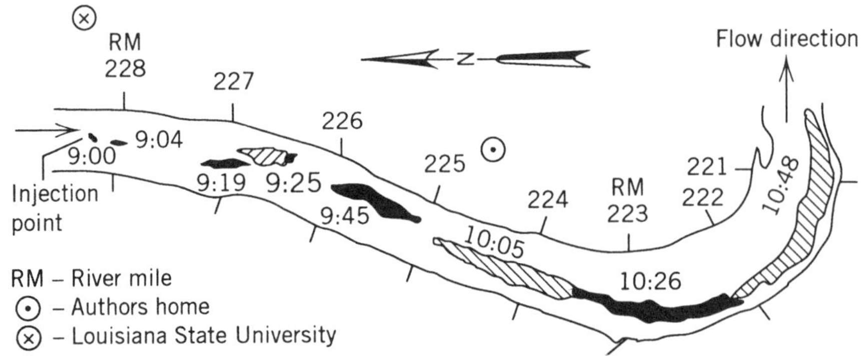

Figure 7.1-20. Tracer cloud experiment on the Mississippi River. (Reprinted with permission. From Ref. 24.)

adjustable diffusion-related parameters, such as D_{2x}, to more fundamental turbulent diffusion quantities such as $\mathscr{D}^{(t)}_{A2x}$, $\mathscr{D}^{(t)}_{A2y}$, and $\mathscr{D}^{(t)}_{A2z}$, with the ultimate goal of developing procedures and unified approaches of estimating these transport coefficients from easily measured parameters. With this goal in mind the following material was assembled for rivers and estuaries.

In 1974, McQuivey and Keefer performed a longitudinal dispersion experiment on the Mississippi River. At 9 A.M. on April 24 a dye solution of Rhodamine WT was released near the center of the channel at river mile 228 near Baton Rouge, Louisiana. Figure 7.1-20 shows the outline of the dye cloud from aerial photographs taken during the early period of the experiment. At the time the flow rate was 22,600 m^3/s, hydraulic radius 17.0 m, $v_* = 5.8$ cm/s, width 870 m, and average velocity 1.48 m/s. The movement of the peak concentration indicates a velocity of 1.4 m/s. The D_{2x} obtained from the experiment was 700 m^2/s. Several other field experiments are presented in detail.[24]

A nontidal river is a truly homogeneous estuary, but this is true of very few, if any, real estuaries. In a truly homogeneous estuary, one in which a uniform density is present at all time, there will be no gravitational convection, but dispersion can still occur. In this case, as with the Mississippi River example, the dispersion is caused by velocity shear due to the friction at the sediment bed. If one considers a patch of dye in a channel, the dye near the bed travels slower than the average and vertical diffusion will mix it to some extent throughout the water column. The dye near the surface travels faster than average but will be mixed downward. Consequently, the dye spreads out longitudinally as the patch moves along the channel, because of the variations in velocity with depth coupled with vertical eddy diffusion. The process is very pronounced in laminar flow, as in the Taylor–Aris dispersion in a capillary tube (see Eq. 3.3-27). Bends in rivers cause secondary circulation called *helical motion* which also aids dispersion.[24] The studies performed by Taylor[25] in 1954 on the dispersion of matter in turbulent flow through pipes is probably the origin of our concepts of longitudinal dispersion. The science is well advanced and Ref. 24 is a good review for nontidal rivers and a source of algorithms to predict traverse mixing and longitudinal dispersion. Reference 27 contains information on vertical dispersion.

Because of the presence of salinity stratification the dispersion in estuaries is different and more complex. Salinity profiles on the Hudson river yield a D_{2x} value of 318 m^2/s based on the data in Problem 7.1R. Another reported observation is 600 m^2/s. The Barataria Bay area, located on the Louisiana Gulf coast, is a large (40 km by 40 km) shallow (1 to 2 m depth) estuary with very little freshwater input but has 25‰ salinity, due to the proximity of the mouth of the Mississippi River. Due to a lack of sufficient data, Trivedi[23] selected dispersion coefficients for the dynamic two-dimensional vertically integrated model of the bay on a basis that gave "'stable and acceptable solutions to the transport equations." For the constituents: coarse detritus, fine detritus, animal biomass, and phytoplankton biomass the value was 5.8 m^2/s, and for dissolved NH_3-N, nitrite-nitrate-N, and organic nitrogen the value was 20 m^2/s. These were $D_{2x} = D_{2z}$ values.

The above are specific examples of estuaries and the ranges of D_{2x} and D_{2z} values encountered. Algorithms exist for estimating these parameters in unstratified streams.[27] Compiled observations on D_{2x} values for 15 tidal rivers and estuaries reported by Lawler et al.[21] range from 15 to 150 m^2/s. The data are displayed as a function of net nontidal flow. Others have observed that it is noticeable how D_{2x} decreases as river flow decreases. Thomann[26] summarized D_{2x} values for 14 estuary experiments and the range covers 6.0 to 720 m^2/s. He presents two formulas, the four-thirds law and the random process law, which correlates D_{2x} with the maximum tidal velocity. Various experimental estimation techniques are used in the field measurements. Thomann notes that the dispersion coefficient, in general, decreases toward the mouth of the estuary. The effect being ascribed to the reduced effects of mixing is caused by steep salinity gradients. Until sufficient field data are available to relate D_{2x} to theoretical models that combine tidal velocity, density gradients, and geometric parameters values of this parameter adjusted to site-specific concentration, field measurements remain the best way to quantify dispersion unique to a particular estuary.

Horizontal Diffusion in Oceans and Large Lakes. Just as in the case of rivers and estuaries, wastewater containing particulates and/or substances in solution will be disposed in oceans and lakes. In this subsection, the ideas and methods developed to predict concentration of selected species are extended and applied to water bodies, where they disperse horizontally into a two-dimensional layer.

The problem of dispersion arises in the context of oceanic disposal of sewage or sludge of domestic or industrial aqueous waste. A usual method of disposal is to convey the aqueous waste through a submarine pipe to a point some distance offshore in a large lake, sea, or ocean and release it there through a system of diffuser ports. The ports are designed to provide considerable mixing

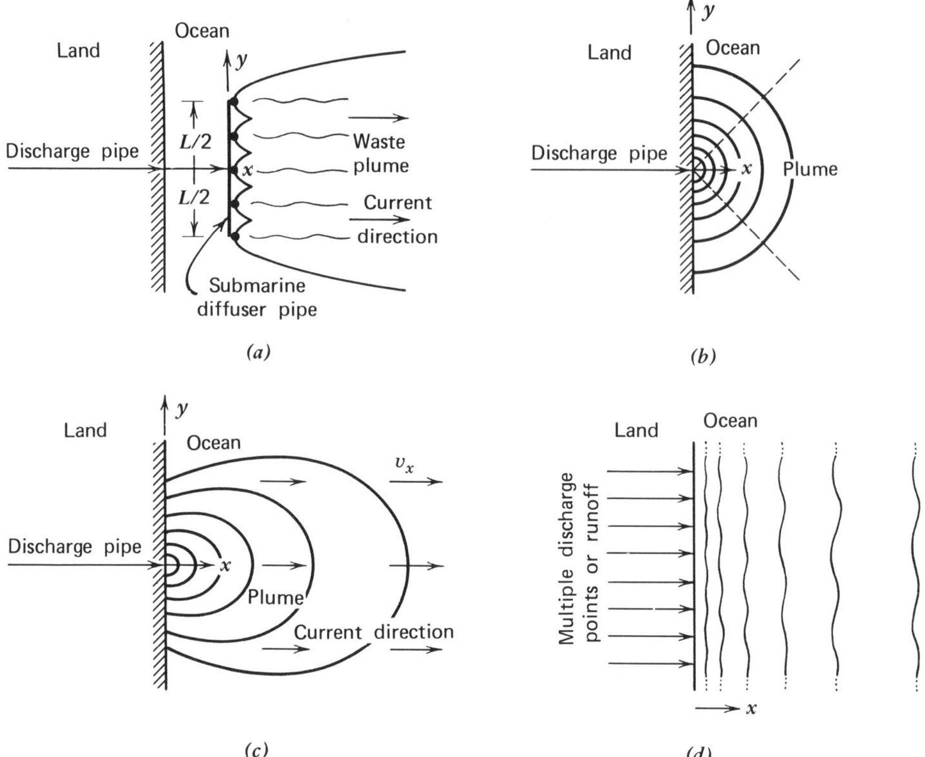

Figure 7.1-21. Pollutant dispersion models for ocean or lake outfalls: (*a*) offshore submerged diffuser; (*b*) point source without advection; (*c*) point source with advection; (*d*) distributed source.

with seawater or lake water (see Fig. 7.1-21). The net result is that the waste is distributed in some initial concentration ρ_{A2}^{0} over a considerable volume, which then drifts along with the lake or ocean current and mixes with its surroundings under the influence of the turbulence naturally present. A theory and mathematical models are desired to predict the dispersion or natural dilution processes of the waste plume.

Diffusion of Aqueous Waste Plumes. The diffusion of aqueous waste plumes occurs mainly in the horizontal direction (i.e., $x-y$ plane). In such applications vertical diffusion in the z direction makes little contribution to the dilution of pollutants. Such is the case, for example, when the initial cloud extends from the sea or lake bottom to the free surface. Even if the depth of water increases in the direction in which the pollutant cloud drifts, so that its vertical size increases, this does not imply vertical mixing. Often, the aqueous waste has some positive buoyancy and "boils" to the surface, effectively distributing itself over an available vertical depth h_2 during the initial mixing phase. The dramatic effect of increased density stratification, or increased buoyancy, in reducing vertical mixing is displayed in Fig. 7.1-8. In other situations a stable density interface, such as the thermocline, stops downward or upward vertical mixing beyond a short initial phase, much as an inversion lid does in the atmosphere. The problems to be addressed and the models presented are therefore concerned with the horizontal spread of pollutants in a surface layer of constant depth h_2 through oceanic or lake turbulence. Means of estimating h_2 will not be covered.

Brooks[16] developed a useful theoretical model of oceanic diffusion arising out of work on aqueous waste disposal at sea. The model assumes that the waste "boils" to the surface after having been mixed with seawater by a diffuser several hundred meters in length lying on the sea bottom (see Fig. 7.1-21*a*). The effluent is forced through many ports of the diffuser, forming a number of jets that mix with the ambient fluid until the momentum is dissipated. The size of the initial waste field is thus fairly large horizontally and is well mixed from surface to bottom.

Since currents near the shore usually follow the depth contours, an adequate model assumes that the depth h_2 of the waste plume remains constant and that its further dilution is caused by lateral diffusion alone. The relevant form of the diffusion equation is

$$v_x \frac{\partial \rho_{A2}}{\partial x} + k_A''' \rho_{A2} = \frac{\partial}{\partial y}\left(\mathscr{D}_{A2y}^{(t)} \frac{\partial \rho_{A2}}{\partial y}\right) \tag{7.1-64}$$

where v_x is the current velocity in the x direction and $\mathscr{D}_{A2y}^{(t)}$ is the eddy diffusivity in the y direction. An advantage of using the equation of continuity is that one may model nonconservative effects, such as die-off of bacteria (in a field of diffusing waste) or flocculation, radioactive decay, and so on. In many such cases the rate of loss of the diffusing substance (e.g., bacteria) is proportional to their concentration (Eq. 1.2-2), the factor of proportionality being a decay constant of dimensions s^{-1} and denoted by k_A'''.

A useful equation capable of mapping the concentration field of chemical A in the waste plume has been presented by Csanady:

$$\rho_{A2} = \frac{1}{2}\rho_{A2}^0 \exp\left(\frac{-k_A''' x}{v_x}\right)\left[\operatorname{erf}\left(\frac{L/2 + y}{\sqrt{2}\sigma_y}\right) + \operatorname{erf}\left(\frac{L/2 - y}{\sqrt{2}\sigma_y}\right)\right] \tag{7.1-65}$$

where ρ_{A2}^0 is the initial concentration of chemical A after mixing by the diffuser and L is the diffuser length shown in Fig. 7.1-21*a*. The eddy diffusivity was converted to a variance by the relation

$$\mathscr{D}_{A2y}^{(t)} = \frac{v_x}{2}\frac{d\sigma_y^2}{dx} \tag{7.1-66}$$

$\mathscr{D}_{A2y}^{(t)}$ and hence σ_y is a function of x and not of y.

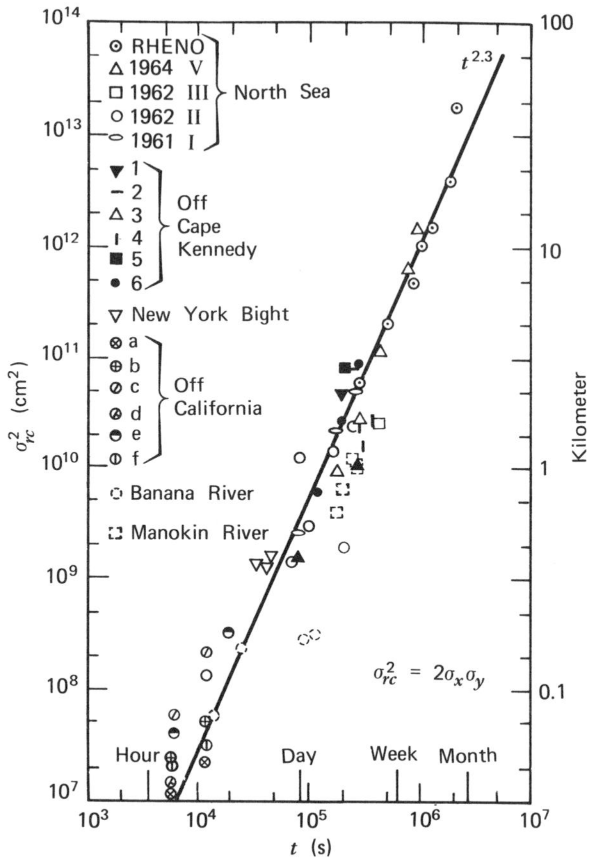

Figure 7.1-22. Diffusion diagram for variance versus diffusion time $\sigma_{rc}^2 \equiv 2\sigma_x\sigma_y$. (Reprinted by permission from Ref. 13.)

Practical estimates of the concentration field may therefore be arrived at if we have data on the decay constant k_A''' and on the point source standard deviation σ_y. In large-scale oceanic diffusion problems we may use Okubo diagrams (Fig. 7.1-22) to estimate the behavior of σ_y. The time dependence of σ_y is obtained in a continuous plume from the simple relation $x = v_x t$.

Point Sources of Pollutants Entering Lakes or Oceans. A significant factor of water quality in lakes and similar bodies of water is the dispersion of chemicals and particulates in the vicinity of a river discharge or a wastewater discharge, as shown in Fig. 7.1-21*b* and (*c*). The constituents in the incoming wastewater are initially diluted by the turbulence of the discharging stream into a portion of the lake water. After this initial phase, further dilution occurs due to the natural mixing processes in the lake, as the incoming mass spreads laterally or longitudinally in the horizontal plane. As pointed out in the previous discussion, the analysis is reduced to a two-dimensional problem because of rapid but limited dispersion in the vertical direction.

POINT SOURCES WITH DIFFUSION AND ADVECTION. When a wind-driven or offshore current is significant, an advective term must be included for point sources. In this case the concentration pattern is distorted by a projection along the axis of the wind velocity as shown in Fig. 7.1-21*c*. Thus elements of fluid containing higher concentration of pollutants are moved greater distances from the shore. To simplify the basic differential equation, the assumption is made that dispersion is negligible along the x axis because of the advection in that direction.

Numerous experiments have been performed to observe the horizontal dispersion in large bodies of water. The usual technique is to release a marked fluid such as rhodamine B dye and follow the spreading process by crisscrossing the plume in a boat and obtaining samples that are subsequently analyzed for the dye content. In a large body of water the horizontal diffusion coefficients are generally governed by the $\frac{4}{3}$ *power law*, that is, the horizontal diffusion coefficient is proportional to the $\frac{4}{3}$ power of the length scale of the diffusing patch or plume:

$$\mathscr{D}_{A2y}^{(t)} = A_L L^{4/3} \tag{7.1-67}$$

where A_L is a dissipation parameter of dimensions $cm^{2/3}/s$ or $ft^{2/3}/sec$ and L is the width of the plume, usually taken to be $4\sigma_y$. The plume illustrated in Problem 3.1F displays this behavior.

In the ocean numerous field experiments have been performed to estimate $\mathscr{D}_{A2y}^{(t)}$. These data are summarized in Fig. 7.1-23. It can be seen that the values of A_L are in the neighborhood of E-2 to E-4 $ft^{2/3}/sec$. It should be pointed out that the field data include effects of shear currents. If these shear currents need to be excluded, the lower value of $A_L =$ E-4 may be more appropriate.

It was first reported by Richardson that the coefficient of lateral diffusion does not remain constant with distance but varies with the scale of the diffusion phenomena. Under these conditions, Eq. 7.1-64 is still applicable. By relating the variation of the lateral diffusion coefficient with the longitudinal dimension in accordance with the $\frac{4}{3}$ power relationships, Brooks[11] integrated the equation for $y = 0$ or the centerline of the plume, obtaining

$$\rho_{A2} = \rho_{A2}^{0} \exp\left(\frac{-k_A''' x}{v_x}\right) \operatorname{erf} \sqrt{\frac{3/2}{[1 + (8\mathscr{D}_{A20}^{(t)} x)/v_x L_0^2]^3 - 1}} \tag{7.1-68}$$

where $\mathscr{D}_{A20}^{(t)}$ is the initial diffusivity of the near-shore well-mixed cloud of outlet width L_0.

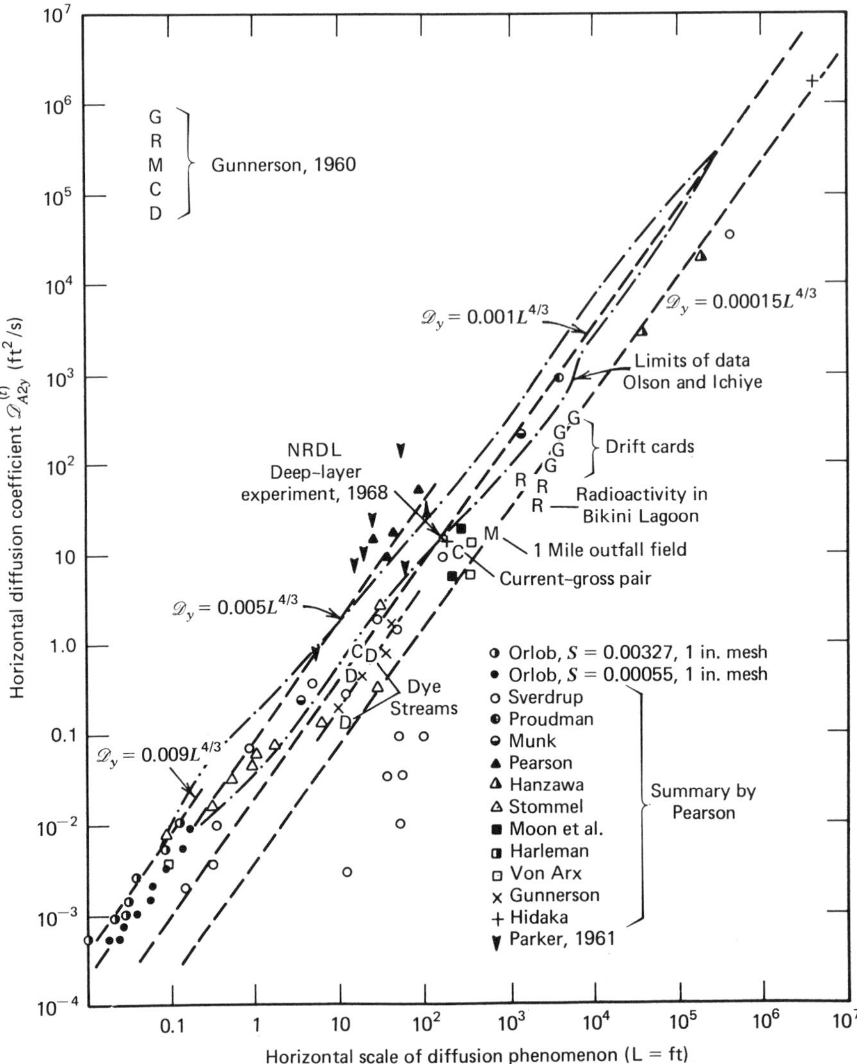

Figure 7.1-23. Horizontal diffusion coefficient as a function of horizontal scale. (From Ref. 7.)

Csanady suggests that in a nearly homogeneous field, plume growth is rather slower than suggested by Fig. 7.1-23 and may be calculated to a satisfactory approximation using a constant eddy diffusivity. He states further that experimental evidence on the behavior of sewage plumes in their early dispersal phase (which is practically the most important part) suggests an eddy diffusivity on the order 1E3 cm^2/s. It would seem to be safer to make such an assumption on the basis of engineering estimates of oceanic dispersal rather than using the more optimistic projections that may be taken from Fig. 7.1-23.

If the diffusion coefficient is assumed constant, Eq. 7.1-64 can be integrated and for $y = 0$ becomes

$$\rho_{A2} = \rho_{A2}^{0} \exp\left(\frac{-k_A'''x}{v_x}\right) \operatorname{erf} \sqrt{\frac{v_x L_0^2}{16\mathscr{D}_{A2y}^{(t)} x}} \tag{7.1-69}$$

where $\mathscr{D}_{A2y}^{(t)}$ is some well-chosen value of the horizontal diffusivity.

POINT SOURCES WITH DIFFUSION AND NO ADVECTION.[11] During some critical periods when there are no surface winds to create a surface current and the net water flow is practically absent, pollutant dispersion is solely by water turbulence. This situation is illustrated in Fig. 7.1-21*b*. Considering the surface plane, the steady-state distribution of concentration of a nonconservative substance may be described by the following equation:

$$0 = \mathscr{D}_{A2x}^{(t)} \frac{\partial^2 \rho_{A2}}{\partial x^2} + \mathscr{D}_{2Ay}^{(t)} \frac{\partial^2 \rho_{A2}}{\partial y^2} - k_A''' \rho_{A2} \tag{7.1-70}$$

The x and y axes are, respectively, perpendicular and parallel to the shore. Dispersion coefficients in both the longitudinal and lateral directions are assumed constant. If it is further assumed that the dispersion coefficients are equal in each direction, Eq. 7.1-70 becomes

$$0 = \mathscr{D}_{A2x}^{(t)} \left(\frac{\partial^2 \rho_{A2}}{\partial x^2} + \frac{\partial^2 \rho_{A2}}{\partial y^2}\right) - k_A''' \rho_{A2} \tag{7.1-71}$$

In addition to being appropriate for conditions in which there is no wind, Eq. (7.1-71) is appropriate for conditions in which upwelling and density difference are negligible.

Since the dispersion process is uniform in the x–y plane, Eq. 7.1-71 may be transformed to polar coordinates to yield

$$\frac{\partial^2 \rho_{A2}}{\partial r^2} + \frac{1}{r}\frac{\partial \rho_{A2}}{\partial r} + \frac{1}{r^2}\frac{\partial^2 \rho_{A2}}{\partial \theta^2} - \frac{k_A''' \rho_{A2}}{\mathscr{D}_{A2x}^{(t)}} = 0 \tag{7.1-72}$$

where r is the radial distance from the source and θ is the angle on either side of the x axis. If it is assumed that ρ_{A2} is constant for a given r, $\partial \rho_{A2}/\partial \theta$ and $\partial^2 \rho_{A2}/\partial \theta^2$ are zero. This assumption does not apply strictly in the vicinity of the shore, where reflection occurs and concentration patterns overlap, but applies to a better degree in a zone radiating from the discharge point and 45° on either side of the x axis. For this assumption the preceding partial differential equation reduces to the following ordinary differential equation:

$$\frac{d^2\rho_{A2}}{dr^2} + \frac{1}{r}\frac{d\rho_{A2}}{dr} - \frac{k_A'''\rho_{A2}}{\mathscr{D}_{A2x}^{(t)}} = 0 \tag{7.1-73}$$

which is a Bessel equation of zero order. The solution is

$$\rho_{A2} = C_1 I_0\left(\sqrt{\frac{k_A''' r^2}{\mathscr{D}_{A2x}^{(t)}}}\right) + C_2 K_0\left(\sqrt{\frac{k_A''' r^2}{\mathscr{D}_{A2x}^{(t)}}}\right) \tag{7.1-74}$$

where $I_0(x)$ and $K_0(x)$ are modified Bessel functions of the first and second kinds and C_1 and C_2 are constants.

To develop boundary conditions, it is assumed that the point source is replaced with an initial cloud of pollutant of concentration ρ_{A2}^0 and radius r_0. In the case of a river discharging into a lake, a reasonable estimate of r_0 is half the effective river width and ρ_{A2}^0 is the concentration of chemical A in the entering river water. The secondary boundary condition is that far away from the source; the concentration is zero. In summary, the two boundary conditions are $\rho_{A2} = \rho_{A2}^0$ at $r = r_0$ and $\rho_{A2} = 0$ at $r = \infty$. The solution of Eq. 7.1-74 is

$$\rho_{A2} = \rho_{A2}^0 \frac{K_0(\sqrt{k_A'''' r^2/\mathscr{D}_{A2x}^{(t)}})}{K_0(\sqrt{k_A''' r_0^2/\mathscr{D}_{A2x}^{(t)}})} \tag{7.1-75}$$

Values of the Bessel function $K_0(x)$ may be obtained from most mathematical handbooks.

Distributed Source with no Advection.[11] A distributed point source may be approximated by placing multiple point sources positioned side by side as shown in Fig. 7.1-21*d*. The ideal distributed source consists of an infinite number of point sources. Such sources arise when chemicals such as pesticides or fertilizers are flushed into a body of water at numerous points along the shoreline by runoff from the land surface and when chemicals enter water bodies by subterranean discharges such as pollutants in groundwater. If the extent of the source is wide enough to minimize lateral gradients, the working differential equation for a nonconservative chemical or substance is

$$0 = \mathscr{D}_{A2x}^{(t)} \frac{d^2\rho_{A2}}{dx^2} - k_A'''\rho_{A2} \tag{7.1-76}$$

At the shoreline the concentration of chemical A is the maximum, and well away from the shoreline the concentration is zero. This yields the boundary conditions $\rho_{A2} = \rho^0_{A2}$ at $x = 0$ and $\rho_{A2} = 0$ at $x = \infty$. The solution of Eq. 7.1-76 is

$$\rho_{A2} = \rho^0_{A2} \exp\left(- \sqrt{\frac{k'''_A x^2}{\mathscr{D}^{(t)}_{A2x}}} \right) \tag{7.1-77}$$

This model may be used to describe the steady-state surface concentration due to a distributed input along a shore. It should be noted that ρ^0_{A2} is the near-shore concentration of A after the source has been mixed with a quantity of lake water and is not necessarily the concentration of the source.

Closure. The models presented in the preceding development are neither necessarily definitive nor final descriptions of the various phenomena. These simple equations do provide a first-order approximation of the dispersion process and provide a backdrop upon which to correlate field data in that they do indicate the significance of the various major processes that include diffusion, advection, and reaction.

Example 7.1-5. Bacterial Distribution in Lake Michigan in the Vicinity of the Indiana Harbor. Brooks et al.[11] obtained field observations of bacterial concentrations in Lake Michigan in the vicinity of the point source of the Indiana Harbor. Three sets of data taken along radii located on the x-axis and 45° on either side of this axis for the months of June, July, and September appear in Fig. E7.1-5.

The diameter of the initial field is taken at 300 ft, which is the order of the outlet width. The dispersion coefficient has a value of 1 mi²/day. Available data on the value of the bacteria decay coefficient k'''_A range from 0.5 to 3.0 per day. Since the wind velocity during this period was less than 1 mi/hr, use the appropriate model and calculate relative bacteria concentrations for the range of decay constants given.

SOLUTION This particular situation can be modeled best with the point source model without advection. The radius of the initial field is 150 ft. Equation 7.1-75 for $\rho^0_{A2} = 100$ and $k'''_A = 0.5$ per day becomes

$$\rho_{A2} = 100 \frac{K_0\left(\sqrt{\dfrac{0.5}{d} \left| \dfrac{r^2\,\text{mi}^2}{} \right| \dfrac{d}{1.0\,\text{mi}^2}} \right)}{K_0\left(\sqrt{\dfrac{0.5}{} \left| \dfrac{150^2}{5280} \right| \dfrac{}{1.0}} \right)}$$

$$= 100 \frac{K_0(\sqrt{0.5r^2})}{K_0(0.02)}$$

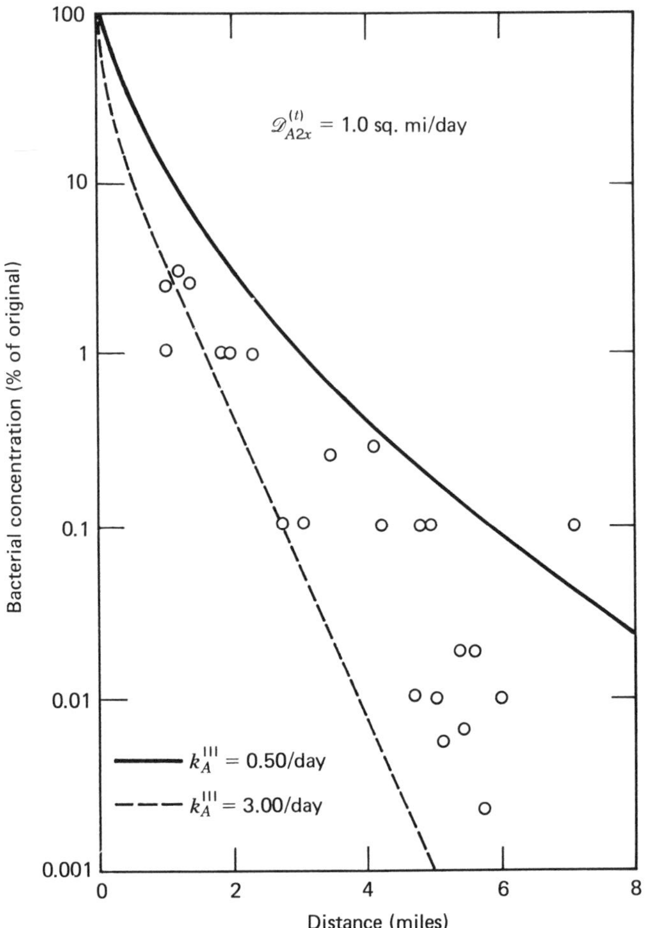

Figure E7.1-5. (From Ref. 11.)

From a table of Bessel functions, $K_0(0.02) = 3.908$. The following calculations appear as a function of distance from the source, r miles.

r (mi)	$\sqrt{0.5r^2}$	$K_0(\sqrt{0.5r^2})$	ρ_{A2}
1	0.707	0.654	16.7
2	1.41	0.240	6.14
4	2.83	0.0425	1.09
6	4.24	0.00852	0.22
8	5.66	0.00181	0.05

For $k_A''' = 3.0$ per day we have the following calculations: $K_0(0.0492) = 3.012$

r (mi)	$\sqrt{3.0r^2}$	$K_0(\sqrt{3r^2})$	ρ_{A2}
1	1.73	0.159	5.29
2	3.46	0.0204	0.68
4	6.93	0.459-3	0.15E-1
6	10.4	0.119E-4	0.40E-3

The model results are shown plotted in Fig. E7.1-5. It is apparent that no significant error is introduced by omitting the advective component, and the range of k_A''' values is sufficient to bracket the observations.

PROBLEMS

7.1A. Thermal Structure of Keowee Reservoir

Keowee Reservoir is an artificial, multipurpose lake located in the southeastern United States. Table 7.1A contains thermal profile data for the reservoir in 1972.

1. Graph the annual development of the temperature profile. Plot 12 months of data on a single sheet and label each month. Place depth on the ordinate (y axis) with zero depth at the top and place temperature on the abscissa (x axis).
2. Study the graph of the thermal history very carefully and choose a date for model time $t = 0$. Also choose a value for T^0.
3. Compute a $\alpha_2^{(t)}$ value of each depth and time except for the months of November, December, January, and February, which are nearly isothermal. Obtain a monthly average. Obtain a yearly average. Report all values in m^2/day.
4. Review the $\alpha_2^{(t)}$ values in the table and explain the likely causes of the variations in a $\alpha_2^{(t)}$ with time and with depth. Do the numerical values of $\alpha_2^{(t)}$ fall within the range for diffusion coefficients reported for the lakes in Fig. 7.1-7 and the data in Fig. 7.1-8.

7.1B. Thermocline

A dictionary of scientific and technical terms gives the following definitions of thermocline: “(1) a temperature gradient in a layer of seawater, in which the

Table 7.1A. Keowee Reservoir Temperatures, 1972

Depth (m)	Temperature (°C) on:											
	Jan 27[a] 27[b]	Feb. 22 53	Mar. 30 89	Apr. 18 108	May 16 136	June 22 173	July 19 200	Aug. 22 234	Sept. 17 262	Oct. 27 300	Nov. 22 326	Dec. 20 354
0	10.4	8.0	12.4	19.1	22.2	24.4	28.4	28.3	26.8	17.8	13.9	10.8
1	10.3	8.0	12.1	18.7	21.8	24.4	27.9	28.3	26.8	17.9	13.9	10.7
2	10.1	8.0	11.8	18.0	21.4	24.2	28.0	28.0	26.6	17.9	13.9	10.5
3	10.0	7.9	11.6	17.4	21.2	24.2	27.9	28.0	26.6	17.7	13.8	10.5
4	9.9	7.9	11.4	17.0	21.1	23.7	27.4	28.0	26.5	17.8	13.8	10.5
5	9.9	7.8	11.2	16.4	20.7	23.1	26.2	27.5	26.4	17.8	13.8	10.5
6	9.8	7.8	11.0	16.0	19.3	21.8	23.6	25.6	25.0	17.8	13.8	10.5
7	9.8	7.8	10.8	15.2	17.1	20.6	21.4	22.2	22.2	17.8	13.8	10.5
8	9.8	7.8	10.6	14.7	15.6	19.2	19.3	20.0	20.1	17.8	13.8	10.5
9	9.7	7.8	10.4	13.7	14.6	17.6	17.9	18.5	18.8	17.6	13.8	10.5
10	9.7	7.8	10.2	12.9	14.1	15.8	16.8	17.6	17.8	17.4	13.7	10.5
11	9.7	7.7	10.0	12.1	13.2	14.8	15.9	16.8	17.3	16.9	13.7	10.5
12	9.7	7.7	9.7	11.6	12.7	14.1	15.0	16.1	16.6	16.5	13.7	10.5
13	9.7	7.7	9.5	11.1	12.1	13.2	14.1	15.0	15.8	16.3	13.7	10.5
14	9.6	7.7	9.2	10.7	11.6	12.2	13.2	14.0	14.8	15.6	13.7	10.5
15	9.6	7.7	8.9	10.4	11.3	12.0	12.4	13.2	14.0	14.9	13.7	10.4
20	9.5	7.7	8.3	9.3	9.9	10.6	10.6	10.9	11.4	11.8	12.5	10.3
25	9.5	7.6	8.2	8.9	9.4	9.7	9.8	9.9	10.7	10.4	10.6	10.3
30	9.5	7.6	8.1	8.7	9.1	9.5	9.3	9.6	9.8	10.0	10.0	10.1
35	9.6	7.6	7.8	8.7	8.8	–	9.1	9.3	9.5	10.0	10.0	10.3

Source: J. H. Elrod, SERT U.S. Government, Clemson, S.C. March 1975, private communication.

[a]Date.

[b]Julian day.

temperature decrease with depth is greater than that of the overlying and underlying water, also known as a metalimnion; (2) a layer in a thermally stratified body of water in which such a gradient occurs."

It is not uncommon to refer to the zone between the hypolimnion and the epilimnion as the thermocline or metalimnion. A more precise definition is needed from a mathematical point of view. A common mathematical definition of the thermocline is the point where the thermal gradient $\partial T/\partial y$ is a maximum. The maximum in the gradient occurs at those points where its derivative is zero; therefore, $\partial^2 T/\partial y^2 = 0$. Apply this criterion to Eq. 7.1-5 to obtain Eq. 7.1-6.

7.1C. Lake Water Column Temperature Profiles via the Semi-infinite Solid Model

Using the result of the semi-infinite solid model, compute the temperature profiles for the following reservoirs and compare them with the actual profiles in Fig. 7.1-7. Prepare a graph of each for comparison purposes. Use the data of Table 7.1C.

Table 7.1C

	Temperatures (°C)		
Lake or reservoir	Surface Water	Bottom Water	Turbulent Diffusivity (cm^2/s)
Tahoe	14.8	5.5	1.14
Castle	19.0	4.5	0.53
Hungry Horse	18.0	4.0	0.99

7.1D. Water Density and Stratification

1. Make a graphical plot of the density of water (g/cm^3) as a function of temperature for the range of 0 to 30°C based on the information in Appendix D.

2. Study the Shagawa Lake thermal profile data for the months of December, January, February, and March of 1973 given in Problem 5.2A. Based on the effect of temperature on water density, explain how a lake can "reverse stratify" in the winter.

3. Study the Hungry Horse Reservoir thermal profile data shown in Fig. 7.1-2. Based on the effect of temperature on water density, explain how a lake can stratify in the summer.

7.1E. Surface Temperature Expression

If Eq. 7.1-3 is used, show that the temperature may be expressed in terms of the surface temperature as follows:

$$T(y, t) = T^0 + \{T(0, t) - T^0\} \exp\left(\frac{-y^2}{4\alpha_2^{(t)} t}\right) - \frac{1}{(\alpha_2^{(t)}\pi)^{1/2} \rho_2 \hat{C}_{p_2}} \cdot \int_0^t \frac{y^2}{4\alpha_2^{(t)} \tau^2} \cdot \exp\left(\frac{-y^2}{4\alpha_2^{(t)} \tau}\right) \left[\int_0^\tau q_{12}(t - \tau') \frac{d\tau'}{\sqrt{\tau'}}\right] d\tau \quad (7.1E)$$

where the positive terms on the right is Eq. 7.1-5 and the negative term is the temperature error committed in its use.

7.1F. Lake Surface Reaeration Coefficients

From the thermal data presented for the lakes in Fig. 7.1-7 and Table 7.1C, estimate the surface reaeration coefficients (cm/h) for the epilimnion region. Do the calculated values agree with the range of values reported in the Table 7.1-2? Discuss pitfalls in using thermal data for mass transfer coefficient estimation.

7.1G. Rate-Limiting Diffusion Processes at Depths in Water Bodies

When the topic of evaporation of chemicals from a lake ecosystem was considered on page 204, it was mentioned that other rate-limiting diffusion processes at depths in water bodies (e.g., the thermocline) may be as significant as or more significant than those near the air–water interface.

1. Show for a stratified lake or ocean that

$$\frac{1}{{}^1K'_{A2}} = \frac{1}{{}^1k'_{A2}} + \frac{c_2}{c_1 H_{Ax}\, {}^2k'_{A1}} + \frac{h_e}{\mathscr{D}^{(t)}_{A2e}} + \frac{h_m}{\mathscr{D}^{(t)}_{A2m}} + \frac{h_h}{\mathscr{D}^{(t)}_{A2h}} \quad (7.1G)$$

where the last three terms are, respectively, the epilimnion (e), metalimnion (m), and hypolimnion (h) resistances. This equation is applicable for the transport of a chemical from the air through the water column to the water–mud interface or for the reverse direction.

2. Table 4.2-1 contains evaporation parameters for various compounds from lakes. Use Eq. 7.1G to evaluate the significance of stratification on the total mass transfer coefficient with the data for the three reservoirs shown in Fig. 7.1-7. Report all resistances in hr/m. Are the other rate-limiting diffusion processes at depths in water bodies significant? Discuss your answer.

7.1H. Movement of Pesticide Through the Thermocline of Reservoir

Employing the problem statement of Example 7.1-2 and the results of model II, reestimate the maximum concentration likely to be observed in the hypolimnion of Beaver Reservoir.

7.1I. Steady-State Phosphorus Profile in Lakes

Phosphorus as orthophosphate is released from the mud on a lake bottom, migrates through the hypolimnion, and is consumed by algal particles in the epilimnion.

1. Develop the model equations for a steady-state phosphorus profile.
2. Calculate concentration (mg/L) as orthophosphate versus distance (m) from the air–water interface for the conditions shown in Table 7.1I.

Table 7.1I

Layer	h (m)	$\mathscr{D}^{(t)}_{A2y}$ (m^2/day)	r_{A2} (g/m$^3\cdot$day)	Other Data
Epilimnion	$h_1 = 5$	1	0.00086	ρ_{A2} at $h_2 = 0.5$ mg/L
Hypolimnion	$h_2 = 20$	0.2	0.0	n_{Ay} at $h_2 = 0.0043$ g/m$^2\cdot$day

7.1J. Vertical Oxygen Profile in Stratified Lake

Consider a stratified lake that contains sufficient phytoplankton nutrients and CO_2 to produce oxygen via respiration at a rate

$$r_{A2} = 0.25 \exp(-y) \tag{7.1D-1}$$

with r_{A2} in g O_2/m$^3\cdot$day and y in meters from the air–water interface. For a water surface temperature of 30°C, compute and graph the steady-state, daylight, vertical oxygen concentration profile as a function of depth. Table 7.1J contains pertinent chemical and physical data for both layers of the lake.

Table 7.1J

Layer	$\mathscr{D}^{(t)}_{A2y}$ (m^2/day)	h (m)	Other Data
Epilimnion	1.0	5	${}^1k'_{A2} = 0.2$ m/day
Hypolimnion	0.2	20	$n_{Ah2} = 0.05$ g O_2/m$^2\cdot$day

7.1K. Vertical Oxygen Profile in Stratified Lake: Simple System

Consider a stratified lake containing little or no phytoplankton that produce oxygen and little or no organic and/or inorganic matter that consumes dissolved oxygen within the water column. For a water surface temperature of

Table 7.1K

Layer	$\mathscr{D}^{(t)}_{A2y}$ (m^2/day)	h (m)	Other Data
Epilimnion	1.0	10	$n_{Ah_1} = 0.2$ g O_2/m$^2 \cdot$day
Hypolimnion	0.4	10	$^1k'_{A2} = 0.3$ m/day

30°C, compute and graph the vertical oxygen concentration profile as a function of depth. Table 7.1K contains pertinent chemical and physical data for both layers of the lake.

7.1L. Low Oxygen Levels in Caminada Pass

This condition is suspected to be the cause of the death of crabs, shrimp, eels, and other marine species near Grand Isle, Louisiana (*Baton Rouge Sunday Advocate*, August 26, 1990). Normally, the dissolved oxygen (DO) is 4 g/m^3 or greater in surface waters. The heat in mid-August combined with several days of relatively calm winds is suspected to lower the DO content. Under these adverse conditions it was measured at 0.0 to 0.2 g/m^3. The conditions biostress many marine species and they die.

1. Select the appropriate steady-state model and use it to obtain a graph of the DO profile for normal conditions. Use conditions at New Orleans; assume a water depth of 3 m and $\mathscr{D}^{(t)}_{A2y} = 0.04$ m^2/h.
2. Graph the DO profile for the adverse condition. Use $T_2 = 35$°C and wind speed $\leqslant 1$ mi/hr.
3. Does the model predict low DO for the adverse condition?
4. Estimate the time required for water column DO depletion.

7.1M. Pritchard Salt Box Model

Based on the estuary illustrated in Fig. 7.1-17*b*, show that the surface and bottom volumetric flows, Q_S and Q_B, respectively, can be estimated by

$$Q_S = \frac{Q_F \bar{\rho}_B}{\bar{\rho}_B - \bar{\rho}_S} \qquad Q_B = Q_S - Q_F \qquad (7.1\text{M-1, 2})$$

where Q_F is the fresh, river water flow and $\bar{\rho}_S$ and $\bar{\rho}_B$ are average, surface, and bottom layer salinities.

7.1N. Steady-State River-Estuary Pollutant Model

1. Using the transformations given in the text following Eq. 7.1-58, develop the dimensionless version of it, including the four boundary condition.
2. Solve the second-order differential equation to arrive at Eq. 7.1-59.

7.1O. Reactive Decay Plus Evaporation of Pollutants in Estuary

Repeat the solution of Example 7.1-4 assuming that the dominant constituent is benzene, with overall ${}^1K'_{A2} = 3.3$ cm/h in addition to being biodegradable with $k'''_A = 0.5$ per day. What fraction of the benzene is lost by evaporation? The average water depth is 2.0 m.

7.1P. Dimensionless Pollutant Concentration Profiles

Verify the dimensionless concentrations, P, shown in Fig. 7.1-19 for χ and N values in the following matrix.

		N		
		0	1	2
χ	−1.0			
	0.0			
	+ 1.0			

7.1Q. Special Cases of River-Estuary Model

Beginning with Eq. 7.1-59:

1. Show the procedure and list the assumptions in arriving at Eq. 7.1-60.
2. Repeat the process for Eq. 7.1-61.

7.1R. Advection and Dispersion of Salt into Rivers

The sea is a source of the salinity found in estuarine streams. Equation 7.1-59 is inappropriate for $A =$ salinity because $k'''_A = 0$; however, Eq. 7.1-58 applies.

1. Develop the boundary conditions for A in a long estuarine stream that enters the sea.

2. Solve the ordinary differential equation to arrive at the following salinity profile:

$$\rho_{A2} = C \exp\left(\frac{xv_x}{D_{Ax}^{(t)}}\right) \tag{7.1R-1}$$

where C is a constant.

3. For summertime salinity values above the battery on the Hudson River are given in Table 7.1R. With 230 m^3/s river flow and 15,000 m^2 cross-sectional area, estimate the longitudinal dispersion coefficient. Answer in m^2/s.

Table 7.1R. Hudson River Salinity versus Distance

ρ_{A2} (‰)	28	16	13	8	6
x (km)	0	8	16	24	32

Source: Ref. 21.

7.1S. Steady-State Biochemical Oxygen Demand Profiles

Do the following for the given conditions on certain estuarine streams:

1. Calculate the maximum BOD and its location given the following: Q(waste) = 37,800 m^3/day, ρ_{A2}(waste) = 20 mg O_2/L, $k_A''' = 0.15$ per day, $D_{2x} = 210\,m^2/s$, Q(river) = $40\,m^3/s$, and $A = 930\,m^2$.
2. Repeat the calculation for part 1 if $D_{2x} = 0$.
3. Repeat the calculation for part 1 if $v_x = 0$.
4. Estimate k_A''' if $A = 930\,m^2$, $D_{2x} = 300\,m^2/s$, and $w_A = 1360$ kg O_2/day for BOD versus distance data (Table 7.1S).

Table 7.1S

x (km)	-16	-8	0	$+8$	$+16$
ρ_{A2}(mg O_2/L)	0.1	0.2	0.4	0.3	0.1
Estuary region	Upstream		Max.	Downstream	

5. Two waste streams $w_{A1} = 450\,kg\,O^2$/day at $x = 0.0$ km and $w_{A2} = 680\,kg\,O_2$/day at $x = +16$ km are entering an estuary with $D_{2x} = 150\,m^2/s$ and $A = 930\,m^2$. Calculate the maximum BODs.
6. If 0.2 mg O_2/L is the maximum BOD concentration allowed for the estuary in part 5, are both waste streams contributing to the exceedence?
7. Reduce each waste load by the same fraction such that the standard is met everywhere in the estuary. Calculate w_{A1} and w_{A2} for this condition.

7.1T. BOD and DO Profiles in Estuary

A point discharge places 760 kg O_2/day demand onto an estuary. Its flow is 40 m^3/s.

1. For $D_{2x} = 210\,m^2/s$, $A = 930\,m^2$, and $k_A''' = 0.15$ per day, compute the concentration profile of BOD on either side at 1-km intervals up to 16 km.
2. For the same estuary compute the DO profile at 1-km intervals: 16 km upstream and 48 km downstream. DO saturation is 8.0 mg/L, ${}^1k'_{02}/h = 0.5$ per day.

7.1U. Bacterial Distribution in Lake Michigan[11]

Additional data were obtained on the distribution of bacteria in Lake Michigan in the vicinity of the Indiana Harbor, and these data appear in Table 7.1U.

Table 7.1U. Bacterial Concentration in Lake Michigan

x (mi)	ρ_{A2}	x (mi)	ρ_{A2}
0	100	2.2	1.2
0.4	80	4.0	1.1
0.5	90	2.8	1.0
0.6	15	6.0	0.8
1.6	15	5.5	0.2
1.8	10	7.2	0.13
3.8	8.5	5.2	0.11
4.3	2.0		

During the periods of observation, the winds were reasonably uniform and constant and the wind direction was along the x axis for the periods. The water velocity was taken as 2.5% of the wind velocity and was 6.0 mi/day. Using the range of bacterial decay constants given in Example 7.1-5, calculate the concentration profile for the following model cases:

1. Point source with diffusion and advection and variable $\mathscr{D}_{A2}^{(t)}$. Use $\mathscr{D}_{A20}^{(t)} = 0.0063$ mi^2/day.
2. Point source with diffusion and advection and constant diffusion. Use $\mathscr{D}_{A2}^{(t)} = 1.0$ mi^2/day.
3. Plot the computed profiles and the data observed in the manner of Fig. E7.1-5.

7.1V. Surface Eddy Diffusivity from Okubo's Diagram

1. Convert the variance versus time function in Figure 7.1-22 to its equivalent eddy diffusivity versus time function.

2. Using the result of part 1, estimate the eddy diffusivity in m^2/s at $t = 5E3$, 1E5, 1E6, and 1E7 seconds.

7.1W. Horizontal Dispersion in Surface Waters

1. Beginning with Eq. 3.2-29, list and justify the assumptions necessary to arrive at Eq. 7.1-64.

2. Equation 7.1-75 is not applicable for $k_A''' = 0$; rederive it for this case choosing realistic boundary conditions.

3. Equation 7.1-77 is also not applicable for $k_A''' = 0$; rederive it.

7.2 CHEMICAL TRANSPORT AND FATE WITHIN ATMOSPHERIC BOUNDARY LAYER

The atmospheric boundary layer (ABL) is the lower part of the troposphere, its bottom contacts the Earth's surface and its top extends 100 to 3000 m into the troposphere depending on local conditions. The processes within and the characteristics of the ABL are not typical of what occurs in the rest of the troposphere or the atmosphere beyond. The main reason for this difference is the dominating influence of the earth's surface on the lowest layers of air. We are familiar with its calm periods, breezes, storms, and other micro- and macroclimates. Within it we feel the warmth of the daytime sun and the chill of the nighttime air. It is here that our crops are grown, our dwellings are constructed, and much of our commerce takes place. As presented in Chapters 4 and 6, the many anthropogenic effluents that enter the atmosphere do so from surface sources, so that the resulting dispersion of the pollutants is influenced by ABL processes.

The major objective of this section is to connect the chemodynamic processes presented in Chapters 4 and 6, having to do with chemical flux and emission rates from water and soil surfaces, respectively, to fate and transport processes that are present in the ABL. Obviously, the reason for studying this connection is because the air in the ABL is a major carrier of chemicals, in both vapor and particulate forms, to humans, other bioreceptors, engineered structures, and soil beyond.

Air-pollution meteorology and agricultural meteorology are applied subjects within the science of boundary layer meteorology.[28] These subjects are quite mature, with numerous handbooks[29,30] and textbooks[31,32] available to the practitioner. This being the present situation, the extent of coverage here on chemical fate and transport (FaT) processes in the ABL will be very limited. The material presented will focus primarily on pollutant sources at the Earth's surface and from an elevated point above the surface using the least complex models. The major goal is to introduce the student to broad outlines of principles and procedures needed in predicting chemical concentration in air associated with these type sources.

Mean Boundary Layer Characteristics

In Section 6.1 the subject of thermal turbulence above the air–soil interface was presented. That material provided the meteorological basis for understanding some key aspects of the ABL very near ($y \lesssim 10\,\text{m}$) the Earth's surface in the constant-flux layer. The diurnal variation of the vertical temperature structure above the Earth (Fig. 6.1-2), the concept of thermal stability, and a means of quantifying it using the Richardson number was presented in enough detail so that it could be used subsequently for estimating chemical flux rates through the lowermost portion of the ABL. The student should review that material in Section 6.1 at this time. What follows completes the brief description of meteorological conditions in the ABL. It is adapted from that given by Stull.[28]

The time evolution of the ABL depth and structure is illustrated in Fig. 7.2-1. It depicts conditions during a high-pressure region over land, which is a very common occurrence. This is a convenient way to start a study of the mean BL characteristics since a well-defined structure evolves with the diurnal cycle. Given the temperature profiles in Fig. 6.1-2, it is useful to relate these to the BL depth and structure in Fig. 7.2-1. The evolutionary history for time flags S1 through S6 follows.

Just before sunset at S1 the ABL is the warmest of the day. The soil surface is very warm transferring heat by natural convection to the air above. The BL is unstable, with vertical mixing occurring, hence the term *convective mixed layer*. The S2 profile is just after sunset. The soil is now losing heat rapidly via radiation to the dark, clear sky. This loss of heat, in turn, cools the air creating a stable layer near the ground. During the night from S2 to S3, the stable (nocturnal) BL continues to cool and increase in height until about 0700 hours. After sunrise at S4 the soil begins to heat up, warming the surface air layers

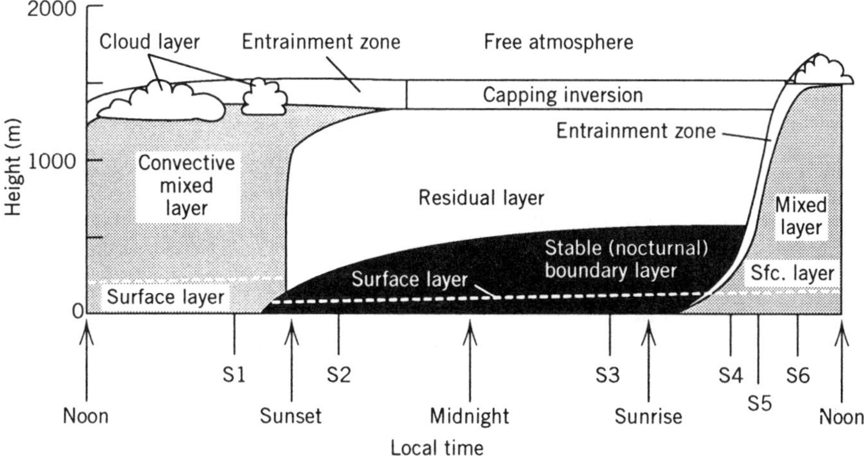

Figure 7.2-1. Height and structure diurnal evolution of the atmospheric boundary layer. Time markers indicated by S1 to S6 correspond to those in Fig. 6.1-2. (Reprinted with permission. Modified from Ref. 28.)

and an inversion is seen at height of about 100 ft. A temperature inversion is a condition where the absolute temperature increases with height. The mixed layer (ML) continues to warm, grows rapidly in height, and the inversion height is about 600 ft at S5. At S6 the ABL is unstable throughout; the mixed BL condition has returned. The ML continues to warm throughout the remainder of the day, reaching maximum temperatures at S1. At this time the diurnal cycle is complete.

The three major components of the ABL depicted in Fig. 7.2-1 are the *mixed layer*, the *residual layer* (RL), and the *stable boundary layer* (SBL). At the bottom is the *surface layer* (SL). This is the region where the turbulent fluxes and stress vary by less than 10% of their magnitude. The bottom approximately 10% of the ABL is called the surface layer, regardless of whether it is part of the ML or SBL. Finally, a thin layer called a *microlayer* or *diffusion layer* has been identified in the lowest few centimeters of air, where molecular transport dominates over turbulent transport.

Mixed Layer. Turbulence in the ML is driven by natural (free) convection, although a nearly well-mixed layer can form in regions of strong wind. Convective sources include heat transfer from a warm ground surface and radiative cooling from the top of the cloud layer. The first situation creates thermals of warm air rising from the ground, while the second creates thermals of cool air sinking from cloud top. Both can occur simultaneously. The resulting turbulence readily transports heat, moisture, and momentum vertically. Water vapor mixing ratios (g H_2O/g air) tend to decrease with height. This reflects the evaporation of soil and plant moisture from below and the entrainment of drier air from above. Water vapor or smoke emitted from elevated stacks into the ML, in the daytime in particular, exhibits a characteristic looping behavior as those portions of the plume are effected in turn by warm-air updrafts and cool-air downdrafts. Wind speed tends to be nearly constant in the upper ML and nearly logarithmic with height in the surface layer.

The ML reaches its maximum depth in the late afternoon. It grows by entraining, or mixing down into it, the less turbulent air from above. A stable layer at the top of the ML acts as a lid to the rising thermals, thus restraining the domain of turbulence. It is called the *entrainment zone* because entrainment of air from the free atmosphere (FA) above occurs there. At times this capping stable layer is strong enough to be classified as a temperature inversion. Since most pollutant sources are near the earth's surface, concentrations can build up in the ML while concentrations in the FA remain low. Trapping of pollutants below such an inversion layer is common in high-pressure regions and sometimes leads to pollution level alerts in large communities.

Residual Layer. As illustrated in Fig. 7.2-1, the ML exists primarily in the daylight hours. About a half hour before sunset thermals cease to form, allowing turbulence to decay in the formerly well-mixed layer. The resulting layer after sunset is sometimes called the RL because its initial mean state variables and concentration variables are the same as those of the recently decayed ML. The RL does not have direct contact with the ground. During the night, the nocturnal stable layer gradually increases in thickness at the expense of the RL. Thus the remainder of the RL is not affected by turbulent transport of surface-related properties and hence does not strictly fall within the definition of a boundary layer. Nevertheless, the RL is included as an exception to the rule.

The RL is neutrally stratified, a result of turbulence that is mechanically generated and of equal intensity in all directions. Visible substances emitted at height into the RL are seen to disperse at equal rates in the vertical and lateral directions, creating a cone-shaped plume (i.e., a coning plume). The RL often exists for a while in the mornings before being entrained into the new ML. When the top of the next day's ML reaches the base of the RL, the ML grows rapidly.

Stable Boundary Layer. As the night progresses, the bottom portion of the RL is transformed by its contact with the ground into a stable boundary layer. The SBL is characterized by statically stable air with weaker, sporadic turbulence. Although the wind at ground level frequently becomes lighter or calm at night, the winds aloft may accelerate in a phenomenon called the *nocturnal jet*. The statically stable air tends to suppress turbulence, while the developing nocturnal jet enhances wind shears that tend to generate turbulence. As a result, turbulence sometimes occurs in relatively short bursts that can cause mixing throughout the SBL. During the nonturbulent periods the RL flow becomes essentially decoupled from the surface.

Pollutants emitted into the SBL disperse relatively little in the vertical. They disperse more rapidly, or fan out, in the horizontal (i.e., a fanning plume). Sometimes at night when winds are lighter, plumes meander left and right as it drifts downwind. Winds exhibit very complex behavior at night. Just above ground level the wind speed often becomes light or even calm. At altitudes on the order of 200 m above ground the wind may reach 10 to 30 m/s (20 to 70 mi/hr) in the nocturnal jet. *A* few hundred meters above, the wind speed drops closer to its geostrophic value.

Visible emissions from elevated stacks entering the RL may disperse to the point where the bottom of the plume hits the top of the NBL. The strong static stability and frequent reduced turbulence in the NBL enhibit downward mixing of the pollutants. The top of the smoke plume sometimes can continue to rise into the neutral air (i.e., a lofting plume). After sunrise a new ML begins to grow, eventually reaching the height of elevated plumes from the previous night. At this time elevated pollutants in either the SBL or RL are mixed downward to the ground by ML entrainment and turbulence in a process called *fumigation*.

Boundary Layer Under Other Pressure Regions. The description of mean ABL characteristics above is for high-pressure regions over land. Over both land and oceans, the general nature of the BL is to be thinner in high-pressure regions than in low-pressure regions. Thunderstorms, although not normally considered to be a boundary layer phenomenon, can modify the BL in a matter of minutes by drawing up boundary layer air into the cloud, or by laying down a carpet of cold downdraft air. Over the oceans, the boundary layer depth varies relatively slowly in time and space. This is because the sea surface temperature changes little over the diurnal cycle because of the large heat capacity of water and good mixing within the top layers. Over land, on days with high or middle overcast conditions, the insolation at ground level is reduced. This, in turn, reduces the intensity of the thermals and the ML exhibits slower growth. It may even become nonturbulent or neutrally stratified if the clouds are thick enough. At night, where the SBL touches the ground, a thin layer (order of a few meters) of katabatic or drainage winds exist. These winds are caused by colder air flowing down mountains or hills under the influence of gravity. Wind speeds of 1 m/s at a height of 1 m are possible. The cold air collects in the valleys and depressions and stagnates there.

These are just a few examples of interesting aspects of BL meteorology. See Stull[28] for more. Coastal meteorology is unique enough that it deserves a treatment of its own; see Hsu.[33].

Dispersion of Pollutants in Atmospheric Boundary Layer

From the material presented in the preceding section it is apparent that air stability and boundary layer depth changes during the course of a day, as does wind speed and wind direction. Models that attempt to make pollutant concentration predictions in the ABL must accommodate the diurnal structure and other peculiarities of the locale. The models used most often are for continuous, steady-state sources in a uniform flow field of constant wind speed. These are appropriate for many applications and include dispersion models such as K-theory, the statistical model, and similarity models. The similarity model was presented in Chapter 6 (see pp. 378–382). Only the Gaussian plume model is presented here; for the others, see Ref. 29.

Gaussian Distribution for Pollutants in Air. The Gaussian frequency distribution describes molecular scale and macroscale random movement processes. The spread of any pollutant in the atmosphere takes place in a flow field that is almost invariably turbulent. In a general way, one may say that matter in the gaseous or liquid state is often subject to two kinds of random movements: one on the molecular scale, the thermal agitation of molecules, and one on a macroscopic scale, turbulence. The main differences between molecular agitation and turbulent movements are (1) a difference of scale, the typical "sweep" of a single turbulent movement being ordinarily very large compared to one molecular free path, and (2) the constraint imposed by continuity. The fairly large parcels of fluid partaking of turbulent movements can move only by displacing other fluids, which eventually has to fill in the space vacated by the moving parcel. Consequently, one may picture turbulent flow as consisting of a number of closed-flow structures of diverse shapes and sizes, called *eddies*, although it would be a mistake to think of these as resembling regular vortices or, indeed, regular flow structures of any kind. For a more detailed introduction to the subject, the reader is referred to Csanady[13] and Slade[14].

Mainly because of these differences, two theoretical models have evolved. The molecular diffusion and Brownian diffusion processes can be adequately quantified by gradient transport models such as Fick's law, whereas the turbulent diffusion processes are best quantified by statistical models. Both the molecular and Brownian processes are Gaussian, and the experimental evidence indicates that the turbulent processes of diffusion are also Gaussian; however, each is characterized by widely different mean-square displacements σ^2. It was first shown by Gauss, for an infinite data set of observations on the quantity z where the variations are random, that the distribution of values about the population mean, $\bar{z}$, is given by

$$f = \frac{\exp(-u^2/2)}{\sqrt{2\pi\sigma^2}} \tag{7.2-1}$$

where $u \equiv (z - \bar{z})/\sigma$ and σ^2 is the variance. Here f is the frequency, or probability of occurrence, of a value of magnitude z. The value f has a maximum at $z = \bar{z}$ and the integral of $f\,dz$ from $z = -\infty$ to $z = +\infty$ is 1.0.

For example, in the case of the continuous point-source plume illustrated in Fig. 7.2-2, f is the frequency the instantaneous plume boundary is displaced a distance z from the center for the downwind position x in wind of constant velocity v_x. At this position, or any other position downwind, 95% of the values of z (i.e., $\int f\,dz \simeq 0.95$), the instantaneous plume distance from the center, are within $\pm 2\sigma$. It is therefore common and convenient to use $\pm$ two standard deviations (σ) as a reasonable measure of plume width.

The problem of turbulent diffusion in the environment has not been uniquely formulated in the sense that a single basic physical model capable of

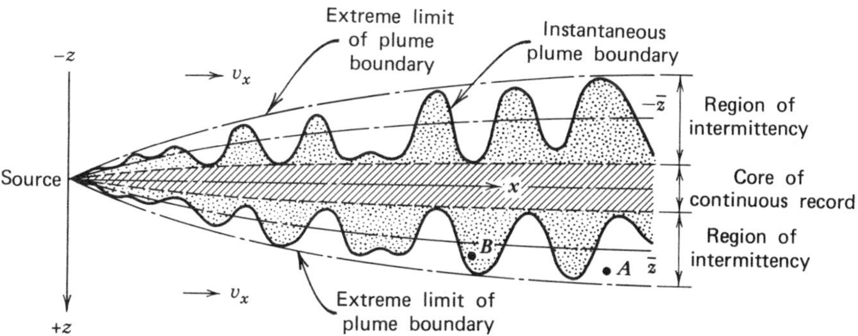

Figure 7.2-2. Continuous-point-source plume.

explaining all the significant aspects of the problem has not yet been proposed. Instead, there are available two alternative approaches, neither of which can be categorically eliminated from consideration since each has areas of utility that do not overlap the other's. The two approaches to diffusion are the gradient transport theory and the statistical theory. Diffusion at a fixed point in the atmosphere, according to the gradient transport theory, is proportional to the local concentration gradient. Consequently, it could be said that this theory is Eulerian in that it considers properties of the fluid motion relative to the spatial fixed-coordinate system. On the other hand, statistical diffusion theories consider motion following fluid particles and thus can be described as Lagrangian. Diffusion theories may be described as either continuous-motion or discontinuous-motion theories, depending on whether the particle motion is postulated to occur continuously or as discrete events. There must necessarily be a close connection among all these approaches to the diffusion problem, since obviously there is only one atmosphere.

Gradient Transport Models. This topic needs little introduction since it has been used extensively through the book in various applications. A prior introduction appears in Section 3.1. For our purposes it is sufficient to regard air as an incompressible fluid. The mathematical statement of Fick's law has (in the three-dimensional case) the form of the classical equation of continuity for species A without reaction:

$$\frac{\partial \rho_{A1}}{\partial t} + \mathbf{v} \cdot \nabla \rho_{A1} = \mathscr{D}_{A1} \nabla^2 \rho_{A1} \tag{7.2-2}$$

where $\mathscr{D}_{A1}$ (in the atmosphere) is the eddy diffusivity coefficient or the molecular diffusivity and ρ_{A1} refers to the value of some conservative air property per unit volume of air. For illustration we discuss here some solutions of particular relevance to atmospheric diffusion problems.

Consider first the idealized one-dimensional case of an instantaneous plane source in a uniform medium at rest ($\mathbf{v} = 0$). The one-dimensional equation (Eq. 3.1-3) describes a phenomenon not dependent on y or z—say, diffusion along the axis of a duct or pipe with conditions uniform across the section or in the atmosphere along the vertical, assuming homogeneous conditions in the horizontal. The solution to the instantaneous plane source problem is

$$\rho_{A1} = \frac{m'_A}{2\sqrt{\pi \mathscr{D}_{A1} t}} \exp\left(\frac{-x^2}{4\mathscr{D}_{A1} t}\right) \tag{7.2-3a}$$

where m'_A is an amount of material per unit area. The asymptotic behavior of ρ_{A1} as $t \to 0$ is

$$\begin{aligned} \rho_{A1} &= 0, \qquad |x| > 0 \\ \rho_{A1} &= \infty, \qquad x = 0 \end{aligned} \tag{7.2-3b}$$

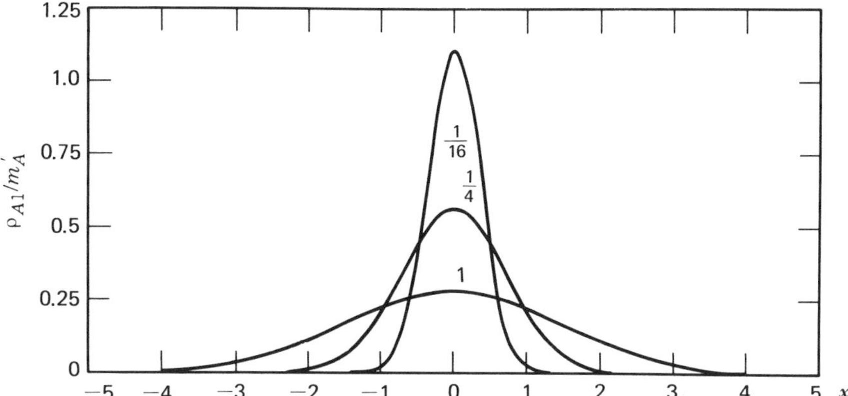

Figure 7.2-3. Concentration–distance curves for an instantaneous plane source. Numbers are values of $\mathscr{D}_{A2}t$. (Reprinted by permission from Ref. 10.)

These properties show that as $t \to 0$, $\bar{\rho}_{A1}$ becomes proportional to the *delta function:*

$$\rho_{A1} = m'_A \delta(x) \qquad (t = 0) \tag{7.2-3c}$$

The delta function is the mathematical description of such idealizations as the concentrated source, such that Eq. 7.2-3*a* describes the diffusion of an amount of material m'_A that was initially in a thin sheet at $x = 0$, in an instantaneous plane source. As time proceeds, the material spreads out as illustrated in Fig. 7.2-3.

Equation 7.2-3*a* is otherwise known as a *Gaussian* or *normal curve*, with the maximum concentration at the center of gravity, which remains at the origin. The movement of inertia or second moment of this distribution may be used to characterize its spread

$$\int_{-\infty}^{\infty} \rho_{A1} x^2 \, dx = 2m'_A \mathscr{D}_{A1} t \tag{7.2-4}$$

If we divide this by the total diffusion material, we obtain a kind of mean-square distance to which the particles have diffused.

$$\sigma^2 = 2\mathscr{D}_{A1} t \tag{7.2-5}$$

The length σ serves as a convenient scale of the width of the distribution and is known as the *standard deviation.* Expressed in terms of this scale the distribution becomes

$$\rho_{A1} = \frac{m'_A}{\sigma\sqrt{2\pi}} \exp\left(\frac{-x^2}{2\sigma^2}\right) \tag{7.2-6}$$

This is perhaps the usual form in which one encounters a Gaussian distribution. It is similar to Eq. 7.2-1. An advantage of this form is that it also applies in turbulent diffusion, where σ is a different function of time since the process is not Fickian.

Similarly, a product of three one-dimensional solutions provides a description of diffusion from an "*instantaneous point source*:

$$\rho_{A1} = \frac{m_A}{(\sqrt{2\pi}\,\sigma)^3} \exp\left(-\frac{x^2}{2\sigma^2} - \frac{y^2}{2\sigma^2} - \frac{z^2}{2\sigma^2} \right) \tag{7.2-7}$$

where $\sigma = \sqrt{2\mathscr{D}_{A1}t}$ is still true for Fickian diffusion and m_A is the amount of material released at the origin at $t = 0$ in grams.

All the previous results pertain to diffusion of a chemical or particle in a uniform medium at rest. From the standpoint of environmental diffusion, a somewhat more realistic model is one in which the medium moves at a constant and uniform velocity v_x in the x-direction. Mathematically, this case differs little from the at-rest medium case in that the coordinate system moving at the wind velocity v_x is still at rest. Applying the coordinate transformation $x' = x - v_x t$, we have at once an equation of the instantaneous point source in a wind:

$$\rho_{A1} = \frac{m_A}{(\sqrt{2\pi}\,\sigma)^3} \exp\left[-\frac{(x - v_x t)^2 + y^2 + z^2}{2\sigma^2} \right] \tag{7.2-8}$$

This equation is a model of a single *puff* (point-source cloud) of material moving downwind and growing in size.

A number of important problems may be modeled by the assumption that a source emits material continuously. The case of a continuous point source with Fickian diffusion in a wind provides a particularly important model (a very crude model of a chimney plume). The concentration field now consists of a series of puffs with their centers stretched out along the x axis as the growing puffs are convected downwind. For a point source maintained indefinitely, let the rate of emission be w_A (g/s), such that in the short interval t to $t + dt$ an amount $w_A dt$ (g) is emitted. Each puff generates its own cloud, and the total concentration is obtained by a summation of contributions from the individual puffs. We obtain the combined concentration field of the many puffs by integration (see Csanady[13]) to obtain

$$\rho_{A1} = \frac{w_A}{4\pi\mathscr{D}_{A1}^{(t)} r} \exp\left[-\frac{v_x}{2\mathscr{D}_{A1}^{(t)}} (r - x) \right] \tag{7.2-9}$$

where $r^2 = x^2 + y^2 + z^2$. The concentration distribution is independent of time. This final result is not generally applicable for modeling all plumes; however, it is applicable to Fickian diffusion in uniform laminar flow and some isotropic turbulent flow cases, such as the center regions of pipes. It is applicable to all cases in which $\mathscr{D}_{A1}^{(t)}$ is constant.

Statistical Theory of Turbulent Diffusion. It is apparent that the loops in the plume shown in Fig. 7.2-2 are not the result of concentration gradients but are due to eddies embedded within the flow structure. An instantaneous sampler located at point A would not detect the diffusing pollutant species, whereas one located at point B would. We recognize at the outset therefore that practically observable concentrations of a pollutant at some location in the environment are, in general, random variables, about which we are only able to make probabilistic predictions. Effectively, the only quantity about which we have adequate evidence, both theoretical and experimental, is the first moment of the concentration probability distribution, that is, its *expectation*, which can also be shown to be equal to the mean concentration of an ensemble of independently diffusing particles. The statistical model results relate to the ensemble-mean concentration field.

At a fixed point in the wake of a continuous release of a tracer at a constant rate into a statistically steady and homogeneous field of turbulence of constant mean velocity, the observable ("instantaneous") concentration varies in a random manner. By the ergodic property of such processes, the ensemble-mean concentration at a given time t is equal to the time-averaged concentration $\bar{\rho}_{A1}$, obtained by averaging the concentration observed in a single experiment over a sufficiently long period of time compared to the lifetime of a typical turbulent eddy.

The time-averaged concentration field at a fixed point may be related to particle displacement probabilities. This leads us to what might be called the *elementary statistical theory of turbulent diffusion*. The importance of this theory rests partly on the fact that a continuous point source in a uniform wind models fairly satisfactorily such prime pollution sources as factory chimneys, although in applying the theory and obtaining the necessary parameters, many approximations have to be made.

The random movements of a diffusing particle in a field of homogeneous and stationary turbulence consist of the movements within eddies and a superimposed Brownian or molecular motion. Both components contribute to dispersion independently, the mean-square displacements along the x axis:

$$\sigma_x^2 = \sigma_{xt}^2 + 2\mathscr{D}_{A1}^{(l)} t \tag{7.2-10}$$

where σ_{xt}^2 is the mean-square displacement due to bulk turbulent motions alone. In most environmental applications the molecular contribution $2\mathscr{D}_{A1}^{(l)} t$ is negligible and $\sigma_x^2 \simeq \sigma_{xt}^2$.

Experimental evidence on diffusion in a homogeneous field shows that the particle-displacement probability distribution, of which the mean-square dispersion σ_x^2 is the second spatial moment, is to a high degree Gaussian. Because its second moments specify a Gaussian distribution completely, the problem of describing the mean concentration field is thus for practical purposes solved. However, the question as to why a Gaussian distribution is observed in experiments is not answered satisfactorily in theory.

On accepting that the probability distribution function of the independently diffusion particles is Gaussian, we may now write down the ensemble-average concentration field of an *instantaneous point source.* Since diffusion by turbulent movement is a linear phenomenon, *continuous sources* may be built up by adding the fields of simple sources. In this instance superposition yields

$$\rho_{A1} = \frac{w_A}{(2\pi)^{2/3}} \int_0^\infty \exp\left[-\frac{(x - v_x t)^2}{2\sigma_x^2} - \frac{y^2}{2\sigma_y^2} - \frac{z^2}{2\sigma_z^2} \right] \frac{dt}{\sigma_x \sigma_y \sigma_z} \tag{7.2-11}$$

where σ_x, σ_y, and σ_z are the mean-square dispersions on the principal axes. This equation cannot be integrated until the functional dependence of the standard deviations σ_x, σ_y, σ_z on diffusion time t is specified.

At short distances from the source the instantaneous-source ensemble average cloud grows according to

$$\sigma_x = \frac{\sqrt{\overline{v_x'^2}}\, x}{v_x} \tag{7.2-12}$$

where $\sqrt{\overline{v_x'^2}}$ is the root-mean-square velocity (see Section 3.3). This equation and similar relations for σ_y and σ_z are a consequence of *Taylor's theorem.* Taylor's theorem is a most important basic result in the theory of diffusion by random movements. The interested reader should consult Csanady[13] for details.

Equation 7.2-11 can be integrated once the form of the σ's is specified. One finds after a few simple approximations,

$$\rho_{A1} = \frac{w_A}{2\pi\sigma_y\sigma_z v_x} \exp\left(-\frac{y^2}{2\sigma_y^2} - \frac{z^2}{2\sigma_z^2} \right) \tag{7.2-13}$$

where we have substituted the time-dependent standard deviations back into the final result. The diffusion takes place along y and z only, and not in the x direction, the mean concentration gradients along x being negligible compared to those along y and z. This equation is exactly as found in molecular diffusion, except that the standard deviations σ_y and σ_z are now more complex functions of diffusion time $t = x/v_x$.

Gaussian Plume Model for Continuous Sources. Beginning with the work of Sutton,[15] the elementary statistical theory of turbulent diffusion was widely utilized as a theoretical framework for the representation of data on atmospheric diffusion. The application of the theory is not straightforward and much empiricism is included. Here we use the elementary statistical theory with whatever empirical inputs are necessary to represent atmospheric observations. Practical predictions of atmospheric diffusion may be made on this basis, mainly because a good deal of empirical information may be absorbed in the functional form of the standard deviations σ_y and σ_z.

Following Sutton and others, we may write for the mean concentration field due to a continuous elevated point source of strength w_A:

$$\rho_{A1} = \frac{w_A}{2\pi v_x \sigma_y \sigma_z} \left\{ \exp\left[-\frac{y^2}{2\sigma_y^2} - \frac{(z-h)^2}{2\sigma_z^2} \right] + \exp\left[-\frac{y^2}{2\sigma_y^2} - \frac{(z+h)^2}{2\sigma_z^2} \right] \right\} \qquad (7.2\text{-}14)$$

where the coordinate origin is vertically below the source at ground level and the source is located at height h, as shown in Fig. 7.2-4. The purpose of the last term is to account for reflection of the plume at the ground (i.e., zero deposition) by assuming an image source at a distance h below the ground surface. Of particular interest is the prediction of concentration at ground level, $z = 0$:

$$\rho_{A1} = \frac{w_A}{\pi v_x \sigma_y \sigma_z} \exp\left(-\frac{y^2}{2\sigma_y^2} - \frac{h^2}{2\sigma_z^2} \right) \qquad (7.2\text{-}15)$$

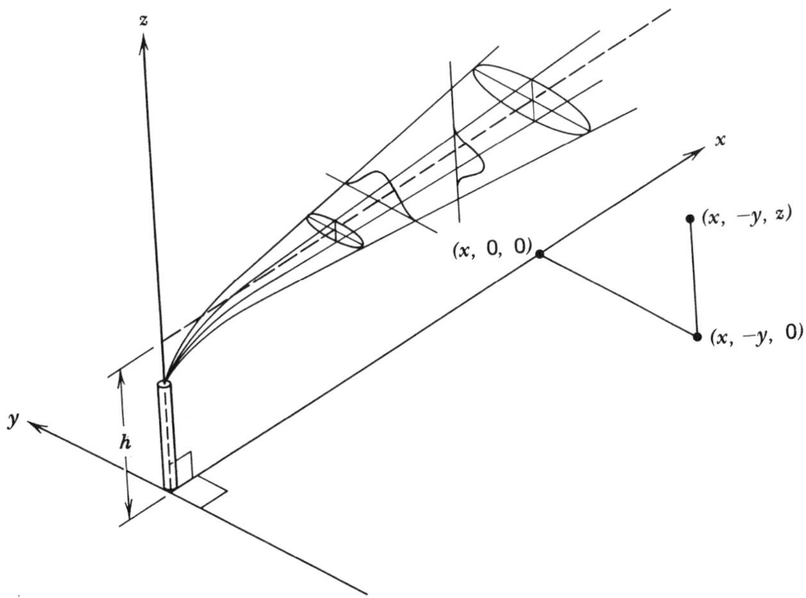

Figure 7.2-4. Coordinate system showing Gaussian distributions in the horizontal and vertical.

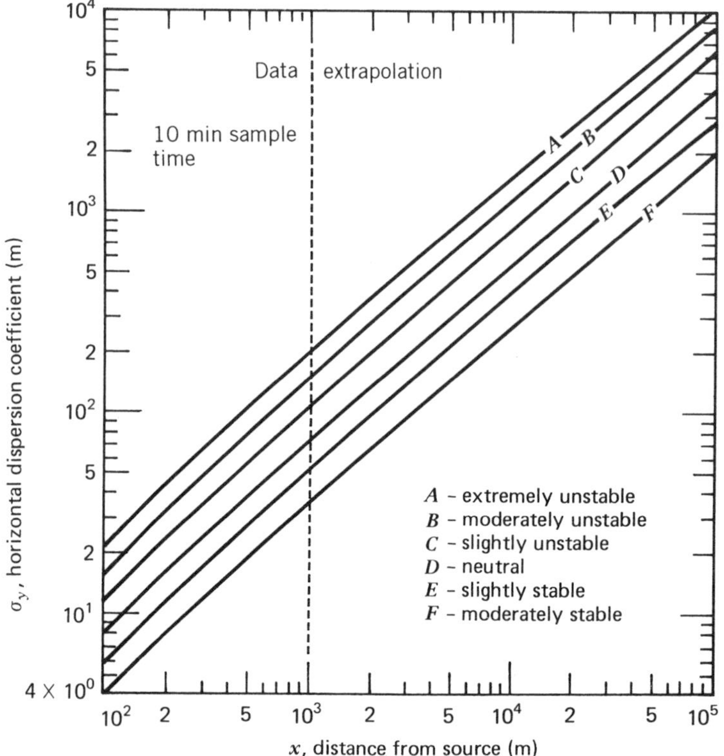

Figure 7.2-5. Horizontal dispersion coefficient as a function of distance from source. (From Ref. 13.)

The worst condition occurs along the axis of the plume, $y = 0$:

$$\rho_{A1} = \frac{w_A}{\pi v_x \sigma_y \sigma_z} \exp\left(-\frac{h^2}{2\sigma_z^2}\right) \tag{7.2-16}$$

For practical use to be made of the dispersion formulas, numerical values of the dispersion coefficients σ_y and σ_z must be determined. These are given as functions of distance downwind, x, and stability and are based on a combination of experimental results and theory. To deal with the resulting wide variations in turbulent properties, meteorologists have introduced stability categories into which atmospheric conditions may be classified. Figures 7.2-5 and 7.2-6 exhibit families of curves for σ_y and σ_z for various stability categories.

The most widely used scheme for stability classes was developed by Pasquill and modified slightly by Turner. Table 7.2-1 contains the criteria for Pasquill's six stability classes, which are in turn based on five classes of surface wind speeds, three classes of daytime insolation, and two classes of nighttime

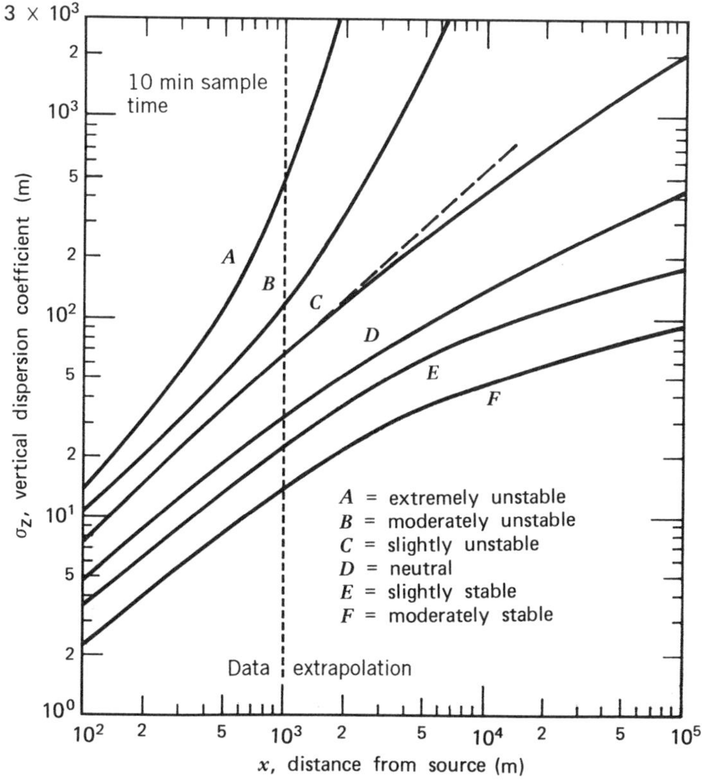

Figure 7.2-6. Vertical dispersion coefficient as function of distance from source. (From Ref. 13.).

cloudiness. The student should appreciate the relationships between Pasquill's stability classes and the stability conditions observed in the atmospheric boundary layer as presented at the beginning of Section 7.2.

The Gaussian model remains popular for performing dispersion calculations for many reasons.[29] It is appealing conceptually and is a solution to the Fickian diffusion equation for constants $\mathscr{D}_{A1}^{(t)}$ and v_x; it produces results that agree with field data as well as any model and is consistent with the random nature of turbulence. The curves in Figs. 7.2-5 and 7.2-6 are based on carefully performed diffusion experiments. The terrain was uniform, releases were from near the ground, and the concentration measurements were at a downwind distance less than 1 km. Note that at distances beyond 1 km, the lines are extrapolated. They may work under certain ideal conditions at greater distances, but there is less basis in observations. All variables are assumed to be averaged over a period of about 10 min. Extensions of the Gaussian plume model to other types of terrain, for other averaging times, for deposition onto the ground, for use with other turbulence measures, and so on, are available in Ref. 29. Turner[30] proposed a virtual upwind point source to model ground-level emission of substances from area sources. Example 7.2-1 illustrates the

Table 7.2-1. Pasquill Stability Categories (They Correspond to Categories in Figs. 7.2-5 and 7.2-6)[a]

Surface Wind Speed (m/s)	Insolation			Night, Mainly Overcast or ⩾4/8 Low Cloud	⩽3/8 Low Cloud
	Strong	Moderate	Slight		
2	A	A–B	B	—	—
2–3	A–B	B	C	E	F
3–5	B	B–C	C	D	E
5–6	C	C–D	D	D	D
6	C	D	D	D	D

[a] A, extremely unstable; B, moderately unstable; C, slightly unstable; D, neutral; E, slightly stable; F, moderately stable.

concept applying the Gaussian plume model.

Example 7.2-1. Emission Rate of Vinyl Chloride from Area Source at Ground Level.[34] A roughly square surface of ground 2.6E8 cm^2 in area, within a hazardous waste landfill in the Los Angeles basin is emitting vinyl chloride (VC). Sample site A is located 175° directly south and 250 m from the center of the area source. Nighttime hillside drainage winds with speeds up to 6 knots fluctuating between 330° and 30° deliver quantities of VC to site A, which is near a residential area. On five days, concentrations of 12, 5, 7, 12, and 9 ppb (volume) were measured. Estimate the emission rate of VC in g/s.

SOLUTION Equation 7.2-16 with $h = 0$ applies to ground-level sources. The virtual point source is located a distance x_{vp} upwind of the area source boundary. This distance, which varies with stability, is chosen to give the virtual point source an initial cross-wind plume size standard deviation at the area source boundary of $\sigma_y = w/4.3$, where w is the width of a side of the area source: $w \simeq (2.6\text{E8 cm}^2)^{1/2} = 161$ m and $\sigma_y = 161/4.3 = 37.4$ m. 6-knot wind is 3.08 m/s.

According to Table 7.2-1, D or E class stability applies. With $\sigma_y = 37.4$ m, D class, Fig. 7.2-5 yields $x_{vp} = 500$ m. The virtual plume source distance is $x = 500 + 250 = 750$ m up-wind of site A. With this as x, Figures 7.2-5 and 6 yields $\sigma_y = 55$ and $\sigma_z = 25$ m for D class at site A.

The average $y_A = 9\text{E-}9$ mol fraction VC, and $M_A = 62.5$ g/mol in air. The mass concentration in air is $\rho_{A1} = y_A M_A p/RT$. At STP RT/p is 22.4 L/mol. This yields $\rho_{A1} = 25.1\text{E-}9$ g/L.

Equation 7.2-16 for emission rate, w_A, is

$$w_A = \pi \rho_{A1} v_x \sigma_y \sigma_z = \pi \left| 25.1\text{E-}6 \, \frac{\text{g}}{\text{m}^3} \right| 3.08 \, \frac{\text{m}}{\text{s}} \left| 55 \text{ m} \right| 25 \text{ m} = 0.33 \text{ g/s}$$

A lower estimate, 0.26 g/s, is obtained for E class stability. Baker and MacKay[34] report 0.20 and 0.28 g/s at site A.

The example above illustrates the ground-level short-range, dispersion case very common in the exposure of humans and other bioreceptors located on and near area sources containing hazardous substances. In addition to landfills, these include particles and/or chemical vapors from automobiles on highways, wastewater treatment facilities, farmland, manufacturing facilities, and so on. Estimating concentrations at locations on the source can be a problem using the Gaussian plume model because the minimum downwind distance is 100 m, as seen in Figs. 7.2-5 and 7.2-6. This data gap has been amended using sulfur hexafloride as a tracer to distances 10 to 30 m at ground level.[35] A model based on horizontal wind fluctuations for lateral dispersion and K-theory for vertical dispersion is used to interpret the data and extend the measurements to other such nonbuoyant plumes.

Closure. The preceding material on chemical transport within the ABL has been abstracted in large part from References 28 through 33 and has been presented as a brief introduction and for completeness. The topic of atmospheric diffusion has enjoyed a disproportionate amount of study compared to other transport topics related to chemodynamics. The serious student should consult the summary works of these authors and/or several other excellent works on this important topic. The problem section that follows is very brief, to save space. The references cited above contain some excellent student exercise problems.

PROBLEMS

7.2A. Sulfuric Acid Vapors

On a clear night with a prevailing wind from the west at 2 mi/hr, a sulfuric acid spill produces vapors in the air. The worst complaints from downwind residents ranged from distinctly unpleasant to unbearable in a location about 200 m. A well-known reference on chemical hazards suggests that humans find sulfuric acid vapors very unpleasant at concentrations of about 10 mg/m^3.

1. Estimate the release rate of sulfuric acid vapors.
2. Find the maximum distance downwind of the source where the concentration would equal the occupational exposure limit of 1 mg/m^3.

(This problem, provided by D. D. Reible, is based on an actual incident.)

7.2B. SO_2 Emission from Refinery

It is estimated that a refinery emits 80 g/s of SO_2 from an average effective height of 60 m. At 0800 on an overcast winter morning, with a surface wind of

6 m/s, what is the ground-level concentration 500 m downwind at a cross-wind distance of 50 m from the centerline? Answer in g/m^3 assuming that a 10-min average concentration is desired. (This problem was provided by James H. Clarke.)

7.2C. SO_2 Maximum Downwind Concentration

A 60-cm-diameter stack discharges 170 m^3/min (STP) air containing 0.5 vol% SO_2 at an effective stack height of 55 m. calculate:

1. The 10-min. average, maximum, ground-level concentration and its location downwind for a 3-m/s wind speed and class D stability. Answer in mg/m^3.
2. The 10-min average concentration at 1500 m downwind on centerline under B, D, and F class stability for a 3-m/s wind. Answer in mg/m^3.

7.3. CHEMICAL TRANSPORT AND FATE WITHIN SUBTERRANEAN MEDIA

This section is included for completeness, being a brief introduction to the subject. Certain chemicals that are inappropriately placed in subterranean media, if mobile, will cause extensive contamination of soil underlying geologic strata, groundwater, and to ground air. Whereas contaminated ground air is usually an insignificant threat to the atmospheric clean air resources, contaminated groundwater can be a significant loss of clean water resources. Other pollution problems will occur if the contaminated soil or geologic strata solids are unearthed and left unattended. However, if the mass of the mobile fraction remains in place, groundwater contamination becomes the environmental chemodynamic problem of major concern. This is the primary topic of Section 7.3. There are textbooks and book chapters devoted to this topic; see Refs. 36 and 37 as good examples.

The section begins with a generic, subterranean media contamination scenario that reflects most of the intricacies of the problem. This will be followed by some information about the nature and types of porous and permeable geologic strata. The next subsection covers groundwater and its behavior in these strata. This is followed by an analysis of the fate and transport of chemical constituents in the mobile mass fraction.

Introduction to Subterranean Media

Contamination of Subterranean Media. Traditionally, groundwaters have been the focus of this environmental chemodynamic problem; however, due to its multiphase, interconnected nature, it is more appropriate to consider the subterranean medium as a whole. Figure 7.3-1 illustrates a generic subterranean medium contamination scenario. This hypothetical chemical mess

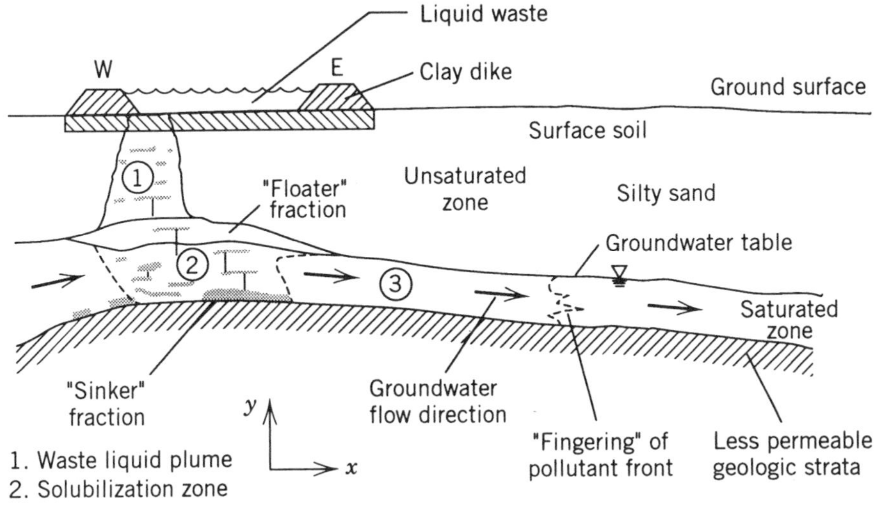

Figure 7.3-1. Subterranian media contamination from waste liquid impoundment.

apparently resulted from a failure in the waste containment barrier. *A* complex liquid waste mixture entered the ground. Its movement through the unsaturated region produced zone 1 which is now a stationary plume containing once-dry surface solids coated with a layer of nonaqueous liquid (NAL), isolated large pockets of NAL (i.e., discontinuous ganglia) in the vadose zone, and substances in the porewater. The zone remains generally unsaturated after downward passage of the chemical mass. Also read the discussion "Some Processes of Soil Contamination" in Section 6.3.

A portion of the chemical mass now resides in zone 2, which is within the saturated region of this silty-sand strata. A mound of liquid containing lighter-than-water fluids floats on the surface of the groundwater. The capillary fringe is omitted in the figure to reduce the clutter (see pp. 31–33, Ref. 38). The dense fractions of the NAL has moved into zone 2 as well. It is shown trapped as inclusions in selected spots within the strata of zone 2. See Refs. 39 to 41 for further information of the behavior of dense NALs. Due to its heavier-than-water nature, the dense material can move by gravity flow upstream counter to the groundwater flow direction.[42] Generally, the plume of zone 2 tilts downstream in the direction of the groundwater flow. Beyond and downstream is zone-3, which contains quantities of the waste material in solution and possibly sorbed to mobile groundwater colloids. The outer limits of the pollutant plume in this zone is shown "fingering" into the virgin groundwater.

Porous Geologic Strata. There is a fair amount of information on this topic in several previous chapters. The student is encouraged to review that material before proceeding. Section 2.1 contains information of the chemical composition of soil and rocks; information on NALs is supplied in Section 2.1 in the discussion "Earthen Solid-Water Equilibrium Occurrences" and Section 2.3 has information of the multicompartment nature of subterranean media. Section 3.3 has a discussion "Porous Media." In addition, Section 3.3 has a brief review of transport processes in porous media. Although specific to bed sediment, the material in Sections 5.2 and 5.3 is directly applicable to the saturated zone. Section 6.3 deals primarily with unsaturated soils. The discussions "Brief Introduction to Soil Structure" and "Some Processes of Soil Contamination" should be reviewed.

Most of the information noted above is concerned with surface soils and the strata downward to 1 to 2 m. The generic subterranean contamination scenario in Fig. 7.3-1 typically involves porous geologic strata to depths of 100 m, more or less. The geologic formations involved with such chemical messes are typically aquifers. An aquifer is a geologic formation, or a stratum, that contains water and permits significant amounts to move through it under ordinary field conditions. An aquiclude, on the other hand, is a geologic stratum that may contain water but is incapable of transmitting significant quantities under ordinary field conditions. A clay layer is an example.

That portion of the rock not occupied by solid matter is the void space (also pore space, pores, interstices, and fissures). This space contains water and/or air. Only connected interstices can act as elementary conduits within the formation. These voids must be distinguished from their interconnecting pathways, which allow fluids to circulate through them exhibiting porous media permeability. The porosity of a aquifer material can generally be grouped in two classes: original interstices, mainly in sedimentary and igneous rocks, created by geologic processes at the time the rock was formed, and secondary interstices, mainly in the form of fissures, joints, and solution passages, developed after the rock was formed.

Granulated Rocks. Most rocks are constituted by solid mineral particles more or less tightly stuck together, forming a skeleton around which empty spaces remain. These are porous media in terms of fluid mechanics. For example, sand and sandstone have a total porosity that may reach 30%. However, even rocks that are generally thought to be solid have a certain porosity; examples are limestone, dolomite (particularly secondary), and even crystalline and metamorphic rocks (from 1 to 5%).

Clays belong to a separate category; their constituent elements resemble thin shavings and are organized into sheets, which are stacked in parallel layers separated by variable intervals where a fluid might lodge. This gives the clay, in particular, the property of swelling in the presence of water. This water is strongly linked to the solid clay. All the same, the percentage of voids may be very high, on the order of 40% and even up to 90%, in unconsolidated marine red clays.

Fractured Rocks. Fracturing is a special case of voids in solid rocks. Because of tectonic movement (e.g., faults, fissures, joints, cracks, and openings along bedding planes) almost all rocks in the earth's crust are fractured. These fractures are generally oriented in at least two (generally three or four) main directions, which cut up the rock into blocks. Some fractures are increased in size by dissolution, resulting in huge limestone caverns.

We then have a network of fractures, more or less interconnected, which may create voids in the rock if the fractures are not sealed by some kind of deposit (clay, calcite, quartz, etc.) In this case we talk about fracture porosity as opposed to the interstitial porosity already mentioned. Moreover, these two types of porosity may coexist (sandstone, limestone, etc.). (The material above was extracted from de Marsily[37] and Bear.[43] These sources contain more details on the nature of aquifers.)

Groundwater And Its Behavior. It is convenient to describe the transport of water and other mobile constituents in porous media in terms of differential equations. However, conceptual difficulties arise from the fact that the notions of porosity and permeability, which are notions concerning points in an equation with partial derivatives, for instance, cannot be defined or measured at single points, since a porous medium is a conglomeration of solid grains and voids. Below a certain scale of volume, porosity, permeability, and other properties have no physical meaning. De Marsily[37] gives two theoretical ways of defining the local properties of a porous medium: the notion of the representative elementary volume and that of random function.

For a porous medium the real fluid velocity in each of the pores, also called *microscopic velocity*, is $\mathbf{v}$. The fluid also has a mass per unit volume, ρ_2, at this scale and point porosity ε ($\varepsilon = 1$ in pore and 0.0 in a grain). At this scale, the continuity equation as Eq. 3.2-1 applies. Based on either of the two ways of defining local properties, the macroscopic quantities or "averages" in the porous media $\langle \mathbf{v} \rangle$, $\langle \rho \rangle$, and $\langle \varepsilon \rangle$ are shown to be, respectively, at a point for all possible realizations of the medium. De Marsily then develops the equation of continuity in porous media equivalent to Eq. 3.2-1:

$$\nabla \cdot [\langle \rho \rangle \langle \mathbf{v} \rangle] + \frac{\partial}{\partial t} [\langle \rho \rangle \langle \varepsilon \rangle] = 0 \tag{7.3-1}$$

where $\langle \cdot \rangle$ designates the average taken and $\langle \mathbf{v} \rangle$ is the fictitious mean velocity, sometimes called the *filtration velocity*. Although expressed at a point, the equation is always established for an elementary volume that is fixed in space. To avoid cumbersome expressions, we shall now dispense with the $\langle \cdot \rangle$ for the properties while remembering that these have been defined by the operation of taking averages. Equations 3.2-1 and 7.3-1 are the consequence of the basic principle of the overall mass balance as expressed by Lavoisier: "Nothing is lost, nothing is created."

Henri Darcy, while studying the fountains in the city of Dijon, France, around 1856, established empirically that the flux of water through a sandy formation, Q, may be calculated by

$$Q = \frac{K_3 A\, \Delta h}{L} \tag{7.3-2}$$

where A is the area of the cross section of the sand formation, L its length, Δh the difference in hydraulic head in the water between the top and bottom of the sand formation, and K_3 a constant that depends on the porous medium, called the *hydraulic conductivity* or sometimes the *coefficient of permeability*. By dividing Q by A the fictitious Darcian velocity, v, or infiltrative velocity is obtained. Bear in mind that this definition of velocity considers the entire section to be open to the flow. The mean pore fluid velocity is obtained by dividing v by the kinematic porosity (i.e., the volume fraction of fluid that can circulate and is less than the total porosity).

From simplifying and integrating the Navier–Stokes equations for schematic porous media, theoretical phenomenological laws can be obtained. For n parallel fractures of width w in an otherwise impermeable rock, de Marsily obtains the steady-state water flux:

$$Q = A\,\frac{\varepsilon w^2}{12}\,\frac{1}{\mu}\,\frac{\Delta P}{L} \tag{7.3-3}$$

For a circular tube of radius r the flux is similarly

$$Q = A\,\frac{\varepsilon r^2}{8}\,\frac{1}{\mu}\,\frac{\Delta P}{L} \tag{7.3-4}$$

where $\Delta P/L$ is the pressure gradient. Thus the Navier–Stokes analysis leads to the conclusion that the fluid viscosity is inversely proportional to Q.

The driving force for fluid displacement must be generalized to include external pressure gradients and gravity forces. Consequently, Darcy's law in generalized form is

$$\mathbf{v} = -\frac{k_3}{\mu}\,(\nabla p + \rho g\, \nabla y) \tag{7.3-5}$$

where y is elevation of the water table and k_3 is the intrinsic or specific permeability that relates to the porous medium regardless of the characteristics of the fluid. It is defined only on the macroscopic scale and has dimensions of L^2. It is often expressed in darcys; 1 darcy is equal to 0.987E-12 m^2. The hydraulic conductivity, K_3, in Eq. 7.3-2 is related to k_3, the intrinsic permeability, by

$$K_3 = \frac{k_3 \rho g}{\mu} \tag{7.3-6}$$

Values of K_3 are given in Table 7.3-1.

Table 7.3-1. Hydraulic Conductivity of Rocks

Unconsolidated Rocks: Medium	K_3 (m/s)	Hard Rocks: Medium	K_3 (m/s)
Coarse gravels	E-1 to E-2	Dolomite limestones	E-3 to E-5
Sands and gravels	E-2 to E-5	Weathered chalk	E-3 to E-5
Fine sands, silt, loess	E-5 to E-9	Unweathered chalk	E-6 to E-9
Clay, shale, glacial till	E-9 to E-13	Limestone	E-5 to E-9
		Sandstones	E-4 to E-10
		Granite, gneiss	E-9 to E-13

Source: Ref. 37.

According to de Marsily, the limit separating permeable rocks from impermeable ones is set arbitrarily at 1E-9 m/s. Clays are impermeable, despite their great total porosity, because their small pores give them very low effective porosity. The intrinsic permeability and hydraulic conductivity values are nonisotropic properties of the porous medium; they are dependent on the orientation in space. Permeability is therefore a tensorial property. For this and other advanced topics on flow in porous media, see de Marsily,[37] Freeze and Cherry,[36] or Bear.[43]

Example 7.3-1. Groundwater Flow in Shallow, Unconfined, and Unconsolidated Silty-Clay Alluvial Deposit. At a hazardous waste site north of Baton Rouge, Louisiana, hexachlorobenzene and hexachlorobutadiene exist in the soil and groundwater similar to the case depicted in Fig. 7.3-1. Steady recharge and a line of on-demand, low-volume, extraction wells downgradient maintain the water table elevation at $y = ix + b$ in zone 3.

(a) Beginning with Eqs. 7.3-1 and 7.3-5, list all assumptions to obtain $Q = AK_3i$ for the flux, where i is the water table hydraulic gradient in L/L.

(b) If the water table gradient is 20 ft per 800 ft and K_3 is estimated to be 1.1E-4 to 4.2E-4 cm/s, estimate the velocity in the saturated zone in m/yr.

SOLUTION (a) For water at steady-state flow in the x-direction only, Eq. 7.3-1 reduces to $dv_x/dx = 0$, therefore, constant velocity in the x-direction. The aquifer is unconfined; therefore, the pressure is atmospheric throughout, so $\bar{\nabla}p = 0$. For the x-direction, $dy = i\,dx$ and Eq. 7.3-5 reduces to $v_x = k_3\rho g i/\mu$. Using Eq. 7.3-6, the flux is $Q = AK_3i$.

(b) For the range of K_3 values given and $i = 0.025$, $v_x = K_3i = 0.87$ to 3.3 m/yr.

Subterranean Transport and Fate of Nonreactive Substances

Clay Layer for Waste Containment. A properly designed, installed, operated, and maintained recompacted clay layer can provide an effective, low-leakage rate barrier to aqueous waste in an impoundment of the type shown in Fig. 7.3-1, thereby protecting the groundwater from pollution. This is possible for selected substances that do not harm the liner provided that the combined chemical transport mechanisms of advection plus molecular diffusion can be minimized by engineering liner transmissivity. Layers a meter in thickness are common and the pore spaces become saturated with water very quickly. Some earthen containers for the disposal of solid waste (i.e., landfills) are also fitted with liners of a similar design, and they too contain water, probably originating from rainwater infiltration.

An investigation into some of the chemodynamics within the clay liner begins with Eq. 5.2-42, the diagenetic equation for trace chemicals in bed sediment. As to the assumptions: bioturbation is absent, no reactions are occuring and no advection or colloids are present; the familiar Fick's second law results. If the semi-infinite solid boundary and initial conditions are used, the following solution analogous to Eq. 5.2-2 applies:

$$\frac{\rho_{A2} - \rho^0_{A2}}{\rho^*_{A2} - \rho^0_{A2}} = 1 - \operatorname{erf}\left(\frac{y}{2\sqrt{D_{A3}t/R_{A3}\varepsilon}}\right) \tag{7.3-7}$$

where ρ^0_{A2} is the initial concentration of A in the porewater, ρ^*_{A2} the assumed constant high concentration of A in the aqueous impoundment water, and ρ_{A2} the concentration at distance y into the layer from the water interface and at time t after placement of the aqueous waste. R_{A3} is the *retardation factor*, defined as $R_{A3} \equiv 1 + K^*_{A32}\rho_3(1 - \varepsilon)/\varepsilon$. The D_{A3} is given by Eq. 5.1-34.

Assume that $\rho^0_{A2} = 0$ and the chemical breaks through or emerges from the liner bottom when $\rho_{A2} = Y\rho^*_{A2}$, where, for example, Y= 1E-4, 1 part in 10,000. From Table B.1, erf$[\phi] = Y$, then $[\phi] = 2.8$ and solving the bracketed term for the breakthrough time τ_{bt} yields

$$\tau_{bt} \simeq \frac{\left(\dfrac{y}{2.8}\right)^2 R_{A3}\varepsilon}{4D_{A3}} \tag{7.3-8}$$

Due to the semi-infinite slab constraint this equation gives a crude estimate. Other values of Y reflecting better or poorer groundwater protection may be selected.

In the case of pure advection, similar assumptions and simplifications of Eq. 5.2-42 yield

$$\frac{\rho_{A2}}{\partial t}\bigg/\frac{\partial\rho_{A2}}{\partial y} = \frac{v_y}{R_{A3}\varepsilon} \tag{7.3-9}$$

where v_y is the y component of the Darcian velocity. Upon inspection the left-hand side is interpreted to be the chemical velocity defined as v_{yA}. The pure advection case result is therefore

$$v_{yA} = \frac{v_y}{R_{A3}\varepsilon} \tag{7.3-10}$$

and the breakthrough time is

$$\tau_{bt} = \frac{y}{v_{yA}} = \frac{yR_{A3}\varepsilon}{v_y} \tag{7.3-11}$$

where y is the layer thickness or distance traveled. Note that if $R_{A3} > 1$, the chemical velocity through the porous layer is less than the water velocity. As in gas or liquid chromatographic separations, due to species sorption differences on the media, their appearance at the detector lags behind that of the solvent (i.e., water), hence is "retarded."

Superimposing diffusion on advection[45] for the initial conditions $y < 0$, $\rho_{A2} = \rho_{A2}^*$ and $y \geqslant 0$, $\rho_{A2} = \rho_{A2}^0$ yields

$$\frac{\rho_{A2} - \rho_{A2}^0}{\rho_{A2}^* - \rho_{A2}^0} = \frac{1}{2}\left[1 - \operatorname{erf}\left(\frac{y - v_y t/R_{A3}\varepsilon}{2\sqrt{D_{A3}t/R_{A3}\varepsilon}}\right)\right] \tag{7.3-12}$$

and recall that $\operatorname{erf}(-\Phi) = -\operatorname{erf}(\Phi)$ (see Table B.1). The point where the left-hand side is $\frac{1}{2}$ moves with the fluid's average velocity. The semi-infinite slab condition applies here also.

Information on the operating conditions for a particular wastewater impoundment containing traces of phenol and A-1254, a polychlorinated biphenyl, are given in Problem 7.3A. Based on these conditions the estimated diffusion-only and advection-only breakthrough times are shown in Table 7.3-2. When the processes are combined, as by Eq. 7.3-12, the estimated times for breakthrough are less. It appears that under the proper conditions, a clay liner can be an effective barrier. The student is encouraged to work Problem 7.3A and verify the numerical results.

Table 7.3-2. Chemical Breakthrough Time (τ_{bt}) in years Estimate for a Clay Liner 0.91 m Thick[a]

Chemical Species	Retardation Factor, R_{A3}	Molecular Diffusion, τ_{bt}	Advection, τ_{bt}	Diffusion Plus Advection, τ_{bt}
Phenol	1.3	14.9	4.71	2.86
A-1254	4,500	52,000	16,200	9,900

[a]See Problem 7.3A.

Alternative Models for Retarded Diffusion and Advective Transport in Packed Beds. Laboratory columns packed with porous, natural materials such as sand, silt, clay particles, and glass beads serve a convenient pilot-scale simulators of chemical behavior for the above-described waste impoundment liner. Flow experiments performed on uniform beds of solids are also described by simple mathematical models such as Eq. 7.3-12. In combination, key parameters such as chemical diffusion coefficients, D_{A3}, or dispersion coefficients $\mathscr{D}^{(t)}_{A2x}$, can be determined. Typically, these experiments are conducted with chemical tracers that are introduced at the head of the column as a step increase in concentration. The initial conditions are $\rho_{A2} = \rho^0_{A2}$ at $x > 0$ for $t = 0$ and $\rho_{A2} = \rho^*_{A2}$ at $x < 0$ for $t = 0$. If no dispersion occurs in the flow stream entering the bed but it does occur in the bed, the boundary condition at $x = 0$ is

$$-\mathscr{D}^{(t)}_{A2x}\left(\frac{\partial \rho_{A2}}{\partial x}\right)_{x>0} + v_x(\rho_{A2})_{x>0} = v_x\rho^*_{A2} \tag{7.3-13}$$

for $t > 0$. Smith[49] outlines the solution, which is analogous to that given by Eq. 7.3-12 for the y direction. The term in parentheses in this equation can be written, for the x direction, as

$$\frac{\sqrt{\text{Pe}}}{2}\,\frac{1-\theta_A}{\sqrt{\theta_A}}$$

where $\theta_A \equiv v_x t/\varepsilon R_{A3} x$. The dimensionless time, θ_A, when equal to unity is the plug flow breakthrough time, t, for Darcian velocity v_x, distance x "retarded" by R_{A3} for the chemical species. The other dimensionless number is the bed Peclet number, $\text{Pe} \equiv v_x x/\mathscr{D}^{(t)}_{A2x}$. For a bed of length l, Eq. 7.3-12 is graphed in Fig. 7.3-2. Plug flow behavior is shown for $\text{Pe}^{-1} = 0$, which occurs when $\mathscr{D}^{(t)}_{A2x} = 0$.

Alternatively, the *tank-in-series model* can be used to describe the combined processes in a porous material. The bed is assumed to be a parallelepiped of total volume Al, where A is the uniform cross-sectional area and l is the length of the porous bed. The flow orientation is in the x direction parallel to the l dimension. The bed is divided into n equal subvolume elements each of length $\Delta x = l/n$.

As the flow moves through the bed it is envisioned to proceed from subvolume to subvolume, in which the chemical concentration changes abruptly, each subvolume being characterized by an average concentration ρ^n_{A2}. Students familiar with the study of fluid mixing in open, tubular flow reactors appreciate that for the case of a packed bed the subvolume average concentration replaces the uniformly mixed concentration within a single tank of the equivalent *series-of-stirred-tanks model.* See Smith[49] for the development of the latter model.

In the case of chemical behavior in packed beds a useful result of the

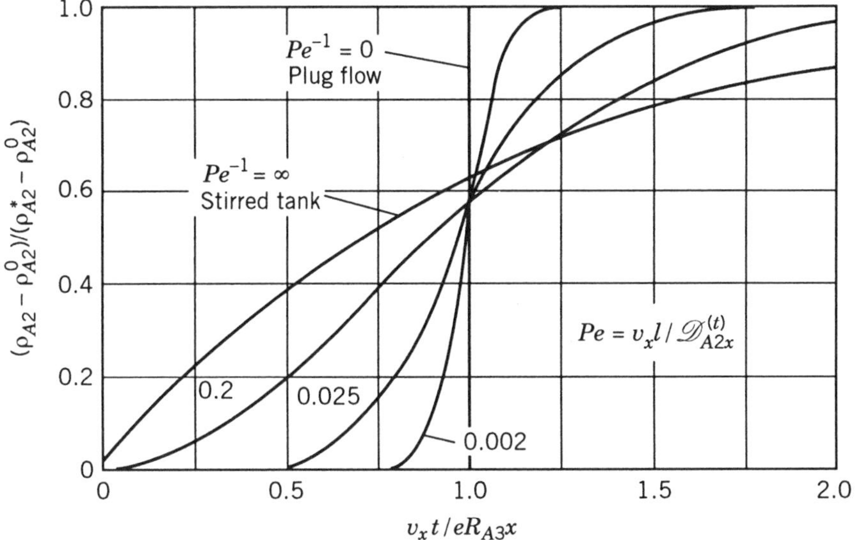

Figure 7.3-2. Response curves for dispersion model. (Reprinted with permission. Adapted from Ref. 49.)

tanks-in-series model is obtained by imposing a step-concentration increase upon the fluid entering the first subvolume, $n = 1$. This pulse of chemical of concentration, ρ_{A2}^*, then enters the second subvolume $n = 2$ and the subsequent ones being diluted in the process. The output concentration, ρ_{A2}^n, from the nth subvolume of the bed is

$$\frac{\rho_{A2}^n - \rho_{A2}^0}{\rho_{A2}^* - \rho_{A2}^0} = 1 - \exp\left(\frac{-nt}{\tau_c}\right)\left[1 + \frac{nt}{\tau_c} + \left(\frac{nt}{\tau_c}\right)^2 \Big/ 2! + \cdots + \left(\frac{nt}{\tau_c}\right)^{n-1} \Big/ (n-1)!\right] \tag{7.3-14a}$$

where ρ_{A2}^0 in the initial concentration of species A in the porewater of the bed and

$$\tau_c \equiv \frac{\varepsilon R_{A3} l}{v_x} \tag{7.3-14b}$$

being the plug flow breakthrough time for the bed.

Figure 7.3-3 is a plot of Eq. 7.3-14a for various values of n. The similarity to Fig. 7.3-2 suggests that the longitudinal dispersion model, which is the name given to Eq. 7.3-12, and the tanks-in-series model give the same general output shape of concentration response to the step-input increase concentration. The analogy is exact for $n = 1$, which models the bed as a single volume element of length $\Delta X = l$ and infinite dispersion coefficient $\mathrm{Pe}^{-1} = \infty$. Agreement is also exact at the other extreme, the plug flow case with $n = \infty$ and $\mathrm{Pe}^{-1} = 0$. The shapes of the curves for the two alternative models are more nearly the same for larger values of n.

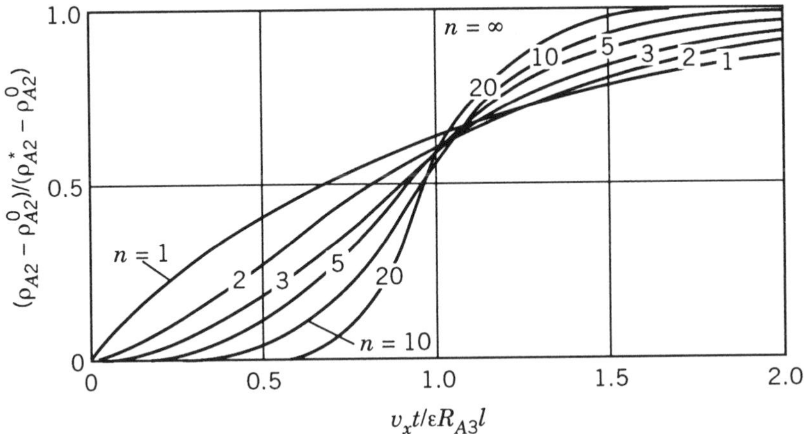

Figure 7.3-3. Response curves for series of stirred tanks. (Reprinted with permission. Adapted from Ref. 49.)

Equating the variance or second moment of the dispersion in the bed for each model Fogler[51] showed that

$$\frac{1}{n} = \mathrm{Pe}^{-1}(2 + 8\mathrm{Pe}^{-1}) \tag{7.3-15}$$

With a measured value Pe^{-1} this equation is used to obtain the nearest integer n, the number of tanks in series; however, it gives incorrect results for $n \lesssim 3$. When calibrated with a numerical value of n or Pe, either model may be used to estimate the concentration in a packed bed as a function of time, t, and location, l.

Some Chemodynamic Processes in Zone 1. Improper design, construction, or operation probably caused the catastrophic liner failure under the east dike resulting in the subterranean contamination mess illustrated in Fig. 7.3-1. The remainder of this section is devoted to some of the chemodynamic processes occurring in zones 1, 2, and 3 shown in the figure.

Aspects of the creation of zone 1 were mentioned in beginning of Section 7.3, but more details are needed. The discussion "Some Processes of Soil Contamination" in Section 6.3 and the discussion "Nonaqueous Liquids in Contact with Earthen Solids" in Section 2.1 should be reviewed now; they describe aspects of the formation of zones 1 and 2. In what follows it will be assumed that the events leading to the NAL downward infiltration have occurred and the NAL movement has ceased. As will be seen, the time scale of the contamination event described above is comparatively short (i.e., months to years) considering the time scale of the subsequent chemical dissolution process. Although the NAL mass is stabilized, the infiltration of rainwater through zone 1 will continue to deliver small amounts of soluble material to zone 2.

Attention is now focused on some fate and transport of the soluble, nonreactive substances in zones 2 and 3 of the silty sand, unconfined aquifer illustrated in the figure.

Some Chemodynamic Processes in Zone 2. As illustrated, zone 2 is a complex region containing two NALs, a heavy *sinker* fraction and a lighter-then-water, *floater* fraction. The groundwater flows from left to right mobilizing pooled quantities in excess of the residual saturation and soluble constituents to locations down-gradient in the aquifer. The presence of NAL quantities, maldistributed within the aquifer, causes the groundwater flow to be redirected around these areas. Due to the reverse slope on the left-hand side of the underlying less permeable rock, quantities of the sinker may migrate up-gradient from right to left.

When the groundwater moves through zone 2 the chemical solubilization process occurs from at least four places: (1) from underneath the floating NAL layer, (2) from the NAL in the disconnected ganglia, (3) from quantities of NAL that have moved easily into the high-porosity pockets or sand lenses, and (4) from quantities sorbed onto aquifer solids. If the porewater contains colloidal matter with a chemical sorptive capacity, the solubilization potential is enhanced. Once solubilized, the substances move with the groundwater and via molecular diffusion. Downstream sorption from solution onto clean media solids occurs, retarding the chemical velocity. The process of dispersion, introduced in Section 3.3, also occurs, tending to transport the solutes beyond the average advection velocity front. Clearly, the physicochemical processes occurring in zone 2 are complex and poorly quantifiable at this time, so that only very basic vignette models illustrating selected aspects are presented here.

Chemical mass distribution in the subterranean media. It is instructive to analyze a finite volume element within zone 2 below the floater fraction depicted in Fig. 7.3-1. The element has length Δx in the groundwater flow direction and cross-sectional area A. The following is a mass balance on species A within the volume element where the exit concentration is ρ^1_{A2}:

$$\rho^0_{A2} v_x A - \rho^1_{A2} v_x A = \frac{d}{dt}(m_{A3}) \tag{7.3-16}$$

and ρ^0_{A2} is the inlet concentration. This balance is for the first of n volume elements used to characterize the entire length, $l = n\,\Delta x$, of zone 2. The mass of species A is distributed as

$$m_{A3} = [S_{43}\rho_A \omega_{A4}\varepsilon + \omega_{A3}\rho_3(1-\varepsilon) + \rho_{A2}\varepsilon + \omega_{AC}\rho_{C2}\varepsilon]A\,\Delta x \tag{7.3-17}$$

where the terms in brackets represent the fractions: residual saturation, sorbed onto solids, and solution in porewater and on colloids in porewater, respectively. In general, the magnitudes of the mass quantity terms in Eq. 7.3-17 decrease from left to right. Notice that the presence of colloids in the porewater of concentration, ρ_{C2}, effectively enhances the soluble fraction by a factor of $(1 + \rho_{C2}K^{*}_{AC2})$.

The residual saturation, defined by Eq. 2.1-15, accounts for the NAL in the disconnected ganglia and other forms, such as coating on particles. Wilson et al[44] report S_{43} values for numerous substances; these fall within the range 0.25 to 0.35 cm^3/cm^3. Using large physical models, Tissa Illangasecare (personal communication, 1994) has dicovered that these S_{43} values are probably formation averages, in that the NAL saturation can vary from 0.1 to about 1.0 at specific locales. He further observed that if $S_{43} \simeq 1.0$, the formation is virtually plugged at this locale and the NAL itself has created a tight heterogenetic zone which is both nonporous and nonpermeable; being so, it forces the flowing ground water to be diverted away from this portion of the NAL.

Nonaqueous Liquid Dissolution.[46] The maldistribution, noted above, may preclude that the groundwater flow is uniform and contacts the NAL mass uniformly. However, for the sake of vignette modeling, uniform contact will be assumed. It will also be assumed that the NAL fraction of A must dissolve before any amount of the sorbed fraction does; then a simple relation for the dissolution time, t_d, results. In the case where the NAL is a mixture of substances such that $\omega_{A4} + \omega_{B4} + \cdots = 1$ and the dissolution rate of each species is different, due to their solubilities, the mass fraction distribution, ω_{A4}, changes with time. This case requires that S_{43} be subdivided into its constituent fractions, a mass balance performed for each, and an equilibrium thermodynamic law used to relate the concentrations of each in the adjacent porewater (see the discussion "Equilibrium Occurrences Between Water and Other Liquids" in Section 2.1). Where the NAL is a pure substance, $\omega_{A4} = 1$, $S_{43} = S_{A3}$, and $\rho^{1}_{A2} = \rho^{*}_{A2}$, its solubility. The dissolution time is simply

$$\tau_d = \frac{S_{A3}\rho_4\varepsilon\,\Delta x}{v_x(\rho^{*}_{A2} - \rho^{0}_{A2})} \tag{7.3-18}$$

The volume element length, Δx, is unspecified. As was shown above, it can be some integer fraction of the length of zone 2, which is l. However, replacing Δx by l and letting $\rho^{0}_{A2} = 0$ yields a result that gives an estimate of the minimum time for NAL dissolution.

See Problem 7.3D for the time versus concentration behavior of a NAL mixture undergoing dissolution. The actual times for NAL dissolution will be much larger than the predictions based on the models above. This occurs for several reasons; the models are not mass transfer (i.e., kinetics) based; equilibrium between the NAL and porewater is assumed and is actually a poor assumption.[47]. In addition, A desorbs from the solid media in the upstream NAL–free volume elements, and therefore $\rho^{0}_{A2} > 0$ for much of the porewater.

If the flow is nonuniform, such as with some NAL held in clay lens within the silty-sand aquifer material, the dissolution time will probably be extended considerably.[46]

Desorption–Dissolution of the Solid-Phase Fraction. From the experimental evidence that has accumulated to date[48] it appears that the solid-phase fraction of A on the desorption cycle behaves much different than when it is on the adsorption cycle. Possibly due to extremely slow desorption kinetics and/or irreversible bound quantities remaining on/in the solid phase, a point set (ρ_{A2}, ω_{A3}) is path dependent, as illustrated in Fig. 7.3-4. The absorption and desorption isotherms shown do not reflect an equilibrium state. Heuristically, this fact should be acknowledged by distinguishing between adsorption partition coefficients, K^a_{A32}, and desorption partition coefficients, K^d_{A32}. The adsorption part of the pathway was covered in Chapter 2 and presented as an equilibrium condition! See the section "Partition Coefficients for Sediment–Water and Soil–Water Systems" in Section 2.1.

As the process proceeds down the desorption path from point 1, a portion of the quantity on the solid, indicated by $-\Delta\omega_{A3}$ on the ordinate, is released to the water fairly quickly, in days to a few weeks. However, considerable time, sometime years, is required to further desorb an equivalent fraction. A desorption cycle starting at point 2 also appears to have a similar loosely bound fraction, followed by a more resilient fraction. The dashed line shown in the figure is an attempt at representing this tightly bound fraction, the slope of which is equivalent to a numerically large desorption partition coefficient. In the case of naphthalene on a sandy riverbed sediment,[48] K^a_{A32}, = 2.4 L/kg,

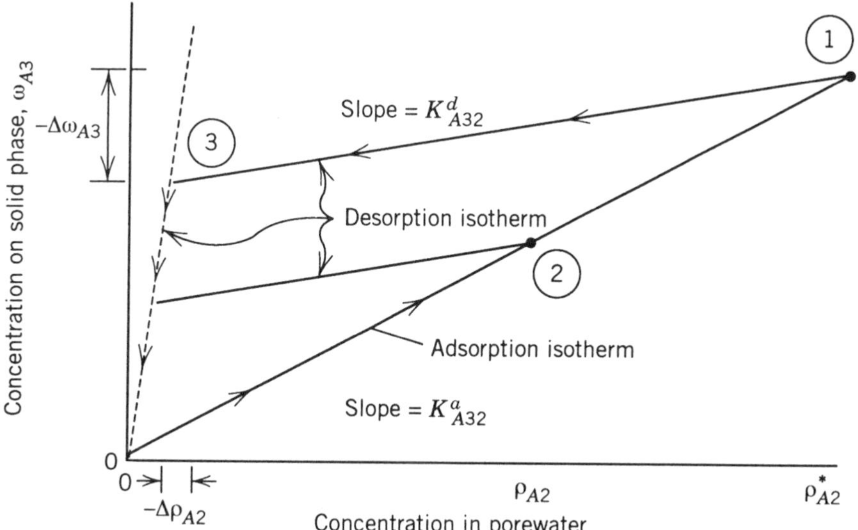

Figure 7.3-4. Adsorption and desorption pathways for the solid and aqueous phases.

$K^d_{A32} \simeq 0.5$ L/kg for the loosely bound fraction, and $k^d_{A32} \simeq 30$ L/kg for the tightly bound fraction.

Once free of the NAL contribution, the instantaneous desorption rate of A from the subterranean media, r_{des}, involves time measurement on the last three terms in Eq. 7.3-17:

$$r_{\text{des}} = \frac{d}{dt}\left[\omega_A \rho_3(1-\varepsilon) + \rho_{A2}(1 + \rho_{C2}K^*_{AC2})\varepsilon\right] \tag{7.3-19}$$

Starting at point 3 in Fig. 7.3-4, the pathway-derived slope of the upper curve is $K^d_{A32} = (-\Delta\omega_{A3}/-\Delta\rho_{A2})$. Differentiating the above and using the chain rule assuming that $\omega_{A3} = f_1(\rho_{A2})$ and $\rho_{A2} = f_2(t)$ yields

$$r_{\text{des}} = \left\{\rho_3(1-\varepsilon)\left[\frac{d\omega_{A3}}{d\rho_{A2}}\right] + (1 + \rho_{C2}K^*_{AC2})\right\}\frac{d\rho_{A2}}{dt} \tag{7.3-20}$$

where K^d_{A32} is the term in brackets. This treatment demonstrates that the solids desorption–dissolution process can be handled by an analytical methodology that tracks the chemical concentration in porewater identical to absorption but with a different set of partition coefficients and related mass fractions.

As done above, for the mass of A sorbed onto the aquifer solids, it is mathematically convenient to assume that the volume of contaminated zone 2, Al, can be subdivided into a series of n subvolumes of equal size $A\,\Delta x$, where $\Delta x = l/n$. The concentration of A in the subvolumes is assumed to be an average value. For $n = 1$ the species mass balance, neglecting the colloid fraction, is

$$\rho^0_{A2}v_x A - \rho^1_{A2}v_x A = \frac{d}{dt}\left\{A\,\Delta x[\varepsilon + K^d_{A32}\rho_3(1-\varepsilon)]\rho^1_{A2}\right\} \tag{7.3-21}$$

where ρ^0_{A2} and ρ^1_{A2} are the entering and exiting concentrations, respectively. Integration yields

$$\rho^1_{A2} - \rho^0_{A2} = (\rho^*_{A2} - \rho^0_{A2})\exp\left(-\frac{nt}{\tau_c}\right) \tag{7.3-22}$$

where $\tau_c \equiv [\varepsilon + K^d_{A32}(1-\varepsilon)]l/v_x \equiv R_{A3}\varepsilon l/v_x$ is the chemical residence timescale for total contaminated volume Al.

For the second subvolume element, $n = 2$, the exiting concentration is

$$\rho^2_{A2} - \rho^0_{A2} = (\rho^*_{A2} - \rho^0_{A2})\exp\left(\frac{-nt}{\tau_c}\right)\left(1 + \frac{nt}{\tau_c}\right) \tag{7.3-23}$$

The concentration in the nth subvolume is

$$\rho^n_{A2} - \rho^0_{A2} = (\rho^*_{A2} - \rho^0_{A2}) \exp\left(\frac{-nt}{\tau_c}\right)\left[1 + \frac{nt}{\tau_c} + \left(\frac{nt}{\tau_c}\right)^2 \bigg/ 2! + \cdots + \left(\frac{nt}{\tau_c}\right)^{n-1} \bigg/ (n-1)!\right] \tag{7.3-24}$$

With this equation a lengthy contaminated zone of uniform cross section can be modeled in space and time by a series of n subvolumes that allows the concentrations to be estimated when species A is desorbing from the aquifer solid media.

Example 7.3-2. Hexachlorobutadiene Dissolution in Zone 2 of Shallow Aquifer. Pure HCB has contaminated zone 2 of the aquifer described in Example 7.3-1.

(a) For the NAL fraction, determine the bulk concentration and estimate the minimum dissolution time. For HCB use: density 1.68 g/cm^3 and 4.0 mg/L solubility. For the aquifer, use: residual saturation 0.25, porosity 0.5, and 10 m zone length.
(b) For the sorbed-porewater fraction determine the bulk concentration and the zone chemical time scale. Use particle density 2.1 g/cm^3; only $K^a_{A32} = 18$ l/kg is available.
(c) For the sorbed-porewater fraction estimate the HCB concentration in the zone exit waters at 100 years assuming that $n = 1$, 3, and 5.

SOLUTION

(a) Use the first term in Eq. 7.3-17 with $\Delta x = l$.

$$\frac{m_A}{Al} = S_{43}\rho_4\omega_{A4}\varepsilon = 0.25 \left| 1.68 \frac{\text{kg}}{\text{L}} \right| 1.0 \left| 0.5 = 0.21 \frac{\text{kg } A}{\text{L}}\right.$$

For τ_d use Eq. 7.3-18 with $\rho^0_{A2} = 0$.

$$\tau_d = \left(0.21 \frac{\text{kg}}{\text{L}} \bigg| 10\,\text{m}\right) \bigg/ \left(3.3 \frac{\text{m}}{\text{yr}} \bigg| 4\text{E-}6 \frac{\text{kg}}{\text{L}}\right) = 1.6\text{E5 yr}$$

The exit concentration is 4.0 mg/L for this time period.

(b) Use the two middle terms in Eq. 7.3-17.

$$R_{A3} = 1 + \rho_3(1 - \varepsilon)K^a_{A32}/\varepsilon = 38.8$$

$$\frac{m_A}{Al} = R_{A3}\varepsilon\rho^*_{A2} = 38.8(0.5)4\text{E-}6 = 7.8\text{E-}5 \frac{\text{kg } A}{\text{L}}$$

(c) For τ_c use Eq. 7.3-22: $\tau_c = 38.8(0.5)10/3.3 = 59$ yr. Use Eq. 7.3-24 with $\rho^0_{A2} = 0$, $t/\tau_c = 100/59 = 1.69$.

$$n = 1: \quad \frac{nt}{\tau_c} = 1.69(1) = 1.69, \quad \frac{\rho^1_{A2}}{\rho^*_{A2}} = \exp(-1.69) = 0.185$$

$$n = 3 \quad \frac{nt}{\tau_c} = 1.69(3) = 5.07, \quad \frac{\rho^3_{A2}}{\rho^*_{A2}} = \exp(-5.07)\left(1 + 5.07 + \frac{5.07^2}{2}\right) = 0.119$$

$$n = 5 \quad \frac{nt}{\tau_c} = 1.69(5) = 8.45, \quad \frac{\rho^5_{A2}}{\rho^*_{A2}} = \exp(-8.45)(359) = 0.0768$$

$\rho^*_{A2} = 4.0$ m/L; therefore, concentrations are, respectively, 0.74, 0.48, and 0.31 mg/L. The models above are generally restricted to aquifers of homogeneous porous media, uniform flow, and constant cross-sectional area.

Equation 7.3-24 gives the concentration at the exit from n tanks in series for a step decrease in the input concentration to tank 1. Figure 7.3-3 is an analogous graphical solution for a step increase in the input concentration to tank 1. To use the graphical solution for Eq. 7.3-24, subtract the ordinate value from 1.00.

Kinematic Dispersion. The subject of mechanical and hydrodynamic dispersion in porous media was introduced in Section 3.3; see specifically the discussion "Porous Media." An ideal model for the hydrodynamic dispersion using bundles of capillary tubes is presented along with a generalized correlation. De Marsily[37] develops the subject in more detail.

In a single uniform-diameter, capillary with the fluid in laminar flow, it moves faster in the center region than in the wall region. Molecular diffusion between these regions produces a progressive spreading of the transported substance, and the coefficient of dispersion is proportional to the mean velocity squared (i.e., Eq. 3.3-27). In the case of the pores between the grains, differences of aperture and travel distances from one pore to another create a difference in mean velocities. The fluids traveling by each of the paths mix with each other, causing a dilution of concentration. It also causes a spreading of the substance at right angles to the main direction of flow. Disregarding molecular diffusion the dispersion coefficient is proportional to mean velocity.[43] Other large-scale heterogenetic features, such as lenses, layered deposits, and broken or fractured zones, behave with respect to the velocity field in the same way, causing the substance transported to mix and spread in all directions of space.

It appears that the dispersion coefficient in uniform porous media is a function of mean velocity. De Marsily observes that the division of the transport into a convective term quantified by a fictitious mean Darcy velocity representing the mean displacement and a dispersive term, integrating the effects of heterogeneities, is quite arbitrary. However, the mathematical form given to the kinematic dispersion coefficient has both a theoretical and an experimental origin. The mathematical form suggested is similar to Fick's law with a dispersion coefficient $\mathbf{D}_{A3}^{(t)}$, which is (1) a symmetrical second-order tensor; (2) has its principal component in the flow direction and the two others at right angles; and (3) each is a function of flow velocity. Expressed in these principal directions of anisotropy it is

$$\mathbf{D}_{A3}^{(t)} \equiv \begin{vmatrix} \mathscr{D}_{A3x}^{(t)} & 0 & 0 \\ 0 & \mathscr{D}_{A3y}^{(t)} & 0 \\ 0 & 0 & \mathscr{D}_{A3z}^{(t)} \end{vmatrix} \tag{7.3-25}$$

where the x component is the longitudinal dispersion coefficient in the direction of flow and the other two are the y and z components, the traverse dispersion coefficients, which are numerically equal. This ideal behavior of $\mathbf{D}_{A3}^{(t)}$ is valid provided that the porous medium is uniform in the x, y, and z directions. *Uniform* here means that bedding planes and other sedimentary-derived features, for example, do not exist. *A* truly uniform porous medium rarely exists in nature and is difficult to create even in the laboratory. The superscript (t) on the dispersion coefficients is used to denote the fluid mixing and spreading process described above rather than to imply that the flow within the pores is turbulent.

The coefficients are related to the mean pore or Darcy velocity. In describing this functionality, the media Peclet number is used; it is defined as $\text{Pe}_m = ud/\mathscr{D}_{A2}$, where u is the fictitious mean convective velocity; it is the Darcy velocity divided by the porosity, $u \equiv v/\varepsilon$. The molecular diffusion coefficient is $\mathscr{D}_{A2}$ and d is a characteristic length of the porous medium, such as mean diameter, pore size, lens size, or fracture block size. Depending on the size of the Peclet number, five flow regimes have been defined.[37] The laboratory data shown in Fig. 7.3-5, giving the dispersion coefficient-molecular diffusivity ratio as a function of Pe_m for one-dimensional flow, covers this range of flow regimes. However, in the domain of useful velocities, $\text{Pe}_m > 10$, the following relations are generally used:

$$\mathscr{D}_{A3x}^{(t)} = \alpha_L v_x \tag{7.3-26a}$$

and displayed in the figure. For the traverse coefficients

$$\mathscr{D}_{A3y}^{(t)} = \mathscr{D}_{A3z}^{(t)} = \alpha_T v_x \tag{7.3-26b}$$

The α_L and α_T parameters have dimensions of length and are known as intrinsic dispersion coefficients or dispersivities.

Laboratory-derived coefficients are of little use in forecasting a real migration of substances in the field, where the scale of heterogeneities is different and the coefficients are much larger. Consequently, they have to be measured by tracer experiments, which are interpreted by analytical or numerical methods. When measured in the laboratory on a column of sand, α_L is on the order of a few centimeters. In the field it is on the order of a meter or hundreds of

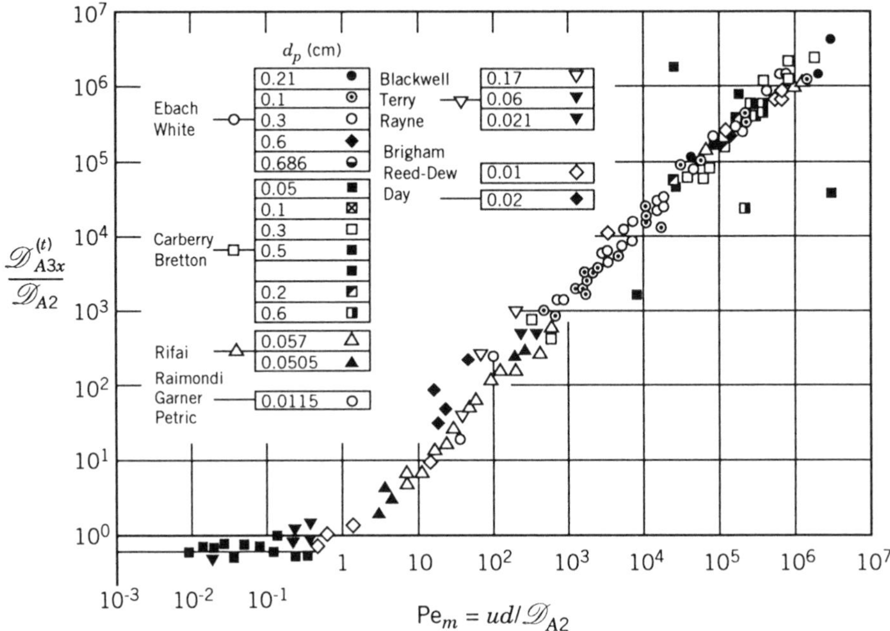

Figure 7.3-5. $\mathscr{D}^{(t)}_{A3x}/\mathscr{D}_{A2}$ versus Pe_m correlation. (Adapted from Ref. 50.)

meters, depending on the degree of heterogeneity of the formation. However, α_T is much smaller, between $\frac{1}{3}$ and $\frac{1}{100}$ of α_L, according to de Marsily.

There are other properties of dispersivity, including the *scale effect*. It is to be appreciated that the α's are not a function of the properties of the porous media alone, as the intrinsic permeability k_3, is, but also a function of the velocity field. For these and other details, see de Marsily.[37]

Analytical Solution of Dispersion Equation. If we choose a one-dimensional case and study the movement of a trace contaminant in a semi-infinite medium, there is a useful analytical solution. The one-dimensional tracer dispersion equation is

$$\mathscr{D}^{(t)}_{A3x}\frac{\partial^2 \rho_{A2}}{\partial x^2} - v_x\frac{\partial \rho_{A2}}{\partial x} = \varepsilon R_{A3}\frac{\partial \rho_{A2}}{\partial t} \tag{7.3-27}$$

For boundary conditions represented by the step-function input concentration and described mathematically by

$$\rho_{A2}(x, 0) = \rho^0_{A2} \qquad x \geqslant 0 \tag{7.3-28a}$$

$$\rho_{A2}(0, t) = \rho^*_{A2} \qquad t \geqslant 0 \tag{7.3-28b}$$

$$\rho_{A2}(\infty, t) = \rho^0_{A2} \qquad t \geqslant 0 \tag{7.3-28c}$$

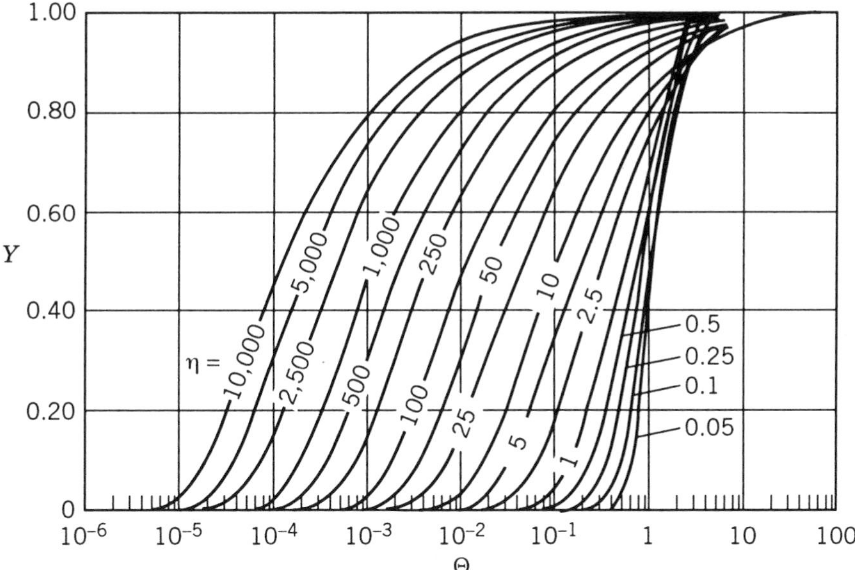

Figure 7.3-6. One-dimensional solution of the dispersion equation, Eq (7.3-29). (Reprinted with permission. Adapted from Ref. 37.)

the solution is

$$\frac{\rho_{A2} - \rho_{A2}^{0}}{\rho_{A2}^{*} - \rho_{A2}^{0}} = \frac{1}{2}\left\{\operatorname{erfc}\left[\frac{x - (v_x/\varepsilon R_{A3})t}{2\sqrt{\mathscr{D}_{A3x}^{(t)}t/\varepsilon R_{A3}}}\right] + \exp\left(\frac{v_x x}{\mathscr{D}_{A3x}^{(t)}}\right)\operatorname{erfc}\left[\frac{x + (v_x/\varepsilon R_{A3})t}{2\sqrt{\mathscr{D}_{A3x}^{(t)}t/\varepsilon R_{A3}}}\right]\right\} \tag{7.3-29}$$

where erfc is the complementary error function (see Appendix B). *A* graphical solution of Eq. 7.3-29 is given in Fig. 7.3-6. The abscissa, ordinate, and parameter are the following dimensionless groups of variables, respectively:

$$\Theta \equiv \frac{v_x t}{\varepsilon R_{A3} x} \qquad Y \equiv \frac{\rho_{A2} - \rho_{A2}^{0}}{\rho_{A2}^{*} - \rho_{A2}^{0}} \qquad \eta \equiv \frac{\mathscr{D}_{A3x}^{(t)}}{v_x x} \tag{7.3-30}$$

where Θ and Y are dimensionless time and concentration, respectively. Note that η is the reciprocal of the formation Peclet number. It is not to be confused with the media Peclet number. This equation is restricted to trace quantities of chemicals in porous media, constant cross-sectional area, and uniform flow. *A* typical desired solution of Eq. 7.3-29 or Fig. 7.3-6 is as follows. For a step-input concentration ρ_{A2}^{*} into an aquifer initially containing ρ_{A2}^{0} quality porewater, the solution yields the concentration ρ_{A2}, at a distance x after time t. The parameters v_x, ε, R_{A3}, and $\mathscr{D}_{A3x}^{(t)}$ must be specified. The figure is most useful for chemical arrival times less than the plug flow time at $\Theta = 1$.

Some Chemodynamic Processes in Zone 3. Return to the subterranean contamination problem illustrated in Fig. 7.3-1. Ideally, during the dissolution time period while NAL exists in zone 2 the soluble material enters zone 3, contaminating it. The exit concentration from zone 2 was assumed to be ρ_{A2}^*, a constant, and in the case of pure NAL it may be as high as its solubility. It is not unrealistic, then, to assume that a steplike function on the flow field of this concentration is imposed on zone 3, which begins at $x = 0$ and $t = 0$. In this case, Eq. 7.3-29 can be used to estimate the concentration of trace substance A throughout zone 3 at position x and the time t after input from zone 2 begins. The ρ_{A2}^0 is the initial concentration in zone 3; if uncontaminated initially, it is zero.

In a similar manner the solution given by Eq. 7.3-29 can be used in reverse. That is, if clean water of concentration $\rho_{A2}^* = 0.0$ is injected at the head of zone 3 (i.e., $x = 0$) with an initial level of contamination ρ_{A2}^0 throughout, the left-hand side of Eq. 7.3-29 becomes

$$Y \equiv \frac{\rho_{A2}^0 - \rho_{A2}}{\rho_{A2}^0 - \rho_{A2}^*} \tag{7.3-31}$$

The other parameters remain the same in Eq. 7.3-30, so Fig. 7.3-6 can also be used to obtain $\rho_{A2}(x, t)$ under contaminant dissolution conditions.

Example 7.3-3. HCB Dissolution in Zone 2 of Shallow Aquifer. For the aquifer of Examples 7.3-1 and 7.3-2 estimate the exit concentration of HCB from zone 2 assuming that it is initially contaminated uniformly at 4 mg/L and then flushed with fresh water for 100 years. Assume that the zone is 10 m in length, $\eta = 0.1$, and use the data in the previous examples as needed.

SOLUTION Use Eq. 7.3-29 with Y of Eq. 7.3-31. $v_x = 3.3$ m/yr.

$$\frac{v_x}{\varepsilon R_{A3}} t = 17.0 \text{ m} \qquad \mathscr{D}_{A3x}^{(t)} = \eta v_x x = 3.3 \text{ m}^2/\text{yr} \qquad 2\sqrt{\frac{\mathscr{D}_{A3x}^{(t)} t}{\varepsilon R_{A3}}} = 8.25 \text{ m}$$

Substituting into Eq. 7.3-29 yields, from Appendix B,

$$\frac{\rho_{A2}^0 - \rho_{A2}}{\rho_{A2}^0} = \frac{1}{2}\left[\operatorname{erfc}\left(\frac{10 - 17}{8.25}\right) + \exp(10)\operatorname{erfc}\left(\frac{10 + 17}{8.25}\right)\right] = 0.897$$

$$\rho_{A2} = \rho_{A2}^0(1 - 0.897) = 4.0(0.10) = 0.40 \text{ mg/L}$$

Now using Fig. 7.3-6 with $\Theta = 1.70$ and $\eta = 0.1$ gives $Y \simeq 0.9$, which gives approximately the same exit concentration.

At this junction the reader should appreciate that the exit concentration of HCB computed above is within the range of values of those computed for zone 2 using the tanks-in-series model. See part (c) of Example 7.3-2.

Closure. The intent of this section was to introduce the student to a few major processes that produce ground contamination and spread chemical pollution about. The emphasis was on the capture and retention of contaminants within media solids and the subsequent translocation by moving groundwaters. Vignette mathematical models that assume chemical equilibrium between the NAL or solid phase and the porewater, restricted in dimensionality and media complexity, were used in quantifying some basic concepts of dissolution in porous media. The example problems reveal that natural recovery by dissolution alone of low-solubility constituents will involve extremely long time periods for chemicals with moderate to low solubility in water. However, the solubilization process often must precede processes that destroy or otherwise eliminate the pollution source.

The subject of fate and transport as presented here is incomplete. In some cases, in situ chemical reactions driven by biotic and abiotic processes are capable of converting the hazardous constituents to harmless products or other hazardous constituents. Other in situ processes that may be of importance at sites include physical filtration and entrapment or release of colloids, ion exchange, substance behavior in fractured media, radioactive waste, and the associated heat effects on porous media, including salt deposits. These and related factors are receiving intense research attention aimed primarily at aquifer restoration. Despite this enormous effort and cost, the prospect of effective and timely aquifer restoration (with exceptions) remains extremely elusive.

PROBLEMS

7.3A. Chemical Movement Through Recompacted Clay Liner

An aqueous solution containing the substances listed in Table 7.3A is to be placed in the surface impoundment illustrated in Fig. 4.1-1. A 91.4-cm liner, 40% porosity, 1.5 g/cm^3 particle density, and the equivalent of 1.68% natural organic matter is to be used.

Table 7.3A. Retardation Factors for Selected Chemicals[a]

Substance	K^*_{AC2} (L/kg)	R_{A3}
A-1254	120,000	4500
Kepone	4,800	180
Phenol	8.4	1.3
HTO	0.0	1.0

[a] $\omega_c = 0.0168$, $\varepsilon = 0.4$, $\rho_3 = 1.5$ g/cm^3.

1. Assuming a water depth of 2.0 m, estimate the Darcian velocity through the liner in cm/yr. The hydraulic conductivity of the recompacted clay is 1E-7 cm/s.
2. Based on a 0.5-mm-diameter particle, estimate the hydrodynamic dispersion coefficient for an assumed 1E-6 cm^2/s molecular diffusivity.
3. Confirm the retardation factors in Table 7.3A.
4. Confirm the breakthrough time for each substance in Table 7.3-2 if diffusion, advection, and combined transport is occurring through the liner. Assume that $Y = 1E\text{-}4$.

7.3B. Waste Injection Legacy

An aqueous waste containing primarily phenol was injected into the lower sandstone strata as shown in Fig. 7.3B. Piezometer P_1 indicates 61 m, $P_2 = 20$ m water pressure, and $K_3 = 8E\text{-}8$ cm/s for the shale with 0.2 effective porosity.

1. Estimate the Darcian velocity in cm/yr.
2. Over a 10-year period, phenol has penetrated about 1 m into the shale, the concentration being 20% of the concentration in the lower sandstone layer. Estimate the longitudinal diffusion coefficient and the dispersivity of the shale. In this case the piezometric pressure is the height of water rise from the bottom of the borehole.

(Problem contributed by Sarah Sharygi.)

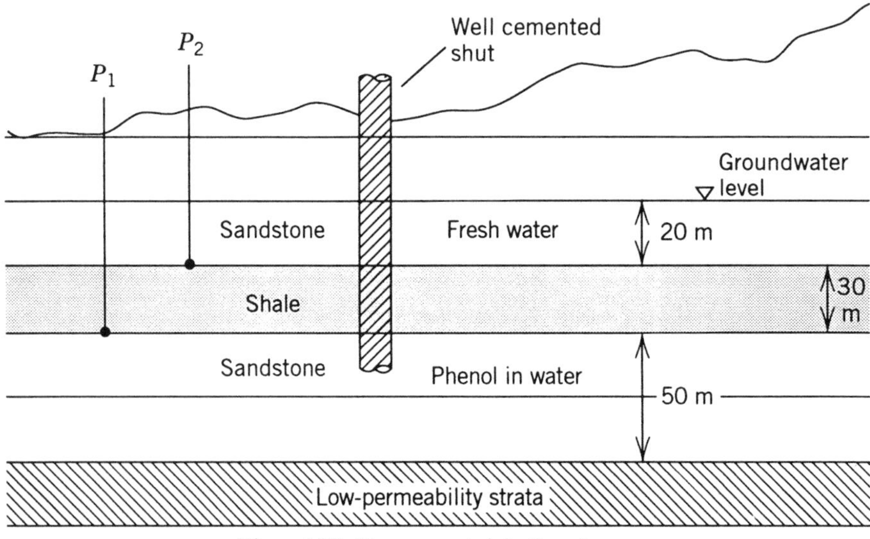

Figure 7.3B. Former waste injection site.

7.3C. Leaky Landfill by River

A landfill is leaking a hazardous leachate into an aquifer and endangering the nearby river 3.9 km down-gradient. The unconfined aquifer has a hydraulic conductivity of 210 m/day and 0.2 effective flow porosity. Two monitoring wells placed 910 m apart have groundwater depths of 19.2 and 18.3 m, respectively.

1. What is the Darcian velocity, and how long will it take the leachate–water to reach the river?

2. For a nonadsorbing tracer, estimate the percent leachate mixture concentration at the river after each year of landfill operation. Assume that $\alpha_L = 10\,\text{m}$.

7.3D. Dissolution Time for Species *A* in NAL Mixture

Rederive the dissolution model for the case of a NAL consisting of a mixture of two or more miscible liquids and show that for fresh water entering zone 2 the dissolution time versus mass fraction *A* is

$$t = \frac{2S_{43}^{0}\rho_4\varepsilon\,\Delta x}{v_x\rho_{A2}^{*}}\,\frac{(1 - \omega_{A4}/\omega_{A4}^{0})M_A}{\gamma_{A4}M_4} + \frac{\varepsilon\Delta x}{v_x}\ln\frac{\omega_{A4}^{0}}{\omega_{A4}} \tag{7.3D}$$

(*Hint*: Assume variation of residual saturation tracks ω_{A4} such that $S_{43} = S_{43}^{0}\omega_{A4}/\omega_{A4}^{0}$, where the superscript zero denotes the initial values.)

7.3E. Series-of-Stirred-Tanks Model

Show the necessary steps in arriving at Eqs. 7.3-22 to 7.3-23.

REFERENCES

1. G. T. Orlob, L. A. Roesner, and W. R. Norton, "Mathematical Models for the Prediction of Thermal Energy Changes in Impoundments," 16130EXT12/69, Water Pollution Control. Research Series, Washington, D.C., 1969.
2. L. J. Thibodeaux, "Semi-infinite Solid Model for Prediction of Temperature in Deep Reservoirs and Lakes," *Water Resources Bull.*, **11**(3), 449–454 (1975).
3. H. S. Carslaw and J. C. Jaeger, *Conduction of Heat in Solids*, 2nd ed., Oxford University Press, London, 1958, p. 76.
4. F. L. Parker, B. A. Benedict, and C. Tsai, "Evaluation of mathematical Models for Temperature Prediction in Deep Reservoirs," EPA-660/3-75-38, U.S. EPA, Corvallis, Oreg., 1975.
5. D. N. Lapedes, *Encyclopedia of Enviromental Science*, McGraw-Hill, New York, 1974, pp. 619–620.

6. G. Neumann and W. J. Pierson, Jr., *Principles of Physical Oceanography*, Prentice Hall, Englewood Cliffs, N. J., 1966, pp. 445–450.
7. C. Y. Koh and L.-N. Fan, "Mathematical Models for the Prediction of Temperature Distributions Resulting from the Discharge of Heated Water into Large Bodies of Water," 16130DW010/70, Water Pollution Control Research Series, Washington, D.C., 1970.
8. C. Eckart, *Hydrodynamic of Oceans and Atmosphere*, Pergamon Press, Elmsford, N.Y., 1960, pp. 57–71.
9. A. Lerman, "Time to Chemical Steady-States in Lakes and Oceans," in J. D. Hem, Ed., *Non-equilibrium Systems in Natural Water Chemistry*, Advances in Chemistry Series 106, American Chemical Society, Washington, D.C., 1971, pp. 30–76.
10. J. Crank, *The Mathematics of Diffusion*, Oxford University Press, Oxford, 1956, pp. 9–61, 121–146.
11. D. J. O'Conner, R. V. Thomann, D. M. Ditoro, and N. H. Brooks, "Mathematical Modeling of Natural Systems," Summer Institute in Water Pollution Control, Manhattan College, New York, 1974.
12. R. V. Thomann, *Systems Analysis and Water Quality Management*, McGraw-Hill, New York, 1972, p. 104.
13. G. T. Csanady, *Turbulent Diffusion in the Environment*, D. Reidel, Boston, 1973.
14. D. H. Slade, Ed., *Meterology and Atomic Energy*, U.S. Atomic Energy Commission, Oak Ridge, Tenn., 1968.
15. O. G. Sutton, *Micrometeorology*, McGraw-Hill, New York, 1953.
16. N. H. Brooks, in *Proceedings of the 1st International Conference on Waste Disposal in the Marine Environment*, Pergamon Press, New York, 1960, p. 246.
17. L. R. Dodd, personal communication, 1988.
18. S. A. Hsu, *Coastal Meteorology*, Academic Press, San Diego, Calif., 1988, p. 221.
19. S. W. Zison, W. B. Mills, D. Deimer, and C. W. Chen, "Rates, Constants and Kinetics Formulations in Surface Water Quality Modeling," *EPA-600/3-78-105*, U.S. EPA, Athens, Ga., Dec. 1978.
20. K. R. Dyer, *Estuaries: A Physical Introduction*, Wiley, New York, 1973, p. 140.
21. J. P. Lawler, G. Apicella, P. J. Lawler, and T. Vanderbeek, *Water Quality Modeling and Waste Load Allocation Workshop*, manual for American Petroleum Institute, New Orleans, La., Mar. 1985.
22. J. P. Lawler, "Mathematical Aspects of Estuarine Pollution," in *Engineering Aspects of Marine Waste Disposal*, Environmental Health Sciences and Engineering Training Course Manual, R. A. Taft Sanitary Engineering Center, Cincinnati, Ohio, 1965.
23. H. K. Trivedi, "Transport Phenomena in a Bay-Marsh System," Ph.D. dissertation, Louisiana State University, Baton Rouge, La., 1976.
24. E. R. Holley, and G. H. Jirka, 1986, "Mixing in Rivers," *Technical Report E-86-11*, U.S. Army Corps of Engineers Waterways Experiment Station, Vicksburg, Miss.
25. G. I. Taylor, "The Dispersion of Matter in Turbulent Flow Through a Pipe," *Proc. Roy. Soc.*, **A223**, 446–468 (1954).
26. R. V. Thomann, *System Analysis and Water Quality Management*, McGraw-Hill, New York, 1972, p. 165.

27. J. L. Schnoor, C. Sato, D. McKechnie, and D. Sahoo, "Processes, Coefficients and Models for Simulating Toxic Organics and Heavy Metals in Surface Waters," *Technical Report EPA/600/3-87/015*, U.S. EPA, Athens, Ga., 1987.

28. R. B. Stull, *An Introduction to Boundary Layer Meteorology*, Kluwer Academic Publishers, Dordrecht, The Netherlands, 1988.

29. S. R. Hanna, G. A. Briggs, and R. P. Hosker, Jr., *Handbook on Atmospheric Diffusion*, U.S. Department of Commerce, National Technical Information Center, Springfield, Va., 1982.

30. D. B. Turner, *Workbook of Atmospheric Dispersion Estimates*, Publ. AP-26, U.S. EPA, Washington, D.C., 1971.

31. J. H. Seinfeld, *Atmospheric Chemistry and Physics of Air Pollution*, Wiley, New York, 1986.

32. R. A. Dobbins, *Atmospheric Motion and Air Pollution*, Wiley, New York, 1979.

33. S. A. Hsu, *Coastal Meteorology*, Academic Press, San Diego, Calif., 1988.

34. L. W. Baker and K. P. MacKay, "Screening Models for Estimating Toxic Air Pollution Near Hazardous Waste Landfills," *J. Air Pollut. Control Assoc.*, **35**(11), 1190 (1985).

35. N. P. Chitgopekar, D. D. Reible, and L. J. Thibodeaux, "Modeling Short Range Air Dispersion from Area Sources of Non-buoyant Toxics," *J. Air Waste Manage. Assoc.*, **40**, 1121–1128 (1990).

36. R. A. Freeze and J. A. Cherry, *Groundwater*, Prentice Hall, Englewood Cliffs, N.J., 1979.

37. G. de Marsily, *Quantitative Hydrogeology*, Academic Press, Orlando, Fl., 1986.

38. F. Schwille, *Dense Chlorinated Solvents in Porous and Fractured Media*, translated by J. F. Pankow, Lewis Publishers, Chelsea, Mich., 1988.

39. J. Clarke, D. D. Reible, and R. Mutch, "Contaminant Transport and Behavior in the Subsurface", Chapter 1 in D. J. Wilson and A. Clarke, Ed., *Hazardous Waste Soil Remediation; Theory and Application of Innovative Technologies*, Marcel Dekker, New York, 1993, pp. 1–49.

40. D. D. Reible and T. H. Illangasekare, "Subsurface Processes of Non-aqueous Phase Liquids," in D. Allen, Y. Cohen, and I. Kaplan, Eds., *Intermedia Pollutant Transport: Modeling and Field Measurements*, Plenum Press, New York, 1989, pp. 237–254.

41. D. D. Reible, T. H. Illangasekare, D. V. Doshi, and M. E. Malhiet, "Infiltration of Immiscibe Contaminants in the Unsaturated Zone," *Ground Water*, Aug.–Sept. 1990, pp. 685–692.

42. P. Combes, P. Goblet, and G. de Marsily, "Analyse d'une pollution au pyralene de l'aquifère de cusset près de Lyon," LHM/RD/86/72 CIG Ecole Nationale Superieure Des Mines De Paris, Fontainbleau, France, 1986.

43. J. Bear, *Dynamics of Fluids in Porous Media*, Dover Publication, New York, 1972.

44. J. L. Wilson, S. H. Conrad, W. R. Mason, W. Peplinski, and E. Hogan, "Laboratory Investigation of Residual Liquid Organics," *EPA/600/6- 90/004*, U.S. EPA, Roberts S. Kerr Environmental Research Laboratory, Office Research and Development, Ada, Okla., Apr. 1990, p. 143.

45. D. U. Von Rosenberg, "Mechanics of Steady-State Single Phase Fluid Displacement from Porous Media," *J. Am. Inst. Chem. Eng.*, **2**, 55–58 (1956).

46. D. M. Mackey and J. A. Cherry, "Groundwater Contamination: Pump-and-Treat Remediation," *Environ. Sci. Technol.*, **23**(6), 630–636 (1989).
47. J. L. Smith, "An Experimental Study of the Transport and Fate of a NAPL in a Heterogeneous Soil and Its Removal by Vacuum Extraction," M.S. thesis, Louisiana. State University, Baton Rouge, La. 1994.
48. A. T. Kan, G. Fu, and M. B. Tomson, "Adsorption/Desorption Hysteresis in Organic Pollutant and Soil/Sediment Interaction," *Environ. Sci. Technol.*, **28**(5), 859–867 (1994).
49. J. M. Smith, *Chemical Engineering Kinetics*, 3rd ed., McGraw-Hill, New York, 1981, pp. 279–285.
50. H. O. Pfankuch, *Rev. Inst. Fr. Petrol.*, **18**, 215 (1963).
51. H. S. Fogler, *Elements of Chemical Reaction Engineering*, 2nd, ed., Prentice Hall, Englewood Cliffs, N.J., 1992, p. 765–771.

APPENDIX A

METRIC SYSTEM OF MEASUREMENT AND CONVERSION TABLE

Interpretation and Modification of the International System of Units for the United States

The International System of Units (SI) is constructed from seven base units for independent quantities plus two supplementary units for plane angle and solid angle, as listed in Table 1.1-1. Units of all other quantities are derived from these units. In Table A.1 are listed 19 SI derived units with special names that were derived from the base and supplementary units in a coherent manner. All other SI derived units, such as those in Table A.2, are similarly derived in a coherent manner from the 28 base, supplementary, and special-name SI units. For use with the SI units there is a set of 16 prefixes (see Table A.3) to form multiples and submultiples of these units. It is important to note that the kilogram is the only SI unit with a prefix.

Table A.1. Derived (SI) Units with Special Names

	SI Unit		
Quantity	Unit	Symbol	Expressed as:
Frequency	hertz	Hz	s^{-1}
Force	newton	N	$kg \cdot m/s^2$
Pressure, stress	pascal	Pa	N/m^2
Energy, work, quantity of heat	joule	J	$N \cdot m$
Power, radiant flux	watt	W	J/s
Quantity of electricity, electric charge	coulomb	C	$A \cdot s$
Electronic potential, potential difference, electromotive force	volt	V	W/A
Electrical capacitance	farad	F	C/V
Electrical resistance	ohm	Ω	V/A
Electrical conductance	siemens	S	A/V
Magnetic flux	weber	W	$V \cdot s$
Magnetic flux density	tesla	T	Wb/m^2
Inductance	henry	H	Wb/A

Table A.1. ***(Continued)***

Quantity	SI Unit		
	Unit	Symbol	Expressed as:
Celsius temperature	degree Celsius	°C	K
Luminous flux	lumen	lm	cd · sr
Illuminance	lux	lx	lm/m^2
Activity of a radionuclide	becquerel	Bq	s^{-1}
Absorbed dose	gray	Gy	J/kg
Dose equivalent	sievert	Sv	J/kg

Table A.2. Some Common Derived Units of SI

Quantity	Unit	Symbol
Absorbed dose rate	gray per second	Gy/s
Acceleration	meter per second squared	m/s^2
Angular acceleration	radian per second squared	rad/s^2
Angular velocity	radian per second	rad/s
Area	square meter	m^2
Concentration (of amount of substance)	mole per cubic meter	mol/m^3
Current density	ampere per square meter	A/m^2
Density, mass	kilogram per cubic meter	kg/m^3
Electric charge density	coulomb per cubic meter	C/m^3
Electric density	joule per cubic meter	J/m^3
Electric field strength	volt per meter	V/m
Electric flux density	coulomb per square meter	C/m^2
Entropy	joule per kelvin	J/K
Exposure (x and gamma rays)	coulomb per kilogram	C/kg
Heat capacity	joule per kelvin	J/K
Heat flux density irradiance	watt per square meter	W/m^2
Molar energy	joule per mole	J/mol
Molar entropy	joule per mole kelvin	J/(mol · K)
Molar heat capacity	joule per mole kelvin	J/(mol · K)
Moment of force	newton meter	N · m
Power density	watt per square meter	W/m^2
Specific energy	joule per kilogram	J/kg
Specific entropy	joule per kilogram kelvin	J/(kg · K)
Specific heat capacity	joule per kilogram kelvin	J/(kg · K)
Specific volume	cubic meter per kilogram	m^3/kg
Surface tension	newton per meter	N/m
Thermal conductivity	watt per meter kelvin	W/(m · K)
Velocity	meter per second	m/s
Viscosity, dynamic	pascal second	Pa · s
Viscosity, kinematic	square meter per second	m^2/s
Volume	cubic meter	m^3
Wave number	1 per meter	1/m

Table A.3. SI Prefixes

Factor	Prefix	Symbol
10^{18}	exa	E
10^{15}	peta	P
10^{12}	tera	T
10^{9}	giga	G
10^{6}	mega	M
10^{3}	kilo	k
10^{2}	hecto	h
10^{1}	deka	da
10^{-1}	deci	d
10^{-2}	centi	c
10^{-3}	milli	m
10^{-6}	micro	μ
10^{-9}	nano	n
10^{-12}	pico	p
10^{-15}	femto	f
10^{-18}	atto	a

Certain units not part of the SI which are used so widely that they are accepted for continued use in the United States with the International System are listed in Table A.4. In those cases where their use is already well established, use, for a limited time, of the units in Table A.5 is accepted subject to future review.

Table A.4. Units in Use with the International System

Name	Symbol	Value in SI Unit
Minute (time)	min	1 min = 60 s
Hour	h	1 h = 60 min = 3600 s
Day		1 day = 24 h = 86,400 s
Degree (angle)	$^\circ$	$1^\circ = (\pi/180)$ rad
Minute (angle)	$'$	$1' = (1/60)^\circ = (\pi/10{,}800)$ rad
Second (angle)	$''$	$1'' = (1/60)' = (\pi/648{,}800)$ rad
Liter	L	1 L = 1 $dm^3 = 10^{-3}$ m^3
Metric ton	t	1 t = 10^3 kg
Hectare (land area)	ha	1 ha = 10^4 m^2

Table A.5. Units in Use with SI Temporarily

Quantity	Unit	Symbol	Definition
Energy	kilowatthour	kWh	1 kWh = 3.6 MJ
Cross section	barn	b	1 b = 10^{-28} m^2
Pressure	bar	bar	1 bar = 10^5 Pa
Activity (of a radionuclide)	curie	Ci	1 Ci = 3.7×10^{10} Bq
Exposure (x and gamma ray)	roentgen	R	1 R = 2.58×10^{-4} C/kg
Absorbed dose	rad	rd	1 rd = 001 Gy

Conversion Table A.6 and Directions for Its Use

- Avoid use of prefixes in denominators (except kg).
- The use of hecto, deka, deci, and centi prefixes should be avoided except when used in areas and volumes.
- SI symbols are not capitalized unless the unit is derived from a proper name (e.g., Hz for H. R. Hertz). Unabbreviated units are not capitalized (e.g., hertz, newton, kelvin). Only the T, G, and M prefixes are capitalized.
- Except at the end of a sentence, SI units are not to be followed by periods.
- Four or more digits in a group should be separated in groups of three with no comma (e.g., 1 983 212.322 7, not 1,983,212.3227).
- With derived unit abbreviations, use the center dot to denote multiplication and a slash for division (e.g., newton-second/meter^2 = $N \cdot s/m^2$).

Table A.6. Unit Conversion Factors

To Convert from:	To:	Multiply by[a]:
angstrom	meter (m)	1.000 000*E-10
atmosphere (normal)	newton/meter^2 (N/m^2)	1.013 250*E + 05
barrel (for petroleum: 42 gal)	meter^3 (m^3)	1.589 873E-01
British thermal unit (International Table)	joule (J)	1.055 056E + 03
Btu/ft^2-hr-°F (heat transfer coefficient)	joule/meter^2-second-kelvin ($J/m^2 \cdot s \cdot K$)	5.678 264E + 00
Btu/ft^2-hr (heat flux)	joule/meter^2-second ($J/m^2 \cdot s$)	3.154 591E + 00
Btu/ft-hr-°F (thermal conductivity)	joule/meter-second-kelvin ($J/m \cdot s \cdot K$)	1.730 735E + 00
Btu/h	watt (W)	2.930 711E-01
Btu/lb_m-°F (c, heat capacity)	joule/kilogram-kelvin ($J/kg \cdot K$)	4.186 800*E + 03
cal/g-°C	joule/kilogram-kelvin ($J/kg \cdot K$)	4.186 800*E + 03
calorie (International Table)	joule (J)	4.186 800*E + 00

Table A.6. ***(Continued)***

To Convert from:	To:	Multiply by[a]:
centipoise	pascal · second (Pa · s) (N · s/m^2)	1.000 000*E-03
centistoke	meter^2/second (m^2/s)	1.000.000*E-06
degree Celsius	kelvin (K)	$t_K = t_C + 273.15$
degree Fahrenheit	kelvin (K)	$t_k = (t_F + 459.67)/1.8$
degree Rankine	kelvin (K)	$t_K = t_R/1.8$
dyne	newton (N)	1.000 000*E-05
erg	joule (J)	1.000 000*E-07
fluid ounce (U.S.)	meter^3 (m^3)	2.957 353E-05
foot	meter (m)	3.048 000*E-01
foot^2/hour	meter^2/second (m^2/s)	2.580 640*E-05
foot-pound-force	joule (J)	1.355 818E + 00
foot^2/second	meter^2/second (m^2/s)	9.290 304*E-02
gallon (U.S. liquid)	meter^3 (m^3)	3.785 412E-03
horsepower (550 ft · lb$_f$/s)	watt (W)	7.456 999E + 02
inch	meter	2.540 000*E-2
inch of mercury (60°F)	pascal (Pa) (N/m^2)	3.376 85E + 03
inch of water (60°F)	pascal (Pa) (N/m^2)	2.488 4E + 02
knot (international)	meter/second (m/s)	5.144-444E-01
micron	meter (m)	1.000.000*E-06
mil	meter (m)	2.540 000*E-05
mile (U.S. statute)	meter (m)	1.609 344*E + 03
mile/hour	meter/second (m/s)	4.470 400*E-01
millimeter of mercury (0°C)	pascal (Pa) (N/m^2)	1.333 224E + 02
ohm (International of 1948)	ohm (Ω)	1.000 495E + 00
ohm-mass (avoirdupois)	kilogram (kg)	2.834 952E-02
ounce (U.S. fluid)	meter^3 (m^3)	2.957 353E-05
pint (U.S. liquid)	meter^3 (m^3)	4.731 765E-04
poise (absolute viscosity)	pascal · second (Pa · s) (N · s/m^3)	1.000 000*E-01
pound-force (lb$_f$ avoirdupois)	newton (N)	4.448 222E + 00
pound-mass (lb$_m$ avoirdupois)	kilogram (kg)	4.535 924E-01
pound-mass/foot^3	kilogram meter^3 (kg/m^3)	1.601 846E + 01
psi	pascal (Pa) (N/m^2)	6.894 757E + 03
quart (U.S. liquid)	meter^3 (m^3)	9.463 529E-04
slug	kilogram (kg)	1.459 390E + 01
stoke (kinematic viscosity)	meter^2/second (m^2/s)	1.000.000*E-04
ton (long, 2240 lb$_m$)	kilogram (kg)	1.016 047E + 03
ton (short, 2000 lb$_m$)	kilogram (kg)	9.071 847E + 02
torr (mmHg, 0°C)	pascal (Pa) (N/m^2)	1.333 22E + 02
yard	meter (m)	9.144.000*E-01

Source: Abstracted from *Am. Inst. Chem. Eng. J.*, **17**(2), 511–512 (Mar. 1971).

[a] An asterisk after the sixth decimal place indicates that the conversion factor is exact and all subsequent digits are zero.

APPENDIX B

PHYSICAL CONSTANTS AND MATHEMATICAL TABLE

Physical Constants

- Universal gas constants: 0.0821 atm·L/mol·K, 8.319 J/mol·K, 0.7302 atm·ft^3/lb mol·R, 1.987 cal/mol·K
- Acceleration of gravity: 980.7 cm/s^2
- Avogadro's number: 6.023E23 molecules/mol
- Electron charge: 1.6022E-19 C
- Faraday constant: 9.64867E4 C/mol
- Planck's constant: 6.624E-34 J/s
- Stefan–Boltzmann constant: 5.673E-12 J/s·cm^2 K^4
- Boltzmann's constant: (=universal gas constant/Avogadro's number) 1.3805E-23 J/molecule K
- Heat of fusion of water at 1 atm, 0°C: 334 J/g
- Heat of vaporization of water at 1 atm, 100°C: 2260 J/g
- Molecular weight of dry air: 28.97 g/mol
- One mole of an ideal gas at 0°C, 1 atm occupies: 22.4 L
- One atmosphere of pressure is: 760 mmHg = 1.013E5 N/m^2

Table B.1. Error Function

ϕ	erf ϕ	ϕ	erf ϕ	ϕ	erf ϕ
0.00	0.00000 00000	0.74	0.70467 80779	1.48	0.96365 40654
0.02	0.02256 45747	0.76	0.71753 67528	1.50	0.96610 51465
0.04	0.04511 11061	0.78	0.73001 04313	1.52	0.96841 34969
0.06	0.06762 15944	0.80	0.74210 09647	1.54	0.97058 56899
0.08	0.09007 81258	0.82	0.75381 07509	1.56	0.97262 81220
0.10	0.11246 29160	0.84	0.76514 27115	1.58	0.97454 70093
0.12	0.13475 83518	0.86	0.77610 02683	1.60	0.97634 83833
0.14	0.15694 70331	0.88	0.78668 73192	1.62	0.97803 80884
0.16	0.17901 18132	0.90	0.79690 82124	1.64	0.97962 17795
0.18	0.20093 58390	0.92	0.80676 77215	1.66	0.98110 49213
0.20	0.22270 25892	0.94	0.81627 10190	1.68	0.98249 27870
0.22	0.24429 59116	0.96	0.82542 36496	1.70	0.98379 04586

Table B.1. ***(Continued)***

ϕ	erf ϕ	ϕ	erf ϕ	ϕ	erf ϕ
0.24	0.26570 00590	0.98	0.83423 15043	1.72	0.98500 28274
0.26	0.28689 97232	1.00	0.84270 07929	1.74	0.98613 45950
0.28	0.30788 00680	1.02	0.85083 80177	1.76	0.98719 02752
0.30	0.32862 67595	104	0.85864 99465	1.78	0.98817 41959
0.32	0.34912 59948	1.06	0.86614 35866	1.80	0.98909 05016
0.34	0.36936 45293	1.08	0.87332 61584	1.82	0.98994 31565
0.36	0.38932 97011	1.10	0.88020 50696	1.84	0.99073 59476
0.38	0.40900 94534	1.12	0.88678 78902	1.86	0.99147 24883
0.40	0.42839 23550	1.14	0.89308 23276	1.88	0.99215 62228
0.42	0.44746 76184	1.16	0.89909 62029	1.90	0.99279 04292
0.44	0.46622 51153	1.18	0.90483 74269	1.92	0.99337 82251
0.46	0.48465 53900	1.20	0.91031 39782	1.94	0.99392 25709
0.48	0.50274 96707	1.22	0.91553 38810	1.96	0.99442 62755
0.50	0.52049 98778	1.24	0.92050 51843	1.98	0.99489 20004
0.52	0.53789 86305	1.26	0.92523 59418	2.00	0.99532 22650
0.54	0.55493 92505	1.28	0.92973 41930	2.1	0.997021
0.56	0.57161 57638	1.30	0.93400 79449	2.2	0.998137
0.58	0.58792 29004	1.32	0.93806 51551	2.3	0.998857
0.60	0.60385 60908	1.34	0.94191 37153	2.4	0.999311
0.62	0.61941 14619	1.36	0.94556 14366	2.5	0.999593
0.64	0.63458 58291	1.38	0.94901 60353	2.6	0.999764
0.66	0.64937 66880	1.40	0.95228 51198	2.7	0.999866
0.68	0.66378 22027	1.42	0.95537 61786	2.8	0.999925
0.70	0.67780 11938	1.44	0.95829 65696	2.9	0.999978
0.72	0.69143 31231	1.46	0.96105 35095	3.0	1.000000

The error function (Table B.1) is defined as the integral

$$\mathrm{erf}(x) = \frac{2}{\sqrt{\pi}} \int_0^x e^{-y^2}\, dy$$

The limiting values of erf(x) are erf(0) = 0, erf(∞) = 1, erf($-\infty$) = -1. The complementary of the error function is

$$\mathrm{erf}(x) = 1 - \mathrm{erf}(x)$$

The limits are erfc(0) = 1 and erfc(∞) = 0 and the following relationships hold:

$$\mathrm{erfc}(-x) = -\mathrm{erf}(x)$$

$$\mathrm{erfc}(-x) = \begin{cases} 1 - \mathrm{erf}(-x) \\ 1 + \mathrm{erf}(x) \\ 2 - \mathrm{erfc}(x) \end{cases}$$

APPENDIX C

CHEMICAL DATA

Table C.1. Relative Atomic Weights for the Chemical Elements (1971)[a]

Element	Symbol	Atomic Number	Atomic Weight	Element	Symbol	Atomic Number	Atomic Weight
Actinium	Ac	89		Mercury	Hg	80	200.5_9
Aluminum	Al	13	26.98154	Molybdenum	Mo	42	95.9_4
Americium	Am	95		Neodymium	Nd	60	144.2_4
Antimony	Sb	51	121.7_5	Neon	Ne	10	20.17_9
Argon	Ar	18	39.94_8	Neptunium	Np	93	237.0482
Arsenic	As	33	74.9216	Nickel	Ni	28	58.7_1
Astatine	At	85		Niobium	Nb	41	92.9064
Barium	Ba	56	137.3_4	Nitrogen	N	7	14.0067
Berkelium	Bk	97		Nobelium	No	102	
Beryllium	Be	4	9.01218	Osmium	Os	76	190.2
Bismuth	Bi	83	208.9804	Oxygen	O	8	15.999_4
Boron	B	5	10.81	Palladium	Pd	46	106.4
Bromine	Br	35	79.904	Phosphorus	P	15	30.97376
Cadmium	Cd	48	112.40	Platinum	Pt	78	195.0_9
Calcium	Ca	20	40.08	Plutonium	Pu	94	
Californium	Cf	98		Polonium	Po	84	
Carbon	C	6	12.011	Potassium	K	19	39.09_8
Cerium	Ce	58	140.12	Praseodymium	Pr	59	140.9077
Cesium	Cs	55	132.9054	Promethium	Pm	61	
Chlorine	Cl	17	35.453	Protactinium	Pa	91	231.0359
Chromium	Cr	24	51.996	Radium	Ra	88	226.0254
Cobalt	Co	27	58.9332	Radon	Rn	86	
Copper	Cu	29	63.54_6	Rhenium	Re	75	186.2
Curium	Cm	96		Rhodium	Rh	45	102.9055
Dysprosium	Dy	66	162.5_0	Rubidium	Rb	37	85.467_8
Einsteinium	Es	99		Ruthenium	Ru	44	101.0_7
Erbium	Er	68	167.2_6	Samarium	Sm	62	150.4
Europium	Eu	63	151.96	Scandium	Sc	21	44.9559
Fermium	Fm	100		Selenium	Se	34	78.9_6
Fluorine	F	9	18.99840	Silicon	Si	14	28.08_6
Francium	Fr	87		Silver	Ag	47	107.868
Gadolinium	Gd	64	157.2_5	Sodium	Na	11	22.98977
Gallium	Ga	31	69.72	Strontium	Sr	38	87.62
Germanium	Ge	32	72.5_9	Sulfur	S	16	32.06
Gold	Au	79	196.9665	Tantalum	Ta	73	180.947_9
Hafnium	Hf	72	178.4_9	Technetium	Tc	43	98.9062
Helium	He	2	4.00260	Tellurium	Te	52	120.6_0
Holmium	Ho	67	164.9304	Terbium	Tb	65	158.9254

Table C.1. *(Continued)*

Element	Symbol	Atomic Number	Atomic Weight	Element	Symbol	Atomic Number	Atomic Weight
Hydrogen	H	1	1.0079	Thallium	Tl	81	204.3$_7$
Indium	In	49	114.82	Thorium	Th	90	232.0381
Iodine	I	53	126.9045	Thulium	Tm	69	168.9342
Iridium	Ir	77	192.2$_2$	Tin	Sn	50	118.6$_9$
Iron	Fe	26	55.84$_7$	Titanium	Ti	22	47.9$_0$
Krypton	Kr	36	83.80	Tungsten	W	74	183.8$_5$
Lanthanum	La	57	138.905$_5$	Uranium	U	92	238.029
Lawrencium	Lr	103		Vanadium	V	23	50.941$_4$
Lead	Pb	82	207.2	Xenon	Xe	54	131.30
Lithium	Li	3	6.94$_1$	Ytterbium	Yb	70	173.0$_4$
Lutetium	Lu	71	174.97	Yttrium	Y	39	88.9059
Magnesium	Mg	12	24.305	Zinc	Zn	30	65.38
Manganese	Mn	25	54.9380	Zirconium	Zr	40	91.22
Mendelevium	Md	101					

Source: Reprinted by permission of Pergamon Press Ltd.

[a]Based on an assigned relative atomic weight for $^{12}C = 12$. The values apply to the elements as they exist in materials of terrestial origin and to certain artificial elements. Weights are reliable to ± 3 if the last digit is in small type. See *Pure Appl. Chem.*, **21**(1), 105 (1970).

Table C.2. Dissolved-Oxygen Solubility Data (mg O_2/L; 760 mmHg Total Pressure)

	Temperature (°C)						
Water Type	0	5	10	15	20	25	30
Fresh[a]	14.81	12.79	11.25	10.04	9.07	8.27	7.50
Sea[b] (33,000 mg/L NaCl)	11.32	10.01	8.98	8.14	7.42	6.74	6.13

[a]From formula by Montgomery et al.: $O_2 = 468/(31.6 + T°C)$. From S. C. McCutcheon, *Water Quality Modeling*, Vol. 1, *Transport and Surface Exchange in Rivers*, CRC Press, Boca Raton, Fla., 1989, p. 244.

[b]From G. C. Whipple and M. C. Whipple, "Solubility of Oxygen in Sea Water," *J. Am. Chem. Soc.*, **33**, 362 (1911).

Benchmark Chemicals

The chemicals itemized in Fig. C.1 and Table C.3, a so-called "benchmark" group, which represent 1300 well-defined organic structures, are also a partial list of toxic substances. Use of these in exercise and example problem calculations demonstrates the importance of chemical structure and other properties on chemodynamics behavior.

CH_3CCl_3
Methylchloroform

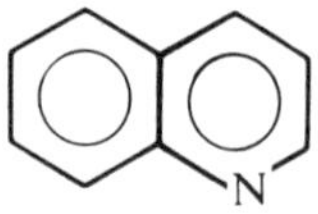

Quinoline

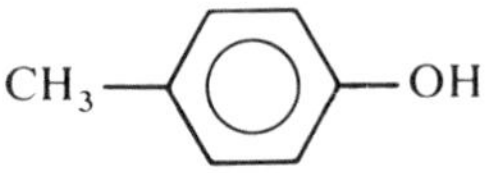

CH_3—⟨benzene⟩—OH

p-Cresol

CCl_3F
Trichlorofluoromethane

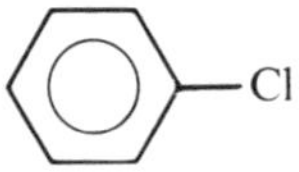

Monochlorobenzene

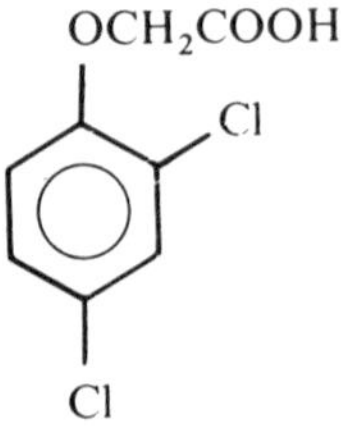

OCH_2COOH, Cl, Cl

2, 4-D

Cl, Cl, Cl

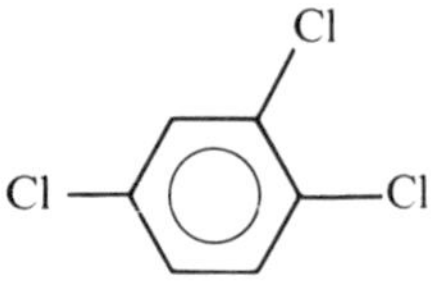

1,2,4-Trichlorobenzene

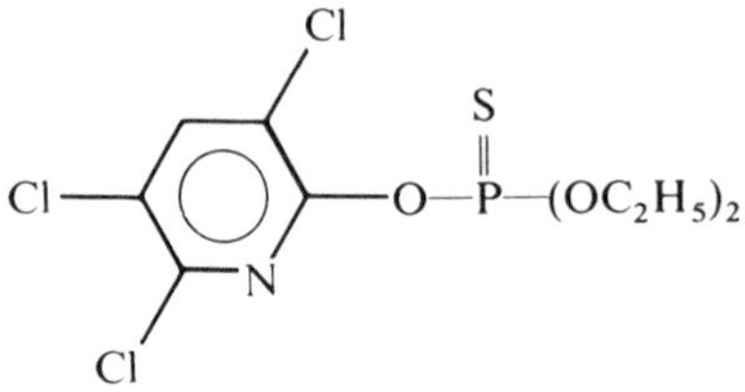

Cl, Cl, Cl, N, O—P—$(OC_2H_5)_2$, S

Chlorpyrifos

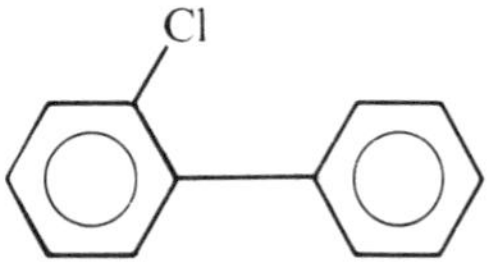

Cl

2-Chlorobiphenyl

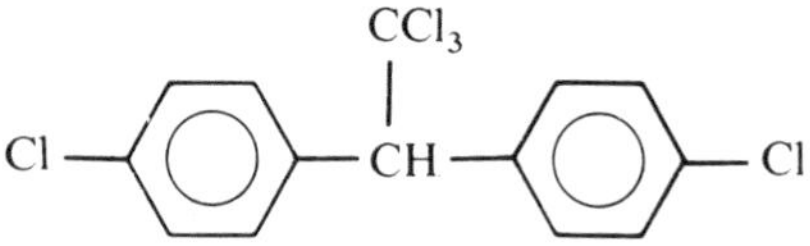

CCl_3, Cl, CH, Cl

DDT

Cl, Cl, Cl, Cl

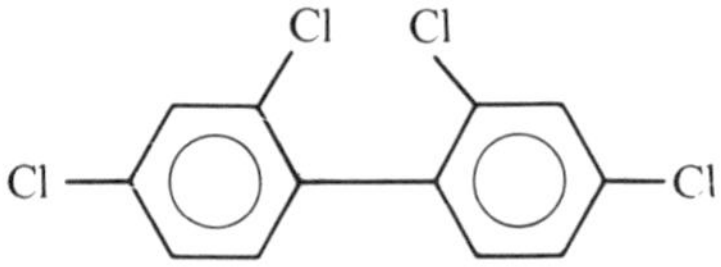

2,4,2′,4′-Tetrachlorobiphenyl

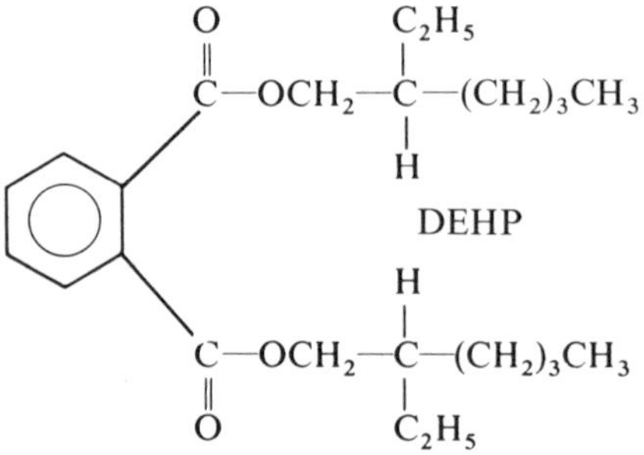

O, C_2H_5, C—OCH_2—C—$(CH_2)_3CH_3$, H

DEHP

H, C—OCH_2—C—$(CH_2)_3CH_3$, O, C_2H_5

Figure C.1

Table C.3. Recommended Properties for the Benchmark Chemicals[a]

Chemical	Molecular Weight (g/mol)	Density[b] (g/cm^3)	Melting Point (°C)	Boiling Point (°C)	Vapor Pressure at 25°C (atm)	Water Solubility at 25°C (mg/L)	log K_{ow}	log K_{oc}
Methylchloroform	133.4	1.325 (26°)	−32.5	74.1	0.163	1334	2.48	2.11
Trichlorofluoromethane	137.4	1.476 (25°)	−111	23.7	1.05[c]	1240	2.53	2.13
Chlorobenzene	112.6	1.106 (20°)	−45	131.7	1.55×10^{-2}	503	2.84	2.46
1,2,4-Trichlorobenzene	181.4	1.574 (10°)	17	213.5	5.2×10^{-4}	48.8	4.10	3.69
2-CBP[d]	188.7	—	34	547	1.1×10^{-5}	5.0	4.54	4.32
2,4,2′,4-TetraCBP[d]	291.9	—	41	—	1.13×10^{-7}	0.068	6.31	4.78
Quinoline	129.2	1.095 (20°)	−19.5	237.7	1.2×10^{-5}	6110	2.04	2.1
p-Cresol	108.1	1.034 (20°)	34.7	201.9	1.16×10^{-4}	1840	1.95	1.67
Chlorpyrifos	350.6	—	42	—	2.46×10^{-8}	0.87	4.90	4.24
2,4-D	221.0	1.565 (30°)	138	215 decomp.	7.9×10^{-10}	690	2.8	1.29
DDT	354.5	—	108.5	—	2.6×10^{-10}	3.29×10^{-3}	6.36	5.39
DEHP[e]	390.6	0.983 (25°)	52.5	—	2.6×10^{-10}	2.49	5.3	5.0

Source: W. B. Neely and G. E. Blau, *Environmental Exposure from Chemicals*, Vol. 1, CRC Press, Boca Raton, Fla., 1985, pp. 8–9.

[a] Average values, with the exception of molecular weight.

[b] Temperature is in parenthesis.

[c] Vapor pressure estimated using Antoine equation.

[d] CBP, chlorobiphenyl.

[e] Di(2-ethylhexyl)phthalate.

Table C.4. Physical and Chemical Properties of Selected Substances[a]

Name	Molecular Weight (g/mol)	Density (20/4°) (kg/L)	Melting Point (°C)	Boiling Point at 760 mm (°C)	Vapor Pressure at 25°C (atm)	Water Solubility at 25°C (mg/L)	log K_{oc} (L/kg)
1. Pesticides							
Aldrin	365	1.70	105	145 (2 mm)*	3.0E-8	0.017	2.61
γ-Chlordane	410	—	104	175	E-4, E-5	E2	5.7
p,p′-DDT	354	1.56	108	193	2.0E-10(*)	0.005(*)	5.4(*)
p,p′-DDE	319	—	89	—	8.5E-9 (30°)	0.07(*)	5.7(*)
Dieldrin	381	1.75	175	—	2.4E-10	0.20(*)	4.3(*)
Endrin	381	—	200	—	6.0E-10(*)	0.23	3.92
Heptachlor	373	1.66	96	145 (1.5 mm)	5.3E-7	0.12	4.34
Lindane	291	1.57	113	323	8.8E-8	8.0(*)	3.1(*)
Toxaphene	414	1.6	85	120	3 E-1(?)	0.9(*)	3.18
2. Aliphatic and aromatic hydrocarbons							
n-Butane	58.1	$0.60^{(0.4)}$	−135	−0.6	2.40	61.4	—
n-Hexane	86.1	0.66	−94	69	0.205	9.5	—
n-Nonane	128	0.72	−54	151	5.64E-3	0.22	—
Tetradecane	198	0.77	5.5	253	1.26E-4	2.2E-2	—
Benzene	78.1	0.88	5.5	80.1	0.125	1780	1.94
Toluene	92.1	0.87	−95	111	0.0374	515	2.1
Ethylbenzene	106	0.88	−95	136	0.0125	152	2.2
o-Xylene	106	0.88	−25	144	8.71E-3	175	2.1
Naphthalene	128	1.14	80.2	218	1.14E-4	34.4	3.1
Biphenyl	—	—	—	—	7.45E-5	7.48	—
Anthracene	178	1.24	216	340	1.4E-7(*)	0.075(*)	4.3
Phenanthrene	178	0.98	100	340	4.53E-6	1.18	4.4(*)
Pyrene	202	1.27	150	400	1.6E-7(*)	0.15	4.8(*)
Benzo[*a*]pyrene	252	1.35	179	495	7.0E-12	0.004	6.0(*)

3. Halogenated hydrocarbons							
Trichloromethane	119	—	−63.5	61.7	0.20 (20)	8200 (20)	1.97(0W)
Tetrachloromethane	154	—	−22.9	76.5	0.12 (20)	785 (20)	2.64(0W)
1,2-Dichloroethane	99.0	—	−97.0	57.3	0.24 (20)	5500 (20)	1.79(0W)
Chloroethene	62.5	—	−154	−13.4	3.5	1.1	0.60(0W)
Hexachlorobutadiene	261	1.55	−21	215	2 E-4 (20°)	4.0	3.67
Bromoform	253	2.89	8.3	150	7.4E-3	3100	2.3
Hexachlorobenzene	285	2.05	230	332	1.4E-8 (20°)	5 E-3	3.4(*)
4. Polychlorinated biphenyls and related compounds							
2-Chloronaphthalene	163	1.27	61	—	2.2E-5	6.74	4.12(0W)
Aroclor 1016	258 (ave.)	1.33	—	—	5.3E-7	0.049	4.70
Aroclor 1242	261 (ave.)	1.39	—	—	5.3E-7	0.24	3.71
Aroclor 1254	327 (ave.)	1.51	—	—	1.0E-7	0.057	5.61
Aroclor 1260	370 (ave.)	1.57	—	—	5.3E-8	0.080	6.42
2,3,7,8-TCDD	322	1.83	300	412	1.8E-12	0.0193	6.66
5. Miscellaneous compounds							
Phenol	94.1	1.06	41	182	4.6E-4	8 E4(*)	1.3(*)
2-Chlorophenol	129	1.26	9.0	175	1.9E-3	2 E4(*)	2.56
Pentachlorophenol	266	1.98	185	310	2.0E-3	25	2.95
p-Chloro-*m*-cresol	143	—	65	235	7 E-5(?)	3850	2.89
Benzidine	184	1.25	~120	401	1 E-3	500	1.60
Acrylonitrile	53	0.806	—	77	0.15	8 E4	−1.13
Mercury	201	13.5	−38.9	358	1.6E-6	0.081 (30°)	—

Source: J. H. Montgomery and L M. Welkom, *Groundwater Chemicals Desk Reference*, Lewis Publishers, Chelsea, Mich., 1990; M. A. Callahan et al., "Water-Related Environmental Fate of 129 Priority Pollutants," Vols. I and II, *EPA-440/4-79-029*, U.S. EPA, Washington, D.C., Dec. 1979; D. Mackay and W.-Y. Shin, *Chemistry and Physics of Aqueous Gas Solutions*, ASTM, Philadelphia, 1974, pp. 104–108.

[a]These data are for student use only. Numerical values denoted by (*) are only approximate, often being the midrange of diverse values reported in the source material. Practitioners are advised, when selecting data, to use original sources.)

Table C.5. Solubility of Selected Hydrocarbons in Water and Seawater (g/m^3)

Hydrocarbon	Water	Seawater	Ratio (%)
n-Pentene	38.5	27.6	71.5
Dodecane	0.0037	0.0029	78
Tetradecane	0.0022	0.0017	77.5
Benzene	1780	1391	78
Toluene	515	402	78

Table C.6. Diffusivities and Schmidt Numbers of Gases in Air

Substance	Temperature (°C)	$\mathscr{D}_{A1}$ (cm^2/s)	Sc = $\mu_1/\rho_1\mathscr{D}_{A1}$
Acetic acid	25	0.133	1.16
Acetone	0	0.109	
Ammonia	25	0.28	0.78
n-Amyl alcohol	25	0.07	2.21
sec-Amyl alcohol	30	0.072	
Amyl butyrate	0	0.040	
Amyl formate	0	0.0543	
i-Amyl formate	0	0.058	
Amyl isobutyrate	0	0.0419	
Amyl propinate	0	0.046	
Aniline	25	0.072	2.14
Anthracene	0	0.0421	
Benzene	25	0.088	1.76
Benzidine	0	0.0298	
Benzyl chloride	0	0.066	
Biphenyl	0	0.0610	
Bromine	25	0.100	
i-Butyl acetate	0	0.0612	
n-Butyl acetate	0	0.058	
i-Butyl alcohol	0	0.0727	
n-Butyl alcohol	25.9	0.097	1.72
Butyl amine	25	0.101	1.53
i-Butyl amine	0	0.0853	
i-Butyl butyrate	0	0.0468	
i-Butyl formate	0	0.0705	
i-Butyl isobutyrate	0	0.0457	
i-Butyl propionate	0	0.0529	
i-Butyl valerate	0	0.0424	
Butyric acid	0	0.067	
i-Butyric acid	25	0.081	1.91

Table C.6. ***(Continued)***

Substance	Temperature (°C)	$\mathscr{D}_{A1}$ (cm^2/s)	$Sc = \mu_1/\rho_1\mathscr{D}_{A1}$
Caproic acid	0	0.050	
i-Caproic acid	25	0.060	2.58
Carbon dioxide	25	0.164	0.94
Carbon disulfide	25	0.107	1.45
Carbon monoxide	25	0.203	
Carbon tetrachloride	25	0.0828	
Chlorine	0	0.093	1.42
Chlorobenzene	0	0.062	2.13
2-Chlorobiphenyl	25	0.0594	
Chloroform	0	0.091	
Chloropicrin	25	0.088	
Chloropyrifos	25	0.0455	
Chlorotoluene	25	0.065	2.38
m-Chlorotoluene	0	0.054	
o-Chlorotoluene	0	0.059	
p-Chlorotoluene	0	0.051	
2,4-Cresol	25	0.0603	
p-Cresol	25	0.0777	
Cyanogen chloride	0	0.111	
Cyclohexane	45	0.086	
DDT	25	0.0468	
Diethylamine	25	0.105	1.47
Diphenyl	25	0.068	2.28
Ethane	0	0.108	1.22
Ether (diethyl)	0	0.0778	
Ethyl acetate	30	0.089	
Ethyl alcohol	25	0.119	1.30
Ethyl benzene	25	0.077	2.01
Ethyl *n*-butyrate	0	0.0579	
Ethyl *i*-butyrate	0	0.0591	
Ethyl ether	25	0.093	1.66
Ethyl formate	0	0.0840	
Ethyl propionate	0	0.068	
Ethyl valerate	0	0.0512	
Ethylene dibromide	25	0.0704	
Eugenol	0	0.0377	
Formic acid	25	0.159	0.97
Hexane	21	0.080	
Hexyl alcohol	25	0.059	2.60
Hydrogen	25	0.410	0.22
Hydrogen chloride	25	0.173	
Hydrogen cyanide	0	0.173	
Hydrogen peroxide	60	0.188	

Table C.6. ***(Continued)***

Substance	Temperature (°C)	$\mathscr{D}_{A1}$ (cm^2/s)	Sc $= \mu_1/\rho_1\mathscr{D}_{A1}$
Hydrogen sulfide	25	0.166	
Iodine	0	0.07	
Mercury	25	0.131	
Mesitylene	25	0.067	2.31
Methane	0	0.16	0.84
Methyl acetate	0	0.084	
Methyl alcohol	25	0.159	0.97
Methyl butyrate	0	0.0633	
Methyl *i*-butyrate	0	0.0639	
Methyl chloroform	25	0.0832	
Methyl formate	0	0.0872	
Methyl propionate	0	0.0735	
Methyl valcrate	0	0.0569	
Monochlorobenzene	25	0.0810	
Naphthalene	0	0.0513	
Nitrogen	0	0.13	0.98
Nitrogen oxide (NO)	25	0.204	
Nitrous oxide (N_2O)	25	0.155	
n-Octane	25	0.060	2.58
Oxygen	25	0.206	0.75
n-Pentane	21	0.071	
Phosgene	0	0.095	
Propane	0	0.088	1.51
Propionic acid	25	0.099	1.56
Propyl acetate	0	0.067	
i-Propyl alcohol	30	0.101	
n-Propyl alcohol	25	0.100	1.55
n-Propyl benzene	25	0.059	2.62
i-Propyl benzene	0	0.0489	
n-Propyl bromide	25	0.105	1.47
i-Propyl bromide	0	0.0902	
Propyl butyrate	0	0.0530	
Propyl formate	0	0.0712	
i-Propyl iodide	0	0.0802	
n-Propyl iodide	25	0.096	1.61
i-Propyl isobutyrate	0	0.059	
n-Propyl isobutyrate	0	0.0549	
Propyl propionate	0	0.057	
Propyl valerate	0	0.0466	
Quinoline	25	0.0723	
Safrol	0	0.0434	
i-Safrol	0	0.0455	
Sulfur dioxide	0	0.103	1.28
2,4,2′,4′-Tetrachlorobiphenyl	25	0.0524	
Toluene	30	0.088	

Table C.6. ***(Continued)***

Substance	Temperature (°C)	$\mathscr{D}_{A1}$ (cm²/s)	Sc = $\mu_1/\rho_1\mathscr{D}_{A1}$
1,2-4-Trichlorobenzene	25	0.0676	
Trichlorofluoromethane	25	0.0902	
Trimethyl carbinol	0	0.087	
i-Valeric acid	0	0.0544	
n-Valeric acid	25	0.067	2.31
Water	25	0.256	0.60
Xylene	25	0.071	2.18

Source: Compiled from: C. O. Bennett and J. E. Myers, Momentum, Heat and Mass Transfer, 2nd ed., McGraw-Hill, New York, 1974, pp. 787–788; T. K. Sherwood and R. L. Pigford, *Absorption and Extraction*, 2nd ed., McGraw-Hill, New York, 1952, p. 20; C. J. Geankoplis, *Mass Transport Phenomena*, Holt, Rinehart & Winston, New York, 1972, pp. 22, 70, 478; J. H. Perry, Ed., *Chemical Engineers' Handbook*, 4th ed., McGraw-Hill, New York, 1964, pp. 14–22 and 14–23; C. A. I. Goring and J. W. Hamaker, *Organic Chemicals in the Soil Environment*, Vol. 1, Marcel Dekker, New York, 1972, p. 348; W. B. Neely and G. E. Blau, *Environmental Exposure from Chemicals*, Vol. I, CRC Press, Boca Raton, Fla., 1985, p. 77.

Table C.7. Diffusivities and Schmidt Numbers of Chemicals in Water

Substance	Temperature (°C)	$\mathscr{D}_{A2} \times 10^5$ (cm²/s)	Sc = $\mu_2/\rho_2\mathscr{D}_{A2}$
Acetamide	25	1.19	
Acetic acid	20	0.88	1140
Acetonitrile	25	1.66	
Acetylene	20	1.56	645
Allyl alcohol	20	0.93	1080
Ammonia	20	1.76	570
i-Amyl alcohol	25	1.0	
Bromine	20	1.2	840
n-Butanol	25	0.96	
Butanol	20	0.77	1310
Caffeine	25	0.63	
Carbon dioxide	20	1.77	559
Carbon monoxide	25	2.17	
Chloral hydrate	25	0.77	
Chlorine	25	1.45	617
2-Chlorobiphenyl	25	0.648	
Chlorpyrifos	25	0.467	
p-Cresol	25	0.871	
2,4-Cresol	25	0.649	
DDT	25	0.485	
Ethanol	20	1.00	1005
Formic acid	25	1.37	
Glucose	25	0.69	

Table C.7. ***(Continued)***

Substance	Temperature (°C)	$\mathscr{D}_{A2} \times 10^5$ (cm^2/s)	Sc = $\mu_2/\rho_2\mathscr{D}_{A2}$
Glycerol	20	0.72	1400
Hydrogen	25	5.85	
Hydrogen sulfide	25	1.36	657
Hydroquinone	20	0.77	1300
Lactose	20	0.43	2340
Maltose	20	0.43	2340
Mannitol	20	0.58	1730
Mercury	23.6	2.9	
Methanol	20	1.28	785
Methyl chloroform	25	0.921	
Monochlorobenzene	25	0.909	
Nitric acid	20	2.6	390
Nitrogen	25	2.0	448
Nitrogen dioxide (NO_2)	25	2.13	
Nitrogen oxide (NO)	25	2.55	
Nitrous oxide (N_2O)	25	2.57	
Oxygen	25	2.35	381
Phenol	20	0.84	1200
n-Propanol	20	0.87	1150
Pyridine	25	0.76	
Pyrogallol	20	0.70	1440
Quinoline	25	0.812	
Radon	25	1.37	
Ratlinose	20	0.37	2720
Resorcinol	20	0.80	1260
Saccharose	25	0.49	
Sodium chloride	20	1.35	745
Sodium hydroxide	20	1.51	665
Succinic acid	25	0.94	
Sucrose	20	0.45	2230
Sulfur dioxide	25	1.9	
Sulfuric acid	20	1.73	580
Tartaric acid	25	0.80	
2,4,2′,4′-Tetrachlorobiphenyl	25	0.552	
1,2-4-Trichlorobenzene	25	0.757	
Trichlorofluoromethane	25	0.102	
Urea	20	1.06	946
Urethane	20	0.92	1090

Source: Compiled from: C. O. Bennett and J. E. Myers, *Momentum, Heat and Mass Transfer*, 2nd ed., McGraw-Hill, New York, 1974, pp. 787–788; R. A. Horne, *Marine Chemistry*, Wiley-Interscience, New York, 1969; J. H. Perry, Ed., *Chemical Engineers' Handbook*, 4th ed., McGraw-Hill, New York, 1964, pp. 14–25 and 14–26; A. Lerman, *Geochemical Processes*, Wiley, New York, 1979, pp. 73–121; W. B. Neely and G. E. Blau, *Environmental Exposure from Chemicals*, Vol. 1, CRC Press, Boca Raton, Fla., 1985, p. 77.

Table C.8. Diffusivities in the Solid State

System	T^a (°C)	$\mathscr{D}_{A3}$ (cm²/s)
Ar in feldspars	25[a]	3.6E-34
Ar in mica	25	2.7E-43
Ca in quartz	25	1.4E-45
Ca in zeolites	25	4E-16
CO_2 in calcite	25	1.4E-46
He in Pyrex	20	4.5E-11
He in Pyrex	500	2E-8
He in SiO_2	20	2.4-5.5E-10
H_2 in SiO_2	500	0.6-2.1E-8
H_2O in zeolites	25	E-8 to E-7
Na in feldspars	25	2.1E-38
Na in quartz	25	1.1E-15
Na in zeolites	25	E-13 to E-12
O_2 in feldspars	25	5.6E-28
Elements in glass (borosilicate)[b]	100	1.0-100.0E-14

Source: A. Lerman, *Geochemical Processes*, Wiley, New York, 1979, pp. 73–121; R. M. Barrer, *Diffusion in and Through Solids*, Macmillan, New York, 1941, pp. 22, 141, 275.

[a] Values at 25°C are extrapolated from higher temperatures.
[b] G. de Marsily, E. Ledoux, and J. Margat, "Nuclear Waste Disposal: Can Geologists Guarantee Isolation," *Science*, **197**, Aug. 5, 1977, p. 197.

Table C.9. Tracer Diffusion Coefficients of Ions in Water at Infinite Dilution and 25°C

Cation	$\mathscr{D}_{i2} \times 10^5$ (cm²/s)	Anion	$\mathscr{D}_{i2} \times 10^5$ (cm²/s)
Na^+	1.33	Cl^-	2.03
Cs^+	2.07	Br^-	2.01
NH_4^+	1.98	I^-	2.00
$Cu(OH)^+$	0.83	HS^-	1.73
$Zn(OH)^+$	0.854	SO_4^{2-}	1.07
Ca^{2+}	0.793	SeO_4^{2-}	0.946
Sr^{2+}	0.794	NO_3^-	1.90
Ra^{2+}	0.889	$H_2PO_4^-$	0.846
Cu^{2+}	0.733	HPO_4^{2-}	0.734
Zn^{2+}	0.715	PO_3^{3-}	0.612
Cd^{2+}	0.717	$H_2AsO_4^-$	0.905
Pb^{2+}	0.945	$H_2SbO_4^-$	0.825
Cr^{3+}	0.594	CrO_4^{2-}	1.12

Source: A. Lerman, *Geochemical Processes*, Wiley, New York, 1979, pp. 73–121.

Table C.10. Solubility of Gases in Seawater

Chlorinity (Cl‰)	Carbon Dioxide (mol/10,000 L)			Nitrogen (mL N_2/L H_2O, STP)		
	0°C	10°C	26°C	0°C	10°C	26°C
15	674	472	299	19.3	15.5	12.0
20	640	452	287	18.0	14.6	11.4

Source: R. A. Horne, *The Chemistry of Our Environment*, Wiley Interscience, New York, 1978, p. 734.

Table C.11. Henry's Law Constants for Gases in Water ($H \times 10^{-4}$)[†]

T, °C	CH_4	C_2H_4	C_2H_6	H_2	He	N_2	O_2	CO	CO_2	H_2S
0	2.24	0.552	1.26	5.79	12.9	5.29	2.55	3.52	0.0728	2.68
10	2.97	0.768	1.89	6.36	12.6	6.68	3.27	4.42	0.104	3.67
30	4.49	1.27	3.42	7.29	12.4	9.24	4.75	6.20	0.186	6.09

Source: National Research Council, "International Critical Tables," Vol. III, McGraw-Hill, New York, 1929.

[†] $p_A = Hx_A$, p_A in atm and x_A in mole fraction.

APPENDIX D

PHYSICAL PROPERTY DATA

Table D.1. Physical Properties of Dry Air at Atmospheric Pressure

Temperature (°C)	Mass Density, ρ (kg/m^3)	Kinematic Viscosity, ν (m^2/s)	Specific Heat, C_p (J/kg·K)	Thermal Diffusivity, α (m^2/s)
0	1.292	13.3E-6	1.005E-3	18.9E-6
5	1.269	13.7E-6	1.006E-3	19.5E-6
10	1.246	14.2E-6	1.006E-3	20.2E-6
15	1.225	14.6E-6	1.006E-3	20.8E-6
20	1.204	15.1E-6	1.006E-3	21.5E-6
25	1.183	15.5E-6	1.006E-3	22.2E-6
30	1.164	16.0E-6	1.006E-3	22.8E-6
35	1.146	16.4E-6	1.006E-3	23.5E-6
40	1.128	16.9E-6	1.007E-3	24.2E-6
45	1.110	17.4E-6	1.007E-3	24.9E-6

Table D.2. Physical Properties of Pure Water

Temperature (°C)	Mass Density, ρ (kg/m^3)	Kinematic Viscosity, ν $(=\mu/\rho)$ (10^{-6} m^2/s)	Absolute Vapor Pressure (kPa)	Specific Heat (J/kg·K)	Thermal Diffusivity, α (10^{-8} m^2/s)
0	999.8	1.785	0.61	4217.4	13.4
5	1000.0	1.519	0.87	—	—
10	999.7	1.306	1.23	4191.9	13.8
15	999.1	1.139	1.70	—	—
20	998.2	1.003	2.34	4181.6	14.2
25	997.0	0.893	3.17	—	—
30	995.7	0.800	4.24	4178.2	14.6
40	992.2	0.658	7.38	4178.3	15.2
50	988.0	0.553	12.33	4180.4	—

Table D.3. Comparison of Properties of Pure Water and Seawater at 1 atm

	Pure Water		Seawater (salinity 35/‰)	
Name (Units)	0°C	20°C	0°C	20°C
Thermal conductivity (W/cm · °C)	0.00566	0.00599	0.00563	0.00596
Kinematic viscosity (cm^2/s)	0.01787	0.01004	0.01826	0.01049
Thermal diffusivity (cm^2/s)	0.00134	0.00143	0.00138	0.00149
Specific volume (cm^3/g)	1.0000	1.0017	0.9842	0.9868
Prandtl number, $Pr = \nu/\alpha$	13.3	7.0	13.1	7.0

Source: R. B. Montgomery in D. E. Gray, Eds., *American Institute of Physics Handbook*, AIS, New York, 1957. Used with permission of McGraw-Hill Book Co.

Table D.4. Physical Properties of Solids at 20°C

Material	Density (g/cm^3)	Specific Heat (J/kg · K)	Thermal Conductivity (J/m · s · K)	Porosity (cm^3/cm^3)
Clay, in water	1.28	—	—	—
Diatomaceous earth, powder	0.224	837	0.52	—
Earth, mud, flowing	1.7	—	—	—
Earth, mud, packed	1.8	—	—	—
Earth's crust	2.67	—	1.67	—
River mud, in water	1.44	—	—	—
Rock				
Granite	2.59–2.76	804	0.167–4.06	0.004–0.0384
Marble	2.64–2.87	879	0.502–2.09	0.004–0.021
Limestone	1.87–2.80	904	—	0.011–0.310
Sandstone	1.91–2.69	921	1.00–2.51	0.019–0.273
Slate	2.69–2.88	—	1.97	0.001–0.017
Soil[a]				
Sandy loam, 4% H_2O	1.66	—	—	0.43
Sandy loam, 10% H_2O	1.94	—	—	—
Clay, clay loam, silt loam	1.00–1.60	—	0.99	0.51
Sands and sandy loam	1.20–1.80	—	1.98–2.44	0.35–0.50
Compact subsoils	$\geqslant 2.0$	—	1.74	—
Soil minerals, dry	2.60–2.75	837	—	—

Sources: Compiled primarily from: A. R. Junikis, *Thermal Geotechnics*, Rutgers University Press, New Brunswick, N.J., 1977; J. H. Perry, Ed., *Chemical Engineers' Handbook*, 3rd ed., McGraw-Hill, New York, 1950.

[a]See Table 6.4-1 for more properties of soils.

Specific Gravity of Water Versus Temperature

$$\begin{aligned}\text{Specific gravity} - 1 = &-0.2260569\text{E-}5 + 0.1546919\text{E-}5T + 0.2141968\text{E-}5\,T^2 \\ &-0.6508630\text{E-}6\,T^3 + 0.1975524\text{E-}7\,T^4 \\ &-0.1894802\text{E-}9\,T^6\end{aligned}$$

where T= temperature in °C, $0° \leqslant T \leqslant 50°$, sp. gr. = ρ_2/ρ_2 at 4°C (from Joseph W. Dewitt, Civil Engineer, U.S. Army Corps of Engineers District Office, Savannah, Ga.).

APPENDIX E

ENVIRONMENTAL DATA

Manning's formula for stream flow is

$$V = \frac{1}{n} R^{2/3} S^{1/2}$$

where V is the mean velocity (m/s), R the hydraulic radius (m), S the hydraulic gradient (m/m), and n the coefficient of roughness (s/m$^{1/3}$) (see Table E.1).

Table E.1. Manning's Roughness Coefficients for Natural Constructed Earthen Channels

Condition	n
Rivers and earth canals in fair condition, some growth	0.025
Winding natural streams and canals in poor condition, considerable moss growth	0.035
Mountain streams with rocky beds and rivers with variable sections and some vegetation along banks	0.040–0.050
Alluvial channels, sand bed, no vegetation	
tranquil flow, Fr < 1	
Plane bed	0.014–0.02
Ripples	0.018–0.028
Dunes	0.018–0.035
Washed-out dunes or transition	0.014–0.024
Plane bed	0.012–0.015
Rapid flow, Fr > 1	
Standing waves	0.011–0.015
Antidunes	0.012–0.020
Earth-lined constructed	
Regular surface in good condition	0.020
In ordinary condition	0.0225
With stones and weeds	0.025
In poor condition	0.035
Partially obstructed with debris or weeds	0.050

Source: R. J. Chorley, Ed., *Water, Earth and Man*, from Chapter 7, "Channel Flow" by D. B. Simmons, Methuen, London, 1969, p. 309; C. V. Davis, Ed., *Handbook of Applied Hydraulics*, McGraw-Hill, New York, 1952, p. 1204. Reprinted by permission.

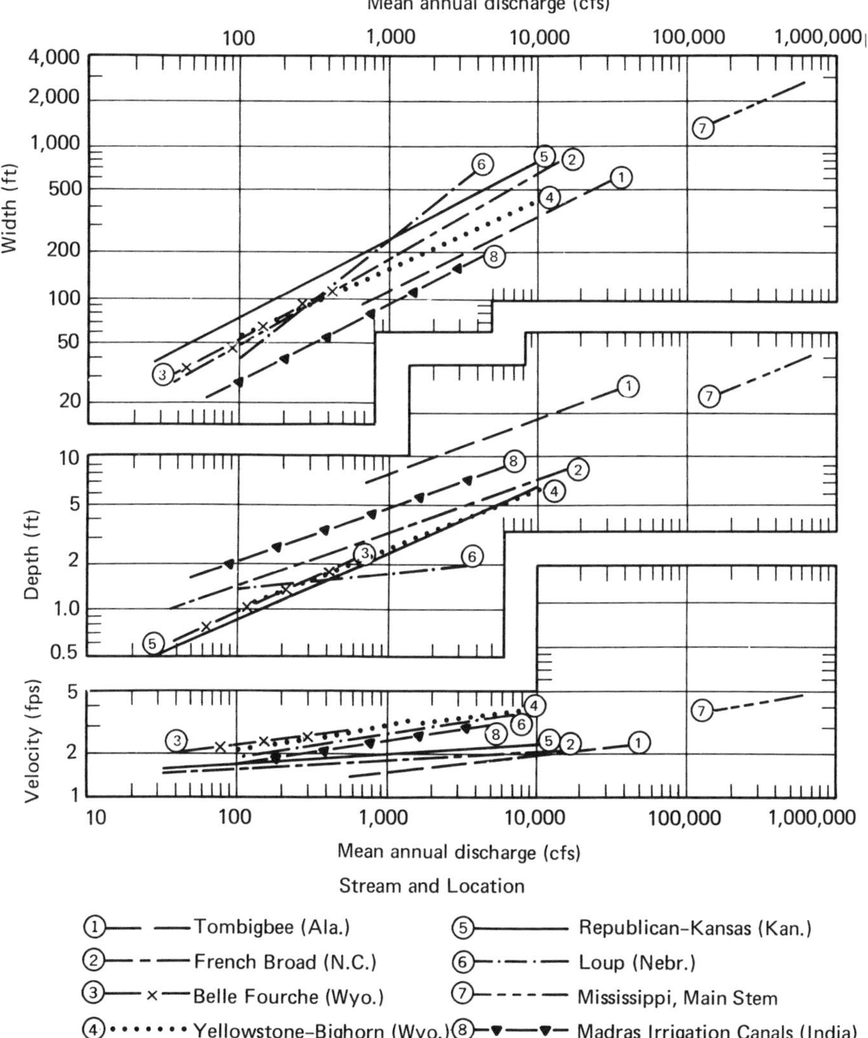

Figure E.1. River systems: width, depth, and velocity. [From L. B. Leopold, "Rivers," *Am. Sci.*, **50**(4), 511 (1962).]

Table E.2. Local Climatological Data in Selected U.S. Cities

Code	City, State	Climatic Region[a]
1	Janeau, Alaska	Subartic
2	Phoenix, Arizona	Desert
3	Los Angeles, California	Mediterranian or dry summer subtropical
4	Denver, Colorado	Steppe
5	Honolulu, Hawaii	Tropical rainforest
6	Chicago, Illinois	Humid continental, warm summer
7	New Orleans, Louisiana	Humid subtropical
8	Portland, Maine	Humid continental, cool summer

[a] *Goodes' World Atlas*, E. B. Espenshade, Jr., Ed., Rand McNally, New York, 1987, p. 8.

Table E.2a. Monthly and Annual Average Relative Humidity at 0700/1300 Hours (%)

Code	Record length (yr)	Jan.	Feb.	Mar.	Apr.	May	June	July	Aug.	Sept.	Oct.	Nov.	Dec.	Annual
						(Upper value at 0700 and lower value at 1300 h)								
1	28	80/77	83/77	79/70	74/64	74/62	75/63	81/70	84/73	88/78	87/81	86/82	82/81	81/73
2	11	67/44	61/39	59/33	46/24	37/18	36/18	47/29	57/36	55/33	53/28	61/38	70/49	54/32
3	12	66/53	71/56	76/58	78/60	81/64	85/70	86/68	85/68	82/64	78/58	75/59	70/55	78/61
4	11	62/44	68/45	70/43	68/37	71/39	73/40	72/36	71/37	72/39	65/36	70/44	67/45	69/40
5	16	80/80	78/78	77/75	75/71	75/69	74/69	74/69	75/70	74/70	76/72	77/75	77/76	76/73
6	8	70/64	70/60	73/58	72/55	71/51	74/54	77/55	79/55	80/55	76/52	77/63	78/70	75/58
7	23	86/67	85/63	84/60	88/60	89/60	90/62	91/66	91/66	89/65	87/59	86/59	86/67	88/63
8	31	78/62	75/60	76/59	74/55	75/57	78/59	80/59	84/59	86/60	86/60	85/64	80/62	80/60

Table E.2b. Monthly and Annual Average Precipitation (inches)

Code	Jan.	Feb.	Mar.	Apr.	May	June	July	Aug.	Sept.	Oct.	Nov.	Dec.	Annual
1	4.00	3.06	3.27	2.87	3.24	3.39	4.49	5.02	6.67	8.33	6.06	4.22	54.62
2	0.73	0.85	0.66	0.32	0.13	0.09	0.77	1.12	0.73	0.46	0.49	0.85	7.20
3	2.66	2.88	1.79	1.05	0.13	0.05	0.01	0.02	0.17	0.39	1.09	2.39	12.63
4	0.55	0.69	1.21	2.11	2.70	1.44	1.53	1.28	1.13	1.01	0.69	0.47	14.81
5	3.76	3.30	2.89	1.31	0.99	0.33	0.44	0.89	0.99	1.84	2.16	2.99	21.89
6	1.86	1.60	2.74	3.04	3.73	4.07	3.37	3.16	2.73	2.78	2.20	1.90	33.18
7	3.74	3.99	5.34	4.55	4.38	4.43	6.92	5.34	5.03	2.84	3.34	4.10	53.90
8	4.37	3.80	4.34	3.73	3.41	3.18	2.86	2.46	3.52	3.20	4.17	3.85	43.85

Table E.2c. Monthly and Annual Average Temperature (°F)

Code	Jan.	Feb.	Mar.	Apr.	May	June	July	Aug.	Sept.	Oct.	Nov.	Dec.	Annual
1	25.1	26.8	30.4	38.0	45.6	52.3	55.3	54.1	48.9	41.6	34.3	28.4	40.1
2	49.7	53.5	59.0	67.2	75.0	83.6	89.8	87.5	82.2	40.7	58.1	51.6	69.0
3	54.4	55.2	57.0	59.4	62.0	64.8	69.1	69.1	68.3	64.9	61.1	56.9	61.9
4	28.5	31.5	36.4	46.4	56.2	66.5	72.9	71.5	63.0	51.4	37.7	31.6	49.5
5	72.5	72.4	72.8	74.2	75.6	77.9	78.8	79.4	79.4	78.5	75.9	73.6	75.9
6	26.0	27.7	36.3	49.0	60.0	70.5	75.6	74.2	66.1	55.1	39.9	29.1	50.8
7	54.6	57.1	61.4	67.9	74.4	80.1	81.6	81.9	78.3	70.4	60.0	55.4	68.6
8	21.8	22.8	31.4	42.8	53.0	62.1	68.1	66.8	58.7	48.6	38.1	25.8	45.0

Table E.2d. Monthly and Annual Average Wind Speed (mi/hr)

Code	Record length (yr)	Jan.	Feb.	Mar.	Apr.	May	June	July	Aug.	Sept.	Oct.	Nov.	Dec.	Annual
1	28	8.5	8.6	9.0	8.9	8.4	7.9	7.9	7.7	8.2	8.8	8.9	9.5	8.6
2	26	4.9	5.5	6.2	6.5	6.6	6.6	6.9	6.3	6.6	5.5	5.0	4.8	5.9
3	23	6.7	7.3	7.9	8.4	8.2	7.8	7.5	7.4	7.1	6.7	6.6	6.6	7.4
4	23	9.3	9.3	10.0	10.4	9.4	9.0	8.5	8.2	8.2	8.2	8.7	9.0	9.0
5	22	9.9	10.6	11.2	11.8	12.1	12.8	13.5	13.6	11.7	10.7	11.1	11.2	11.7
6	29	11.5	11.7	11.9	11.8	10.5	9.3	8.3	8.1	9.0	9.8	11.4	11.2	10.4
7	23	9.6	10.2	10.2	9.6	8.3	6.9	6.3	6.2	7.5	7.8	8.9	9.2	8.4
8	31	9.2	9.5	10.0	9.9	9.2	8.2	7.7	7.5	7.8	8.5	8.8	9.0	8.8

Table E.2e. Monthly and Annual Average Possible Sunshine (%)

Code	Record length (yr)	Jan.	Feb.	Mar.	Apr.	May	June	July	Aug.	Sept.	Oct.	Nov.	Dec.	Annual
1	26	33	31	37	39	38	35	30	30	24	19	23	20	31
2	76	78	70	83	88	93	94	84	85	89	88	84	77	86
3	31	71	72	73	69	66	65	82	83	79	73	74	71	73
4	22	72	71	71	67	64	70	71	72	75	74	66	68	70
5	19	63	65	69	69	71	73	75	77	75	67	61	60	69
6	29	44	47	51	53	61	67	69	68	64	61	41	40	57
7	46	49	51	57	65	69	67	61	63	64	72	62	48	61
8	31	55	59	58	57	57	61	65	65	62	58	46	54	59

Source: U.S. National Oceanic and Atmospheric Administration, local climatological data.

INDEX

ENVIRONMENTAL SCIENCE AND TECHNOLOGY

A Wiley-Interscience Series of Texts and Monographs

Edited by JERALD L. SCHNOOR, *University of Iowa*
ALEXANDER ZEHNDER, *Swiss Federal Institute for Water Resources and Water Pollution Control*

PHYSICOCHEMICAL PROCESSES FOR WATER QUALITY CONTROL
Walter J. Weber, Jr.

ph and pION CONTROL IN PROCESS AND WASTE STREAMS
F. G. Shinskey

AQUATIC POLLUTION: An Introductory Text
Edward A. Laws

INDOOR AIR POLLUTION: Characterization, Prediction, and Control
Richard A. Wadden and Peter A. Scheff

PRINCIPLES OF ANIMAL EXTRAPOLATION
Edward J. Calabrese

SYSTEMS ECOLOGY: An Introduction
Howard T. Odum

INTEGRATED MANAGEMENT OF INSECT PESTS OF POME AND STONE FRUITS
B. A. Croft and S. C. Hoyt, Editors

WATER RESOURCES: Distribution, Use and Management
John R. Mather

ECOGENETICS: Genetic Variation in Susceptibility to Environmental Agents
Edward J. Calabrese

GROUNDWATER POLLUTION MICROBIOLOGY
Gabriel Bitton and Charles P. Gerba, Editors

CHEMISTRY AND ECOTOXICOLOGY OF POLLUTION
Des W. Connell and Gregory J. Miller

SALINITY TOLERANCE IN PLANTS: Strategies for Crop Improvement
Richard C. Staples and Gary H. Toenniessen, Editors

ECOLOGY, IMPACT ASSESSMENT, AND ENVIRONMENTAL PLANNING
Walter E. Westman

CHEMICAL PROCESSES IN LAKES
Werner Stumm, Editor

INTEGRATED PEST MANAGEMENT IN PINE-BARK BEETLE ECOSYSTEMS
William E. Waters, Ronald W. Stark, and David L. Wood, Editors

PALEOCLIMATE ANALYSIS AND MODELING
Alan D. Hecht, Editor

BLACK CARBON IN THE ENVIRONMENT: Properties of Distribution
E. D. Goldberg

GROUND WATER QUALITY
C. H. Ward, W. Giger, and P. L. McCarty, Editors

TOXIC SUSCEPTIBILITY: Male/Female Differences
Edward J. Calabrese

ENERGY AND RESOURCE QUALITY: The Ecology of the Economic Process
Charles A. S. Hall, Cutler J. Cleveland, and Robert Kaufmann

AGE AND SUSCEPTIBILITY TO TOXIC SUBSTANCES
Edward J. Calabrese

ECOLOGICAL THEORY AND INTEGRATED PEST MANAGEMENT PRACTICE
Marcos Kogan, Editor

AQUATIC SURFACE CHEMISTRY: Chemical Processes at the Particle Water Interface
Werner Stumm, Editor

RADON AND ITS DECAY PRODUCTS IN INDOOR AIR
William W. Nazaroff and Anthony V. Nero, Jr., Editors

PLANT STRESS–INSECT INTERACTIONS
E. A. Heinrichs, Editor

INTEGRATED PEST MANAGEMENT SYSTEMS AND COTTON PRODUCTION
Ray Frisbie, Kamal El-Zik, and L. Ted Wilson, Editors

ECOLOGICAL ENGINEERING: An Introduction to Ecotechnology
William J. Mitsch and Sven Erik Jorgensen, Editors

ANTHROPOD BIOLOGICAL CONTROL AGENTS AND PESTICIDES
Brian A. Croft

AQUATIC CHEMICAL KINETICS: Reaction Rates of Processes in Natural Waters
Werner Stumm, Editor

GENERAL ENERGETICS: Energy in the Biosphere and Civilization
Vaclav Smil

FATE OF PESTICIDES AND CHEMICALS IN THE ENVIRONMENT
J. L. Schnoor, Editor

ENVIRONMENTAL ENGINEERING AND SANITATION, Fourth Edition
Joseph A. Salvato

TOXIC SUBSTANCES IN THE ENVIRONMENT
B. Magnus Francis

CLIMATE-BIOSPHERE INTERACTIONS
Richard G. Zepp, Editor

AQUATIC CHEMISTRY: Chemical Equilibria and Rates in Natural Waters, Third Edition
Werner Stumm and James J. Morgan

PROCESS DYNAMICS IN ENVIRONMENTAL SYSTEMS
Walter J. Weber, Jr., and Francis A. DiGiano

ENVIRONMENTAL CHEMODYNAMICS: Movement of Chemicals in Air, Water, and Soil, Second Edition
Louis J. Thibodeaux